Complex Analytic Desingularization

José Manuel Aroca • Heisuke Hironaka
José Luis Vicente

Complex Analytic Desingularization

Springer

José Manuel Aroca
Catedrático de Geometría y Topología
Universidad de Valladolid
Valladolid, Spain

Heisuke Hironaka
Professor Emeritus
Harvard University
Cambridge, MA, USA

José Luis Vicente
Catedrático de Álgebra
Universidad de Sevilla
Sevilla, Spain

ISBN 978-4-431-70218-4 ISBN 978-4-431-49822-3 (eBook)
https://doi.org/10.1007/978-4-431-49822-3

Library of Congress Control Number: 2018959845

This Springer imprint is published by the registered company Springer Japan KK part of Springer Nature.
The registered company address is: Shiroyama Trust Tower, 4-3-1 Toranomon, Minato-ku, Tokyo 105-6005, Japan

Foreword

The main ideas of the proof of resolution of singularities of complex-analytic spaces presented here were developed by Heisuke Hironaka in the late 1960s and early 1970s. The difficulty of presenting with sufficient clarity these subtle and original ideas, as well as the other interests of the co-authors, is one of the causes of the unusually long delay before the book's publication. During this period, a number of proofs of resolution of singularities in all dimensions by compositions of permissible blowing-ups, all inspired by Hironaka's general approach, have appeared, the validity of some of them extending beyond the complex-analytic case. The proof has been so streamlined that, while in the 1960s and 1970s, it was considered to be one of the most difficult proofs produced by mathematics; it can now be the subject of an advanced university course. In the bibliography below, one finds a small sample of proofs and expositions.

So, setting apart the epilogue on singularities of foliations, is this book of historical interest only? We think that it is not at all the case, and that reading and perusing it will be very rewarding for any mathematician interested in the subject of resolution of singularities and more generally in singularity theory. Unlike most of the recent proofs, the redaction is not oriented toward a proof of a canonical or algorithmic resolution of singularities or even toward a most "economical" proof.

What is presented in this book is in fact a masterly study of the infinitely near "worst" singular points of a complex-analytic spaces obtained in finite sequences of permissible blowing-ups and of the way to tame them using subspaces of the ambient space having "maximal contact" with the singular space at a given point. This taming proves by an induction on the dimension that there exist finite sequences of permissible blowing-ups at the end of which the worst infinitely near points have disappeared, and this is essentially enough to obtain resolution of singularities.

Here "permissible" means that the center of blowing-up is non-singular and contained in a "Samuel stratum" of X, where the Hilbert–Samuel stratum of the local rings of the points of X is largest (in the lexicographic order). For infinitely near points, "worst" is taken as meaning that the Hilbert–Samuel function is the same as that of the worst points on X. The totality of all such infinitely near worst points is of a nature similar to the Zariski–Riemann manifold or, more relevant in this case, the *voûte étoilée* (starry vault), the analogue in complex-analytic geometry of the Zariski–Riemann manifold created by Hironaka. The fact that the structure of this extremely complicated object can be described, at least locally, by resolutions of the singularities of an *idealistic exponent*, which is slightly more general than a singular space, living in a non-singular space of lower dimension with maximal contact with the given space, is the very foundation of Hironaka's method in his 1964 proof for algebraic varieties and still appears, under various guises since maximal contact does not exist as such, in most attempts to prove resolution in positive characteristics. It appears here in a very purified form which, we believe, makes it much easier to understand the fundamental mechanisms at work in all proofs of birational resolution in high dimensions.

In the complex-analytic case, in comparison to the algebraic case, there is an essential new difficulty due to the fact that permissible local centers of blowing-up do not globalize, even at the price of adding singularities of Zariski closures and complicating the induction. The recent proofs use some form of uniqueness in the local construction, for example, precise conditions on the centers of blowing-ups using the exceptional divisors created by previous blowing-ups, which imply, according to a classical scheme, that locally defined centers will glue up. Hironaka's method is completely different and may serve as a template for solving certain globalization problems where there is no natural local uniqueness, by creating new objects (in this case, *groves* and *polygroves*) which contain more information and glue up naturally.

If we add that this book contains an elegant presentation of all the prerequisites of complex-analytic geometry, including basic definitions and theorems needed to follow the development of ideas and proofs, and that the epilogue presents the use of similar ideas in the resolution of singularities of complex-analytic foliations, we have given our reasons to find this book so useful and interesting. We think it will be a great help to the members of the younger generation who wish to understand one of the most fundamental results in algebraic and analytic geometry and invent possible extensions and applications of the methods created to prove it.

Bibliography

1. Bierstone, E., Milman, P.: Canonical desingularization in characteristic zero by blowing up the maximum strata of a local invariant. Invent. Math. **128**(2), 207–302 (1997)
2. Cutkosky, S.: Resolution of Singularities. Graduate Studies in Mathematics, vol. 63. American Mathematical Society, Providence (2004)
3. Hauser, H.: The Hironaka theorem on resolution of singularities. Bull. A.M.S. **40**(3), 323–403 (2003)
4. Hironaka, H.: Bimeromorphic smoothing of a complex-analytic space. Acta Math. Vietnam. **2**(2), 103–168 (1977)
5. Kollár, J.: Lectures on Resolution of Singularities. Annals of Mathematics Studies, vol. 166. Princeton University Press, Princeton (2007)
6. Lipman, J.: MR 1440306, Math. SciNet review of [1]

7. Villamayor, O.E.: Constructiveness of Hironaka's resolution. Ann. Sci. É.N.S. 4ème Série **22**(1), 1–32 (1989)
8. Włodarczyk, J.: Simple Hironaka resolution in characteristic zero. J. Amer. Math. Soc. **18**(4), 779–822 (2005)
9. Włodarczyk, J.: Resolution of singularities of analytic spaces. In: Proceedings of Gökova Geometry-Topology Conference 2008, pp. 31–63. Gökova Geometry/Topology Conference (GGT), Gökova (2009)

Paris, France
March 2018

Bernard Teissier

Preface

This book presents a complete and self-contained proof of the theorem of desingularization for complex-analytic varieties. Since I proposed a proof of the theorem [70–72], there have been many alterations and improvements, starting from the basic inductive techniques employed in my earlier paper on the resolution of singularities of algebraic varieties over a field of characteristic zero [63]. In this direction, one of the most notable improvements in the elegance of the style of the proof was seen in the paper of Bierstone and Milman [10]. Technically speaking, the first completely new approach, since my paper, was proposed by de Jong [75] including desingularization problems in all characteristics and even for arithmetic varieties. Although the end results of his work are somewhat weaker in terms of the constructive birationality between the given singular variety and the resulting smooth one, the approach was indeed remarkable in its breaking through the essential barriers to the zero characteristic world.

The approach taken in this work may be considered to be a return to the old one in my earlier papers [70, 71]. It may appear to be not so elegant as that of Bierstone and Milman, but it revives the notions that are not only geometrically interesting by themselves but also have some possibility of broader implications. The end result of this work is confined to the complex-analytic case, and many of the techniques employed are based upon the characteristic zero nature of complex numbers. Therefore, unlike the work of de Jong [75], the applicability beyond the world of characteristic zero of the methods presented here may appear slight.

Now, technically speaking, the essential features of this work are two notions, maximal contact and gardening, of which I now give brief accounts.

Maximal contact [50, 51, 71, 83]: We define a notion of maximal contact between a singular variety and a smooth variety in the same ambient space, which seems to have possibilities to be generalized between two singular varieties, though such generalizations are not explored in this work. In terms of its effective use for the desingularization problems, the following property is essential. If a smooth variety H has maximal contact with a singular variety X at a point x of X, then H contains all the infinitely near singularities of X at x which are not better than that of X itself at x. Infinitely near singularities at x mean those of the permissible transforms of X at the points corresponding to x. Here the permissible transforms are those obtained by the sequences of successively permissible blow-ups, where a blow-up is said to be permissible if its center is a smooth subvariety along which the singular variety has a constant Hilbert–Samuel function. (If the singular variety is a hypersurface in a smooth ambient space, the Hilbert–Samuel constancy means simply the equimultiplicity.) The maximal contact then implies that every sequence of permissible blow-ups of X to eliminate singularities as bad as X at x is essentially happening in H. What actually takes place in the desingularization procedure is that we find some singular data Y confined in H such that the question of improving the singularities of X to be better than the given is in fact equivalent to the desingularization of the data Y. The existence of H with smaller dimension than the original ambient space is the essence of the inductive nature of the proof of the complex-analytic desingularization theorems. The existence of such an H uses nothing but the characteristic zero property of the field of complex numbers, and it is indeed false for an algebraic variety of positive characteristics [51].

Gardening of infinitely near singularities [71, 73]: The idea here is to consider the totality of all the finite sequences of permissible local blow-ups for the given singular variety. In a sense, it is an analogy to the sheaf theory. Let us say that a singular variety X is given in a smooth ambient space Z. Then for each open set U of Z, we consider the set $G(U)$ of all sequences of successively permissible local blow-ups for the restriction of X to U. Here the adjective "local" to blow-ups means that the blow-up may be taken after restrictions to smaller open sets at every step of the succession. Our interest lies in the loci of those singular points of the transforms of X contained $hG(U)$ which are as bad as the worst singularity of X within U. Here "the worst" means having the greatest values of Hilbert–Samuel functions at singular points. Paying attention to the loci of those worst singular points in $G(U)$, called infinitely near worst singularities of X at x, we call $G(U)$ the grove of the infinitely near worst singularities of X within U. For various open sets U in Z, it is a sheaf of sets with singularity data, which are the transforms of X and the loci of the worst singular points in them. The most important fact about the groves, called the globality theorem [7, 72], is that the Hilbert–Samuel function at the worst singularity in X is not quite determined by $G(U)$, but it indeed is determined by $G(U \times \mathbb{C}^e)$ for $X \times \mathbb{C}^e$ in $Z \times \mathbb{C}^e$ for various positive integers e. In other words, flattening of the grove data enables us to recover the essential characters of the given singularities. This is a very useful fact in understanding the relation between "local" and "global" in a manner of sheaf theory.

The idea of writing this book began to surface quite a long time ago. My original thoughts began in the late 1960s when I conceived a clear idea for the globalization of complex-analytic desingularization, a few years after the publication of my *Annals* paper [63] in which the local desingularization was already done for general complex-analytic singularities. The idea became clearer when I encountered a group of young and brilliant mathematicians in Paris, who had strong interests in topics in singularities in algebraic and analytic geometry. They were Monique Lejeune, Bernard Teissier, Lê Dung Tráng, and several others. I received many useful suggestions and advice from them with regard to analytic desingularization and on many other closely related topics, especially as to the style of presentation for complex-analytic desingularization. I was much indebted particularly to Lejeune and Teissier. Later, in the early 1970s, I encountered two young and energetic Spanish mathematicians, José Manuel Aroca and José Luis Vicente, who are now the co-authors of this book. From that time, I became accustomed to paying several visits to Spain to talk with them about writing a book on complex-analytic desingularization. Preliminary outcomes of the plan were achieved with their help; these were the three volumes of lecture notes published in the Memorias de Matemática del Instituto Jorge Juan thanks to the support of the Consejo Superior de Investigaciones Científicas of Spain. The current book is essentially a rewriting of those three volumes. Added as an epilogue is a survey of applications of the methods exposed in this book to differential equations mainly developed by J.M. Aroca and his collaborators.

Cambridge, USA Heisuke Hironaka

Contents

Prologue

Singularities are ubiquitous. In daily life, they are almost everywhere. They are mostly pleasant to our senses, sometimes quite pretty and useful. For instance, personal signatures on official documents or personal accounts would be much less distinct or significant if they were written without singularities, nodes, and cusps. Cartoonists for magazines and newspapers should be particularly sensitive about singularities, perhaps beside other technical things, in drawing caricatures of celebrities or personalities in order to make them instantly recognizable. Writers should also try to search for or to create singularities of events in their works. Without singularities, this world would be far too uninteresting for artists and novelists, and indeed it would be very boring for normal human beings to live and work in.

It is the same with science or mathematics. Singularities appear in the shape of physical objects, in the sudden changes of movement, in the catastrophic behavior in the growing of quantities or in their destruction, and so on. Mathematically, all kinds of singularities naturally arise in geometry, in the theory of functions, in differential equations, etc. Singularities are often central research themes in the study of deformations and collapsing of geometric objects, in kinetic or statistical models, in asymptotic behavior of functions and degeneration of transformations, in analysis of scattering and propagations, in catastrophe, and even in chaos. Examples are indeed too many and too diverse to enumerate. Many hard mathematical problems exist and are left open for future research.

For many researchers in mathematics, if not all, singularities are interesting because of their ubiquity and complexity in general. They find it challenging to investigate singularities, for many nontrivial problems (sometimes extremely hard ones) arise in association with the presence of singularities, and thus, a good possibility exists for new research directions to open up.

In this book, the authors modestly admit that the scope of interest is quite limited in the face of the vast uncivilized jungle of singularities. Our interest will be restricted to the type of singularities which appear in the geometry of the zero point set of a system of complex-analytic functions in a complex number space. Moreover, our aim will be focused to a single theme, that is, the desingularization or resolution of singularities in complex-analytic varieties. Yet in higher dimensions the singularities are far from trivial, and the task of desingularization is not the kind of thing that is easily done after a few clever ideas.

Readers may ask: Why should we resolve singularities if they are so precious in many ways? What is the point of eliminating singularities which are often beautiful and sometimes useful? Is it only because the task of resolution in its generality to higher dimensions is far from any sort of triviality and quite challenging?

Our answer is partially yes. It is generally true that many mathematicians, if not all, have a common inclination, or a strong karma in some cases, to challenge any difficult problem for the sake of its difficulty. Even outsiders could imagine, we hope, that solving difficult problems in pure mathematics can sometimes produce a great unexpected reward of opening up a new road to deeper insights and better understandings of human endeavors in general, even if nothing of direct application or immediate use is apparent in any of the physical sciences or in technological production. The authors of this book, however, have a firm belief that our theory of complex-analytic desingularization has better than usual applicability at least within the territory of pure mathematics and perhaps more. Even just the end result of desingularization has been known to be quite useful in quantitative calculus as well as in qualitative analysis of data given with complex-analytic singularity, because the task can be carried out far more easily after pulling back the data to the smooth variety obtained by resolution of all singularities. What is more important in the authors' minds is that the techniques developed and the approach chosen to achieve the end result will provide better and deeper understanding of the properties of complex-analytic singularities in general, which happens now with differential equations and differential Galois theory, where our techniques have been successfully applied.

This book presents proofs of desingularization theorems for complex-analytic varieties by means of bimeromorphic transformations of their ambient complex-analytic manifolds. The main theorems of this work are stated below as four theorems, from Theorems 3 to 6, in general terms and in precise formulation. Needless to say, however, what may be of most interest to some readers is how we prove them rather than how we state them. Such readers might ask what the ideas behind our proofs are. Setting aside all the rigorous details for the moment, we first want to describe how the primary ideas of our proofs came about and grew into the current state of our presentation.

Idea 1

Consider the surface X defined by $f(x) = x_1^3 - x_2^2 x_3^4 = 0$ in the ambient space $\mathbb{C}^3 = Z$ with coordinates $x = (x_1, x_2, x_3)$. The singular locus of X is defined by the vanishing of the gradient $grad(f) = (3x_1^2, -2x_2x_3^4, -4x_2^2x_3^3)$. It is the union of two lines $L_2 = \{x_1 = x_2 = 0\}$ and $L_3 = \{x_1 = x_3 = 0\}$ (see Fig. P.1) intersecting at the origin $\xi = (0, 0, 0)$. Let us take the blow-up of Z with center L_2, which means the closure $Z' \subset Z \times \mathbb{P}^1$ of the graph of the map from (x_1, x_2, x_3) to the ratio $(x_1 : x_2)$, wherever this is well defined. Z' is covered by the following two copies of $\mathbb{C}^3$:

(1) $Z(1)$ with coordinate system $(x_1, x_1^{-1}x_2, x_3)$
(2) $Z(2)$ with coordinate system $(x_2^{-1}x_1, x_2, x_3)$

Since the order of f along L_2 is 2, we have that the strict transform X' of X is defined by $x_1^{-2}f = x_1 - (x_1^{-1}x_2)^2x_3^4 = 0$ within $Z(1)$, which has no singular point. On the other hand, X' is defined by $x_2^{-2}f = x_2(x_2^{-1}x_1)^3 - x_3^4 = 0$ within $Z(2)$, which has multiplicity 4 at the origin ξ' of the above coordinate system of $Z(2)$. This is greater than the multiplicity 3 of X at its image point ξ. In this example, the multiplicity of X at ξ was 3, while it was 2 at the other points of the center L_2 of the blow-up.

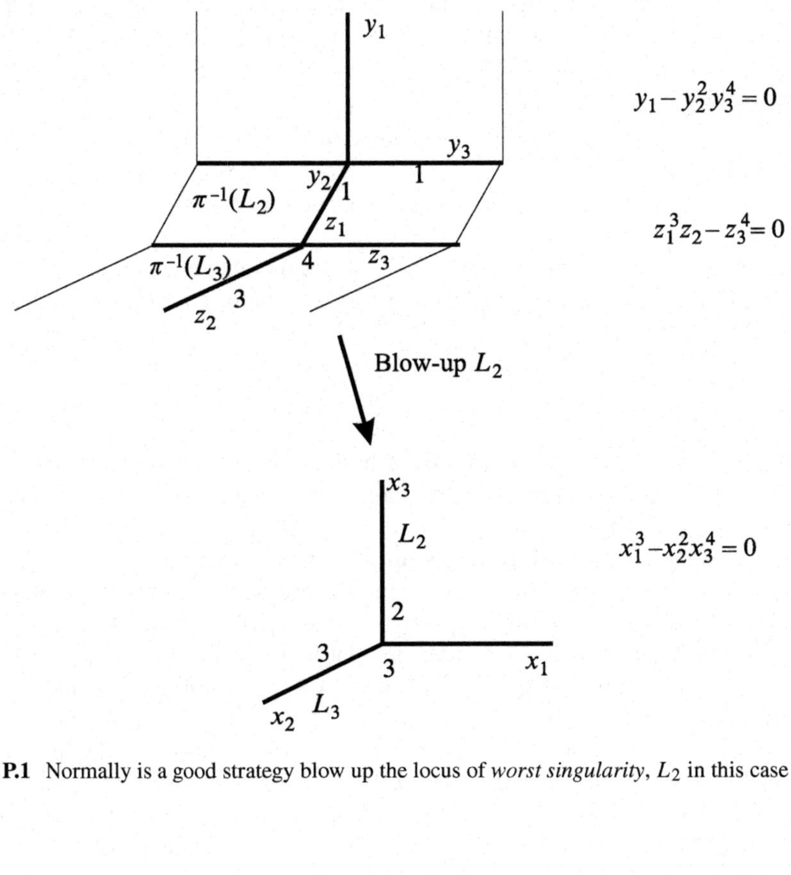

Fig. P.1 Normally is a good strategy blow up the locus of *worst singularity*, L_2 in this case

The lesson we learnt is that X should have a constant multiplicity along a center of choice. A blow-up or its center will be said to be *permissible* if the multiplicity is constant along the center. Indeed this permissibility condition (combined with the smoothness of the center by itself) assures us that the multiplicities of the transform are always bounded by the multiplicities at the image points. Thus, in our theory, we have decided to make use of only *permissible* blow-ups. In view of the above example, for instance, the locus of *worst singularity* seems to be a good candidate for a permissible center in this sense. This is *normally* a good strategy for desingularization, but only *normally*, as will be shown in *Idea 2*.

Idea 2

Let us next consider the surface X defined by the equation $f(x) = x_1^2 - x_2^2 x_3^2 = 0$ in $\mathbb{C}^3 = Z$. The singular locus of X is the union of two lines $L_2 = \{x_1 = x_2 = 0\}$ and $L_3 = \{x_1 = x_3 = 0\}$ intersecting at ξ. It is clear that ξ is the worst kind of singular point of X, that is, *geometrically speaking*. Take the blow-up Z' of Z with center ξ (see Fig. P.2), which is the closure of the graph of the map $(x_1, x_2, x_3) \mapsto (x_1 : x_2 : x_3)$. Z' is covered by the following three copies of $\mathbb{C}^3$:

$Z(1)$ with coordinate system $(x_1, x_1^{-1}x_2, x_1^{-1}x_3)$,
$Z(2)$ with coordinate system $(x_2^{-1}x_1, x_2, x_2^{-1}x_3)$ and
$Z(3)$ with coordinate system $(x_3^{-1}x_1, x_3^{-1}x_2, x_3)$.

In $Z(1)$, the strict transform of X is defined by

$$x_1^{-2} f(x) = 1 - x_1^2(x_1^{-1}x_2)^2(x_1^{-1}x_3)^2 = 0,$$

which has no singular point. For $i = 2, 3$, we let the above coordinate system of $Z(i)$ be called $x' = (x_1', x_2', x_3')$ in the ordering given there. Then the strict transform of X in $Z(i)$ is defined by

$$x_i^{-2} f(x) = x_1'^2 - x_2'^2 x_3'^2 = 0,$$

which is exactly the replacement of x by x' in the original equation. In other words, the blow-up with center ξ creates two singular points which are locally isomorphic to the original singular point ξ.

This gives us a lesson that the blow-up with the center of the worst singularity may serve no good outcome toward desingularization. In this example, the quickest way to desingularize X is as follows: Firstly, blow up Z choosing the center to be one of the two lines, say L_2, so that the singular locus of the strict transform $X_2 \subset Z_2$ of $X \subset Z$ is exactly the strict transform L_3' of L_3. Secondly, blow up Z_2 with center L_3' and find that the strict transform $X_{2,3} \subset Z_{2,3}$ of $X_2 \subset Z_2$ has no singular point. Under certain "special circumstances," we should choose a blow-up having a center from among the *largest permissible* ones instead of any one contained in the locus of *worst singularity*. The question of what makes "special circumstances" will be made precise later, but here we simply point out that the second term $x_2^2 x_3^2$ of f is *monomial*. Geometrically speaking, this is the case of *normal crossings* type.

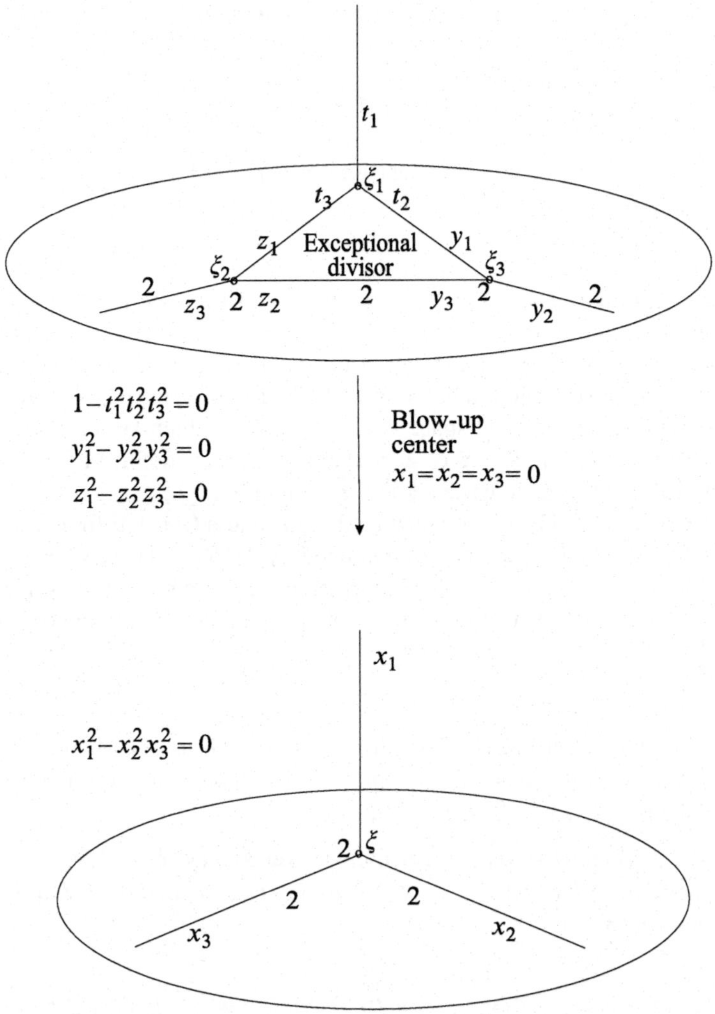

Fig. P.2 The blow-up with center in the worst singularity (0, 0, 0) does not improve the singularity

As a matter of fact, the same principle as above applies to an equation $f(y, x)$ of the following type where (y, x) is a system of $n = m + 1$ variables with $x = (x_1, \ldots, x_m)$. Namely, locally about the origin ξ of the coordinate system (y, x), we have a convergent power series factorization $f(y, x) = U(y, x)P(y, x)$ with $U(0, 0) \neq 0$ where $P(y, x)$ is as follows:

$$P(y, x) = y^d + \sum_{1 \le j \le d} h_j(y, x) y^{d-j} \tag{1}$$

$$\textit{where } \exists\, \alpha \in \mathbb{Z}_0^n \textit{ and } \exists\, k, 2 \le k \le d \textit{ such that}$$

$$h_k(y,x) = u(y,x)x^{\alpha} \quad \textit{with } u(0,0) \neq 0$$
$$h_1(y,x)^k = v(y,x)x^{\alpha} \quad \textit{with } v(0,0) = 0$$
$$\textit{and } \; h_j(y,x)^k \textit{ is divisible by } x^{j\alpha} \quad \textit{for } j, 1 \leq j \leq d$$

Of course, we are only interested in the case when $f(y,x)$ has order d, which means $|\alpha| = \sum_{1\leq i\leq m} \alpha_i \geq d$.

Idea 3

For an equation $f(y,x) = U(y,x)P(y,x)$ of the type Eq. (1), we can find an explicit algorithm to reduce its order by a finite sequence of permissible blow-ups (locally about the origin ξ). In fact it is a combinatoric process determined by α alone. For the sake of notational simplicity (with no serious loss of generality), let us assume that our ambient space is $\mathbb{C}^{m+1} = Z$ of which (y,x) is the standard global coordinate system. Let $X \subset Z$ be the hypersurface defined by $f(y,x) = 0$. For a subset $A \subset [1,m]$, we write $\sigma(A)$ for the number $\sum_{i\in A} \alpha_i$.

Now pick any $B \subset [1,m]$ such that $\sigma(B) \geq d$ and $\sigma(A) < d$ for every $A \subsetneq B$. Let $D(B)$ be the center of the blow-up defined by $D(B) = \{y = x_j = 0, j \in B\}$. Let $\pi : Z' \longrightarrow Z$. Then there is always a definite improvement of the singularity in the strict transform X' of X at every point of X' corresponding to the origin $\xi \in Z$ (see Fig. P.3).

Idea 4

In the desingularization presented in this book as well as elsewhere, Weierstrass theorems (preparation, division, etc.) play the fundamental role in the *proof by induction*. There the key is the following:

Lemma 1 *Let us consider a convergent power series $f(y,x) \in \mathbb{C}\{y,x\}$ where y is a single variable and $x = (x_1,\ldots,x_m)$ is a system of m variables. Let us assume that it is written as*

$$f(y,x) = U(y,x)P(y,x) \textit{ with } P(y,x) = y^d + \sum_{j=2}^{d} h_j(x)y^{d-j} \tag{2}$$

where $U(0,0) \neq 0$ and $h_j \in \mathbb{C}\{x\}, \forall j$. Then for every prime ideal $I \subset \mathbb{C}\{y,x\}$ defining a smooth subvariety D through the origin ξ, if $f \in I^d$, then we have

(1) $y \in I$, i.e., y vanishes on D and
(2) $h_j \in (x)^j\mathbb{C}\{x\}$ for every j.

The important point of this assertion is that the ideal I has no a priori relation with the choice of the coordinates (y,x).

Proof Let ∂_y be the derivation with respect to the variable y (with reference to the coordinate system (y,x)). We have $\partial_y^{d-1}P = d!y$, which is contained in $\partial_y^{d-1}I^d \subset I$. Hence $y \in I$. Next $\partial_y^{d-j}P - ((d-j)!)h_j$ is equal to $\binom{d}{d-j}y^j +$

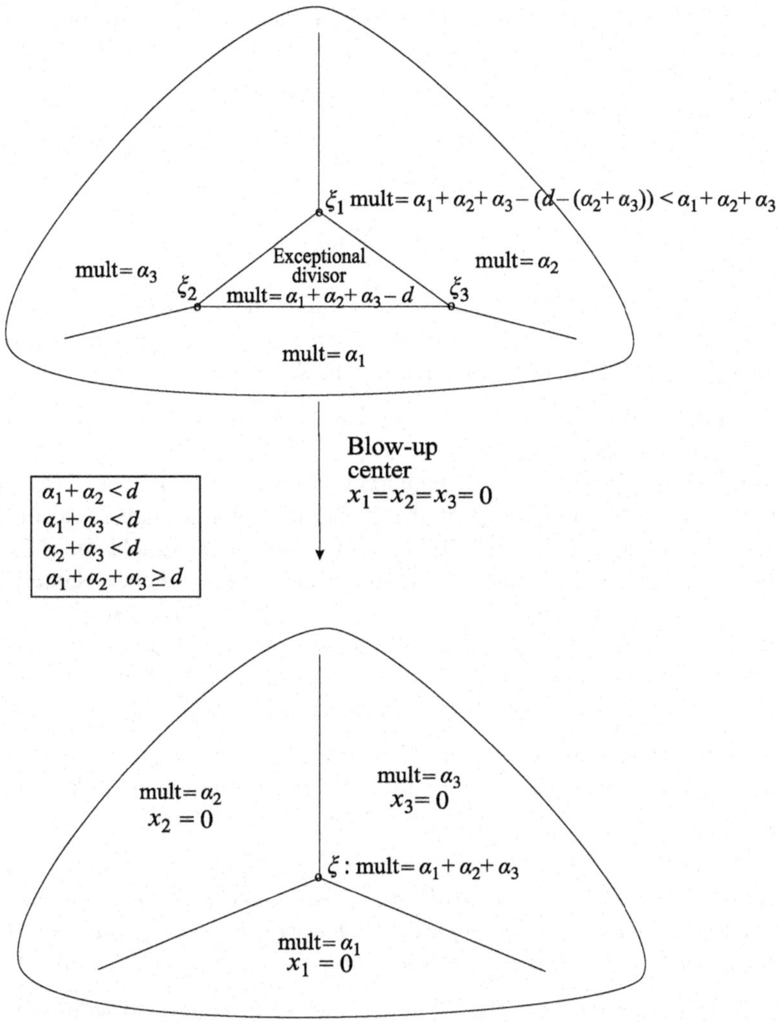

Fig. P.3 Idea 3 for $B = \{1, 2, 3\}$

$\sum_{2\leq i<j}\binom{d-i}{d-j}h_i y^{j-i}$. Hence, having $y \in I$ and $\partial_y^{d-j}P \in I^i$, from the induction on j starting from $j = 2$ to $j = d$ step by step, we can deduce that $h_j \in I^j$ for all j.□

Incidentally the assumption Eq. (2) can be always realized for any convergent power series $f \in \mathbb{C}\{y, x\}$ and the integer $d = ord_{(0)}(f)$ by means of a suitable $\mathbb{C}$-linear transformation of the variables and then by a suitable translation of y by an element of $\mathbb{C}\{x\}$.

The following corollary of Lemma 1 was indeed the very initial idea of our *proof by induction on the ambient dimensions*, i.e., dimensions of the ambient spaces (rather than dimensions of given singular data) as it asserts an *equivalence relation* between $m+1$-variables problems with m-variables problems.

Corollary 2 *The problem of reducing the multiplicity of $X : f(y,x) = 0$ down to less than d is equivalent to eliminating all the points of multiplicities $\geq b$ of the following ideal $J \subset \mathbb{C}\{x\}$ with respect to the following number b:*

$$J = \left(h_j^{\frac{b}{j}}, 2 \leq j \leq d\right)\mathbb{C}\{x\} \text{ and } b = d! \tag{3}$$

However, with respect to the *equivalence relation*, it should be kept in mind that the nature of problems is somewhat different from $m+1$ variables to m variables according to our inductive correspondences. The differences are:

(a) We should deal with a system of equations or an ideal (in m variables) rather than a single equation (in $m+1$ variables).
(b) We must keep in mind the denomination number b. The number b is usually greater than the multiplicity of the ideal. It is not enough to resolve the singularity of the maximal multiplicity of the ideal (in m variables).
(c) The transformation law is dictated by the denomination number. Namely, each time we take a blow-up, we must factor out b times the exceptional ideal (which means the principal ideal defining the inverse image of the center of the blow-up). Naturally this produces a *divisorial gap* between the transform with denomination and the usual *strict transform*.

At any rate, from now on, we consider a pair (J, b) with an ideal J defining the singular data and with the denomination b for targeting multiplicity reduction.

Idea 5

Theorem 3 *Given a coherent ideal sheaf $J \neq (0)$ on a complex manifold M, there exists a complex manifold $\tilde{M}$ and a proper bimeromorphic map $\tilde{\pi} : \tilde{M} \longrightarrow M$ such that the pullback $\tilde{J}$ of J by $\tilde{\pi}$ is locally normal crossings everywhere in $\tilde{M}$.*

Remark 1 Here *complex manifold M* is assumed to be connected with no loss of generality. We moreover assume that M is countable at infinity, i.e., it is a union of countably many compact subsets. We let $dim_{\mathbb{C}} M = dim_{\mathbb{C}} \tilde{M} = n$. The term *local normal crossings* will mean that for every point $\tilde{\xi} \in \tilde{M}$ we have a local coordinate system $x = (x_1, \ldots, x_n)$ of $\tilde{M}$ at $\tilde{\xi}$ such that $\tilde{J}$ is generated by a monomial in x within a neighborhood of $\tilde{\xi}$. If $n = 1$, then $\tilde{\pi}$ must be an isomorphism and J itself is *local normal crossings* from the beginning.

As a matter of fact, our result will be stronger than it is stated above. Indeed, we have the following:

Theorem 4 *The $\tilde{\pi}$ of Theorem 3 is locally obtained by a finite sequence of blow-ups with smooth centers. To be precise, let K be any compact subset of M. Then there exists an open neighborhood Z_0 of K in M such that $\tilde{\pi}\big|_{Z_0}$ is decomposed into a finite sequence of blow-ups*

$$\pi_j : Z_{j+1} \longrightarrow Z_j, 0 \leq j \leq r,$$

where the center D_j of π_j is such that:

(1) *D_j is closed and nowhere dense in Z_j and*
(2) *D_j is smooth by itself.*
(3) *Let $U_k, k = 1, 2,$ be two open subsets of Z_0. Assume that we have an isomorphism $t : U_1 \longrightarrow U_2$ such that $J\big|_{U_1}$ is the pullback of $J\big|_{U_2}$. Then for every $j \geq 0$, we have an isomorphism $t(j) : \pi(j)^{-1}(U_1) \longrightarrow \pi(j)^{-1}(U_2)$ which makes a commutative diagram:*

$$\begin{array}{ccc} \pi(j)^{-1}(U_1) & \xrightarrow{t(j)} & \pi(j)^{-1}(U_2) \\ \Big\downarrow \pi(j) & & \Big\downarrow \pi(j) \\ U_1 & \xrightarrow{t} & U_2 \end{array}$$

where $\pi(j)$ denotes the composition of π_i for all $i \leq j$.

Remark 2 It follows from Theorem 4 that the center D_j must be contained in the zero set of the pullback J_{j-1} for all j because any pair of points outside the zero set of J have isomorphic neighborhoods in which J induces the unit ideals.

Theorem 5 *Following the notation of Theorem 4, let us start with*

$$J_0^\sharp = J_0^\flat = J\big|_{Z_0},$$

which denotes the restriction of J to Z_0 and successively define $J_{j+1}^\sharp$ to be the pullback of $J_j^\sharp$ by π_j and $J_{j+1}^\flat$ by letting

$$J_{j+1}^\flat = P_{j+1}^{-d_j} J_j^\flat,$$

with the ideal P_{j+1} of the new exceptional divisor $\pi_j^{-1}(D_j) \subset Z_{j+1}$ and with the order of vanishing d_j of $J_j^\flat$ along the center $D_j \subset Z_j$. Furthermore, imposing an additional condition that the order $ord_\xi(J_j^\flat)$ is constant for all ξ in each of the connected components of D_j, we also have the existence of such a sequence as of Theorem 4 in which the final transform $J_{r+1}^\flat$ is the unit ideal everywhere in $\tilde{Z}$.

Remark 3 Under the additional condition of Theorem 5, $J_{j+1}^\flat$ is called the *strict transform* of $J_j^\flat$ by π_j. For instance, if J is the ideal of an irreducible reduced hypersurface X in Z, then there exists an integer $j_0, 0 \leq j_0 \leq r$, such that J_{j_0} is the ideal of a smooth hypersurface X_{j_0} in Z_{j_0} which is a desingularization of X. It is then true that for every j, $j_0 \leq j \leq r$, X_j is also a desingularization of X provided that it is not empty. It is important to note that making of the normal crossings of J_{r+1} is usually much stronger than a desingularization of X that is the smoothness of a strict transform X_j of X for some j. Incidentally, the imposed additional condition in Theorem 5 means that the multiplicity of the hypersurface X_j is constant in each of the connected components of D_j. The desingularization for non-hypersurface cases is stated in the theorem below.

The next desingularization theorem appears somewhat different in nature in comparison with the previous theorems. However, there is a general preparative technique of reducing a proof of the next theorem to that of Theorem 5.

Theorem 6 *For a complex-analytic subvariety, that is, a reduced complex-analytic subspace, $X \subset M$, there exists a complex manifold $\tilde{M}$ with a proper bimeromorphic map $\tilde{\pi} : \tilde{M} \longrightarrow M$ which is isomorphic outside the singular locus of X and such that the strict transform $\tilde{X}$ of X by $\tilde{\pi}$ is smooth. Moreover, we may require that for every compact subset K of M, there exists an open neighborhood Z_0 of K in M and $\tilde{\pi}\big|_{Z_0}$ is decomposed into a finite sequence of blow-ups*

$$\pi_j : Z_{j+1} \longrightarrow Z_j, 0 \leq j \leq r,$$

which satisfy the following conditions:

(1) *The center D_j of π_j is smooth by itself,*
(2) *D_j is closed and nowhere dense in Z_j, and*
(3) *D_j is contained in the singular locus of the strict transform X_j of X in Z_j.*
(4) *Let $U_k, k = 1, 2$, be two open subsets of Z_0 and assume that we have an isomorphism $t : U_1 \longrightarrow U_2$ which induces an isomorphism $X\big|_{U_1} \longrightarrow X\big|_{U_2}$. Then for every $j \geq 0$, we have an isomorphism $t(j) : \pi(j)^{-1}(U_1) \longrightarrow \pi(j)^{-1}(U_2)$, inducing an isomorphism $X_j\big|_{\pi(j)^{-1}(U_1)} \longrightarrow X_j\big|_{\pi(j)^{-1}(U_2)}$, which makes a commutative diagram:*

$$\begin{array}{ccc} X_j\big|_{\pi(j)^{-1}(U_1)} & \longrightarrow & X_j\big|_{\pi(j)^{-1}(U_2)} \\ \downarrow \pi(j) & & \downarrow \pi(j) \\ X_j\big|_{U_1} & \longrightarrow & X_j\big|_{U_2} \end{array}$$

where $\pi(j)$ denotes the composition of π_i for all $i \leq j$.
(5) *X_j is normally flat along D_j for every j.*

Remark 4 The last condition *normal flatness* means that at every point $\zeta \in D_j$ the tangent cone of X_j at ζ is isomorphic to the product of the following two cones in a natural way:

(1) the tangent cone of D_j at ζ which is a vector space because D_j is smooth and
(2) the fiber at ζ of the normal cone of X_j along D_j where the normal cone is a fiber space over the base space D_j.

This condition of normal flatness turns out to be technically useful for the proof of the above desingularization theorem more than the strengthening of the theorem.

Incidentally, the normal flatness is equivalent to saying that the Hilbert–Samuel function of X_j at ζ is constant for points ζ of each connected component of D_j. This constancy is equivalent to that of multiplicity for the case of hypersurfaces as was considered in Remark 3, but it is stronger than the multiplicity constancy in general for non-hypersurface cases.

We now give some guiding comments which we hope will be of help to readers before starting this book.

Chapter 1 This is an introduction to complex-analytic spaces where the readers are assumed to be familiar with basic facts in the theory of commutative algebras and to have seen the notion of sheaves of rings and ideals on a topological space. It would be helpful to the readers if a certain degree of familiarity has been already gained about holomorphic functions in several complex variables. Not much more is needed even for novice rearers to understand the contents of this chapter.

To those who are familiar with the theory of several complex variables and complex spaces, singular or smooth, almost nothing new will be found in this chapter except for our special way of presentation. New terminology of this book is chosen according to our specific objective and needs associated with it. Let us explain them here.

A complex space (or variety, or manifold) is defined in a usual way, that is, to build up a *global object* by piecing together a bunch of *local objects*. A *local object* means the one defined by firstly picking an open subset U in a complex number space $\mathbb{C}^n$ and then a set of complex-analytic equations within U:

$$f_\lambda(z_1, \ldots, z_n) = 0, \ 1 \le \lambda \le m.$$

Those f_λ's generate an ideal, let us call it J. Here we have two different ways of viewing things: one (strictly geometric way) in which our attention is focused to the set of zeros of the equations and another (rather algebraic way) in which the ideal J itself is viewed important and so is the open set U carrying J. In this book, we take the second way. The notion of *complex-analytic element* (Definition 1.2.1, Chap. 1) is exactly for the reason to emphasize this point of view all the way through this book.

We use certain new terms which may sound a bit unusual, such as *horizontal* and *vertical*. A morphism (complex-analytic map, complex conjugation map, etc.) is said to be *horizontal* (see Definition 1.2.2, Chap. 1) if it is locally isomorphic everywhere (as ringed spaces morphism). The reader may be more comfortable with such terms as *etalè* or *unramified*, which mean more or less the same. In the process of our desingularization, horizontal morphisms are abundantly present in the basic objects such as complex-analytic spaces.

When we consider a complex-analytic space, we add to the structure many of the local isomorphisms among its open pieces (or more generally local horizontal morphisms) as built-in furniture. In other words, in our way of dealing with a complex-analytic space (not always but quite often), we add some suitable pseudo-group of local isomorphisms as a part of its structure. This is exactly why we introduce the notion of $\mathbb{C}$-situations (see Definition 1.2.3, Chap. 1). The point is that our desingularization in the sense of $\mathbb{C}$-situations is invariant by any a priori chosen pseudo-group of local isomorphisms. For instance, when a complex-analytic space is given with an action of a set of automorphisms (finite, infinite, or continuous), its desingularization is carried out invariant for those automorphisms as a matter of *structural* consequence. For instance, such an automorphism could be the complex conjugation, and then our desingularization process is automatically real-analytic in every one of the steps.

In contrast to horizontal morphisms, we will define and make use of the term *vertical morphism* (see Definition 1.4.1, Chap. 1) which is roughly speaking a blow-up (see Definition 1.3.4 and the preceding paragraphs in the same section of Chap. 1) followed by an open restriction. There a *vertical morphism* is defined for $\mathbb{C}$-element. Namely, it is applied to the ambient space as well as to the possibly singular subvariety, and the center is required to be non-singular so that the resulting ambient space is held to be smooth. A *tree* means a finite sequence of vertical morphisms. This term is chosen to fit our later discussion on the *gardening of singularities*.

We would like to add one important comment at this point. Our theory is not searching for a special desingularization process, minimal or canonical, but rather for all possible desingularizations by successions of elementary transformations (subject to certain permissibility conditions), among which readers may choose the one of their liking.

Chapter 2 The chapter is about a generalization of the classical Weierstrass preparation theorem. This generalization is formulated and proved here as an important technical tool for desingularization of general complex-analytic singularities. A singular variety X may be locally defined in a complex space by a set of more than one holomorphic equations, i.e., may be defined as the intersection of a certain set, not unique, of hypersurfaces (each one locally defined by a single holomorphic equation), say:

$$X = \bigcap_{j=1}^{m} H_j.$$

Among all such expressions, we know how to choose a *good one* for the purpose of desingularization of X. Namely:

(1) A blow-up of the ambient space is permissible for X if an only if it is so for every one of the H_j's.
(2) A finite sequence of permissible blow-ups *improves* the singularity of X if and only if it does at least one of the H_j's.
(3) Consequently the *worst* singular locus of X is exactly the intersection of the *worst* singular loci of those H_j's.

Here for X, the permissibility is in the sense of *normal flatness* (see Definition 2.4.1, Chap. 2), which is equivalent to the constancy of the Hilbert–Samuel function along the center. This condition turns out to be equivalent, in the hypersurface case, to the *equimultiplicity*, because in this case the Hilbert–Samuel function is determined by the multiplicity and conversely. Moreover, the *improvement* means to get a certain change in the Hilbert–Samuel function (see the theorem of B. Bennet, Theorem 2.5.1, Chap. 2), which turns out to be equivalent to a decrease in multiplicity in the hypersurface case. There, the important point is the if and only if in the above statements.

To be more precise with definitions and theorems in regard to this matter, the reader must refer to their geometric version in Sect. 3.3 of the next chapter, definition of *suitedness* (see Definition 3.3.1, Chap. 3) and definitions of *tangential intersection* and *total suitedness* (see Definition 3.3.2, Chap. 3).

Furthermore, the notion of *idealistic exponent* is needed and introduced in Sect. 3.3 of Chap. 3 (see Definition 3.3.3). The permissibility of a blow-up for an idealistic exponent and the transform of the idealistic exponent by a permissible blow-up are all built in the definitions of *grove* and *polygrove*, which are presented later in Sect. 4.3 of Chap. 4 (Remark 4.3.1, Chap. 4).

Now after these references are made, we can state:

- A sequence of blow-ups over the ambient space is *permissible* for X in regard to successive strict transforms if and only if it is permissible for every one of the hypersurfaces H_j's in the same regard.

This is the merit of an expression $X = \bigcap_j H_j$, which is chosen to be tangential and totally suited to X.

Our generalization of the Weierstrass theorem is needed to prove the existence (and indeed to construct) an expression $X = \bigcap_i^m H_j$ which possesses all these properties and hence the above quoted property (and indeed more!).

At any rate, the desingularization problem (or rather its crucial inductive step) for an arbitrary singular variety X can be interpreted *equivalently* as the same problem for a system of hypersurfaces $\{H_j\}_{1 \leq j \leq m}$, related to X by $X = \bigcap_i^m H_j$. We prefer to express this by saying that the desingularization of X can be interpreted as the same for a certain idealistic exponents (written in terms of the ideal generated by suitable powered defining equations of the H_j). Thus, from that point on (mainly in Chaps. 4 and 5), we have only to deal with idealistic exponents.

Chapter 3 This is the theory of maximal contact. Given a singular point ξ of a complex-analytic variety X in an ambient space Z, we consider various smooth subvarieties (in particular, smooth hypersurfaces) of Z intersecting X at ξ and look for the one among them which has *maximal contact* with X at ξ. The notion of *maximal contact* is rather easily explained in the case of a smooth curve L with a singular plane curve X. It can be defined in terms of the intersection numbers of L and the analytic branches of X, together with the first characteristic exponents of these branches (see Sect. 3.1, Chap. 3). The notion in the general case is defined by means of various test curves, essentially reducing the problem to the case of smooth curves (see Definition 3.2.5, Chap. 3). The existence of maximal contact is proved in Theorem 3.2.3 (Chap. 3), which plays a key role in the inductive proof of desingularization theorems (see Chap. 5) with respect to the dimensions of the ambient spaces. If it is only for the sake of a proof by induction, there exists a simpler (though a little weaker) notion of *maximal contact* proposed by Jean Giraud [50], which is enough for the proof of desingularization theorems. However, we decided to present in this book our definition of maximal contact for we knew that ours is indeed stronger than that of Giraud. Moreover, we believe that the existence theorem (see Theorem 3.4.1, Chap. 3) in our sense of maximal contact is significant in its own right. At any rate, the *smooth* maximal contact in any good sense is strictly a *characteristic zero* phenomenon, meaning that there is not good analogy in the case of *positive characteristics* (here we are speaking of the characteristics of the base fields in algebraic geometry).

Chapters 4 and 5 From Chap. 4 on, what we call *gardening of infinitely near singularities* begins to develop. Infinitely near singular points mean those singular points which appear after a finite sequence of permissible blow-ups and are found to be as *bad* as the original point. So a *grove* is an abstract formulation of the totality of all such infinitely near singular points. In our formulation, we choose to take the totality of all sequences of permissible blow-ups, which are called *trees*, for a given idealistic exponent.

One point that deserves a special attention is that the extension from the notion of *groves* (Definition 4.1.1, Chap. 4) to that of *polygrove* (Definition 4.1.3, Chap. 4) is absolutely indispensable in our theory of infinitely near singularities.

For the sake of an imagery, the reader may take the simple case in which the idealistic exponent is the pair (J, b), where J is the ideal generated by an equation of a plane curve X, say analytically irreducible at a singular point ξ, and b is the multiplicity of X at ξ. Here the *badness* at an infinitely near point means that the multiplicity b remains the same. In this case, we can clearly state that, at least locally at ξ, the *grove* means the integral part $[r]$ of the *first characteristic exponent* r of X at ξ, while the *polygrove* means the rational number r itself. The difference between the two is great, isn't it?

However, the notion of *polygrove* is introduced by the need for the globalization of desingularization processes, and it is essential in higher dimensions (though it appears useless for the case of curve singularities). Generally speaking, we may start with an *idealistic exponent* given globally, but later on, the locally derived idealistic exponents (especially the ones derived by cutting down the dimensions of the ambient spaces by means of *maximal contact*) may not piece together in any sense of ideals. On the other hand, its *groves* and *polygroves* piece together thanks to our definition of trees, which are successions of local blow-ups. Roughly speaking, *geometry* is friendly to globalization, while *algebra* resists. Or the *equality* is much easier to handle than the *equivalence*. In our methodology, we consider the *equivalence relation* among idealistic exponents by means of the *equality* of associated polygroves, and then we see that the equivalence classes automatically piece together globally. Respect should be paid, however, to the power that *algebra* possesses in dealing with local problems on singularities. A golden rule is a good collaboration of *algebra* and *geometry*. This is exactly what our idealistic gardening of infinitely near singularities is all about.

There is one more point of extreme importance. It is about dealing with *normal crossings*. It concerns a question, explained above in this introduction, of choosing a strategy in attacking singularities in order to eliminate them. Generally speaking, it sounds like a good tactic to start attacking the part where singularities are *worst* in a certain sense. In the language of stratification with respect to certain measures, the first thought is to start attacking the smallest strata. It is indeed a wise decision, except when we achieved the state of *normal crossings*, as we saw above in the explanation of the basic ideas of desingularization; when we are in this state, we should change our tactics by 180 degrees. Namely, we should begin to blow up the largest possible strata.

The above observation led us to introduce the notion of *allées* in our gardening (see Definition 4.5.1, Chap. 4), which means nothing but normal crossings. They are meant to be paid special attention and respect.

Having the notions of *polygroves* and *allées*, we can start walking into a *forest* which may look wild at first sight and we then will learn to enjoy gardening all the way through to the end. The end result is our inductive proof of complex-analytic desingularization in all dimensions, which is done in the last chapter.

The authors would like to express their gratitude to Bernard Teissier for his suggestions, comments, and corrections. The authors would like to acknowledge the help and patience of those at Springer who made this book possible. Our very sincere thanks go to Ms. Chino Hasebe, without whose support this book would not have become a reality.

Chapter 1
Complex-Analytic Spaces and Elements

1.1 Complex-Analytic Spaces

Definition 1.1.1 A **ringed space** is a pair $(X, \mathscr{O})$, where X is a topological space (called the underlying topological space) and $\mathscr{O}$ is a sheaf of commutative rings with multiplicative unity (called the structure sheaf). Given two ringed spaces $(X, \mathscr{O})$ and $(X', \mathscr{O}')$, a **morphism of ringed spaces** between them is a pair (f, f^*), where f is a continuous map from X to X' and f^* is an f-homomorphism from $\mathscr{O}'$ to $\mathscr{O}$, i.e., a collection of ring homomorphisms (mapping unity to unity) $f^*{}_{U'} : \mathscr{O}'(U') \to \mathscr{O}(f^{-1}(U'))$, one for each open subset U' of X', such that for every $U'_1 \supset U'_2$ the diagram

$$
\begin{array}{ccc}
\mathscr{O}(f^{-1}(U'_1)) & \xleftarrow{f^*{}_{U'_1}} & \mathscr{O}'(U'_1) \\
\Big\downarrow \text{res} & & \Big\downarrow \text{res} \\
\mathscr{O}(f^{-1}(U'_2)) & \xleftarrow{f^*{}_{U'_2}} & \mathscr{O}'(U'_2)
\end{array}
$$

is commutative, where the vertical arrows are the restriction maps given in the sheaves $\mathscr{O}$ and $\mathscr{O}'$.

Definition 1.1.2 A **local-ringed space** is a ringed space $(X, \mathscr{O})$ such that the stalks $\mathscr{O}_x$, $x \in X$, are all local rings, i.e., commutative rings with unique maximal ideals. A **morphism of local-ringed spaces** from $(X, \mathscr{O})$ to $(X', \mathscr{O}')$ is a morphism of ringed

J. M. Aroca et al., *Complex Analytic Desingularization*,
https://doi.org/10.1007/978-4-431-49822-3_1

spaces (f, f^*) such that f^* induces local homomorphisms in the stalks, i.e., for each $x \in X$ the induced homomorphism of stalks

$$\mathscr{O}'_{f(x)} = \varinjlim_{f(x)\in U'} \mathscr{O}'(U') \xrightarrow{f^*} \varinjlim_{f(x)\in U'} \mathscr{O}(f^{-1}(U')) = \varinjlim_{x\in U} \mathscr{O}(U) = \mathscr{O}_x$$

maps the maximal ideal of $\mathscr{O}'_{f(x)}$ into that of $\mathscr{O}_x$.

Definition 1.1.3 Let K be a ring. A **K-local-ringed space** is a local-ringed space $(X, \mathscr{O})$ together with a K-algebra structure p in $\mathscr{O}$. A K-algebra structure p in $\mathscr{O}$ means a collection of ring homomorphisms $p_U : K \to \mathscr{O}(U)$, one for each open $U \subset X$, such that for every $U_1 \supset U_2$ the diagram

$$\begin{array}{ccc} & \overset{p_{U_1}}{\nearrow} & \mathscr{O}(U_1) \\ K & & \big\downarrow \text{res} \\ & \underset{p_{U_2}}{\searrow} & \mathscr{O}(U_2) \end{array}$$

is commutative. As usually if $z \in \mathscr{O}(U)$ and $a \in K$, we will write az instead of $p_U(a)z$.

Let $(X', \mathscr{O}')$ with p' be another K-local-ringed space. Then a **morphism** of K-local-ringed spaces, or K-**morphism** for short, from $(X, \mathscr{O})$ to $(X', \mathscr{O}')$ is a morphism (f, f^*) of local-ringed spaces such that

$$f^*{}_{U'} \cdot p'{}_{U'} = p_{f^{-1}(U')}$$

for every open subset U' of X'.

We shall almost always consider case $K = \mathbb{C}$ and only occasionally the case $K = \mathbb{R}$.

Example 1.1.1 The complex number space

$$\mathbb{C}^n = \{(\xi_1, \ldots, \xi_n) \mid \xi_i \in \mathbb{C} \text{ for } 1 \le i \le n\}$$

will be viewed as a $\mathbb{C}$-local-ringed space with the natural structure as follows: The underlying topological space is $\mathbb{C}^n$ with the standard metric topology, and the structure sheaf $\mathscr{O}$ is the sheaf of holomorphic functions. Namely, for each open subset U of $\mathbb{C}^n$, $\mathscr{O}(U)$ is the $\mathbb{C}$-algebra of all holomorphic functions on U, where the elements of $\mathbb{C}$ are identified with constant functions. Moreover, an open subset V of $\mathbb{C}^n$ will also be viewed as a $\mathbb{C}$-local-ringed space in which the structure sheaf is the restriction of $\mathscr{O}$ to V. This restriction will be denoted by $\mathscr{O}_V$.

Example 1.1.2 Let V be an open subset of $\mathbb{C}^n$ and V' an open subset of $\mathbb{C}^{n'}$. Let $f = (f_1, \ldots, f_{n'})$ be any system of holomorphic functions on V such that $f(\boldsymbol{\xi}) \in V'$ for all $\boldsymbol{\xi} \in V$. Then f defines a morphism of local-ringed spaces from V to V' in such a way that $f^*_{U'}(g) = g \cdot f$, where U' is any open subset of V' and g is any holomorphic function on U'. In short, f^* is the pull-back of functions by the map f. This pair (f, f^*) is a $\mathbb{C}$-morphism.

Example 1.1.3 Let V and V' be the $\mathbb{C}$-local-ringed spaces of Example 1.1.2. Not all the morphisms of local-ringed spaces from V to V' are obtained in the way described in Example 1.1.2. For instance, consider the case in which $n' = n$ and $V' = \{\bar{\boldsymbol{\xi}} \mid \boldsymbol{\xi} \in V\}$, where $\boldsymbol{\xi} = (\xi_1, \ldots, \xi_n)$ and $\bar{\boldsymbol{\xi}} = (\bar{\xi}_1, \ldots, \bar{\xi}_n)$; here the bars denote complex conjugates. Let $\sigma : V \to V'$ be the map defined by $\sigma(\boldsymbol{\xi}) = \bar{\boldsymbol{\xi}}$.

For any open subset U' of V' and any holomorphic function g on U', define

$$\sigma^*_{U'}(g)(\boldsymbol{\xi}) = \overline{g(\bar{\boldsymbol{\xi}})}, \quad \text{for } \boldsymbol{\xi} \in \sigma^{-1}(U').$$

Then $\sigma^*_{U'}(g)$ is a holomorphic function on $\sigma^{-1}(U')$ and $\sigma^*_{U'}$ is a ring homomorphism from $\mathscr{O}_{V'}(U')$ to $\mathscr{O}_V(\sigma^{-1}(U'))$. In fact, $(\sigma, \sigma^*) : V \to V'$, with $\sigma^* = \{\sigma^*_{U'}\}$, is a morphism of local-ringed spaces that is not a $\mathbb{C}$-morphism. It is called the **complex conjugation morphism**. Note that σ^* is not the pull-back of functions by the map σ. In particular, the complex number field $\mathbb{C}$ being identified with the set of constant functions on V and V', $\sigma^*_{U'}$ induces the complex conjugation in $\mathbb{C}$, not the identity.

Remark 1.1.1 Let V and V' be the same as in Example 1.1.2. Let

$$(f, f^*) : V \to V'$$

be any morphism of local-ringed spaces. If f^* induces the identity in $\mathbb{C}$ (which we identify with the sets of constant functions on V and on V'), then f is given by a system of holomorphic functions $(f_1, \ldots, f_{n'})$ as in Example 1.1.2 and f^* is the pull-back of functions by f. In fact, let $(y_1, \ldots, y_{n'})$ be the coordinate functions on $\mathbb{C}^{n'}$ and let $f_i = f^*_{V'}(y_i)$ for $1 \le i \le n'$. Then, for each $\boldsymbol{\xi} \in V$, if $\boldsymbol{\xi}' = f(\boldsymbol{\xi})$, then $y_i - \xi'_i$ belongs to the maximal ideal of $\mathscr{O}_{V',\boldsymbol{\xi}'}$ and hence $f_i - \xi'_i$ belongs to that of $\mathscr{O}_{V,\boldsymbol{\xi}}$ for $1 \le i \le n'$. This means that $f_i(\boldsymbol{\xi}) - \xi'_i = 0$ or $f_i(\boldsymbol{\xi}) = \xi'_i$ for $1 \le i \le n'$.

In other words, the map f is given by the system of holomorphic functions $(f_1, \ldots, f_{n'})$ as was noted. Moreover, if U' is any open subset of V' and g is any holomorphic function on U', then for $\boldsymbol{\xi} \in f^{-1}(U')$ we have

$$f^*_{U'}(g - g(f(\boldsymbol{\xi}))) = f^*_{U'}(g) - g(f(\boldsymbol{\xi})),$$

which must be in the maximal ideal of $\mathscr{O}_{V,\boldsymbol{\xi}}$ because $g - g(f(\boldsymbol{\xi}))$ is in the maximal ideal of $\mathscr{O}_{V',f(\boldsymbol{\xi})}$. Hence

$$f^*_{U'}(g)(\boldsymbol{\xi}) = g(f(\boldsymbol{\xi})) \quad \text{for all } \boldsymbol{\xi} \in f^{-1}(U'),$$

so that $f^*_{U'}(g) = g \cdot f$.

Remark 1.1.2 Let V and V' be again as above. Assume that V is non-empty and connected in $\mathbb{C}^n$, $n \geq 0$. If $(f, f^*) : V \to V'$ is any morphism of local-ringed spaces, f^* maps the constant functions on V' to constant functions on V, i.e., it is an endomorphism of the field $\mathbb{C}$, the latter being identified with the set of constant functions. We check this as follows. First, the constant 1 on V' is mapped to the same constant 1 on V. By elementary algebra, therefore, every rational number is mapped to itself, and $\sqrt{-1}$ is mapped to either $\sqrt{-1}$ or $-\sqrt{-1}$. Thus f^* induces an automorphism of the field $\mathbb{Q}(\sqrt{-1})$. Now, pick any $c \in \mathbb{C}$ and let $c_{V'}$ denote the constant function on V' with the value c. Suppose $f^*(c_{V'})$ were not a constant function on V. Then we must have $n > 0$ and, since V is connected, $f^*(c_{V'})$ would be even locally non-constant. Hence the image of $f^*(c_{V'})$ would contain a non-empty open subset of $\mathbb{C}$. This implies that there are infinitely many $b \in \mathbb{Q}(\sqrt{-1})$ such that $f^*(c_{V'}) - b_V$ vanishes at some point of V, or, equivalently, such that it is not invertible in $\mathcal{O}_V$. For each b with this property, let $b' \in \mathbb{Q}(\sqrt{-1})$ be such that $b_V = f^*(b'_{V'})$. We deduce that there are infinitely many $b' \in \mathbb{Q}(\sqrt{-1})$ such that $c_{V'} - b'_{V'}$ is not invertible in $\mathcal{O}_{V'}$. This is absurd, because $c_{V'} - b'_{V'} = (c - b')_{V'}$ is invertible except for one possible case in which $c = b'$. We conclude that $f^*(c_{V'})$ is a constant function. In other words, f^* induces an endomorphism of the field $\mathbb{C}$ that is an automorphism of the subfield $\mathbb{Q}(\sqrt{-1})$.

Remark 1.1.3 With the same notation as above, let $(f, f^*) : V \to V'$ be a morphism of local-ringed spaces and assume that:

(1) V is non-empty and connected, and
(2) there exists a holomorphic function g on V' such that $f^*(g)$ is non-constant on V.

Then we claim that f^* induces either the identity or the complex conjugation in $\mathbb{C}$ (identified with the sets of constant functions on V and V').

In fact, first of all, f^* induces an endomorphism of the field $\mathbb{C}$, as was proved in Remark 1.1.2. Call this endomorphism φ. We shall prove that φ is continuous. If this is done, the density of $\mathbb{Q}(\sqrt{-1})$ in $\mathbb{C}$ implies that φ is either the identity (if $\varphi(\sqrt{-1}) = \sqrt{-1}$) or the complex conjugation (if $\varphi(\sqrt{-1}) = -\sqrt{-1}$). We shall make use of the function g of assumption (2). Without any loss of generality, we may assume that we have $v \in V$ such that $g(f(v)) = 0$: just replace g by $g - g(f(v))$. Since $f^*(g)$ is also a non-constant holomorphic function on V, we can choose a holomorphic map

$$h : D = \{z \in \mathbb{C} \mid |z| < 1\} \to V$$

such that $h(0) = v$ and $f^*(g) \cdot h$ is also non-constant in D. Let f_1 denote the composition map $g \cdot f \cdot h$, which is obviously continuous. By Example 1.1.2, g and h are naturally viewed as morphisms of local-ringed spaces. We let $f_1^* = h^* \cdot f^* \cdot g^*$, so that (f_1, f_1^*) is a morphism of local-ringed spaces from D to $\mathbb{C}$. We have $f_1(0) = 0$, and for every rational $\varepsilon > 0$ there exists a rational $\delta > 0$ such that $D_\delta \subset f_1^{-1}(D_\varepsilon)$, where D_a denotes the open disk of radius a in $\mathbb{C}$ whose center is the origin. If $c \in \mathbb{C}$ satisfies $|c| < \delta$, then

$$\lambda = \sum_{i=0}^{\infty} \left(\frac{cz}{\delta\varepsilon}\right)^i \in \mathcal{O}_{\mathbb{C}}(D_\varepsilon).$$

Since f_1^* induces a local homomorphism of local rings at the origin, it is automatically continuous with respect to the adic topology in the local rings. Hence

$$f_1^*(\lambda) = \sum_{i=0}^{\infty} \left(\frac{f_1^*(c) f_1^*(z)}{f_1^*(\delta) f_1^*(\varepsilon)}\right)^i \in \mathcal{O}_{D_1}(D_\varepsilon).$$

Note that $f_1^*(\delta) = \delta$ and $f_1^*(\varepsilon) = \varepsilon$ because δ and ε are rational numbers; moreover, $f_1^*(c) = \varphi(c)$ because h^* and g^* induce the identity in $\mathbb{C}$. Then

$$f_1^*(\lambda) = \sum_{i=0}^{\infty} \left(\frac{\varphi(c) f_1^*(z)}{\delta\varepsilon}\right)^i \in \mathcal{O}_D(D_\delta).$$

The convergence of $f_1^*(\lambda)$ on D_δ implies that

$$\left|\frac{\varphi(c)\delta}{\delta\varepsilon}\right| \leq 1$$

and hence $|\varphi(c)| \leq \varepsilon$. We have proved that $|c| \leq \delta$ implies $|\varphi(\varepsilon)| \leq \varepsilon$. This implies that φ is continuous.

Remark 1.1.4 Let $(f, f^*) : V \to V'$ be a morphism of local-ringed spaces such that f^* induces the complex conjugation in $\mathbb{C}$. Let $V_1 = \{\bar{\xi} \mid \xi \in V\}$ and let (σ, σ^*) be the complex conjugation morphism from V_1 to V as was defined in Example 1.1.3. The morphism $V_1 \to V$ obtained by composing the two morphisms above is **holomorphic**, that is, given by a system of holomorphic functions as described in Example 1.1.2. (For the proof, see Remark 1.1.1.) Similarly, if (σ', σ'^*) is the complex conjugation morphism from V' to $V_1' = \{\bar{\xi} \mid \xi \in V'\}$, the composition from V to V_1' is also holomorphic. The pair (f, f^*) is said to be **antiholomorphic** if f^* induces complex conjugation in $\mathbb{C}$.

Example 1.1.4 Let V be an open subset in $\mathbb{C}^n$ and let $(h_1, \ldots, h_m)$ be a finite system of holomorphic functions in V. Let

$$S = \{\xi \in V \mid h_i(\xi) = 0,\ 1 \leq i \leq m\},$$

with the induced topology from $\mathbb{C}^n$. The quotient sheaf of $\mathcal{O}_V$ by the ideal generated by the system $(h_1, \ldots, h_m)$ is concentrated on S; we will denote by $\mathcal{O}_S$ the restriction of this sheaf to S. Then we obtain a $\mathbb{C}$-local-ringed space $(S, \mathcal{O}_S)$ together with ρ which is nothing but the identification of constants with constant functions. Incidentally, the quotient sheaf $\mathcal{O}_S$ is uniquely characterized by the following properties:

(1) It is a sheaf of $\mathbb{C}$-algebras on S.
(2) There is a homomorphism of sheaves of $\mathbb{C}$-algebras on S,

$$g : \mathcal{O}_V \to i^*\mathcal{O}_S,$$

where $i^*\mathcal{O}_S$ is the extension by zero of $\mathcal{O}_S$ with respect to the inclusion $i^* : S \to V$, that is, for every open subset U of V, we have $i^*\mathcal{O}_S(U) = \mathcal{O}_S(U \cap S)$. This is called the natural homomorphism.
(3) For every point $x \in V$, the local homomorphism $\mathcal{O}_{V,x} \to (i^*\mathcal{O}_S)_x$ induced by g is surjective.
(4) The kernel of this local homomorphism $\mathcal{O}_{V,x} \to (i^*\mathcal{O}_S)_x$ is the ideal generated by the germs at x of $(h_1, \ldots, h_m)$.

It is easy to prove that, if $\mathcal{O}'$ is a sheaf of $\mathbb{C}$-algebras on S and $g' : \mathcal{O}_V \to i^*\mathcal{O}'$ is a homomorphism satisfying the above condition, there exists a unique isomorphism $\varphi : \mathcal{O}' \to \mathcal{O}_S$ (as sheaves of $\mathbb{C}$-algebras) such that $g = i^*(\varphi) \cdot g'$ and $g' = i^*(\varphi^{-1}) \cdot g$.

Definition 1.1.4 A **complex-analytic space**, or $\mathbb{C}$-**space** for short, is a $\mathbb{C}$-local-ringed space $(X, \mathcal{O}_X)$ such that every point of X admits an open neighborhood W in X such that the restriction $(W, \mathcal{O}_X|_W)$ of $(X, \mathcal{O}_X)$ to W is $\mathbb{C}$-isomorphic to some $(S, \mathcal{O}_S)$ of the form described by Example 1.1.4.

Remark 1.1.5 The $\mathbb{C}$-local-ringed space $(S, \mathcal{O}_S)$ of Example 1.1.4, defined by means of $V \subset \mathbb{C}^n$ and $(h_1, \ldots, h_m)$, is of course an example of a $\mathbb{C}$-space. Note that, for instance, if the h_i are replaced by some of their powers, we have the same topological space S but a different structure sheaf, so that the $\mathbb{C}$-space $(S, \mathcal{O}_S)$ is changed. The difference is in nilpotent elements in $\mathcal{O}_S$. In general, if $(X, \mathcal{O}_X)$ is a $\mathbb{C}$-space, the nilpotent elements in $\mathcal{O}_X$ form a coherent ideal sheaf in $\mathcal{O}_X$, say $\mathrm{nil}(\mathcal{O}_X)$. Hence $(X, \mathcal{O}_X/\mathrm{nil}(\mathcal{O}_X))$ is again a $\mathbb{C}$-space. $(X, \mathcal{O}_X)$ is said to be **reduced** if $\mathrm{nil}(\mathcal{O}_X) = (0)$.

Remark 1.1.6 Throughout this work, we will often use the same symbol X for a $\mathbb{C}$-space $(X, \mathcal{O}_X)$ as well as for its underlying topological space, provided no confusion should arise from the context. It is important, however, that when X is called a $\mathbb{C}$-space, a specific structure sheaf $\mathcal{O}_X$ is meant to be attached to X.

Remark 1.1.7 By definition, over a $\mathbb{C}$-space there is given a specific $\mathbb{C}$-local-ringed space structure, not just a local-ringed space structure. It is, however, sometimes useful to consider morphisms from a $\mathbb{C}$-space to another regarding the domain and range only as local-ringed space, that is, morphisms that are not necessarily $\mathbb{C}$-morphisms. A typical example is a complex conjugation morphism as was defined in Example 1.1.3. For this reason, we introduce the following special definition.

Definition 1.1.5 Let X and X' be $\mathbb{C}$-spaces. Then a morphism f from X to X', where X and X' are regarded only as local-ringed spaces, is said to be **holomorphic** (or a **holomorphism**) if f^* induces the identity in $\mathbb{C}$ with reference to the structural maps from $\mathbb{C}$ into $\mathscr{O}_X$ and $\mathscr{O}'_X$. The morphism f is said to be **antiholomorphic** (or an **antiholomorphism**) if f^* induces the complex conjugation in $\mathbb{C}$. Compare Remark 1.1.4.

Remark 1.1.8 A morphism from X to X', where X and X' are regarded as local-ringed spaces, may fail to be either holomorphic or antiholomorphic, even locally about a point of X. For instance:

(1) Pick any point ξ' of X', and let f map all the points of X to the single point ξ'. Pick any local homomorphism φ from $\mathscr{O}_{X',\xi'}$ to $\mathbb{C}$ (i.e., any isomorphism from the residue field of $\mathscr{O}_{X',\xi'}$ to a subfield of $\mathbb{C}$), and define f^* by

$$\mathscr{O}_{X'}(U') \xrightarrow{\lambda_{U'}} \mathscr{O}_{X',\xi'} \xrightarrow{\varphi} \mathbb{C} \xrightarrow{p} \mathscr{O}_X(X)$$

for every open neighborhood U' of ξ', where $\lambda_{U'}$ and p are the canonical homomorphisms. Then (f, f^*) is a morphism of local-ringed spaces. Moreover, f^* can be neither surjective nor continuous in the subfield $\mathbb{C}$ of $\mathscr{O}_{X',\xi'}$.

(2) Suppose we have an element $g \in \mathscr{O}_X(X)$ such that $g \neq 0$ but $g^2 = 0$. Moreover, suppose we have a non-trivial derivation of $\mathbb{C}$ that extends to a derivation ∂ of $\mathscr{O}_X$ into itself. Define an endomorphism of $\mathscr{O}_X$ by $h \to h + g\partial(h)$, and call it f^*. Then (f, f^*), with the identity in the underlying topological space X, is a morphism of X to itself as local-ringed spaces. But f^* maps some constants to non-constants.

Remark 1.1.9 Morphisms such as those described in Remark 1.1.8 are outside our interest throughout this work. We shall be only interested in morphisms that are locally either holomorphic or antiholomorphic. Let X_1, X_2, X_3 be $\mathbb{C}$-spaces. Let $f_1 : X_1 \to X_2$ and $f_2 : X_2 \to X_3$ be morphisms. If both f_1 and f_2 are holomorphic, or if both are antiholomorphic, then $f_2 \cdot f_1 : X_1 \to X_3$ is holomorphic. If one of the two is holomorphic and the other is antiholomorphic, then $f_2 \cdot f_1$ is antiholomorphic.

Remark 1.1.10 Let X be a $\mathbb{C}$-space and let $p : \mathbb{C} \to \mathscr{O}_X$ be the given $\mathbb{C}$-algebra structure in X. Let $\sigma : \mathbb{C} \to \mathbb{C}$ be the complex conjugation map. Then we obtain a new $\mathbb{C}$-local-ringed space $\bar{X}$ that is the same as X as a local-ringed space (that is, has the same underlying topological space and the same structure sheaf) but in which the $\mathbb{C}$-algebra structure is changed by replacing p by $p\cdot\sigma$. Then $\bar{X}$ is also a $\mathbb{C}$-space. In fact, pick a $\mathbb{C}$-isomorphism from $X|_W$ to $(S, \mathscr{O}_S)$ according to Definition 1.1.4. Following Example 1.1.4 in conjunction with Example 1.1.3, set $V' = \{\bar{\xi} \mid \xi \in V\}$ and $h'_i(z) = h_i(\bar{z})$ for $1 \leq i \leq m$, so that the h'_i are holomorphic functions on V'. Let $(S', \mathscr{O}_{S'})$ be the $\mathbb{C}$-space where $S' = \{\xi \in V' \mid h'_i(\xi) = 0 \text{ for } 1 \leq i \leq m\}$ and $\mathscr{O}_{S'}$ is the quotient sheaf of $\mathscr{O}_{V'}|_{S'}$ by the ideal sheaf generated by $(h'_1, \ldots, h'_m)$. Then it is easy to show that $\bar{X}|_W$ is $\mathbb{C}$-isomorphic to $(S', \mathscr{O}_{S'})$. This proves that $\bar{X}$

is a $\mathbb{C}$-space. Moreover, the identity map from X to $\bar{X}$ is an isomorphism of local-ringed spaces that is antiholomorphic.

Definition 1.1.6 The $\mathbb{C}$-space $\bar{X}$ of Remark 1.1.10 is called the **conjugate $\mathbb{C}$-space** of X, and the identity map from X to $\bar{X}$ regarded as local-ringed spaces is called the **canonical antiholomorphism**.

Remark 1.1.11 Let $f : X \to X'$ be a morphism from one $\mathbb{C}$-space to another. The following conditions are equivalent:

(a) f is antiholomorphic.
(b) $f \cdot \sigma^{-1}$ is holomorphic, where $\sigma : X \to \bar{X}$ is the canonical antiholomorphism.
(c) $\sigma' \cdot f$ is holomorphic, where $\sigma' = X' \to \bar{X}'$ is the canonical antiholomorphism.

Remark 1.1.12 Let X be a $\mathbb{C}$-space and let $f_i \in \mathscr{O}_X(X)$, $1 \le i \le n'$. Let $\mathbf{z} = (z_1, \dots, z_{n'})$ be the coordinate system in $\mathbb{C}^{n'}$. Then there exists a unique holomorphic map f from X to $\mathbb{C}^{n'}$ (viewed as $\mathbb{C}$-space) such that $f^*(z_i) = f_i$, $1 \le i \le n'$. This is the same as Example 1.1.2 if X is an open subset of $\mathbb{C}^n$. The existence and uniqueness of f can be easily checked by means of Definition 1.1.4 and Example 1.1.4. Conversely, every holomorphism from X to $\mathbb{C}^n$ is obtained as above. Moreover, for a given system $(f_1, \dots, f_{n'})$ as above, there exists a unique antiholomorphism $\bar{f}$ from X to $\mathbb{C}^{n'}$ such that $\bar{f}^*(\bar{z}_i) = f_i$ for $1 \le i \le n'$. In fact, let $\bar{X}$ be the conjugate $\mathbb{C}$-space of X and let $\tau : \bar{X} \to X$ be the canonical antiholomorphism. Then $(\tau^*(f_1), \dots, \tau^*(f_{n'}))$ defines a unique holomorphism $f : \bar{X} \to \mathbb{C}^{n'}$ such that $\tau^*(f_i) = f^*(z_i)$ for $1 \le i \le n'$. Set $\bar{f} = f \cdot \tau^{-1}$, which is an antiholomorphism from X to $\mathbb{C}^{n'}$ such that $\bar{f}^*(\bar{z}_i) = f_i$ for $1 \le i \le n'$. The uniqueness of $\bar{f}$ follows from that of f. Again, conversely, every antiholomorphism from X to $\mathbb{C}^{n'}$ is given as above.

Example 1.1.5 (Complex-analytic subspace) Let $(X, \mathscr{O}_X)$ be a $\mathbb{C}$-space. Assume that we are given an open subset W of X and a sheaf of ideals $\mathbf{J}$ in $\mathscr{O}_X|_W$ which is coherent. Then we obtain a $\mathbb{C}$-space $(Y, \mathscr{O}_Y)$ deduced from W and $\mathbf{J}$ as follows. The underlying topological space Y is

$$Y = \{\boldsymbol{\xi} \in W \mid \mathbf{J}_{\xi} \subset \max(\mathscr{O}_{X,\xi})\},$$

which is necessarily a closed subset of W; and the structure sheaf $\mathscr{O}_Y$ is the quotient sheaf of $\mathscr{O}_X|_Y$ by the ideal sheaf $\mathbf{J}|_Y$. Note that this $(Y, \mathscr{O}_Y)$ is in fact a $\mathbb{C}$-space. The important point is that $\mathbf{J}$ is locally finitely generated. In fact, by Definition 1.1.3, it is enough to consider the case in which $(X, \mathscr{O}_X)$ is of the form $(S, \mathscr{O}_S)$ of Example 1.1.1, because the question is local. By restricting to a smaller neighborhood of any given point, we may even assume that there exists a system of holomorphic functions $(g_1, \dots, g_e)$ on W that induces a system of generators of $\mathbf{J}$, by Example 1.1.4(4), and that $S = W$. Then

$$Y = \{\boldsymbol{\xi} \in W \mid h_i(\boldsymbol{\xi}) = 0 \text{ for } 1 \le i \le m \text{ and } g_j(\boldsymbol{\xi}) = 0 \text{ for } 1 \le j \le e\},$$

and $\mathscr{O}_Y$ is the quotient of the restriction $\mathscr{O}_W|_Y$ by the ideal sheaf generated by $\{h_1, \ldots, h_m, g_1, \ldots, g_e\}$. We say that $(Y, \mathscr{O}_Y)$ is the **complex-analytic subspace**, or $\mathbb{C}$-**subspace**, of $(X, \mathscr{O}_Y)$ defined by $\mathbf{J}$ in W. Note that we have a canonical holomorphism from $(Y, \mathscr{O}_Y)$ to $(X, \mathscr{O}_X)$, composed of the inclusion $Y \xrightarrow{i} X$ and the natural quotient homomorphism $\mathscr{O}_X \to i^*(\mathscr{O}_Y)$ whose kernel is $\mathbf{J}$ in W. We call this the **inclusion morphism** from the $\mathbb{C}$-subspace $(Y, \mathscr{O}_Y)$ into the ambient $\mathbb{C}$-space $(X, \mathscr{O}_X)$. If Y is a closed subset of X, then $(Y, \mathscr{O}_Y)$ is called a **closed** $\mathbb{C}$-subspace of $(X, \mathscr{O}_X)$. In this case, the ideal sheaf $\mathbf{J}$ can be extended to a coherent ideal sheaf on X so that $\mathbf{J}|_{X \smallsetminus Y} = \mathscr{O}_X|_{X \smallsetminus Y}$. By doing so, we can define Y by a coherent ideal sheaf on all of X (that is, for $W = X$).

Example 1.1.6 Let $f : X \to X'$ be a morphism from a $\mathbb{C}$-space X to another $\mathbb{C}$-space X'. Let Y' be a closed $\mathbb{C}$-subspace of X', say defined by a coherent ideal sheaf $\mathbf{J}'$ in $\mathscr{O}_{X'}$. Then the **inverse image** $f^{-1}(Y')$ in X is defined to be the closed $\mathbb{C}$-subspace of X whose ideal sheaf is the one generated by $f^*(\mathbf{J}')$ in $\mathscr{O}_X$. Since f is a morphism of local-ringed spaces, we have the underlying topological space of $f^{-1}(Y')$ exactly equal to the inverse image of that of Y'. The notion of inverse image can be extended to an arbitrary $\mathbb{C}$-subspace, not necessarily closed, by means of the restriction of f to an appropriate open subset of X'.

Definition 1.1.7 A $\mathbb{C}$-space is said to be **smooth** at a point if its restriction to a suitable neighborhood of the point is $\mathbb{C}$-isomorphic to a $\mathbb{C}$-space $(V, \mathscr{O}_V)$ of the type described in Example 1.1.1. A smooth $\mathbb{C}$-space is a $\mathbb{C}$-space that is smooth at every point. A $\mathbb{C}$-space is said to be **singular** at a point if it is not smooth at the point. Such a point is often called a **singular point** of the $\mathbb{C}$-space.

Example 1.1.7 Let X be the closed $\mathbb{C}$-subspace of $\mathbb{C}^2$ whose ideal sheaf is generated by $z_1^2 - z_2^3$. Then X has only one singular point, the origin $(0, 0) \in \mathbb{C}^2$. By contrast, the closed $\mathbb{C}$-subspace of $\mathbb{C}^2$ whose ideal sheaf is generated by $(z_1^2 - z_2^3)^2$ has no smooth points; it is singular at every point.

Definition 1.1.8 A morphism f from a $\mathbb{C}$-space Y into a $\mathbb{C}$-space X is called an **immersion** if every point $\mathbf{y} \in Y$ admits an open neighborhood W in Y such that $f|_W : Y|_W \to X$ can be given by an isomorphism of local-ringed spaces from $Y|_W$ to a $\mathbb{C}$-subspace of X. It is called an **embedding** if it is an immersion and induces a homeomorphism from the underlying topological space of Y to its image with the induced topology from X. An immersion or embedding is called a $\mathbb{C}$-immersion or $\mathbb{C}$-embedding if it is a $\mathbb{C}$-morphism, that is, a holomorphism.

1.2 Elements, Horizontal Morphisms and $\mathbb{C}$-Situations

As was seen in Example 1.1.4, a local version of a $\mathbb{C}$-space such as $(S, \mathscr{O}_S)$ is given together with a closed $\mathbb{C}$-embedding into a smooth $\mathbb{C}$-space $(V, \mathscr{O}_V)$, where V is an open subset of some $\mathbb{C}^n$. In this way, the local-ringed space $(S, \mathscr{O}_S)$ is described by an ideal sheaf, or by a system of equations, in a smooth ambient space. In our work, as will become clear later, it is more convenient to deal with equations than with the point-set.

Definition 1.2.1 A **complex-analytic element**, or $\mathbb{C}$-**element** for short, is a closed $\mathbb{C}$-embedding of $\mathbb{C}$-spaces $\mathbf{s} : (s)_0 \to (s)^0$ such that $(s)^0$ is smooth. We call $(s)^0$ the **ambient space** of the $\mathbb{C}$-element $\mathbf{s}$. A **morphism** of $\mathbb{C}$-elements $\mathbf{a} : \mathbf{s}' \to \mathbf{s}$ is a pair $\mathbf{a} = ((a)_0, (a)^0)$, where $(a)_0 : (s')_0 \to (s)_0$ and $(a)^0 : (s')^0 \to (s)^0$ are morphisms of local-ringed spaces, making the following diagram commute:

$$\begin{array}{ccc} (s')_0 & \xrightarrow{(a)_0} & (s)_0 \\ \Big\downarrow{\scriptstyle \mathbf{s}'} & & \Big\downarrow{\scriptstyle \mathbf{s}} \\ (s')^0 & \xrightarrow{(a)^0} & (s)^0 \end{array}$$

We say that $\mathbf{a}$ is a $\mathbb{C}$-morphism if $(a)^0$, and hence also $(a)_0$, is a $\mathbb{C}$-morphism.

Example 1.2.1 If $\mathbf{s}$ is a $\mathbb{C}$-element and U is an open set in $(s)^0$, one may construct the $\mathbb{C}$-element

$$\mathbf{s}\,|_U : (s)_0 \cap \mathbf{s}^{-1}(U) \to U,$$

where we consider on both spaces the structure induced by $(s)_0$ and $(s)^0$ respectively. This $\mathbb{C}$-element is called the restriction of $\mathbf{s}$ to U.

Let X be a $\mathbb{C}$-space.

(1) A $\mathbb{C}$-**element** of X is a $\mathbb{C}$-element $\mathbf{s} : (s)_0 \to (s)^0$ in which $(s)_0$ is the restriction of X to an open subset of its underlying topological space.
(2) If we set $(s)_0 = (W, \mathscr{O}_X|_W)$ and $(s)^0 = (V, \mathscr{O}_V)$ with the notation in Example 1.1.4 and Definition 1.1.4, the $\mathbb{C}$-embedding $\mathbf{s} : (s)_0 \to (s)^0$ given by the isomorphism $(W, \mathscr{O}_X|_W) \to (S, \mathscr{O}_S)$ of 1.1 is a $\mathbb{C}$-element. Such a $\mathbb{C}$-element $\mathbf{s}$ will be called a **local** $\mathbb{C}$-**element** of the $\mathbb{C}$-space X.

In later chapters, morphisms of $\mathbb{C}$-elements are among the objects that we will most often speak of. Two kinds of such morphisms will interest us most: **horizontal** morphisms are used for pasting together various $\mathbb{C}$-elements to create certain global objects, and **vertical** morphisms are used for towering up successively created $\mathbb{C}$-elements by means of blow-ups, which we define later. In this section, we shall only talk about horizontal morphisms, which we define as follows.

Definition 1.2.2 A morphism of $\mathbb{C}$-elements $\mathbf{a} : \mathbf{s}' \to \mathbf{s}$ is said to be **horizontal** if the following conditions are satisfied:

(1) $(a)^0$ is an embedding of $\mathbb{C}$-spaces (not necessarily closed) that is locally either holomorphic or antiholomorphic.

(2) $\left((a)^0\right)^{-1}((s)_0) = (s')_0$.
(3) $(a)_0$ is a local isomorphism of local-ringed spaces, that is, for any point $x' \in (s')_0$ exists an open neighborhood V' in $(s')_0$ and an open neighborhood V of $(a)_0(x')$ in $(s)_0$ such that $(a)_0$ induces an isomorphism $(s')_0|_{V'} \to (s)_0|_V$ as local-ringed spaces.

Definition 1.2.3 Denote by ((el)) the category of $\mathbb{C}$-elements and morphisms and by ((el | hor)) the category of $\mathbb{C}$-elements in which the objects are the $\mathbb{C}$-elements and the morphisms are the horizontal morphisms (not necessarily $\mathbb{C}$-morphisms) of $\mathbb{C}$-elements. A **complex-analytic situation**, or $\mathbb{C}$-**situation** for short, is a covariant functor

$$G : [G] \to ((\mathrm{el} \,|\, \mathrm{hor})),$$

where $[G]$ is an abstract category, called the **index category** of the $\mathbb{C}$-situation G.

Example 1.2.2 Let X be a $\mathbb{C}$-space; we are going to construct a situation attached to X. First we needed the index category $[G]$:

Define ob$[G]$ to be the set of all $\mathbb{C}$-elements of X; then G : ob$[G] \to$ ob((el | hor)) will be the obvious map. Next, we want to define the set mor$[G]$ so that $[G] = \{\mathrm{ob}[G], \mathrm{mor}[G]\}$ becomes a category, and to extend G to mor$[G]$ and mor((el | hor)) so that $G : [G] \to ((\mathrm{el} \,|\, \mathrm{hor}))$ becomes a functor and hence a $\mathbb{C}$-situation. There are three obvious ways of doing this (among many others, of course):

(1) For $g, g' \in \mathrm{ob}[G]$, Hom(g, g') may be the set of all those horizontal $\mathbb{C}$-morphisms from $G(g)$ to $G(g')$ such that $(G(g))_0 \to (G(g'))_0$ is the inclusion morphism. (Note that $(G(g))_0$ and $(G(g'))_0$ are restrictions of X to open subsets of the underlying topological space of X.)
(2) Hom(g, g') may be the set of all the horizontal $\mathbb{C}$-morphisms (not necessarily inducing the inclusion morphisms).
(3) Hom(g, g') may be the set of all the horizontal morphisms (not necessarily $\mathbb{C}$-morphisms).

In each case, the extension of G to a functor $G \to ((\mathrm{el} \,|\, \mathrm{hor}))$ is the obvious one.

As will become clear later, each of these three $\mathbb{C}$-situations associated with the same $\mathbb{C}$-space X has its own significance in terms of transformations and eventual resolution of singularities. Case (2) is useful when we want to treat in the same way any two open pieces of X that are isomorphic as $\mathbb{C}$-spaces. The resolution, for instance, based upon the $\mathbb{C}$-situation of type (2) will be the same process when restricted to such a pair of $\mathbb{C}$-isomorphic open pieces of X. In particular, the resolution based upon the $\mathbb{C}$-situation of type (2) would yield complex conjugate processes when restricted to a pair of open pieces of X that are mutually antiholomorphically isomorphic.

Example 1.2.3 A **real-analytic space**, or $\mathbb{R}$-space for short, is by definition an $\mathbb{R}$-local-ringed space $(T, \mathscr{O}_T)$ such that, for every point $\mathbf{t}$ of T, there exists an open neighborhood N of $\mathbf{t}$ in the underlying topological space of T and an $\mathbb{R}$-isomorphism from the restriction $(N, \mathcal{O}_T|_N)$ to some $\mathbb{R}$-local-ringed space $(M, \mathcal{O}_M)$ of the following type: For some open subset L of $\mathbb{R}^n$ and some system of real-valued real-analytic functions $(g_1, \ldots, g_m)$ on L, we have

$$M = \{\boldsymbol{\xi} \in L \mid g_i(\boldsymbol{\xi}) = 0, 1 \leq i \leq m\}$$

with the induced topology from $\mathbb{R}^n$, and $\mathcal{O}_M$ is the restriction to M of the quotient sheaf of $\mathcal{O}_L$ by the ideal sheaf generated by $(g_1, \ldots, g_m)$, where $\mathcal{O}_L$ denotes the sheaf of real-valued real-analytic function on L.

In this work we do not deal with real-analytic spaces in any explicit way. But, as we will see below, they represent an interesting special class of the more general $\mathbb{C}$-situations that we shall work on in later chapters.

Remark 1.2.1 The $\mathbb{R}$-local-ringed space $(M, \mathcal{O}_M)$ of Definition 1.2.3 is of course an example of a real-analytic space. From this we can easily derive a so-called **complexification** as follows. Identifying $\mathbb{R}^n$ with the subspace of real points in $\mathbb{C}^n$, we can find an open subset V of $\mathbb{C}^n$ such that:

(1) $V \cap \mathbb{R}^n = L$ and g_i extends to a holomorphic function h_i in V, $1 \leq i \leq m$, and
(2) the complex conjugation in $\mathbb{C}^n$ maps V into itself and transforms h_i into itself for all i.

Then let $(S, \mathcal{O}_S)$ be the $\mathbb{C}$-space defined by V and $(h_1, \ldots, h_m)$, as in Example 1.1.4. It possesses an **antiholomorphic involution**, i.e., an automorphism σ of local-ringed spaces that induces the complex conjugation in $\mathbb{C}$ and has square σ^2 equal to the identity. This σ is induced by the complex conjugation morphism of $(V, \mathcal{O}_V)$, say τ, which was defined in Example 1.1.3. We call $(S, \mathcal{O}_S)$ together with σ a **complexification** of $(M, \mathcal{O}_M)$. On the other hand, if $(S, \mathcal{O}_S)$ is any $\mathbb{C}$-space with an antiholomorphic involution σ, we obtain a $\mathbb{R}$-local-ringed space $(M, \mathcal{O}_M)$ from them as follows:

(1) M is the set of fixed points by σ in S, with the induced topology;
(2) $\mathcal{O}_M$ is the subsheaf of $\mathcal{O}_S|_M$ that consists of all σ^*-invariant sections.

We call $(M, \mathcal{O}_M)$ the **real part** of $(S, \mathcal{O}_S)$ with respect to σ. Finally, we remark that we have a $\mathbb{C}$-element $\mathbf{s} : (s)_0 \to (s)^0$ and an **antiholomorphic involution a** of $\mathbf{s}$, where $(s)_0 = (S, \mathcal{O}_S)$, $(s)^0 = (V, \mathcal{O}_V)$, $\mathbf{a} = (\sigma, \tau)$ and $\mathbf{s}$ is the canonical embedding.

Example 1.2.4 ($\mathbb{C}$-situation associated with a real-analytic space) Let T be a real-analytic space. Consider $(N, \lambda, \mathbf{s}, \mathbf{a})$ such that:

(1) $\mathbf{s}$ is a $\mathbb{C}$-element and $\mathbf{a} = (\sigma, \tau)$ is an antiholomorphic involution of $\mathbf{s}$, i.e., $\mathbf{a}$ is a morphism from $\mathbf{s}$ to itself such that σ and τ are antiholomorphic involutions of $(s)_0$ and $(s)^0$, respectively, and
(2) N is an open subset of the underlying topological space of T and λ is an $\mathbb{R}$-isomorphism from $T|_N$ to the real part of $(s)_0$ with respect to σ.

Define ob$[G]$ to be the set of all $(N, \lambda, \mathbf{s}, \mathbf{a})$ as above and, for $g_i = (N_i, \lambda_i, \mathbf{s}_i, \mathbf{a}_i) \in \text{ob}[G]$ for $i = 1, 2$, define $\text{Hom}(g_1, g_2)$, to be the set of those horizontal morphisms $\mathbf{b}$ from $\mathbf{s}_1$ to $\mathbf{s}_2$ such that $\mathbf{a}_2 \cdot \mathbf{b} = \mathbf{b} \cdot \mathbf{a}_1$ and $(b)_0 \cdot \lambda_1 = \lambda_2 \cdot i$, where $i : T|_{N_1} \to T|_{N_2}$ is the inclusion morphism. Note that if N_2 does not contain N_1 then $\text{Hom}(g_1, g_2)$ is empty, and also that if $\mathbf{b} \in \text{Hom}(g_1, g_2)$ as above then $\mathbf{a}_2 \cdot \mathbf{b} = \mathbf{b} \cdot \mathbf{a}_1 \in \text{Hom}(g_1, g_2)$. (One of these two elements of $\text{Hom}(g_1, g_2)$ is holomorphic and the other is antiholomorphic.) By setting $\text{mor}[G] = \{\mathbf{b} \in \text{Hom}(g_1, g_2) \mid g_1, g_2 \in \text{ob}[G]\}$, we obtain a category $[G]$ and a functor $G : [G] \to ((\text{el} \mid \text{hor}))$ that maps $(N, \lambda, \mathbf{s}, \mathbf{a})$ to $\mathbf{s}$ and morphisms to themselves. This G is the $\mathbb{C}$-**situation** associated with the given real-analytic space X. It is easy to see that we can recover X from the situation G by putting together $(G(g))_0$ with $g \in \text{ob}[G]$ by means of $(G(\alpha))_0$ with $\alpha \in \text{mor}[G]$. (Warning: When we talk about a $\mathbb{C}$-situation G, $[G]$ is an abstract category and we disregard how its elements are obtained.)

1.3 Blow-Ups

Before we can speak of "vertical" morphisms in the category of $\mathbb{C}$-elements, we need to introduce the notion of a blow-up.

The complex number space $\mathbb{C}^{n+1}$ can be viewed as a complex vector space. In particular, we have the scalar multiplication:

$$\begin{array}{ccc} \mathbb{C} \times \mathbb{C}^{n+1} & \to & \mathbb{C}^{n+1} \\ (a, (\xi_0, \ldots, \xi_n)) & \mapsto & a, (\xi_0, \ldots, \xi_n) = (a\xi_0, \ldots, a\xi_n). \end{array}$$

Thus, the multiplicative group $\mathbb{C}^* = \mathbb{C} \setminus \{0\}$ acts on $\mathbb{C}^{n+1} \setminus \{\mathbf{0}\}$ and the quotient space by this scalar multiplication is the complex projective space:

$$\mathbb{P}^n_{\mathbb{C}} = (\mathbb{C}^{n+1} \setminus \{\mathbf{0}\})/\mathbb{C}^*$$

or $\mathbb{P}^n$ for short. We can do the same in the context of $\mathbb{C}$-spaces. $\mathbb{C}^{n+1}$ and hence $\mathbb{C}^{n+1} \setminus \{\mathbf{0}\}$ are viewed as $\mathbb{C}$-spaces by Example 1.1.1.

If $(z_0, \ldots, z_n)$ denotes the standard coordinate system in $\mathbb{C}^{n+1}$, and $\forall i,\ 0 \le i \le n,\ U_i = \mathbb{C}^{n+1} \setminus \{z_i = 0\}$ the identification:

$$U_i/\mathbb{C}^* \equiv \mathbb{C}^n$$

$$(z_0, \ldots, z_n).\mathbb{C}^* \equiv (z_0/z_i, \ldots, z_{i-1}/z_i, z_{i+1}/z_i, \ldots, z_n/z_i)$$

provides $\mathbb{P}^n$ with a unique structure of $\mathbb{C}$-space such that U_i is open in its underlying topological space and $\mathbb{P}^n|_{U_i}$ is $\mathbb{C}$-isomorphic to $\mathbb{C}^n$ by the above identification for every $i, 0 \le i \le n$. Clearly we have a $\mathbb{C}$-morphism

$$\pi_0 : \mathbb{C}^{n+1} \setminus \{\mathbf{0}\} \to \mathbb{P}^n,\ \pi_0(z_0, \ldots, z_n) = (z_0, \ldots, z_n).\mathbb{C}^*.$$

The graph of π_0 in $(\mathbb{C}^{n+1} \setminus \{\mathbf{0}\}) \times \mathbb{P}^n$ has a closure Z' in $\mathbb{C}^{n+1} \times \mathbb{P}^n$, which is a smooth $\mathbb{C}$-space of dimension $n+1$.

In fact, Z' can be explicitly obtained as follows. For each i fixed, $0 \le i \le n$, let $(w_0^{(i)}, \ldots, \widehat{w_i^{(i)}}, \ldots, w_n^{(i)})$ be the coordinate system in $\mathbb{P}^n|_{U_i}$, where $w_j^{(i)}$ corresponds to z_j/z_i for $j \neq i$ and for any letter w, $\widehat{w}$ means that w is omitted. Then $Z'|_{\mathbb{C}^{n+1}\times U_i}$ is the $\mathbb{C}$-subspace of $\mathbb{C}^{n+1} \times (\mathbb{P}^n|_{U_i})$ whose defining ideal sheaf is generated by the elements $z_j - z_i w_j^{(i)}$, where $0 \le j \le n$ and $j \neq i$. We have a $\mathbb{C}$-isomorphism

$$Z'|_{\mathbb{C}^{n+1}\times U_i} \to \mathbb{C}^{n+1}$$

defined by the system of holomorphic functions

$$(w_0^{(i)}, \ldots, w_{i-1}^{(i)}, z_i, w_{i+1}^{(i)}, \ldots, w_n^{(i)}).$$

Now let $\pi : Z' \to Z = \mathbb{C}^{n+1}$ be the $\mathbb{C}$-morphism induced by the first projection $\mathbb{C}^{n+1} \times \mathbb{P}^n \to \mathbb{C}^{n+1}$. Note that π restricted to $\mathbb{C}^{n+1} \times U_i$ is defined by

$$(w_0^{(i)} z_i, \ldots, w_{i-1}^{(i)} z_i, z_i, w_{i+1}^{(i)} z_i, \ldots, w_n^{(i)} z_i).$$

Definition 1.3.1 The map $\pi : Z' \to Z$ is called the **quadratic transformation** of Z with center $\mathbf{0}$, or the **blow-up** of Z with center $\mathbf{0}$.

Remark 1.3.1 The map π induces an isomorphism away from the center, that is, a $\mathbb{C}$-isomorphism

$$Z' \setminus \pi^{-1}(\mathbf{0}) \to Z \setminus \mathbf{0}.$$

This is the simplest and very basic case of a general blow-up, which we are going to define. A slightly more general case is as follows.

Definition 1.3.2 Set $Z = \mathbb{C}^{n+1} = \mathbb{C}^s \times \mathbb{C}^{r+1}$, where $n = s + r$. Let $Y = \mathbb{C}^s \times (\mathbf{0})$, considered as a $\mathbb{C}$-subspace of Z. Let $\varphi = W' \to \mathbb{C}^{r+1}$ be the quadratic transformation with center $\mathbf{0}$.

Set $Z' = \mathbb{C}^s \times W'$ and let $\pi : Z' \to Z$ be the $\mathbb{C}$-morphism that is the product of the identity in $\mathbb{C}^s$ and φ in W'. Then π is called the **monoidal transformation** with center Y, or the **blow-up** of Z with center Y. Note that π induces an isomorphism outside Y.

To state an implicit characterization of this blow-up and more general ones defined later, we need a notion of an **invertible** sheaf. Let X be any local-ringed space, and let $\mathscr{F}$ be a sheaf of $\mathscr{O}_X$-modules. We say that $\mathscr{F}$ is **invertible** if it is locally isomorphic to $\mathscr{O}_X$ itself as an $\mathscr{O}_X$-module. An ideal sheaf $\mathbf{H}$ in $\mathscr{O}_X$ is said to be invertible if it is invertible as an $\mathscr{O}_X$-module, that is, if and only if for every point

$x \in X$ there exist an open neighborhood N of x in X and $g \in \mathcal{O}_X(N)$ such that $\mathbf{H}|_N = (g)\mathcal{O}_X|_N$ and g is not a zero-divisor in the local ring $\mathcal{O}_{X,\xi}$ for any $\xi \in N$.

These are the most important properties of the monoidal transformation:

Lemma 1.3.1 *Let $\pi : Z' \to Z$ be the blow-up with center Y, as given in Definition 1.3.2. Let $\mathbf{I}$ be the ideal sheaf in $\mathcal{O}_Z$ defining Y as a $\mathbb{C}$-subspace of Z. Then:*

(1) *The ideal sheaf $\mathbf{I}\mathcal{O}_{Z'}$, generated in $\mathcal{O}_{Z'}$ by the pull-back of $\mathbf{I}$ with respect to π^*, is invertible as an $\mathcal{O}_{Z'}$-module.*
(2) *If $t : T \to Z$ is any $\mathbb{C}$-morphism of $\mathbb{C}$-spaces such that $\mathbf{I}\mathcal{O}_T$ is invertible as an $\mathcal{O}_T$-module, then there exists one and only one $\mathbb{C}$-morphism $g : T \to Z'$ such that $t = \pi \cdot g$.*

Proof Let $\mathbf{y} = (y_1, \ldots, y_s)$, $\mathbf{z} = (z_0, \ldots, z_r)$, be the standard coordinate systems in $\mathbb{C}^s$ and $\mathbb{C}^{r+1}$ respectively. Write y'_j for $\pi^*(y_j)$, z'_i for $\pi^*(z_i)$, and set $\mathbf{y}' = (y'_1, \ldots, y'_s)$. Now Z' is covered by $r+1$ open subsets, say V'_i for $0 \leq i \leq r$, such that $Z'|_{V'_i}$ is $\mathbb{C}$-isomorphic to $\mathbb{C}^{n+1}$ by

$$(\mathbf{y}', w_0^{(i)}, \ldots, w_{i-1}^{(i)}, z'_i, w_{i+1}^{(i)}, \ldots, w_r^{(i)})$$

and $\pi|_{V'_i}$ is given by

$$(\mathbf{y}', w_0^{(i)}z'_i, \ldots, w_{i-1}^{(i)}z'_i, z'_i, w_{i+1}^{(i)}z'_i, \ldots, w_r^{(i)}z'_i).$$

The ideal sheaf $\mathbf{I}$ is generated by $\mathbf{z}$; it is clear that $\mathbf{I}\mathcal{O}_{Z'}|_{V'_i}$ is generated by a single element, in fact by z'_i, and that this generator is not a zero-divisor. This being true for all i, we have (1).

To prove (2), we observe that the question is local in T. In fact, if we construct locally a collection of g's, they automatically patch together to a global morphism by the uniqueness in the assertion. So we may assume that there exists an index i such that $\mathbf{I}\mathcal{O}_T$ is globally generated by $t^*(z_i)$. This not being a zero-divisor, we find a unique section h_j of $\mathcal{O}_T$ such that $t^*(z_j) = h_j t^*(z_i)$, for $0 \leq j \leq r$. Then the system

$$(t^*(\mathbf{y}), h_0, \ldots, h_{i-1}, t^*(z_i), h_{i+1}, \ldots, h_r)$$

defines a $\mathbb{C}$-morphism $g : T \to Z'|_{V'_i}$ by means of the coordinate system in $Z'|_{V'_i}$ described above. It is clear that $t = \pi \cdot g$. To prove uniqueness, let $g' : T \to Z'$ be any morphism such that $t = \pi \cdot g'$. First of all, we shall prove that the image of g' must be in V'_i. Say $\eta \in T$ and $g'(\eta) \in V'_j$. Then, in the local ring $\mathcal{O}_{T,\eta}$, we must have

$$t^*(z_i) = g'^*(w_i^{(j)})t^*(z_j)$$

because $t = \pi \cdot g'$. Since $t^*(z_i)$ generates $\mathbf{I}\mathcal{O}_T$, the element $g'^*(w_i^{(j)})$ must be a unit in $\mathcal{O}_{T,\eta}$. Hence $w_i^{(j)}$ must be a unit in $\mathcal{O}_{Z',g'(\eta)}$. This means that $g'(\eta) \in V_j' \cap V_i'$. We have proved that $\mathrm{Im}(g') \subset V_i'$. Again we have

$$t^*(z_j) = g'^*(w_j^{(i)})t^*(z_i) \quad \text{for } 0 \le j \le r,$$

and $t^*(z_i)$ is not a zero-divisor. Hence we must have $g'^*(w_j^{(i)}) = h_j$ for all j. It follows that $g' = g$. □

With the same $\pi : Z' \to Z$ of Definition 1.3.2, let X be any $\mathbb{C}$-subspace of Z. There exists, by definition, an open subset V of Z and a coherent ideal sheaf $\mathbf{J}$ in $\mathcal{O}_V = \mathcal{O}_Z|_V$ that defines X as a closed $\mathbb{C}$-subspace of $Z|_V$.

Definition 1.3.3 The **strict transform** X' of X by the blow-up $\pi : Z' \to Z$ with center Y is, by definition, the smallest $\mathbb{C}$-subspace X' of Z' having the following properties:

(1) If V is any open subset of Z such that X is a closed $\mathbb{C}$-subspace of $Z|_V$, then X' is a closed $\mathbb{C}$-subspace of $Z'|_{\pi^{-1}(V)}$.
(2) The map π induces an isomorphism of $\mathbb{C}$-spaces from $X' \setminus \pi^{-1}(Y)$ to $X \setminus Y$.

Lemma 1.3.2 *The strict transform X' of X by $\pi : Z' \to Z$ with center Y always exists. Moreover, the ideal sheaf of X' in $\mathcal{O}_{V'} = \mathcal{O}_{Z'}|_{V'}$ with $V' = \pi^{-1}(V)$ is equal to*

$$\mathbf{J}' = \bigcup_{\nu=1}^{\infty} (\mathbf{J}\mathcal{O}_{V'} : \mathbf{I}^\nu \mathcal{O}_{V'}),$$

which is coherent.

Proof The union of an increasing sequence of coherent ideal sheaves in an analytic space is coherent (Grauert [56] Satz 8, §2). Hence $\mathbf{J}'$ is coherent. Clearly we have $(\mathbf{J}' : \mathbf{I}^\nu \mathcal{O}_{V'}) = \mathbf{J}'$ for all $\nu > 0$. Let X' be the closed $\mathbb{C}$-subspace of $Z'|_{V'}$ defined by this ideal sheaf $\mathbf{J}'$. We want to prove that this X' is the strict transform of X. Property (2) is clear because $\mathbf{I}$ is the unit ideal outside Y and hence $\mathbf{I}^\nu \mathcal{O}_{V'}$ is also the unit ideal outside $\pi^{-1}(Y)$. Hence $\mathbf{J}'$ and $\mathbf{I}\mathcal{O}_{V'}$ coincide on $Z' \setminus \pi^{-1}(Y)$, which implies (2). To prove the minimality of X', pick any X'' satisfying conditions (1) and (2). Let $\mathbf{J}''$ be the ideal sheaf of X'' in $\mathcal{O}_{V'}$. By (2) we have $\mathbf{J}''|_{V' \setminus \pi^{-1}(Y)} = \mathbf{J}'|_{V' \setminus \pi^{-1}(Y)}$. So if $\mathbf{H} = (\mathbf{J}'' + \mathbf{J}')/\mathbf{J}'$, its support is contained in $\pi^{-1}(Y)$. Since $\pi^{-1}(Y)$ is defined by the ideal $\mathbf{I}\mathcal{O}_{Z'}$, it is annihilated locally everywhere by suitable powers of $\mathbf{I}\mathcal{O}_{Z'}$. Namely,

$$\mathbf{J}'' + \mathbf{J}' \subset \bigcup_{\nu=1}^{\infty} (\mathbf{J}' : \mathbf{I}^\nu \mathcal{O}_{V'}).$$

But this last ideal is $\mathbf{J}'$ itself. Thus $\mathbf{J}'' \subset \mathbf{J}'$, i.e., $X'' \supset X'$. □

Lemma 1.3.3 *Let $p : X' \to X$ be the morphism induced by π, where X' is the strict transform of X by $\pi : Z' \to Z$ with center Y. Then:*

(1) *The ideal $\mathbf{I}\mathcal{O}_{X'}$ is invertible as an $\mathcal{O}_{X'}$-module, where $\mathbf{I}$ is the defining ideal of Y in $\mathcal{O}_{Z'}$.*
(2) *If $t : T \to X$ is any $\mathbb{C}$-morphism of $\mathbb{C}$-spaces such that $\mathbf{I}\mathcal{O}_T$ is invertible as an $\mathcal{O}_T$-module, there exists a unique morphism $g = T \to X'$ such that $t = p \cdot g$.*

Proof By Lemma 1.3.1(1), $\mathbf{I}\mathcal{O}_{Z'}$ is invertible as an $\mathcal{O}_{Z'}$-module. In particular, it is locally principal, that is, locally generated by a single element. It follows that $\mathbf{I}\mathcal{O}_{X'}$ is also locally principal. To prove part (1) of Lemma 1.3.3, it is therefore enough to show that $((0)\mathcal{O}_{X'} : \mathbf{I}\mathcal{O}_{X'}) = (0)\mathcal{O}_{X'}$. But X' is defined by the ideal sheaf $\mathbf{J}'$ in $\mathcal{O}_{V'}$; hence it is enough to show that $(\mathbf{J}' : \mathbf{I}\mathcal{O}_{V'} + \mathbf{J}') = \mathbf{J}'$, or, equivalently, that $(\mathbf{J}' : \mathbf{I}\mathcal{O}_{V'}) = \mathbf{J}'$. This is in fact immediate from Lemma 1.3.2.

To prove part (2), let $t' : T \to Z$ be the composition of t and the inclusion morphism from X into Z. Then there exists a unique morphism $g' : T \to Z'$ such that $t' = \pi \cdot g'$, by Lemma 1.3.1(2). So it is enough to show that g' factors through X'. We have $t'^*(\mathbf{J}\mathcal{O}_{V'}) = (0)$, since $t' = \pi \cdot g'$ with $g' : T \to Z'$. For every integer $\nu > 0$, we have

$$t'^*(\mathbf{J}\mathcal{O}_{V'} : I^\nu \mathcal{O}_{V'}) \subset ((0)\mathcal{O}_{X'} : \mathbf{I}^\nu \mathcal{O}_{X'}).$$

The right-hand side is $(0)\mathcal{O}_{X'}$, because $\mathbf{I}^\nu \mathcal{O}_{X'}$ is invertible as an $\mathcal{O}_{X'}$-module, by Lemma 1.3.3(1). Hence $t'^*(\mathbf{J}\mathcal{O}_{V'} : \mathbf{I}^\nu \mathcal{O}_{V'}) = (0)$ for all ν. Hence $t'^*(\mathbf{J}') = (0)$ by Lemma 1.3.2. This means that t' can be factored through a morphism $t : T \to X'$. □

We now consider the general case in which X is any $\mathbb{C}$-space and D any closed $\mathbb{C}$-subspace of X.

Definition 1.3.4 A $\mathbb{C}$-morphism $p : X' \to X$ is called the **blow-up** of X **with center** D if it has the following properties:

(1) If $\mathbf{H}$ is the ideal sheaf in $\mathcal{O}_X$ that defines D, then $\mathbf{H}\mathcal{O}_{X'}$ is invertible as an $\mathcal{O}_{X'}$-module.
(2) If $t : T \to X$ is any morphism of $\mathbb{C}$-spaces, locally either holomorphic or antiholomorphic, and if $\mathbf{H}\mathcal{O}_T$ is invertible as an $\mathcal{O}_T$-module, there exists a unique morphism $g : T \to X'$ such that $t = p \cdot g$.

Remark 1.3.2 In condition (2) of this definition, we may assume that t is holomorphic, or a $\mathbb{C}$-morphism. In fact, by the uniqueness, the condition is local and hence t may be assumed to be either holomorphic or antiholomorphic. If t is antiholomorphic, it can be replaced by a holomorphic map by composing with the canonical antiholomorphism of T (compare Definition 1.1.6).

Remark 1.3.3 Conditions (1) and (2) of Definition 1.3.4 make up a **universal mapping property**, and X' is unique up to canonical isomorphisms over X. Indeed, let $p_1 : X'_1 \to X$ be another $\mathbb{C}$-morphism satisfying (1) and (2). By (2), we have

$g_1 : X'_1 \to X'$ such that $p_1 = p \cdot g_1$ and $g : X' \to X'_1$ such that $p = p_1 \cdot g$. Now $p_1 = p_1 \cdot (g \cdot g_1)$. Hence, by the uniqueness of (2) applied to p_1 itself, $g \cdot g_1$ must be the identity of X'_1. Similarly, $g_1 \cdot g$ must be the identity of X'. Therefore g_1 is a $\mathbb{C}$-isomorphism. Note that it is also unique.

Remark 1.3.4 The notion of the blow-up of X is "local" in X in the following sense. A $\mathbb{C}$-morphism $p : X' \to X$ is the blow-up of X with center D if and only if for every point $x \in X$ there exists an open neighborhood M of x in X such that $p : X'|_{p^{-1}(M)} \to X|_M$ is the blow-up of $X|_M$ with center $D|_M$. The only-if part is immediate from Definition 1.3.4. To prove the converse, it is enough to check property (2) of Definition 1.3.4. Now, for every M as above, we have $g_M : T|_{t^{-1}(M)} \to X'$ with $t|_{t^{-1}(M)} = p \cdot g_M$. But, by the uniqueness of those g_M (and by the above only-if part), they automatically patch together to make a morphism $g : T \to X'$ with $t = p \cdot g$.

Remark 1.3.5 The **strict transform** X' of X by an arbitrary blow-up $\pi : Z' \to Z$ with center Y is defined exactly as in Definition 1.3.3.

Theorem 1.3.1 *Let X be a $\mathbb{C}$-space and let D be a closed $\mathbb{C}$-subspace of X. Then the blow-up of X with center D, say $p : X' \to X$, always exists. Moreover, if X_1 is a $\mathbb{C}$-subspace of X, the strict transform X'_1 of X_1 by p with respect to D always exists and the morphism $p_1 : X'_1 \to X_1$ induced by p is the blow-up with center $D \cap X_1$.*

Note 1.3.1 The meaning of $D \cap X_1$ is as follows. There exists an open subset W of X such that X_1 is a closed $\mathbb{C}$-subspace of $X|_W$. Let $\mathbf{H}$ be the ideal sheaf of D in $\mathscr{O}_X$. Then $D \cap X_1$ is the closed $\mathbb{C}$-subspace of X_1 defined by the ideal sheaf $(\mathbf{H}|_W)\mathscr{O}_{X_1}$.

Proof To prove the existence of the blow-up, note first that the question is local due to uniqueness, by the universal mapping property. Thus we may assume that X is given as a closed $\mathbb{C}$-subspace of an open subset U of $\mathbb{C}^s$ and that there exists a finite system of holomorphic functions $g = (g_1, \dots, g_r)$ in U that generates the ideal sheaf of D in $\mathscr{O}_X$. Let $V = U \times \mathbb{C}^{r+1}$, and let $(z_0, \dots, z_r)$ be the coordinate system in $\mathbb{C}^{r+1}$. The ideal sheaf of X in $\mathscr{O}_U$, together with $z_i - g_i$ for $0 \le i \le r$, generates an ideal sheaf in $\mathscr{O}_V$ that defines a closed $\mathbb{C}$-subspace of V. This is $\mathbb{C}$-isomorphic to X by the projection $V \to U$, and the $\mathbb{C}$-subspace D of X corresponds to the one whose ideal sheaf is generated by $(z_0, \dots, z_r)$. Therefore we may assume that X is given as a closed $\mathbb{C}$-subspace of an open subset V of $\mathbb{C}^s \times \mathbb{C}^{r+1}$ and that $D = X \cap Y$, where $Y = \mathbb{C}^s \times (\mathbf{0})$.

Let $\pi : Z' \to Z = \mathbb{C}^s \times \mathbb{C}^{r+1}$ be the blow-up with center Y as in Definition 1.3.2, and let X' be the strict transform of X by π, which exists by Lemma 1.3.2. Let $p : X' \to X$ be the morphism induced by π. Then, by Lemma 1.3.3, p is the blow-up with center D because the ideal sheaf of D in $\mathscr{O}_X$ is generated by the ideal sheaf $\mathbf{I}$ of Y in $\mathscr{O}_Z$. This completes the existence of the blow-up.

To prove the existence of the strict transform, consider first a local case as above. Then X_1 can be viewed as a $\mathbb{C}$-subspace of Z. By Lemma 1.3.2, the strict transform X'_1 of X_1 by π exists and is obviously contained in X', which is the strict transform of X by π. Now the property of being the smallest for X_1' is clearly the same for (π, Y) and for (p, D). Hence X_1' is the strict transform of X_1 by the blow-up p with center D. If M is an open subset of X_1 and X_1'' is the strict transform of $X_1|_M$ by π, then X_1'' must be equal to the restriction $X_1'|_{\pi^{-1}(M)}$. In fact, obviously X_1'' is a closed $\mathbb{C}$-subspace of $X_1'|_{\pi^{-1}(M)}$ by minimality. On the other hand, by Lemma 1.3.3, the induced morphism $X_1'' \to X_1|_M$ is the blow-up with center $D|_M$. Similarly, $X_1' \to X_1$ is the blow-up with center D. Since the properties of Definition 1.3.4 are preserved by an open restriction in the target, $X_1'|_{\pi^{-1}(M)} \to X_1|_M$ is also the blow-up with center $D|_M$. By the uniqueness of the blow-up, the inclusion $X_1'' \to X_1'|_{\pi^{-1}(M)}$ must be the identity. We have now proved that the strict transform is compatible with open restrictions. It follows immediately that, in the general case, the strict transforms of the restrictions of X_1 to open subsets of X_1 patch together to make up a global strict transform of X_1. Finally, $p_1 : X_1' \to X_1$ is the blow-up of X_1 with center $D \cap X_1$ because this is the case locally in X_1. □

1.4 Vertical Morphisms and $\mathbb{C}$-Trees

Definition 1.4.1 Let $\mathbf{s}'$ and $\mathbf{s}$ be $\mathbb{C}$-elements. A **vertical morphism of $\mathbb{C}$-elements** from $\mathbf{s}'$ to $\mathbf{s}$ is a triple $(\mathbf{a}, U, E)$, where:

(1) $\mathbf{a} : \mathbf{s}' \to \mathbf{s}$ is a morphism of $\mathbb{C}$-elements,
(2) U is an open subset of the ambient space $(s)^0$ of $\mathbf{s}$ and E is a smooth closed $\mathbb{C}$-subspace of $(s)_0|_U$, and
(3) $(a)^0 : (s')^0 \to (s)^0$ is the blow-up $(s')^0 \to (s)^0|_U$ with center E, followed by the inclusion $(s)^0|_U \to (s)^0$.

Remark 1.4.1 It is important that E is in $(s)_0$, not just in $(s)^0$, so that any horizontal morphism from $\mathbf{s}$ to another $\mathbb{C}$-element $\mathbf{s}_1$ induces a local isomorphism in E.

Remark 1.4.2 Given a $\mathbb{C}$-element $\mathbf{s}$, there exists always a **trivial** vertical morphism of $\mathbf{s}$, namely, $(\mathrm{id}, (s)^0, \phi)$, where id denotes the identity morphism from $\mathbf{s}$ to itself. Note that the identity $(s)^0 \to (s)^0$ is the blow-up with the empty center.

Remark 1.4.3 Let $(\mathbf{a}, U, E)$ be a vertical morphism of $\mathbb{C}$-elements from $\mathbf{s}'$ to $\mathbf{s}$. Let U_0 be the image of $(a)^0$ in the underlying topological space of $(s)^0$. Let E_0 be the set of points $\mathbf{x} \in U_0$ such that $(a)^0$ is not locally a isomorphism about $\mathbf{x}$. Then:

(1) The set U_0 is the union of those connected components of U that are not contained in E, so that U_0 is an open subset of $(s)^0$.
(2) The set E_0 is the set of those points in U at which the ideal sheaf of E is not invertible as $\mathcal{O}_U$-module. Clearly E_0 is a closed subset of U_0. Moreover, since E is smooth, E_0 is simply the union of the connected components of E that have codimension ≥ 2. Hence E_0 is a smooth closed $\mathbb{C}$-subspace of $(s)^0|_{U_0}$.
(3) The morphism $(\mathbf{a}, U_0, E_0)$ is a vertical morphism of $\mathbb{C}$-elements.

Note that U and E are not uniquely determined by the morphism $\mathbf{a}$. We can roughly say, however, that U is determined by $\mathbf{a}$ except for those connected components contained in E (and hence locally equal to E) and that E is determined by $\mathbf{a}$ except for those connected components of codimension ≤ 1.

Definition 1.4.2 A **complex-analytic tree**, or **$\mathbb{C}$-tree**, is a finite sequence $A = \{\mathbf{a}_i, U_{i-1}, E_{i-1}\}_{1 \leq i \leq m}$ of vertical morphisms of $\mathbb{C}$-elements, where $\mathbf{a}_i : \mathbf{s}_i \to \mathbf{s}_{i-1}$. We call the $\mathbb{C}$-element $\mathbf{s}_0$ the **soil** of the $\mathbb{C}$-tree A. The sequence of morphisms of $\mathbb{C}$-spaces $\{(a_i)^0 : (s_i)^0 \to (s_{i-1})^0\}_{1 \leq i \leq m}$ is called the **ambient diagram** of A, and $\{(a_i)_0 : (s_i)_0 \to (s_{i-1})_0\}_{1 \leq i \leq m}$ the **singular diagram** of A. The integer m is the **length** of the $\mathbb{C}$-tree A.

Definition 1.4.3 Let

$$A = \{\mathbf{a}_i, U_{i-1}, E_{i-1}\}_{1 \leq i \leq m} \quad \text{and} \quad A' = \{\mathbf{a}'_i, U'_{i-1}, E'_{i-1}\}_{1 \leq i' \leq m'}$$

be $\mathbb{C}$-trees. A **morphism of $\mathbb{C}$-trees** from A to A' is a sequence of morphisms $F = \{f_i\}_{0 \leq i \leq m}$ such that, for $1 \leq i \leq m$:

(1) $\mathbf{f}_i$ is a horizontal morphism from $\mathbf{s}_i$ to $\mathbf{s}'_i$ and the diagram

$$\begin{array}{ccc} \mathbf{s}_i & \xrightarrow{\mathbf{f}_i} & \mathbf{s}'_i \\ {\scriptstyle a_i}\downarrow & & \downarrow{\scriptstyle a_i} \\ \mathbf{s}_{i-1} & \xrightarrow{\mathbf{f}_{i-1}} & \mathbf{s}'_{i-1} \end{array}$$

is commutative, and

(2) $U_i \subset \left((f_i)^0\right)^{-1}(U'_i)$ and $E_i = \left((f_i)^0\right)^{-1}(E'_i) \cap U_i$.

Also, for fixed $\mathbb{C}$-element $\mathbf{s}$, we define an **s-tree** as any $\mathbb{C}$-tree with soil $\mathbf{s}$, and an **s-morphism** between $\mathbf{s}$-trees as any morphism for which f_0 equals the identity of $\mathbf{s}$.

Remark 1.4.4 By this definition, if $m' < m$, there exists no morphism of $\mathbb{C}$-trees from A to A'.

The $\mathbb{C}$-trees and morphisms of trees form a category, which will be denoted by (($\mathbb{C}$-trees)). It contains the category ((el | hor)) as a full subcategory.

Together, $\mathbf{s}$-trees and $\mathbf{s}$-morphisms form a subcategory of (($\mathbb{C}$-trees)), which will be denoted by (($\mathbf{s}$-trees)).

Remark 1.4.5 Since $\mathbf{f}_i$ is horizontal, $(f_i)_0 : (s_i)_0 \to (s'_i)_0$ is a local isomorphism. By definition of vertical morphism, $E'_i \subset (s_i)_0$, so condition (2) of Definition 1.4.3 implies that $(f_i)^0$ induces a local isomorphism from E_i to E'_i.

Remark 1.4.6 If $\mathbf{f}_0$ is holomorphic, each $\mathbf{f}_i$, for $i > 0$, is holomorphic as well. Similarly, if $\mathbf{f}_0$ is anti-holomorphic, so are the other $\mathbf{f}_i$.

Example 1.4.1 Let $A = \{\mathbf{a}_i, U_{i-1}, E_{i-1}\}_{1\leq i\leq m}$ be a $\mathbb{C}$-tree. Let k be an integer, $1 \leq k \leq m$, and let V_{i-1} an open subset of U_{i-1} such that $(a_i)^0(V_i) \subset V_{i-1}$ for $1 \leq i \leq m$. Then $B = \{\mathbf{b}_i, V_{i-1}, F_{i-1}\}_{1\leq i\leq k}$ is a $\mathbb{C}$-tree, where $F_{i-1} = E_{i-1}|_{V_{i-1}}$ and $\mathbf{b}_i : \mathbf{s}_i|_{V_i} \to \mathbf{s}_{i-1}|_{V_{i-1}}$ is the morphism induced by $\mathbf{a}_i$. Moreover, we have a morphism $\mathbf{f} : B \to A$, where the $\mathbf{f}_i : \mathbf{s}_i|_{V_i} \to \mathbf{s}_i$, for $0 \leq i \leq m$, are the obvious inclusion morphisms. Such a morphism $\mathbf{f}$ will be called an **open inclusion morphism of $\mathbb{C}$-trees.**

Let $\mathbf{s}$ be a $\mathbb{C}$-element, V a smooth $\mathbb{C}$-space and $\tau : V \to (s)^0$ a smooth $\mathbb{C}$-holomorphic map. Then the natural $\mathbb{C}$-embedding $\tau^{-1}((s)_0) \to V$ is a $\mathbb{C}$-element, which will be denoted by $\tau^*(\mathbf{s})$.

Now consider an $\mathbf{s}_0$-tree as in the beginning.

(i) Write $\mathbf{s}'_0 = \tau^*(\mathbf{s}_0)$, $U'_0 = \tau^{-1}(U_0)$, $E'_0 = \tau^{-1}(E_0)$.
(ii) For $1 \leq i \leq m$, write

$$(s'_i)^0 = (s_i)^0 \times_{(s_{i-1})^0} (s'_{i-1})^0$$

and denote by $\tau_i : (s'_i)^0 \to (s_i)^0$ and $(a'_i)^0 = (s'_i)^0 \to (s'_{i-1})^0$ the projections.
(iii) For $1 \leq i \leq m$, write $(s'_i)_0 = \tau_i^{-1}((s_i)_0)$ and let $(a'_i)_0$ be the morphism induced by $(a_i)_0$.
(iv) For $1 \leq i \leq m-1$, write $U'_i = \tau_i^{-1}(U_i)$ $E'_i = \tau_i^{-1}(E_i)$.

Then $A' = \{\mathbf{a}'_i, U'_{i-1}, E'_{i-1}\}_{1\leq i\leq m}$ is a $\tau^*(\mathbf{s}_0)$-tree, which will be denoted by $\tau^*(A)$.

In particular, if V is an open set in $(s)^0$ and τ is the inclusion, $\tau^*(A)$ will be denoted by $A|_V$ and called the restriction of A to V.

If V is a smooth $\mathbb{C}$-space and $\tau : (s)^0 \times V \to (s)^0$ is the projection, $\tau^*(\mathbf{s})$ will be denoted by $\mathbf{s} \times V$ and $\tau^*(A)$ by $A \times V$.

Definition 1.4.4 Let $F : A \to A'$ be a morphism of $\mathbb{C}$-trees, as given in Definition 1.4.3. We call F an **ambient change** if $m = m'$, $U_i = \left((f_i)^0\right)^{-1}(U'_i)$ and $(f_i)^0$ is an isomorphism for $0 \leq i \leq m = m'$. We also call A' an **ambient expansion** of A and A an **ambient contraction** of A' by means of the ambient change F.

If $\mathbf{f} : \mathbf{s}_0 \to \mathbf{s}'_0$ is a horizontal morphism and if there exists an ambient change $F : A \to A'$, $F = \{\mathbf{f}_i\}_{0\leq i\leq m}$ such that $\mathbf{f}_0 = \mathbf{f}$, we shall say that A is an ambient change of A' by $\mathbf{f}$. And we shall say that A is **f-equivalent** to A' if there exists an open set U in $(s'_0)^0$ such that $\mathbf{f}$ induces a $\mathbb{C}$-morphism $\mathbf{f}' : \mathbf{s}_0 \to \mathbf{s}'_0|_U$ in such a way that A is an ambient contraction of $A'|_U$.

Throughout this book, our primary interest in $\mathbb{C}$-trees is only up to ambient changes. We say that a $\mathbb{C}$-tree is **S-equivalent** to another if there exists a finite chain of ambient changes, forward or backward, starting from the one and ending at the other. An S-equivalence class of $\mathbb{C}$-trees can be thought of as a "specific singular diagram together with a specific embedding type".

Example 1.4.2 Suppose we are interested in a fixed singular diagram, for instance "four points mapped to a single point". To be precise, let

$$(s_1)_0 = \{\xi_1, \xi_2, \xi_3, \xi_4\}, \quad (s_0)_0 = \{\eta\}, \quad (a_1)_0(\xi_i) = \eta \text{ for } 1 \leq i \leq 4.$$

What are the possibilities for $\mathbb{C}$-trees $A = \{\mathbf{a}_1, U, E\}$ with $\mathbf{a}_1 : \mathbf{s}_1 \to \mathbf{s}_0$, whose singular diagram is $(a_1)_0 : (s_1)_0 \to (s_0)_0$.

We seek this classification only up to ambient changes of $\mathbb{C}$-trees. First of all, we must have $E \subset (s_0)_0$ and hence $E = \eta$. The question being local, we may assume that $(s_0)^0 = U = \mathbb{C}^k$ for some integer $k \geq 2$. (If $k \leq 1$, then $\left((a_i)^0\right)^{-1}(\eta)$ cannot have more than one point.) Next, $(a_1)^0$ is the quadratic transformation of $\mathbb{C}^k$ with center η, and $\left((a_1)^0\right)^{-1}(\eta) = \mathbb{P}^{k-1}$. Now, it is not difficult to check that, up to S-equivalence, the following are the possibilities of A having the fixed $(a_1)_0$:

Case (a) ($k = 2$) The $\mathbb{C}$-tree A is determined up to S-equivalence by the cross ratio of the four points $\xi_1, \xi_2, \xi_3, \xi_4 \in \mathbb{P}^1$.

Case (b) ($k = 3$) There are exactly two S-equivalence classes, other than those in the preceding case. Namely, we have the class of A in which three of the four points lie in a line $\mathbb{P}^1 \subset \mathbb{P}^2$ but not all four, and the other in which the four points are in general position in $\mathbb{P}^2$.

Case (c) ($k \geq 4$) There is exactly one new class, represented by A with $k = 4$ in which the four points span $\mathbb{P}^3$.

Example 1.4.3 Consider a complex-analytic hypersurface X in an open subset U of $\mathbb{C}^n$. A singular point ξ of X is said to be **absolutely isolated** if the strict transform of X by any finite sequence of quadratic transformations has only isolated singular points that correspond to ξ. Now assume that X has a finite number of absolutely isolated singular points and no other singular points. Let $\mathrm{Sing}(X)$ denote the set of singular points in X. Let $(s_0)^0 = U$ and $(s_0)_0 = \mathrm{Sing}(X)$.

Let $(\mathbf{a}_1, (s_0)^0, \mathrm{Sing}(X))$ be the vertical morphism in which $\mathbf{a}_1 : \mathbf{s}_1 \to \mathbf{s}_0$ has $(s_1)_0 = \mathrm{Sing}(X_1)$, where X_1 is the strict transform of X by the blow-up $(a_1)^0$ of $(s_0)^0$ with center $\mathrm{Sing}(X)$. In general, let

$$(\mathbf{a}_j, (s_{j-i})^0, \mathrm{Sing}(X_{j-i}))$$

be the vertical morphism where $(s_j)_0 = \mathrm{Sing}(X_j)$ with the strict transform X_j of X_{j-i} by the blow-up $(a_j)^0$, for $j = 2, 3, \ldots$. In this way, we obtain a $\mathbb{C}$-tree of length m, $A_m = \{(\mathbf{a}_j, (s_{j-1})^0, (s_{j-1})_0), 1 \leq i \leq m\}$, for each integer $m \geq 1$. The most important among those A_m is the one that has $(s_m)_0 \neq \phi$ but $(s_{m+1})_0 = \phi$, if it exists at all. This A_m will be called the **singularity tree** of $X \subset V$. We can find an $X \subset V = \mathbb{C}^n$ for which the singularity tree is S-equivalent to those classified in Example 1.4.2. Namely, with coordinate system $(z_1, \ldots, z_n)$ of $\mathbb{C}^n$, we have:

Case (a) $X : z_3^5 - z_1^2 z_2^2 (z_1 - z_2)^2 (z_1 - z_2)^2 = 0$ in $\mathbb{C}^3$ (where λ is the cross ratio).
Case (b) $X : z_4^3 - z_1 z_2 z_3 (z_1 + z_3) + z_1^6 = 0$ in $\mathbb{C}^4$ and

$$X : z_4^3 - z_1 z_2 z_3 (z_1 + z_2 + z_3) \text{ in } \mathbb{C}^4.$$

Case (c) $X : z_5^7 - z_1^2 z_2^2 z_3^2 z_4^2 = 0$ in $\mathbb{C}^5$.

1.5 Specan and Cone-Fibered Spaces

Let X be a $\mathbb{C}$-space. We denote the underlying topological space by X as well. Let A be an $\mathscr{O}_X$-algebra. This means that A is a sheaf of commutative rings on X and for every open subset U of X there is given a specific ring homomorphism $\rho_U : \mathscr{O}_X(U) \to A(U)$ (making $A(U)$ into an $\mathscr{O}_X(U)$-algebra), and for every $U_1 \subset U_2$ we have a commutative diagram

$$\begin{array}{ccc} \mathscr{O}_X(U_2) & \xrightarrow{\text{res}} & \mathscr{O}_X(U_1) \\ \downarrow \rho_{U_2} & & \downarrow \rho_{U_1} \\ A(U_2) & \xrightarrow{\text{res}} & A(U_1) \end{array}$$

Example 1.5.1 Let $\mathbf{w} = (w_1, \ldots, w_k)$ be a finite system of indeterminates. Then $\mathscr{O}_X[\mathbf{w}]$ denotes the $\mathscr{O}_X$-algebra in which $\mathscr{O}_X[\mathbf{w}](U) = \mathscr{O}_X(U)[\mathbf{w}]$, the polynomial ring of k indeterminates, and $\rho_U : \mathscr{O}_X(U) \to \mathscr{O}_X(U)[\mathbf{w}]$ is the inclusion.

Example 1.5.2 If $\mathbf{g} = (g_1, \ldots, g_m)$ is a finite system of $g_i \in \mathscr{O}_X[\mathbf{w}](X)$, for $1 \le i \le m$, then $\mathscr{O}_X[\mathbf{w}]/(\mathbf{g})\mathscr{O}_X[\mathbf{w}]$ denotes the quotient sheaf by the ideal sheaf generated by $(\mathbf{g})$. It is an $\mathscr{O}_X$-algebra in a natural manner.

Definition 1.5.1 An $\mathscr{O}_X$-algebra A is said to be **locally finitely presented** if every point of X admits an open neighborhood U for which $A|_U$ is isomorphic to an $\mathscr{O}_X|_U$-algebra of the form described in Example 1.5.2 (where $\mathscr{O}_X$ should be replaced by $\mathscr{O}_X|_U$).

Let $t : T \to X$ be a morphism of $\mathbb{C}$-spaces. For an $\mathscr{O}_X$-algebra A and an $\mathscr{O}_T$-algebra B, we shall speak of a t-**homomorphism** (of algebras) $\alpha : A \to B$. This means a collection of ring homomorphisms α_U, one for each open subset U of X, such that for every U the diagram

$$\begin{array}{ccc} \mathscr{O}_X(U) & \xrightarrow{t_U^*} & \mathscr{O}_T(t^{-1}(U)) \\ \downarrow \rho_U & & \downarrow \rho_{t^{-1}(U)} \\ A(U) & \xrightarrow{\alpha_U} & B(t^{-1}(U)) \end{array}$$

is commutative, and such that for every $U_1 \subset U_2$ the diagram

$$\begin{array}{ccc} A(U_2) & \xrightarrow{\alpha_{U_2}} & B(t^{-1}(U_2)) \\ \Big\downarrow \text{res} & & \Big\downarrow \text{res} \\ A(U_1) & \xrightarrow{\alpha_{U_1}} & B(t^{-1}(U_1)) \end{array}$$

is commutative.

We shall denote by $\mathrm{Hom}_t(A, B)$ the set of all t-homomorphisms from A to B.

Remark 1.5.1 Consider the case in which

$$A = \mathscr{O}_X[\mathbf{w}]/(\mathbf{g})\mathscr{O}_X[\mathbf{w}],$$

as given in Example 1.5.2. Let $\mathbf{z} = (z_1, \dots, z_k)$ be the coordinate system on $\mathbb{C}^k$ and, for $1 \le i \le m$, let $g_i(\mathbf{z})$ denote the section of $\mathscr{O}_{X \times \mathbb{C}^k}$ obtained from g_i by exchanging $\mathbf{w}$ with $\mathbf{z}$. Let X' be the closed $\mathbb{C}$-subspace of $X \times \mathbb{C}^k$ defined by the ideal sheaf generated by $(g_1(\mathbf{z}), \dots, g_m(\mathbf{z}))$. There are two maps p and α, characteristic to X', as follows:

$$p : X' \to X, \ \mathbb{C}\text{-morphism induced by the projection,}$$
$$\alpha : A \to \mathscr{O}_{X'}, \ p\text{-homomorphism given by } \alpha(w_i) = z_i \text{ for all } i.$$

Lemma 1.5.1 *The characteristic (p, α) of X' defined in Remark* 1.5.1 *has the following universal property:*

($\star$) *For every $\mathbb{C}$-morphism $t : T \to X$ and every t-homomorphism $\beta : A \to \mathscr{O}_T$, there exists a unique $\mathbb{C}$-morphism $s : T \to X'$ such that $t = p \cdot s$ and $\beta = s^* \cdot \alpha$.*

Proof Given t and β, the system $(\beta(w_1), \dots, \beta(w_k))$ defines a $\mathbb{C}$-morphism $s_0 : T \to \mathbb{C}^k$ such that $s_0^*(z_i) = \beta(w_i)$, for $1 \le i \le k$. Let s be the $\mathbb{C}$-morphism $T \to X \times \mathbb{C}^k$ defined by t and s_0. We have $s^*(g_j(z)) = \beta(g_j) = 0$ for all j, so that s can be viewed as a morphism into X'. Clearly $t = p \cdot s$ and $\beta = s^* \cdot \alpha$. Conversely, the uniqueness of s is nothing but the uniqueness of $s_0 = g \cdot s$, where $g : X \times \mathbb{C}^k \to \mathbb{C}^k$ is the projection. This uniqueness is clear from the fact that $s_0^*(z_i) = \beta(w_i)$, for $1 \le i < k$, because $\beta = s^* \cdot \alpha$. □

Lemma 1.5.2 *Assume that, for a $\mathbb{C}$-morphism $p_1 : X_1' \to X$ and a p_1-homomorphism $\alpha_1 : A \to \mathscr{O}_{X_1'}$, the pair (p_1, α_1) satisfies the universal condition ($\star$) as (p, α) of Lemma* 1.5.1. *Then there exists a unique $\mathbb{C}$-isomorphism $r : X_1' \to X'$ such that $p_1 = p \cdot r$ and $\alpha_1 = r^* \cdot \alpha$.*

Proof We find $r : X_1' \to X'$ by ($\star$) (Lemma 1.5.1) for $t = p_1$ and $\beta = \alpha_1$. Similarly, we find $r' : X' \to X_1'$ by Lemma 1.5.1 for (p_1, α_1) instead of (p, α). Then, by the uniqueness in Lemma 1.5.1, $r \cdot r'$ and $r' \cdot r$ must be the identity on X_1' and X' respectively. □

Proposition 1.5.1 *Given an $\mathcal{O}_X$-algebra A that is locally finitely presented, there exists a $\mathbb{C}$-space X' equipped with a $\mathbb{C}$-morphism $p : X' \to X$ and a p-homomorphism $\alpha : A \to \mathcal{O}_{X'}$ such that condition ($\star$) in Lemma* 1.5.1 *is satisfied. Moreover, such (X', p, α) is unique up to an isomorphism of X' compatible with (p, α).*

Proof It is clear that (X', p, α) is unique if it exists. (See the proof of Lemma 1.5.2.) Hence the existence is a local question in X. Thus, in view of Definition 1.5.1, Lemmas 1.5.1 and 1.5.2 suffice for the proof of the proposition. □

Definition 1.5.2 The $\mathbb{C}$-space X' of Proposition 1.5.1 together with (p, α) will be denoted by Specan(A). We call p : Specan$(A) \to X$ the canonical projection and $\alpha : A \to \mathcal{O}_{\mathrm{Specan}(A)}$ the canonical p-homomorphism. It is important to keep in mind that, whenever A is presented as in Remark 1.5.1, Specan(A) is obtained in the manner described there.

We consider next a **graded** $\mathcal{O}_X$-algebra A, which means an $\mathcal{O}_X$-algebra having a direct sum decomposition

$$A = \bigoplus_{d \in \mathbb{Z}} A_d$$

(as an $\mathcal{O}_X$-module only), where A_d is an $\mathcal{O}_X$-module and $A_d A_e \subset A_{d+e}$ for all $d, e \in \mathbb{Z}$. We call A_d the **homogeneous component of degree** d in A.

Example 1.5.3 With a finite system of indeterminates $\mathbf{w} = (w_1, \ldots, w_k)$, $A = \mathcal{O}_X[\mathbf{w}]$ has a natural structure of graded $\mathcal{O}_X$-algebra in which A_d is the $\mathcal{O}_X$-module generated by the monomials of degree d in t for every $d \in \mathbb{C}$. We have $A_d = (0)$ for all $d < 0$ and $A_0 = \mathcal{O}_X$.

Remark 1.5.2 A **homomorphism of graded $\mathcal{O}_X$-algebras** $A \to A'$ is a homomorphism of $\mathcal{O}_X$-algebras which respects gradings, that is, maps A_d into A'_d for every $d \in \mathbb{Z}$.

Throughout this book, we shall be interested only in a special kind of graded $\mathcal{O}_X$-algebras A, namely those satisfying the following condition:

For every $x \in X$, there exist an open neighborhood N of x in X and a homomorphism of graded $\mathcal{O}_X|_N$-algebras $\mathcal{O}_X|_N[\mathbf{w}] \to A|_N$ with a system of indeterminates $\mathbf{w}$ (of Example 1.5.3) that is surjective and has a finitely generated kernel as an $\mathcal{O}_X|_N[\mathbf{w}]$-module. In particular, for every d, A_d is a coherent $\mathcal{O}_X$-module, $A_0 = \mathcal{O}_X$ and $A_d = 0$ if $d < 0$. Also the kernel is necessarily generated by homogeneous elements.

Definition 1.5.3 If a graded $\mathcal{O}_X$-algebra A satisfies the preceding condition, we say that A is of **LF type** (meaning that A is linearly generated and finitely presented).

For a **graded** $\mathcal{O}_X$-algebra A, say of LF type, the $\mathbb{C}$-space $X' = \mathrm{Specan}(A)$ with canonical (p, α) as in Definition 1.5.2, has an additional structure of scalar multiplication. Let's make this precise. We denote by $\mathcal{O}_X^*$ the sheaf of multiplicative groups on X that consists of invertible elements in $\mathcal{O}_X$. If $g : Z \to X$ is a $\mathbb{C}$-morphism of $\mathbb{C}$-spaces, we define $\mathrm{Aut}(g)$, or $\mathrm{Aut}(Z/X)$, to be the sheaf of groups on X such that, for every open subset U of X, $\mathrm{Aut}(g)(U)$ is the group of those isomorphisms e of $Z|_{g^{-1}(U)}$ into itself such that $g \cdot e = g$.

Definition 1.5.4 Given $g : Z \to X$ as above, a **scalar multiplication** in g is a homomorphism of sheaves of groups $s : \mathcal{O}_X^* \to \mathrm{Aut}(Z/X)$.

Remark 1.5.3 For a graded $\mathcal{O}_X$-algebra A as above, the canonical map

$$p : X' = \mathrm{Specan}(A) \to X$$

admits a natural scalar multiplication. Namely, let U be any open subset of X and let $h \in \mathcal{O}_X^*(U)$. Then h allows us to define an automorphism of the graded $\mathcal{O}_X|_U$-algebra $A|_U$ by letting $v \to h^d v$ for all $v \in A_d$ and all $d \geq 0$. By Lemma 1.5.1, we have a unique $s(h) \in \mathrm{Aut}(p|_U)$ such that $\alpha \cdot h = s(h)^* \cdot \alpha$, where the first h denotes the above automorphism of graded $\mathcal{O}_X|_U$-algebras. It is easy to see that $h \to s(h)$ is a homomorphism of group sheaves $\mathcal{O}_X^* \to \mathrm{Aut}(X'/X)$, which we denote by s and call the **canonical scalar multiplication** in $\mathrm{Specan}(A)$.

Remark 1.5.4 The notation being the same as in Remark 1.5.3, we define an $\mathcal{O}_X$-module $(\mathcal{O}_{X'})_d$ for each $d \geq 0$ as follows by setting

$$(\mathcal{O}_{X'})_d(U) = \{g \in \mathcal{O}_{X'}(p^{-1}(U)) \mid s(h)^*(g) = h^d g \text{ for all } h \in \mathcal{O}_X^*(U)\}$$

for every open subset U of X. Obviously $(\mathcal{O}_{X'})_d$ is an $\mathcal{O}_X$-module and α induces a homomorphism of $\mathcal{O}_X$-modules

$$\alpha_d : A_d \to (\mathcal{O}_{X'})_d.$$

Proposition 1.5.2 *This α_d is an isomorphism for all $d \in \mathbb{Z}$.*

Proof The question is local in X, so we may assume that $A = \mathcal{O}_X[\mathbf{w}]/\mathbf{J}$ and $\mathcal{O}_{X'} = \mathcal{O}_{X \times \mathbb{C}^k}/\mathbf{J}\mathcal{O}_{X \times \mathbb{C}^k}$, where $\mathbf{w}$ is identified with the coordinate system $\mathbf{z}$ in $\mathbb{C}^k$. Every element of $\mathcal{O}_{X'}(p^{-1}(U))$ can be expanded in the variables $\mathbf{w}$, that is, can be written as a (possibly infinite) sum of elements in $\alpha(A_d(U))$, for $d \in \mathbb{Z}$. As is easily seen, it belongs to $(\mathcal{O}_{X'})_d(U)$ if and only if it is in $\alpha(A_d(U))$. Therefore α_d is surjective. That α_d is injective (equivalently, that α in $A_d(U)$) is seen as follows. For each point $\boldsymbol{\xi} \in U$, we have canonical maps

$$
\begin{array}{c}
A_\xi = \mathscr{O}_{X,\xi}[\mathbf{w}] \triangleleft \mathbf{J}_\xi \\
\cap \\
\Big\downarrow b \\
\text{localization of } A_\xi \\
\cap \\
\Big\downarrow c \\
\mathscr{O}_{X',\xi\times\{0\}} = \mathscr{O}_{X\times\mathbb{C}^k,\xi\times\{0\}} \triangleleft \mathbf{J}'_\xi
\end{array}
$$

where the localization is with respect to the maximal ideal in A_ξ generated by the maximal ideal of $\mathscr{O}_{X,\xi}$ and $\mathbf{w}$, and $\mathbf{J}'_\xi$ is the ideal generated by the image of $\mathbf{J}_\xi$. Now, b is injective because $\mathbf{J}_\xi$ is homogeneous, while c is injective because it is flat and local. This flatness is an easy consequence of the fact that, for two sets of variables $\mathbf{x}$ and $\mathbf{w}$, $\mathbb{C}\{\mathbf{x}\}[\mathbf{w}] \to \mathbb{C}\{\mathbf{x}, \mathbf{w}\}$ is flat. □

Remark 1.5.5 Proposition 1.5.2 means that A_d is exactly the sheaf of functions that are homogeneous of degree d with respect to the canonical scalar multiplication s in $\mathrm{Specan}(A)$.

We next consider the following condition, which is stronger than that of Remark 1.5.2:

Condition 1.5.6 In addition to the condition stated in Remark 1.5.2, the kernel of the homomorphism $\mathscr{O}_X[\mathbf{w}] \to A$ is generated by elements in $\sum_{i=1}^{k} \mathscr{O}_X w_i$ (i.e., linear in $\mathbf{w}$).

Definition 1.5.5 If a graded $\mathscr{O}_X$-algebra A satisfies Condition 1.5.6, then we say that A has **additive FL type**.

The point of this definition is that if A has additive FL type, there is given over

$$p : X' = \mathrm{Specan}(A) \to X$$

an **addition morphism** as well as the scalar multiplication, so that its fibers become vector spaces. We now make this precise.

If $\pi_1 : E_1 \to X$ and $\pi_2 : E_2 \to X$ are $\mathbb{C}$-morphisms of $\mathbb{C}$-spaces, we define the **fiber product** $E_1 \times_X E_2$ (with respect to π_1 and π_2) as the closed $\mathbb{C}$-subspace of $E_1 \times E_2$ whose defining ideal sheaf is generated by $(\pi_1 \cdot p_1)^*(h) - (\pi_2 \cdot p_2)^*(h)$ for all $h \in \mathscr{O}_X$, where $p_i : E_1 \times E_2 \to E_i$ denotes the projection. It then follows that $\pi_1 \cdot p_1$ and $\pi_2 \cdot p_2$ induce the same morphism $E_1 \times_X E_2 \to X$, called the **canonical projection to the base space** X and denoted by $\pi_1 \times_X \pi_2$.

Remark 1.5.7 Let $\Pi = \pi_1 \times_X \pi_2$ and let $P_i : E_1 \times_X E_2 \to E_i$ be the morphism induced by the projection p_i, for $i = 1, 2$. Then the fiber product $\{E_1 \times_X E_2, P_1, P_2\}$ has the following property:

($\star$) For every pair of $\mathbb{C}$-morphisms $t_1 : T \to E_1, t_2 : T \to E_2$ such that $\pi_1 \cdot t_1 = \pi_2 \cdot t_2$, there exists a unique morphism $r : T \to E_1 \times_X E_2$ such that $t_1 = P_1 \cdot r$ and $t_2 = P_2 \cdot r$.

The proof of ($\star$) is straightforward. It is also easily seen that ($\star$) is a universal mapping property characterizing $\{E_1 \times_X E_2, P_1, P_2\}$.

Lemma 1.5.3 *Let A_1 and A_2 be $\mathcal{O}_X$-modules that are locally finitely presented. For $i = 1, 2$, let $E_i = \mathrm{Specan}(A_i)$, with canonical maps $\pi_i : E_i \to X$ and $\alpha_i : A_i \to \mathcal{O}_{E_i}$. Let $\pi : E_1 \times_X E_2 \to X$ be the canonical projection and let $\alpha : A_1 \otimes_{\mathcal{O}_X} A_2 \to \mathcal{O}_{E_1 \times_X E_2}$ be the homomorphism defined by α_1 and α_2. Then $E_1 \times_X E_2 = \mathrm{Specan}(A_1 \otimes_{\mathcal{O}_X} A_2)$, with π and α as canonical maps.*

Proof Let $t : T \to X$ be any $\mathbb{C}$-morphism, and let $\beta : A_1 \otimes_X A_2 \to \mathcal{O}_T$ be any t-homomorphism. Then β induces t-homomorphisms $\beta_1 : A_1 \to T$ and $\beta_2 : A_2 \to T$ by means of the obvious inclusions $A_i \to A_1 \otimes_{\mathcal{O}_X} A_2$. By Lemma 1.5.1 combined with Definition 1.5.2, (t, β_i) gives, for $i = 1, 2$, a unique morphism $s_i : T \to E_i$ with $t = \pi_i \cdot s_i$ and $p_i = s_i^* \cdot \alpha_i$. By Remark 1.5.7, there exists, for $i = 1, 2$, a unique $s : T \to E_1 \times_X E_2$ such that $s_i = P_i \cdot s$, where $P_i : E_1 \times_X E_2 \to E_i$ is the canonical projection. So $t = \pi_i \cdot s_i = (\pi_i \cdot P_i) \cdot s = \pi \cdot s$, and moreover $s^* \cdot \alpha$ coincides with β because they do in both A_1 and A_2.

We now go back to a $\mathbb{C}$-space X and a graded $\mathcal{O}_X$-algebra A of LF type as in Definition 1.5.3. The tensor product $A \otimes_X A$ has a natural graded $\mathcal{O}_X$-algebra structure. In particular, its linear part (homogeneous part of degree 1) is isomorphic to $A_1 \oplus A_1$ in a natural manner:

$$(A \otimes_{\mathcal{O}_X} A)_1 = A_1 \oplus A_1.$$

□

Remark 1.5.8 Let $a_{(1)} : A_1 \to A_1 \oplus A_1$ be the homomorphism of $\mathcal{O}_X$-modules defined by $a_{(1)}(g) = g \oplus g$. This will be called the **coaddition** in A. Now, if A is of additive LF type (Definition 1.5.5), the coaddition extends to a homomorphism of graded $\mathcal{O}_X$-algebras from A to $A \otimes_{\mathcal{O}_x} A$. In fact, the question being local due to the obvious uniqueness of such an extension, we may assume that $A = \mathcal{O}_X[\mathbf{w}]/\mathbf{I}$ with an ideal sheaf $\mathbf{I}$ generated by elements of $\sum_i \mathcal{O}_X w_i$ (see Remark 1.5.2). Then

$$A \otimes_{\mathcal{O}_X} A = \mathcal{O}_X[\mathbf{w}', \mathbf{w}'']/(\mathbf{I}(\mathbf{w}'), \mathbf{I}(\mathbf{w}'')),$$

where $\mathbf{w}'$, $\mathbf{w}''$ are two independent copies of $\mathbf{w}$ and $\mathbf{I}(\mathbf{w}')$, $\mathbf{I}(\mathbf{w}'')$ are obtained from I by replacing $\mathbf{w}$ by $\mathbf{w}'$, $\mathbf{w}''$ respectively. We have a homomorphism of graded $\mathcal{O}_X$-algebras from $\mathcal{O}_X[\mathbf{w}]$ to $\mathcal{O}_X[\mathbf{w}', \mathbf{w}'']$, defined by $\mathbf{w} \to \mathbf{w}' + \mathbf{w}''$; we call it e. Clearly e induces the coaddition in A. The question is whether or not it induces the homomorphism $A \to A \otimes_{\mathcal{O}_X} A$; whether it maps $\mathbf{I}$ into the ideal $(\mathbf{I}(\mathbf{w}'), \mathbf{I}(\mathbf{w}''))$ in $\mathcal{O}_X[\mathbf{w}', \mathbf{w}'']$. Now, by the additive LF type assumption, $\mathbf{I}$ is generated by its linear part and the linear part is clearly mapped into $(\mathbf{I}(\mathbf{w}'), \mathbf{I}(\mathbf{w}''))$. So $e(\mathbf{I}) \subset (\mathbf{I}(\mathbf{w}'), \mathbf{I}(\mathbf{w}''))$, as desired, so that the coaddition extends to a homomorphism $A \to A \otimes_{\mathcal{O}_X} A$ as $\mathcal{O}_X$-algebras, which is induced by e. This extension will be called the **coaddition morphism**, or simply the **coaddition**, in A. Note that the coaddition in A is determined by A: there is no other choice for it, so long as it exists at all.

Remark 1.5.9 Assume that A is of additive LF type, so that we have the unique coaddition $\mathrm{ca} : A \to A \otimes_{\mathcal{O}_X} A$. Let $E = \mathrm{Specan}(A)$ with canonical data $\pi : E \to X$ and $\alpha : A \to \mathcal{O}_E$. Let $\Pi : E \times_X E \to X$ be the canonical projection. Then, by Lemmas 1.5.1 and 1.5.3, we have a unique $\mathbb{C}$-morphism $\mathrm{ad} : E \times_X E \to E$ such that $\Pi = \pi \cdot (\mathrm{ad})$ and $\mathrm{ca} = (\mathrm{ad})^* \cdot \alpha$. We shall call ad the **addition morphism**, or simply the **addition**, for E over X.

Remark 1.5.10 The **addition** $\mathrm{ad} : E \times_X E \to E$ and the **scalar multiplication** $s : \mathcal{O}_X^* \to \mathrm{Aut}(E/X)$ have the following basic properties:

(1) **Associativity:** The diagram

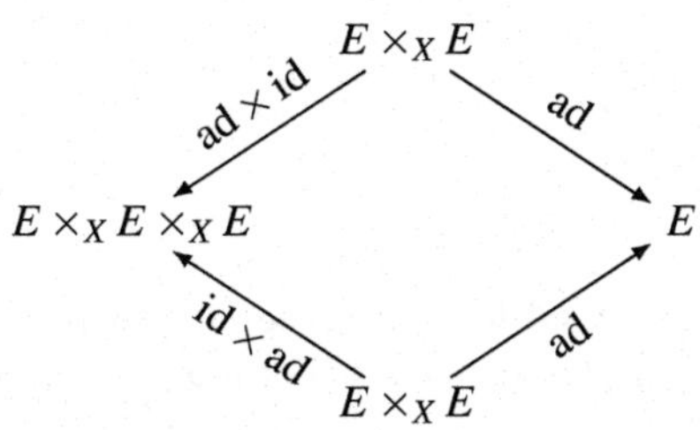

commutes.

(2) **Commutativity:** If σ denotes the automorphism of $E \times_X E$ by permutation of factors, the diagram

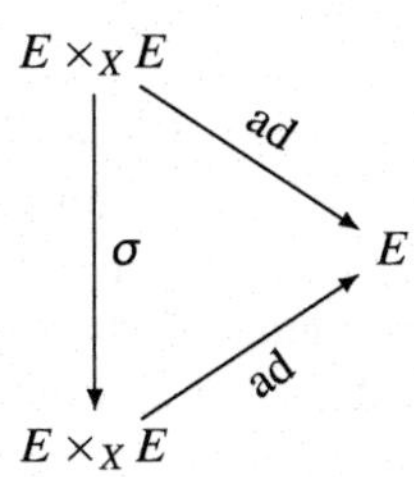

commutes.

(3) **Distributivity:** If $s : \mathscr{O}_X^* \to \operatorname{Aut}(E \times_X E/X)$ denotes the scalar multiplication, where $E \times_X E$ is identified with $\operatorname{Specan}(A \otimes_{\mathscr{O}_X} A)$, then

$$\begin{array}{ccc} E \times_X E & \xrightarrow{\text{ad}} & E \\ \downarrow{\scriptstyle s(g)} & & \downarrow{\scriptstyle s(g)} \\ E \times_X E & \xrightarrow{\text{ad}} & E \end{array}$$

is commutative for all $g \in \mathscr{O}_X^*(X)$, and likewise when all maps are restricted to an open subset in X.

Furthermore, let $\varepsilon : A \to \mathscr{O}_X$ be the homomorphism that maps A_d to 0 for all $d > 0$ and induces the identity in A_0. Let $0 : X \to E$ be the $\mathbb{C}$-morphism defined by ε. (Apply Lemma 1.5.1 to the pair $(\mathrm{id}_X, \varepsilon)$, where id_X denotes the identity on X.) We have $\pi \cdot 0 = \mathrm{id}_X$, and 0 will be called the **zero section** of $\pi : E \to X$. We have:

(4) **Identity law:** The composition

$$E \xrightarrow{(\mathrm{id}_E, 0 \cdot \pi)} E \times_X E \xrightarrow{\text{ad}} E$$

is id_E.

Finally, we have the **additive inverse** morphism $\tau : E \to E$ which is $s(-1)$ where $s : \mathscr{O}_X^* \to \operatorname{Aut}(E/X)$ is the scalar multiplication (Definition 1.5.4). Then we have:

(5) **Additive inverse law:** The composition

$$E \xrightarrow{(\mathrm{id}_E, \tau)} E \times_X E \xrightarrow{\text{ad}} E$$

is $O \cdot \pi$.

Note 1.5.1 These five properties are easily proved by checking the claims on the **linear parts** of the graded $\mathscr{O}_X$-algebras that appear in the statements. Here we may use the following general principle: Let A and B be two graded $\mathscr{O}_X$-algebras of LF type. Let $Z = \operatorname{Specan}(A)$ with canonical data (p, α), and $W = \operatorname{Specan}(B)$ with canonical data (g, β). Then a morphism $r : Z \to W$, subject to $g \cdot r = p$ and commuting with scalar multiplications, is uniquely determined by the effect of r^* on B_1, that is, the composition

$$B_1 \xrightarrow[\beta]{\sim} (\mathscr{O}_W)_1 \xrightarrow[r^*]{} (\mathscr{O}_Z)_1 \xrightarrow[\alpha]{\sim} A_1$$

(see Remark 1.5.4 for the isomorphisms at both ends).

Remark 1.5.11 Let $t : T \to X$ be any fixed $\mathbb{C}$-morphism of $\mathbb{C}$-spaces. Following the assumptions and the notation of Remarks 1.5.8, 1.5.9, and 1.5.10, we let $\mathrm{Hom}_X(T, E)$ denote the set of $\mathbb{C}$-morphisms $h : T \to E$ such that $t = \pi \cdot h$, and we define the following objects:

(1) **Addition:** For $g_1, g_2 \in \mathrm{Hom}_X(T, E)$, the sum $g_1 + g_2$ is the composition

$$T \xrightarrow{(g_1, g_2)} E \times_X E \xrightarrow{\mathrm{ad}} E.$$

(2) **Zero element:** This is the composition

$$T \xrightarrow{t} X \xrightarrow{0} E.$$

(3) **Inverses:** For $g \in \mathrm{Hom}_X(T, E)$, the element $-g$ is the composition

$$T \xrightarrow{g} E \xrightarrow{\tau} E\triangleright$$

(4) **Scalar multiplication:** For each open neighborhood U of $\mathrm{Im}(t)$ in X, we have $s_U : \mathcal{O}_X^*(U) \to \mathrm{Aut}(E/X)(U)$. Therefore, if $k \in \mathcal{O}_X^*(U)$ and $g \in \mathrm{Hom}_X(T, E)$, we define $k\,g$ as the composition

$$T \xrightarrow{g} E \xrightarrow{s_U(g)} E.$$

In particular, $\mathbb{C}^* \subset \mathcal{O}_X^*(U)$ acts on $\mathrm{Hom}_X(T, E)$.

It is straightforward to prove that $\mathrm{Hom}_X(T, E)$ becomes a **vector space** over $\mathbb{C}$ with this addition and scalar multiplication.

Definition 1.5.6 Let $\pi : C \to X$ be a $\mathbb{C}$-morphism of $\mathbb{C}$-spaces. Let

$$s : \mathcal{O}_X^* \to \mathrm{Aut}(C/X)$$

be a homomorphism of group sheaves on X. With such π and s, we define

$$A_d(U) = \{h \in \mathcal{O}_C(\pi^{-1}(U)) \mid s(g)^*(h) = g^d h \text{ for all } g \in \mathcal{O}_X^*(U)\},$$

for each integer $d \in \mathbb{Z}$. Let $A = \bigoplus_{d \in \mathbb{Z}} A_d$; this is obviously a graded $\mathcal{O}_X$-algebra. We also have an obvious π-homomorphism $\alpha = A \to \mathcal{O}_C$. Now, we say that $\pi : C \to X$ is a **cone-fibered space with scalar multiplication** $s : \mathcal{O}_X^* \to \mathrm{Aut}(C/X)$, or simply that C is a **cone-fibered space over** X, if the following conditions are satisfied:

(1) The $\mathcal{O}_X$-algebra A is a graded $\mathcal{O}_X$-algebra of LF type.
(2) The unique morphism $C \to \mathrm{Specan}(A)$ determined by (π, α) according to Lemma 1.5.1 is an isomorphism.
(3) The morphism s is derived from the scalar multiplication in $\mathrm{Specan}(A)/X$ by the isomorphism of (2).

Definition 1.5.7 Let $\pi : C \to X$ and $s : \mathscr{O}_X^* \to \mathrm{Aut}(C/X)$ be as in Definition 1.5.6. Assume we have a $\mathbb{C}$-morphism $a : C \times C \to C$. We say that $\pi : C \to X$ is a **vector-fibered space with addition** a **and scalar multiplication** s, or simply that C is a **vector-fibered space over** X, if the following conditions are satisfied, in addition to those of Definition 1.5.6:

(1) The $\mathscr{O}_X$-algebra A is a graded $\mathscr{O}_X$-algebra of additive LF type.
(2) The morphism $a : C \times_X C \to C$ is derived from the addition morphism for $\mathrm{Specan}(A)$ by the isomorphism of Definition 1.5.6 (2).

Definition 1.5.8 If, in addition to the conditions in Definitions 1.5.6 and 1.5.7, A_1 is a locally free $\mathscr{O}_X$-module (which implies that A is locally of the form $\mathscr{O}_X[\mathbf{w}]$ with a finite system $\mathbf{w}$ of indeterminates), we say that $\pi : C \to X$ is a **vector bundle with addition** a **and scalar multiplication** s, or that C is called a **vector bundle over** X.

1.6 Blowing Cone, Normal Cone and Tangent Cone

Let X be a $\mathbb{C}$-space and D a closed $\mathbb{C}$-subspace. We shall associate various graded algebras with such a pair (X, D), from which we derive the notions of tangent cones, normal cones, blow-ups and differential operators.

Let $\mathbf{I}$ denote the ideal sheaf in $\mathscr{O}_X$ that defines the $\mathbb{C}$-subspace D.

Example 1.6.1 We first define

$$\mathrm{bl}_D(X) = \bigoplus_{d \geq 0} \mathbf{I}^d,$$

where $\mathbf{I}^d$ is the d-th power of $\mathbf{I}$ as ideal sheaves (in particular, $\mathbf{I}^0 = \mathscr{O}_X$). The direct sum $\mathrm{bl}_D(X)$ has a natural graded $\mathscr{O}_X$-algebra structure, in which the multiplication

$$\mathrm{bl}_D(X)_d \times \mathrm{bl}_D(X)_l \to \mathrm{bl}_D(X)_{d+l}$$

is nothing but $\mathbf{I}^d \times \mathbf{I}^l \to \mathbf{I}^{d+l}$ by the multiplication in $\mathscr{O}_X$.

Lemma 1.6.1 *The $\mathscr{O}_X$-algebra* $\mathrm{bl}_D(X)$ *is a graded $\mathscr{O}_X$-algebra of LF type.*

Proof The question being local, we may assume that $\mathbf{I}$ is generated by a finite system $\mathbf{g} = (g_0, \ldots, g_r)$ of global sections g_i in $\mathscr{O}_X$. First pick a corresponding system of indeterminates $\mathbf{t} = (t_0, \ldots, t_r)$ and let

$$\lambda_d = \mathscr{O}_X[\mathbf{t}]_d \to \mathscr{O}_X$$

be the homomorphism of $\mathscr{O}_X$-modules that maps monomials in $\mathbf{t}$ to the corresponding monomials in $\mathbf{g}$, i.e.,

$$t_0^{a_0} t_1^{a_1} \cdots t_r^{a_r} \to g_0^{a_0} g_1^{a_1} \cdots g_r^{a_r},$$

where $a_0 + \cdots + a_r = d$. Note that, by Oka's coherence theorem [97], $\mathrm{Ker}(\lambda_d)$ is locally finitely generated as an $\mathscr{O}_X$-module for every $d \geq 0$.

Now, let's view $\mathbf{t}$ as the standard coordinate system on $\mathbb{C}^{r+1}$ and consider the ideal sheaf in $\mathscr{O}_{X\times\mathbb{C}^{r+1}}$ generated by all the $\ker(\lambda_d)$, for $d \geq 1$. Then this ideal sheaf is locally finitely generated, because it can be expressed as the limit of a growing sequence of coherent ideal sheaves (the ones generated by $\mathrm{Ker}(\lambda_d)$ with bounds on d). Therefore, if any point $x \in X$ is given, we can find an open neighborhood V of $x \times (\mathbf{0})$ in $X \times \mathbb{C}^{r+1}$ in which the ideal sheaf is finitely generated and hence there exists $d_0 > 0$ such that all the $\mathrm{Ker}(\lambda_l)$ with $l > d_0$ are contained by the ideal sheaf generated by the $\mathrm{Ker}(\lambda_d)$, with $d \leq d_0$. Note that V can be chosen to have the form $U \times \mathbb{C}^{r+1}$ with an open neighborhood of x in X, because the ideal sheaf in question is invariant under scalar multiplications on $\mathbf{t}$. Moreover, replacing U by a smaller neighborhood if necessary, we may assume that $\mathrm{Ker}(\lambda_d)$ is finitely generated as an $\mathscr{O}_X$-module for every $d \leq d_0$. Now, replacing X by $X|_U$, we obtain an isomorphism

$$\mathscr{O}_X[\mathbf{t}]/K \to \mathrm{bl}_D(X)$$

of graded $\mathscr{O}_X$-algebras, with $K = (\mathrm{Ker}(\lambda_d), d \leq d_0)\mathscr{O}_X[\mathbf{t}]$, which is finitely generated as an $\mathscr{O}_X[\mathbf{t}]$-module. □

Example 1.6.2 Let D and X be as in Example 1.6.1, and define

$$\mathrm{gr}_D(X) = \bigoplus_{d\geq 0} \mathbf{I}^d/\mathbf{I}^{d+1} = \mathrm{bl}_D(X)/\mathbf{I}\,\mathrm{bl}_D(X),$$

which is viewed as a graded $\mathscr{O}_D$-algebra. (The right-hand side is annihilated by $\mathbf{I}$ and restricted to the $\mathbb{C}$-subspace D defined by $\mathbf{I}$.) By Example 1.6.1 and Lemma 1.6.1, $\mathrm{gr}_D(X)$ is of LF type as a graded $\mathscr{O}_D$-algebra.

Example 1.6.3 Let X be a $\mathbb{C}$-space and let Δ be the diagonal of $X \times X$. To be precise, pick any open subset U of X such that $X|_U$ can be expressed as a closed $\mathbb{C}$-subspace of an open subset V of $\mathbb{C}^n$ for some n. Take all such U and let W be the union of all the $U \times U$ in $X \times X$. Then W is an open subset and Δ is the closed $\mathbb{C}$-subspace of $X \times X|_W$, which is locally defined as follows. $(X|_U) \times (X|_U)$ is embedded in $\mathbb{C}^n \times \mathbb{C}^n$ and $\Delta|_{U\times U}$ is defined by the ideal sheaf generated by the elements $p_1^*(z_i) - p_2^*(z_i)$, for $1 \leq i \leq n$, in the structure sheaf of $(X|_U) \times (X|_U)$, where $p_1, p_2 : \mathbb{C}^n \times \mathbb{C}^n \to \mathbb{C}^n$ are the projections and $(z_1, \ldots, z_n)$ is the coordinate system in $\mathbb{C}^n$. Let $\mathbf{J}$ be the ideal sheaf of Δ in $X \times X|_W$ and let

$$\mathscr{P}_X^{(\nu)} = \mathscr{O}_{X\times X}/J^{\nu+1}$$

for each integer $\nu \geq 0$. We will view each $\mathscr{P}_X^{(\nu)}$ as an $\mathscr{O}_X$-module via the $\mathbb{C}$-isomorphism $\Delta \to X$ induced by the first projection $p_1 : X \times X \to X$. This $\mathscr{O}_X$-module $\mathscr{P}_X^{(\nu)}$ is called the **sheaf of principal parts of order** V on X. Now, let D be any closed $\mathbb{C}$-subspace of X. Then define

$$\mathscr{P}_X^{(\nu)}(D) = \mathscr{P}_X^{(\nu)}/\mathbf{I}\mathscr{P}_X^{(\nu)},$$

where $\mathbf{I}$ denotes the ideal sheaf of D in $\mathscr{O}_X$. This $\mathscr{P}_X^{(\nu)}(D)$ will be viewed as an $\mathscr{O}_D$-module. We have natural homomorphisms

$$\tau_\nu(D) : \mathscr{P}_X^{(\nu)}(D) \to \mathscr{P}_X^{(\nu-1)}(D), \quad \nu \geq 1,$$

all surjective, and we define

$$\tan_D(X) = \bigoplus_{\nu=0}^{\infty} \operatorname{Ker}(\tau_\nu(D)),$$

where $\operatorname{Ker}(\tau_0(D)) = \mathscr{O}_X$. This is again a graded $\mathscr{O}_D$-algebra of LF type, because it is isomorphic to $\operatorname{gr}_{\Delta D}(D \times X)$ viewed as an $\mathscr{O}_D$-algebra via p_1, where ΔD denotes the diagonal of $D \times D$. When $D = X$, we write $\tan(X)$ for $\tan_D(X)$.

Remark 1.6.1 The sheaves $\mathscr{P}_X^{(\nu)}$ are important in defining **differential operators** on X. We define

$$\operatorname{Diff}_X^{(\nu)} = \mathbf{Hom}_{\mathscr{O}_X}(\mathscr{P}_X^{(\nu)}, \mathscr{O}_X),$$

and call it the $\mathscr{O}_X$-module of **differential operators of order** $\leq \nu$ on X. The sections of $\operatorname{Diff}_X^{(\nu)}$ act on the sections of $\mathscr{O}_X$ as follows. For each open subset U of X and $\delta \in \operatorname{Diff}_X^{(\nu)}(U)$, we have

$$\mathscr{O}_X(U) \xrightarrow{p_2^*} \mathscr{P}_X^{(\nu)}(U) \xrightarrow{\delta} \mathscr{O}_X(U),$$

using the second projection $p_2 : X \times X \to X$. We shall often write $\delta(f)$ for $\delta(p_2^*(f))$, so that δ maps $\mathscr{O}_X(U)$ into $\mathscr{O}_X(U)$.

Consider the special case in which $X = \mathbb{C}^n$, with $\mathbf{x} = (x_1, \ldots, x_n)$ as coordinate system. In $\mathbb{C}^n \times C^{n\prime}$, we identify $\mathbf{x}$ with $p_1^*(\mathbf{x})$ and let $t = p_2^*(\mathbf{x}) - p_1^*(\mathbf{x})$. For each $\nu \geq 1$, let $\mathbf{t}(\nu)$ denote the class of t modulo $\mathbf{I}_\Delta^{\nu+1}$. Then $\mathscr{P}_{\mathbb{C}^n}^{(\nu)}$ is a free $\mathscr{O}_{\mathbb{C}^n}$-module, generated by the monomials $\mathbf{t}(\nu)^\alpha$ for $\alpha \in \mathbb{Z}_0^n$ with $|\alpha| \leq \nu$. If $\delta_\alpha \in \operatorname{Diff}_{\mathbb{C}^n}^{(\nu)}(\mathbb{C}^n)$ is the homomorphism defined by $\delta_\alpha(\mathbf{t}(\nu)^\beta) = 0$ for $\beta \neq \alpha$ and $\delta_\alpha(\mathbf{t}(\nu)^\alpha) = 1$, then δ_α operates on holomorphic functions h on $\mathbb{C}^n$ (or on an open subset of $\mathbb{C}^n$) as follows: Expand $h(\mathbf{x} + \mathbf{t})$ as $\sum_{\beta \in \mathbb{Z}_0^n} h_\beta(\mathbf{x})\mathbf{t}^\beta$; then

$$\delta^\alpha(h) = h_\alpha(\mathbf{x}).$$

In other words,

$$\delta_\alpha = \frac{1}{\prod_j \alpha_j!} \frac{\partial^{|\alpha|}}{(\partial x_1)^{\alpha_1} (\partial x_2)^{\alpha_2} \cdots (\partial x_n)^{\alpha_n}},$$

according to Taylor's theorem.

Definition 1.6.1 Following the notation of Example 1.6.1, we now define the **blowing cone of X along D** as the **cone-fibered** space $\mathrm{Specan}(\mathrm{bl}_D(X))$ over X (Definitions 1.5.2 and 1.5.6). We denote this blowing cone by $B_{X,D}$.

Definition 1.6.2 Following the notation of Example 1.6.2, we define **the normal cone of X along D** as the cone-fibered space $\mathrm{Specan}(\mathrm{gr}_D(X))$ over D, which is denoted by $C_{X,D}$. If this happens to be a vector bundle in the sense of Definition 1.5.8, we may call it the **normal bundle** of X along D and denote it by $N_{X,D}$.

Definition 1.6.3 Following the notation of Example 1.6.3, we define the **tangent cone of X along D** as the cone-fibered space $\mathrm{Specan}(\tan_D(X))$ over D, which will be denoted by $C_X(D)$. If this is vector bundle, we call it the **tangent bundle** of X and denote it by $T_X(D)$. If $D = X$, we write C_X and T_X for $C_X(D)$ and $T_X(D)$, respectively.

Remark 1.6.2 Pick a point x of X. We can view x as a closed $\mathbb{C}$-subspace of X, defined by the maximal ideal of the local ring $\mathscr{O}_{X,x}$. Then we have a natural $\mathbb{C}$-isomorphism of $\mathbb{C}$-spaces

$$C_X(x) \to C_{X,x}$$

induced by the second projection $p_2 : X \times X \to X$. In fact, $\mathscr{P}_X^{(\nu)}(x)$ is isomorphic to $\mathscr{O}_{X,x}/M_{X,x}^{\nu+1}$ via the projection p_2, where $M_{X,x}$ denotes the maximal ideal of $\mathscr{O}_{X,x}$, so that

$$M_{X,x}^{\nu}/M_{X,x}^{\nu+1} = \mathrm{Ker}(\mathscr{P}_X^{(\nu)}(x) \xrightarrow{\tau_\nu(x)} \mathscr{P}_X^{(\nu-1)}(x))$$

for all $\nu \geq 0$. We shall make the identification $C_X(x) = C_{X,x}$.

Definition 1.6.4 We call $C_{X,x}$ the **tangent cone of** X at x.

Remark 1.6.3 The tangent cone $C_X(D)$ has canonical projection

$$\Pi : C_X(D) \to D.$$

For a closed $\mathbb{C}$-subspace F of D, we shall write $C_X(D)|_F$ for $\Pi^{-1}(F)$, the inverse image of F by Π (compare Example 1.1.6). We claim that there is a **canonical closed embedding**

$$k : C_X(F) \to C_X(D)|_F.$$

In fact, we have a natural surjection $\mathrm{gr}_{\Delta D}(D \times X) \to \mathrm{gr}_{\Delta F}(F \times X)$ (the two spaces being regarded as $\mathscr{O}_{\Delta D}$-algebras), because the ideal sheaf of ΔD in $\mathscr{O}_{D\times X}$ induces that of ΔF in $\mathscr{O}_{F\times X}$, where ΔD and ΔF denote the diagonal of $D \times D$ and $F \times F$. This surjection maps $\mathbf{I}_F \mathrm{gr}_{\Delta D}(D \times X)$ to (0), where $\mathbf{I}_F$ is the ideal sheaf of F in $\mathscr{O}_D$

and the graded algebras are viewed as $\mathcal{O}_D$-algebras via the first projection. Thus we get the canonical surjection

$$\mathrm{gr}_{\Delta D}(D \times X)/\mathbf{I}_F \mathrm{gr}_{\Delta D}(D \times X) \to \mathrm{gr}_{\Delta F}(F \times X).$$

By Example 1.6.3 and Definition 1.6.3, this gives us the embedding k, as its kernel generates the ideal sheaf of the image of k. Note that k is not in general an isomorphism.

The following example shows that $C_X(\boldsymbol{\xi}) \to C_X|_{\{\boldsymbol{\xi}\}}$ for a point $\boldsymbol{\xi} \in X$ may not be isomorphism.

Example 1.6.4 Let $X = \mathrm{Specan}(\mathbb{C}[z_1, z_2]/(z_1^2 - z_2^3))$. Then we can write

$$X \times X = \mathrm{Specan}(\mathbb{C}[w_1, w_2, z_1, z_2]/(w_1^2 - w_2^3,\ z_1^2 - z_2^3))$$

with a copy (w_1, w_2) of (z_1, z_2). Let $t_1 = w_1 - z_1$ and $t_2 = w_2 - z_2$, so that the ideal $\mathbf{I}_\Delta$ of the diagonal Δ is generated by (t_1, t_2) and

$$X \times X = \mathrm{Specan}(\mathbb{C}[t_1, t_2, z_1, z_2]/(h(\mathbf{t}, \mathbf{z}),\ z_1^2 - z_2^3)),$$

where

$$h(\mathbf{t}, \mathbf{z}) = (t_1 + z_1)^2 - (t_2 + z_2)^3 - (z_1^2 - z_2^3) = (t_1^2 - t_2^3 - 3t_2^2 z_2) + (2t_1 z_1 - 3t_2 z_2^2).$$

It follows that

$$\mathrm{Ker}(\tau_2) = L_2/N_2,$$

where L_2 is a free $\mathcal{O}_X$-module of rank 3, generated by u_{11}, u_{12}, u_{22} (corresponding to $t_1^2, t_1 t_2, t_2^2$), and N_2 is the submodule generated by

$$(2z_1 u_{11} - 3z_2^2 u_{12},\ 2z_1 u_{12} - 3z_2^2 u_{22}).$$

Now, let D be the origin; then

$$\mathrm{Ker}(\tau_2) \otimes_{\mathcal{O}_X} \mathcal{O}_D = \mathrm{Ker}(\tau_2)/M_{X,\mathbf{0}}\, \mathrm{Ker}(\tau_2)$$

is a free $\mathbb{C}$-module of rank 3. However,

$$\mathrm{Ker}(\tau_2(D)) = M_{X,\mathbf{0}}^2/M_{X,\mathbf{0}}^3 = (z_1, z_2)^2 \mathbb{C}[z]/(z_1, z_2)^3 + (z_1^2 - z_2^3),$$

which is free of rank 2. Thus $\mathrm{Ker}(\tau_2) \otimes_{\mathcal{O}_X} \mathcal{O}_D \to \mathrm{Ker}(\tau_2(D))$ is not injective.

Remark 1.6.4 Let D be a closed $\mathbb{C}$-subspace of a $\mathbb{C}$-space X, and let F be a closed $\mathbb{C}$-subspace of D. We recall the definition of $\tan_D(X)$ from Example 1.6.3. Let $\mathbf{J}$ be the ideal sheaf of the diagonal Δ of $X \times X$ (within a suitable open neighborhood of Δ as in Example 1.6.3). Let

$$p_1, p_2 : X \times X \to X$$

be the projections. Let $\mathbf{I}_D$ be the ideal sheaf of D in $\mathcal{O}_X$. Then we have:

$$p_2^*(\mathbf{I}_D)\mathcal{O}_{X\times X} \subset p_1^*(\mathbf{I}_D)\mathcal{O}_{X\times X} + \mathbf{J}\mathcal{O}_{X\times X}.$$

This means that the ideal sheaf of the diagonal ΔD of $D \times D$ in $\mathcal{O}_{D\times X}$ contains the pull-back of $\mathbf{I}_D$ by p_2^*. It follows that we get a natural homomorphism of graded algebras $\mathrm{gr}_D(X) \to \mathrm{gr}_{\Delta D}(D \times X)$ with respect to the second projection p_2. Identifying $\mathrm{gr}_{\Delta D}(D \times X)$ with $\tan_D(X)$ as in Example 1.6.3, we obtain the canonical homomorphism

$$\mathrm{gr}_D(X) \to \tan_D(X) \tag{1.1}$$

of graded $\mathcal{O}_D$-algebras. On the other hand, by applying the same process to the ideal sheaf of the diagonal ΔF of $F \times F$ in $\mathcal{O}_{F\times X}$, we get a natural homomorphism of graded $\mathcal{O}_F$-algebras $\mathrm{gr}_{\Delta F}(F \times X) \to \mathrm{gr}_{\Delta F}(F \times D)$, namely

$$\tan_F(X) \to \tan_F(D) \tag{1.2}$$

via the isomorphism induced by p_2. Applying Specan to both sides of (1.1) we get the canonical $\mathbb{C}$-morphism of cone-fibered spaces

$$C_X(D) \to C_{X,D} \tag{1.3}$$

inducing $C_X(D)|_F \to C_{X,D}|_F$. Applying Specan to (1.2), we get

$$C_D(F) \to C_X(F). \tag{1.4}$$

By combining (1.3) and (1.4), we obtain a sequence

$$C_D(F) \to C_X(F) \to C_{X,D}|_F \tag{1.5}$$

of canonical $\mathbb{C}$-morphisms of cone-fibered spaces, where the last morphism is the composition of the canonical closed $\mathbb{C}$-embedding

$$C_X(F) \to C_X(D)|_F$$

of Remark 1.6.3 with the restriction to F of (1.3).

Definition 1.6.5 The sequence

$$C_D(F) \to C_X(F) \to C_{X,D}|_F$$

of Remark 1.6.4 (1.5) is called the **fundamental tangent-normal sequence** associated with $F \subset D \subset X$.

Example 1.6.5 Take F to be a point $\xi \in D$. Then we have the fundamental sequence

$$C_{D,\xi} \to C_{X,\xi} \to C_{X,D}|_{\{\xi\}}$$

(see Remark 1.6.2). We shall often write $C_{X,D,\xi}$ for $C_{X,D}|_{\{\xi\}}$.

1.7 Projan and Blowing-Up

Let X be a $\mathbb{C}$-space and let A be a graded $\mathscr{O}_X$-algebra of LF type. Let $Z = \mathrm{Specan}(A)$ and let $\pi : Z \to X$ be the canonical projection. Denote by A_+ the ideal sheaf $\bigoplus_{d\geq 1} A_d$ in A, which is in fact generated by A_1. Then A_+ generates a coherent ideal sheaf in $\mathscr{O}_Z$ that defines a closed $\mathbb{C}$-subspace Z_0 of Z. Clearly π induces an isomorphism $Z_0 \to X$. Let $p : Z' \to Z$ be the blow-up with center Z_0 and let X' denote the inverse image of Z_0 by p. Let $g : X' \to X$ be the morphism induced by $\pi \cdot p$.

Definition 1.7.1 The notation being as above, X' is also written $\mathrm{Projan}(A)$ or $\mathbb{P}(Z)$, and is called the **projective $\mathbb{C}$-space associated with the cone-fibered space** Z over X. Note that $\mathrm{Projan}(A)$ has the **canonical projection** $g : X' \to X$.

Theorem 1.7.1 *Let D be a closed $\mathbb{C}$-subspace of a $\mathbb{C}$-space X. Let*

$$X' = \mathrm{Projan}(\mathrm{bl}_D(X))$$

and let $g : X' \to X$ be the canonical projection. Then g is the blow-up of X with center D.

The proof will be given after a few lemmas proved below. Let $\mathbf{I}$ denote the ideal sheaf of D in $\mathscr{O}_X$. Set $A = \mathrm{bl}_D(X) = \bigoplus_{d\geq 0} I^d$, where $A_d = \mathbf{I}^d$ for all $d \geq 0$. Let $\mathbf{I}\mathscr{O}_Z$ denote the ideal sheaf in $\mathscr{O}_Z$ generated by $\pi^*(\mathbf{I})$, and similarly let $\mathbf{I}\mathscr{O}_{Z'}$ be generated by $(\pi \cdot p)^*(\mathbf{I})$. Let $r : Z'' \to Z$ be the blow-up with center $\pi^{-1}(D)$, where $\pi^{-1}(D)$ is the closed $\mathbb{C}$-subspace of Z defined by $\mathbf{I}\mathscr{O}_Z$. We have a canonical π-homomorphism $A \to \mathscr{O}_Z$, by means of which $A_+\mathscr{O}_Z$ makes sense and defines Z_0 in Z. The ideal sheaf $A_+\mathscr{O}_{Z''}$ is constructed by means of r. Note that $A_+\mathscr{O}_Z = A_1\mathscr{O}_Z$ and $A_+\mathscr{O}_{Z''} = A_1\mathscr{O}_{Z''}$.

Lemma 1.7.1 *The ideal sheaves $\mathbf{I}\mathscr{O}_Z$ and $A_1\mathscr{O}_Z$ are invertible as $\mathscr{O}_Z$-modules outside $Z_0 \cap \pi^{-1}(D)$.*

Proof We have $\mathbf{I}^d = A_d$, but to avoid confusion with $\mathbf{I}^d \subset A_0 = \mathscr{O}_X$, we will use capital letters to denote the elements of $\mathbf{I}^d = A_d$. For instance, if $f \in \mathbf{I}^d \subset \mathscr{O}_X$, the corresponding element in A_d will be written F. Now pick any point $\eta \in Z$ and let $\xi = \pi(h) \in X$. For every pair of elements $f, g \in \mathbf{I}_\xi$, we have $fG - gF = 0$ in $(A_1)_\xi$. Therefore $\mathbf{I}\mathcal{O}_{Z,\eta} = f\mathcal{O}_{Z,\eta}$ if and only if $A_1\mathcal{O}_{Z,\eta} = F\mathcal{O}_{Z,\eta}$. Furthermore, for $f \in \mathbf{I}_\xi$ and $H \in (A_e)_\xi$ with $e \geq 0$ we have

$$f^d H = 0 \text{ in } (A_e)_\xi \iff f^d h = 0 \text{ in } \mathbf{I}_\xi^{d+l} \iff F^d H = 0 \text{ in } (A_{e+d})_\xi.$$

Indeed, for a fixed $f \in \mathbf{I}_\xi$, the annihilator of f^d in A_ξ is equal to that of F^d in A_ξ. Since the canonical homomorphism $A \to \mathcal{O}_Z$ is flat, the annihilator of such an element in $\mathcal{O}_Z$ is generated by this annihilator in A. Thus we can conclude that $f \in \mathbf{I}_\xi$ generates $\mathbf{I}\mathcal{O}_{Z,\eta}$ and is not a zero-divisor in $\mathcal{O}_{Z,\eta}$ if and only if $F \in (A_1)_\xi$ generates $A_1\mathcal{O}_{Z,\eta}$ and is not a zero-divisor in $\mathcal{O}_{Z,\eta}$. To complete the proof of Lemma 1.7.1, it is enough to check that $\mathbf{I}\mathcal{O}_Z$ is invertible outside $\pi^{-1}(D)$ and that $A_1\mathcal{O}_Z$ is invertible outside Z_0. But this is obvious. □

Lemma 1.7.2 *The ideal $\mathbf{I}\mathcal{O}_{Z'}$ is invertible as an $\mathcal{O}_{Z'}$-module, and $A_1\mathcal{O}_{Z''}$ is invertible as an $\mathcal{O}_{Z''}$-module.*

Proof We prove the first statement. Pick any point $\eta' \in Z'$ and let $\eta = p(\eta') \in Z$. Let $\xi = \pi(\eta) \in X$. By the definition of $p : Z' \to Z$, $A_1\mathcal{O}_{Z'}$ is invertible as an $\mathcal{O}_{Z'}$-module. Hence there is $F \in (A_1)_\xi$ such that $A_1\mathcal{O}_{Z',\eta'} = F\mathcal{O}_{Z',\eta'}$. Let f be the corresponding element of F in I_ξ. For every $g \in \mathbf{I}_\xi$, we have $fG = gF$ in $(A_1)_\xi$. If $G = \lambda F$ with $\lambda \in \mathcal{O}_{Z',\eta'}$, then $(g - \lambda f)F = 0$ in $\mathcal{O}_{Z',\eta'}$. But F is not a zero-divisor in $\mathcal{O}_{Z',\eta'}$. Hence $g = \lambda f$ in $\mathcal{O}_{Z',\eta'}$. We conclude $f\mathcal{O}_{Z,\eta'} = I\mathcal{O}_{Z',\eta'}$, and that $I\mathcal{O}_{Z'}$ is locally principal everywhere in Z'.

To prove invertibility, it is enough to show that the annihilator of $\mathbf{I}\mathcal{O}_{Z'}$ in Z' is zero. But this annihilator must have its support in $p^{-1}(Z_0)$ because p is an isomorphism outside Z_0 and $\mathbf{I}\mathcal{O}_Z$ is invertible outside Z_0 by Lemma 1.7.1. Therefore it must in turn be annihilated, at least locally everywhere, by sufficiently high powers of the ideal sheaf $A_1\mathcal{O}_{Z'}$ that defines $p^{-1}(Z_0)$ in Z'. This implies its nullity because $A_1\mathcal{O}_{Z'}$ is invertible as an $\mathcal{O}_{Z'}$-module. We now conclude that $\mathbf{I}\mathcal{O}_{Z'}$ is invertible as an $\mathcal{O}_{Z'}$-module.

The proof of the second statement of this lemma is analogous. If $f \in \mathbf{I}_\xi$ is such that $\mathbf{I}\mathcal{O}_{Z'',\eta''} = f\mathcal{O}_{Z'',\eta''}$ with $\eta'' \in Z''$ such that $r(\eta'') = \eta \in Z$ and $\pi(\eta) = \xi \in X$, we can prove that $A_1\mathcal{O}_{Z'',\eta''} = F\mathcal{O}_{Z'',\eta''}$. To prove the nullity of the annihilator of $A_1\mathcal{O}_{Z''}$ in $\mathcal{O}_{Z''}$, we use the fact that $r : Z'' \to Z$ is an isomorphism outside $\pi^{-1}(D)$ and that $A_1\mathcal{O}_Z$ is invertible as an $\mathcal{O}_Z$-module outside $\pi^{-1}(D)$, due to Lemma 1.7.1. So the annihilator has its support in $(\pi \cdot r)^{-1}(D)$. But the ideal sheaf of $(\pi \cdot r)^{-1}(D)$, that is, $\mathbf{I}\mathcal{O}_{Z''}$ in $\mathcal{O}_{Z''}$, defines $(\pi \cdot r)^{-1}(D)$ and is invertible as an $\mathcal{O}_{Z''}$-module. Hence the annihilator should be zero and $A_1\mathcal{O}_{Z''}$ is invertible as an $\mathcal{O}_{Z''}$-module. □

Lemma 1.7.3 *There exists a unique isomorphism $\sigma : Z' \to Z''$ such that $r \cdot \sigma = g$.*

Proof The universal mapping property of blowing-up (see Definition 1.3.4 and Remark 1.3.3) implies that the first part of Lemma 1.7.2 gives us a morphism $\sigma : Z' \to Z''$ and the second gives us $\sigma : Z'' \to Z'$ such that $r \cdot \sigma = g$ and $g \cdot \sigma' = r$. By uniqueness, $\sigma' \cdot \sigma$ is the identity of Z' and $\sigma \cdot \sigma'$ is the identity of Z''. This proves Lemma 1.7.3. □

Lemma 1.7.4 *In general, let X be a closed $\mathbb{C}$-subspace of a $\mathbb{C}$-space Z and assume that we have a $\mathbb{C}$-morphism $\pi : Z \to X$ that induces the identity in X. Let D be any closed $\mathbb{C}$-subspace of X, let $p : Z' \to Z$ be the blow-up with center $\pi^{-1}(D)$, and set $X' = p^{-1}(X)$. Then the morphism $X' \to X$ induced by p is the blow-up with center D.*

Proof The question is local in X and we may assume that Z is given as a closed $\mathbb{C}$-subspace of $X \times \mathbb{C}^s|_U$, where U is an open subset of $X \times \mathbb{C}^s$, in such a way that π is induced by the projection and the inclusion $X \subset Z$ is given by the identification $X = X \times \{\mathbf{0}\}$. Let $g : X'' \to X$ be the blow-up with center D. Then, first of all,

$$g \times \mathrm{id} : X'' \times \mathbb{C}^s \to X \times \mathbb{C}^s$$

is the blow-up with center $D \times \mathbb{C}^s$. This is easily deduced from Definition 1.3.4 (compare Remark 1.3.3), combined with the universal mapping property of the product: $\mathrm{mor}(T, R \times S) = \mathrm{mor}(T, R) \times \mathrm{mor}(T, S)$. By Theorem 1.3.1, Z' can be identified with the strict transform of Z via the blow-up $g \times \mathrm{id}$, and p is induced by $g \times \mathrm{id}$. If $\mathbf{z} = (z_1, \dots, z_s)$ is the coordinate system on $\mathbb{C}^s$, then $(\mathbf{z})\mathscr{O}_Z$ is the ideal sheaf of X. Hence $X' = p^{-1}(X)$ is a closed $\mathbb{C}$-subspace $X'' \times \{\mathbf{0}\}$ and so may be identified as a closed $\mathbb{C}$-subspace of X'' in such a way that p and g induce the same map $X' \to X$. Clearly p is isomorphic outside $D \times \mathbb{C}^s$, so this $X' \to X$ is isomorphic outside D. Similarly g is isomorphic outside D. This implies that the ideal sheaf $\mathbf{N}$ of X' in $\mathscr{O}_{X''}$ has support in $g^{-1}(D)$ and hence is annihilated, at least locally everywhere, by high powers of ideal sheaf $\mathbf{I}\mathscr{O}_{X''}$, where $\mathbf{I}$ denotes the ideal sheaf of D in $\mathscr{O}_X$. But $\mathbf{I}\mathscr{O}_{X''}$ is invertible as an $\mathscr{O}_{X''}$-module by the definition of g, so that $\mathbf{N} = (0)$, that is, $X'' = X'$. □

Proof of Theorem 1.7.1 By Lemma 1.7.3, we may identify Z'' with Z' and assume that $p : Z' \to Z$ is the blow-up with center $\pi^{-1}(D)$. Then, by Lemma 1.7.4, p induces the blow-up $g : X' \to X$ with center D. □

Remark 1.7.1 Let A be a graded $\mathscr{O}_X$-algebra of LF type, where X is a $\mathbb{C}$-space. Let $\mathbf{H}$ be a homogeneous ideal sheaf in A that is locally finitely generated. Then $A/\mathbf{H}$ is again a graded $\mathscr{O}_X$-module of LF type. We have a natural $\mathbb{C}$-morphism

$$c : \mathrm{Projan}(A/\mathbf{H}) \to \mathrm{Projan}(A),$$

which is a closed $\mathbb{C}$-embedding and is often called the **inclusion morphism**. It is obtained as follows. Let $Z = \mathrm{Specan}(A)$ and let Z_0 be the closed $\mathbb{C}$-subspace of Z

defined by $A_+\mathscr{O}_Z$. Similarly, let $W = \mathrm{Specan}(A/\mathbf{H})$ and let W_0 be the $\mathbb{C}$-subspace by $(A/\mathbf{H})_+\mathscr{O}_W$. Since $W_0 = W \cap Z_0$ with respect to the obvious inclusion $W \subset Z$, we have a commutative diagram

$$\begin{array}{ccc} W' & \hookrightarrow & Z' \\ {\scriptstyle r}\downarrow & & \downarrow{\scriptstyle g} \\ W & \hookrightarrow & Z \end{array}$$

where g is the blow-up with center Z_0 and r is the blow-up with center W_0. In fact, W' is simply the strict transform of W by g. The top inclusion induces the inclusion

$$\mathrm{Projan}(A/\mathbf{H}) = r^{-1}(W_0) \subset g^{-1}(Z_0) = \mathrm{Projan}(A).$$

Theorem 1.7.2 *Let A be a graded $\mathscr{O}_X$-algebra of LF type, where X is a $\mathbb{C}$-space. Set $X' = \mathrm{Projan}(A)$ and let $g : X' \to X$ be the canonical projection. Let Y be a closed $\mathbb{C}$-subspace of X and let $\mathbf{I}$ be the ideal sheaf of Y in $\mathscr{O}_X$. Then* $\mathrm{Projan}(A/\mathbf{I}A) = g^{-1}(Y)$ *with respect to the inclusion of Remark* 1.7.1.

Proof The question being local in X, we may assume that $A = \mathscr{O}_X[\mathbf{w}]/\mathbf{H}$, where $\mathbf{w} = (w_0, \dots, w_n)$ is a system of indeterminates and $\mathbf{H}$ is a homogeneous ideal with respect to the natural grading in $\mathscr{O}_X[\mathbf{w}]$. We view $\mathbf{w}$ as the coordinate system in $\mathbb{C}^{n+1}$ so that $\mathrm{Specan}(\mathscr{O}_X[\mathbf{w}]) = X \times \mathbb{C}^{n+1}$. Let $V \to \mathbb{C}^{n+1}$ be the quadratic transformation with center 0. Then $X \times V \to X \times \mathbb{C}^{n+1}$, the product of this transformation by the identity in X, is the blow-up with center $X \times \{\mathbf{0}\}$. We call this map $\mathscr{B}$. We have $\mathscr{B}^{-1}(X \times \{\mathbf{0}\}) = X \times \mathbb{P}^n$. We have a natural projection $V \to \mathbb{P}^n$ whose fibers are the complex lines through $\mathbf{0}$ in $\mathbb{C}^{n+1}$. With this, we define $\Gamma : X \times V \to X \times \mathbb{P}^n$ with identity in X. Our basic set-up is

$$X \;\underset{S}{\overset{\Pi}{\rightleftarrows}}\; X \times \mathbb{C}^{n+1} \xleftarrow{\;\mathscr{B}\;} X \times V \xrightarrow{\;\Gamma\;} X \times \mathbb{P}^n, \tag{1.6}$$

where Π is the projection and S is the section identifying $X \times \{\mathbf{0}\}$ with X by Π. We have the corresponding set-up for $Z = \mathrm{Specan}(A)$:

$$X \;\underset{s}{\overset{\pi}{\rightleftarrows}}\; Z \xleftarrow{\;p\;} Z' \xrightarrow{\;\gamma\;} X'. \tag{1.7}$$

Indeed, π is the canonical projection, s identifies X with the closed $\mathbb{C}$-subspace Z_0 of Z defined by $A_+\mathscr{O}_Z$, and p is the blow-up with center Z_0. Z is presented as a closed subset of $X \times \mathbb{C}^{n+1}$ by the ideal sheaf $\mathbf{H}\mathscr{O}_{X\times\mathbb{C}^{n+1}}$, Z' is identified with the strict transform of Z by $\mathscr{B}$ and p is induced by $\mathscr{B}$. Also $X' = p^{-1}(Z_0) \subset \mathscr{B}^{-1}(X \times \{\mathbf{0}\}) = X \times \mathbb{P}^n$. The scalar multiplication in Z as a cone-fibered space over X is compatible with the scalar multiplication on the second factor of $X \times \mathbb{C}^{n+1}$. This scalar multiplication lifts to $X \times V$ and acts as the scalar multiplication on the fiber lines for Γ. The scalar multiplication in Z leaves Z_0 fixed and hence lifts to Z'. In other words, Z' is invariant under the lifted scalar multiplication in $X \times V$. Thus we must have a closed $\mathbb{C}$-subspace X'' of $X \times \mathbb{P}^n$ such that $Z' \setminus p^{-1}(Z_0) = \Gamma^{-1}(X'') \setminus p^{-1}(Z_0)$. Since Z' is the strict transform, we must have $Z' = \Gamma^{-1}(X'')$. Hence $X'' = X' = p^{-1}(Z_0)$ and $Z' = \Gamma^{-1}(X')$. This Γ induces $Z' \to X'$. We apply the same argument to $A/\mathbf{I}A$ instead of A. Let $Z = \mathrm{Specan}(A/\mathbf{I}A)$, which is a closed $\mathbb{C}$-subspace of Z defined by $\mathbf{I}\mathscr{O}_Z$. Let Z' be the strict transform of Z by p (or, equivalently, by $\mathscr{B}$) so that $p : Z' \to Z$, induced by $\mathscr{P}$, is the blow-up with center Z_0, where Z_0 is defined by $(A/\mathbf{I}A)_+\mathscr{O}_Z$. Then we have

$$Y \underset{s}{\overset{\pi}{\rightleftarrows}} Z \xleftarrow{\ p\ } Z' \xrightarrow{\ \gamma\ } X'. \tag{1.8}$$

This is deduced from (1.6) in the same way as (1.7). Here Z is defined by $\mathbf{I}\mathscr{O}_Z$ and hence Z' is defined by $\mathbf{I}\mathscr{O}_Z$, outside $p^{-1}(Z_0)$. On the other hand, $\gamma^{-1}(g^{-1}(Y))$ is also defined by $\mathbf{I}\mathscr{O}_Z$, outside $p^{-1}(Z_0)$. This is because (first projection)$\cdot\Gamma = \Pi \cdot \mathscr{B}$, so that $g \cdot \gamma = \pi \cdot p$. Thus, Z' being the strict transform of Z, we must have $Z' = \gamma^{-1}(g^{-1}(Y))$. Hence $\mathrm{Projan}(A/\mathbf{I}A) = p^{-1}(Z_0) = Z' \cap p^{-1}(Z_0)$. This is equal to $g^{-1}(Y)$ because $p^{-1}(Z_0)$ is the zero-section of the complex line bundle $\gamma : Z' \to X'$. □

Theorem 1.7.3 *The assumption and notation being the same as in Theorem* 1.7.1, *we have* $g^{-1}(D) = \mathrm{Projan}(\mathrm{gr}_D(X)) = \mathbb{P}(C_{X,D})$. *Namely, the inverse image of the center by a blow-up is the associated projective fiber space of the normal cone.*

Proof We have $\mathrm{gr}_D(X) = \mathrm{bl}_D(X)/\mathbf{I}\,\mathrm{bl}_D(X)$ with the ideal sheaf $\mathbf{I}$ of D in $\mathscr{O}_X$. Thus Theorem 1.7.3 follows immediately from Remark 1.7.1. □

Chapter 2
The Weierstrass Preparation Theorem and Its Consequences

2.1 Filtrations and Graduations

Let Z be a $\mathbb{C}$-space and let $x \in Z$ be a smooth point (see Definition 1.1.7 in Chap. 1). To take a local coordinate system on Z centered at x is, by definition, to consider a specific isomorphism $(\varphi, \varphi^*) : (U, \mathcal{O}_Z/U) \to (V, \mathcal{O}_V)$, where U is an open neighborhood of x in Z, V is an open neighborhood of $\mathbf{0}$ in some $\mathbb{C}^n$, and $\mathcal{O}_V$ is the sheaf of holomorphic functions on V such that $\varphi(x) = \mathbf{0}$. If $h \in \mathcal{O}_V(V')$ is any holomorphic function on the open subset $V' \subset V$, we denote again by h the pull-back function $\varphi^*(h)$. In particular, if $(z_1, \ldots, z_n) = \mathbf{z}$ is the standard coordinate system on $\mathbb{C}^n$, we call the sections $(z_1, \ldots, z_n) \in \mathcal{O}_Z(U)^n$ the (corresponding) local coordinates on Z centered at x. We will also say, more succinctly, that we are taking a local coordinate system $(z_1, \ldots, z_n) = \mathbf{z}$ on Z centered at x (or around x).

With these conditions, the stalk $\mathcal{O}_{Z,x}$ can be identified with the ring of germs of holomorphic functions at $\mathbf{0}$, which is the ring $\mathbb{C}\{\mathbf{z}\}$ of convergent power series in the variables $\mathbf{z}$. We shall identify an analytic function on a connected open neighborhood of x with its germ at x.

Let X be a $\mathbb{C}$-space and let $x \in X$. We shall consistently operate with local $\mathbb{C}$-elements (Definition 1.2.1 in Chap. 1) $\mathbf{s} : (s)_0 \to (s)^0$ of X such that $x \in (s)_0$. Such an object was called a local $\mathbb{C}$-element of X centered at x (or around x). In agreement with the classical terminology, this scene will be often described as follows. Let Z be a smooth $\mathbb{C}$-space, let $x \in Z$, and let X be a closed $\mathbb{C}$-subspace of Z locally around x (here Z and X play the respective roles of some open $(s)^0 \subset \mathbb{C}^n$ and $(s)_0$). In this case, X is defined locally around x (that is, on $(s)^0$) by the vanishing of some holomorphic functions $h_1, \ldots, h_r$ on $(s)^0$. If $\mathbf{z} = (z_1, \ldots, z_n)$ is a local coordinate system on Z centered at x, then X is defined, locally around x, by the vanishing of the functions belonging to the ideal $I = (h_1(\mathbf{z}), \ldots, h_r(\mathbf{z})) \subset \mathbb{C}\{\mathbf{z}\}$.

J. M. Aroca et al., *Complex Analytic Desingularization*,
https://doi.org/10.1007/978-4-431-49822-3_2

Remark 2.1.1 Let X be a $\mathbb{C}$-space, and let $x \in X$ be a smooth point; then $\mathscr{O}_{X,x}$ is a regular local ring, since it is isomorphic to a convergent power series ring, as we have just seen. Conversely, assume that $x \in X$ is such that $\mathscr{O}_{X,x}$ is a regular local ring; we are going to prove that x is a smooth point of X. In fact, take a smooth $\mathbb{C}$-space Z such that $x \in X \subset Z$ locally around x, where X is closed in Z, locally around x. Take also a local coordinate system $\mathbf{z} = (z_1, \ldots, z_n)$ on Z centered at x; then X is defined, locally around x, by an ideal $I \subset \mathbb{C}\{\mathbf{z}\}$, so $\mathscr{O}_{X,x} = \mathbb{C}\{\mathbf{z}\}/I$. Let $\{y_{r+1}+I, \ldots, y_r+I\}$, $y_j \in \mathbb{C}\{\mathbf{z}\}$ be a regular system of parameters in $\mathscr{O}_{X,x}$; then we may write

$$y_j = \sum_{i=1}^{n} a_{ij}\mathbf{z}_i + f_j \qquad \text{for } a_{ij} \in \mathbb{C} \text{ and } f_j \in \mathbb{C}\{\mathbf{z}\} \text{ with } \nu_{\mathbf{0}}(f_j) \geq 2,$$

where $\nu_{\mathbf{0}}$ represents the usual order of a power series. Then a_{ij} has rank $n - r$, so $\{y_{r+1}, \ldots, y_n\}$ can be extended to a regular system of parameters $y = (y_1, \ldots, y_n)$ of $\mathbb{C}\{\mathbf{z}\}$. Then $\mathbb{C}\{\mathbf{z}\} = \mathbb{C}\{\mathbf{y}\}$, and the ideal I is given by $I = (y_1, \ldots, y_r)$, so X is locally isomorphic to $\mathbb{C}^{n-r}$. This proves our assertion.

Remark 2.1.2 Let Z be a smooth $\mathbb{C}$-space, let $x \in Z$, and let W be a smooth closed $\mathbb{C}$-subspace of Z locally around x. We shall say that a local coordinate system $(\mathbf{z}, \mathbf{w})$ on Z centered at x is **associated with** W if W is defined, locally around x, by the ideal $P = (\mathbf{z}) \cdot \mathbb{C}\{\mathbf{z}, \mathbf{w}\}$. If $\mathbf{z} = (z_1, \ldots, z_c)$ and $\mathbf{w} = (w_1, \ldots, w_d)$, the restrictions $w_j|_W$ to W of the coordinate functions w_j form a local coordinate system on W centered at x. Algebraically speaking, this means that the canonical images $w_j + P$ of w_j in $\mathbb{C}\{\mathbf{z}, \mathbf{w}\}/P$ form a regular system of parameters of this ring. We shall often write, by abuse of notation, $w_j + P = w_j|_W$.

Now, by restricting ourselves to suitable neighborhoods of x, we have a map $r : Z \to W$, locally around x, defined by $r(\mathbf{z}, \mathbf{w}) = (\mathbf{w})$. This map is a (local) retraction of the inclusion $W \subset Z$ and corresponds algebraically to the injective homomorphism

$$r_x : \mathbb{C}\{\mathbf{z}, \mathbf{w}\}/P \to \mathbb{C}\{\mathbf{z}, \mathbf{w}\}$$

defined by

$$r_x(w_j + P) = w_j.$$

The fiber $r^{-1}(x)$ is the smooth closed $\mathbb{C}$-subspace of Z defined, locally around x, by the ideal $(\mathbf{w}) \cdot \mathbb{C}\{\mathbf{z}, \mathbf{w}\}$, so $(z_1|_{r^{-1}(x)}, \ldots, z_c|_{r^{-1}(x)})$ is a local coordinate system on it, centered at x.

Conversely, assume we are given a local retraction $r : Z \to W$, around x, of the inclusion $W \subset Z$ (we shall simply say that r is a local retraction around x). A local coordinate system $(\mathbf{z}, \mathbf{w})$ on Z centered at x is said to be **associated with** r if

$(\mathbf{z}, \mathbf{w})$ is associated with W, and r is defined by $r(\mathbf{z}, \mathbf{w}) = \mathbf{w}$. It is not difficult to see that such a local coordinate system always exists. In fact, if $(\mathbf{z}', \mathbf{w}')$ is any local coordinate system on Z, centered at x and associated with W, the homomorphism $r_x : \mathbb{C}\{\mathbf{z}', \mathbf{w}'\}/P \to \mathbb{C}\{\mathbf{z}', \mathbf{w}'\}$ must be defined by equations of the form

$$r_x(w'_j + P) = w'_j + g_j, \quad \text{for } g_j \in (\mathbf{z}') \cdot \mathbb{C}\{\mathbf{z}', \mathbf{w}'\}.$$

Then, if we let $z_i = z'_i$ for $1 \le i \le c$ and $w_j = w'_j + g_j$ for $1 \le j \le d$, we have a local coordinate system $(\mathbf{z}, \mathbf{w})$ associated with r, as required.

Let $r, r' : Z \to W$ be two local retractions around x, and let $(\mathbf{z}, \mathbf{w})$, $(\mathbf{z}', \mathbf{w}')$ be the local coordinate system on Z, centered at x, and associated with r and r', respectively; then $(\mathbf{z}) \cdot \mathbb{C}\{\mathbf{z}, \mathbf{w}\} = (\mathbf{z}') \cdot \mathbb{C}\{\mathbf{z}', \mathbf{w}'\}$. It is easy to see that $r = r'$ if and only if $(\mathbf{w}) \cdot \mathbb{C}\{\mathbf{z}, \mathbf{w}\} = (\mathbf{w}') \cdot \mathbb{C}\{\mathbf{z}', \mathbf{w}'\}$ and $w'_j(\mathbf{z}, \mathbf{w}) = w'_j(\mathbf{0}, \mathbf{w})$ for any $j = 1, \ldots, d$.

Let Z be a smooth complex analytic space, let $x \in Z$, and let W and W' be smooth subspaces of Z containing x. We say that two local retractions $r : Z \to W$ and $r' : Z \to W'$ are equivalent if there exists a local isomorphism $\sigma : W \to W'$ around x such that $r' = \sigma \cdot r$. The equivalence class of r will be called the local projection around x defined by r and will be denoted by $\mathbf{r}$. If $r' \in \mathbf{r}$ and $(\mathbf{z}, \mathbf{w})$, $(\mathbf{z}', \mathbf{w}')$ are local coordinate systems associated with r and r' respectively, then

$$w'_i = w'_i(\mathbf{w}), \quad z'_i = z'_i(\mathbf{z}) + \varphi_i(\mathbf{w}).$$

We have described the conditions that we shall constantly deal with. Therefore, we are led to a thorough study of the ring of convergent powers series $R = \mathbb{C}\{\mathbf{z}, \mathbf{w}\}$, with $\mathbf{z} = (z_1, \ldots, z_c)$, $\mathbf{w} = (w_1, \ldots, w_d)$. Our main tool will be the Weierstrass preparation theorem, which will be stated and proved in the next section. In the rest of this section we will describe some filtrations and graduations associated with R. We shall consistently use the following notation. Every $g \in R$ will be written as

$$g = \sum_{A \in \mathbb{Z}_0^c} g_A \mathbf{z}^A = \sum_{A \in \mathbb{Z}_0^c, B \in \mathbb{Z}_0^d} g_{A,B} \mathbf{z}^A \mathbf{w}^B,$$

where $g_A \in \mathbb{C}\{\mathbf{w}\}$, $g_{A,B} \in \mathbb{C}$, and $\mathbf{z}^A = z_1^{a_1} \cdots z_c^{a_c}$ if $A = (a_1, \ldots, a_c)$ ($\mathbf{w}^B$ is given an analogous meaning). We also write $|A| = a_1 + \cdots + a_c$. The following sets will play a key role in the sequel:

$$\mathscr{E}_{\mathbf{z}}(g) = \{A \in \mathbb{Z}_0^c \mid g_A \neq 0\},$$
$$\mathscr{E}_{\mathbf{z},\mathbf{w}}(g) = \{(A, B) \in \mathbb{Z}_0^c \times \mathbb{Z}_0^d \mid g_{A,B} \neq 0\}.$$

We observe that $\mathscr{E}_{\mathbf{z}}(g)$ is not affected by a coordinate change inside W, that is, by a parameter change among the w_j.

Remark 2.1.3 A positive linear from is a map $L : \mathbb{R}_0^c \times \mathbb{R}_0^d \to \mathbb{R}_0$ given by an equation of the form

$$L(\mathbf{x}, \mathbf{x}') = \sum_{i=1}^{c} l_i x_i + \sum_{i=1}^{d} l'_j x'_j,$$

where $l_i, l'_j \in \mathbb{R}_0$ and not all the l_i are zero. Let Σ be a sub-semigroup of $(\mathbb{R}_0, +)$ containing $\mathbb{Z}_0$; we construct the filtration $\{I(L, \Sigma, \nu)\}_{\nu\in\Sigma}$ on R by writing, for $\nu \in \Sigma$,

$$I(L, \Sigma, \nu) = \sum_{(A,B)\in\mathbb{Z}_0^{c+d},\ L(A,B)\le\nu} \mathbf{z}^A \mathbf{w}^B R.$$

It is easy to see that

$$\bigcap_{\nu\in\Sigma} I(L, \Sigma, \nu) = (0),$$

so, for any non-zero $g \in R$, the number

$$\nu_{L,\Sigma}(g) = \sup\{\nu \in \Sigma \mid g \in I(L, \Sigma, \nu)\}$$

makes sense. We call this number the **order** of g in this filtration. We shall write $\nu_{L,\Sigma}(0) = \infty$.

Now, if we write, for any $\nu \in \Sigma$,

$$I(L, \Sigma, \nu^+) = \sum_{(A,B)\in\mathbb{Z}_0^{c+d},\ L(A,B)>\nu} \mathbf{z}^A \mathbf{w}^B R,$$

we have a graded ring

$$\mathrm{gr}_{L,\Sigma}(R) = \bigoplus_{\nu\in\Sigma} I(L, \Sigma, \nu)/I(L, \Sigma, \nu^+) = \bigoplus_{\nu\in\Sigma} \mathrm{gr}^{\nu}_{L,\Sigma}(R),$$

where we have written

$$\mathrm{gr}^{\nu}_{L,\Sigma}(R) = I(L, \Sigma, \nu)/I(L, \Sigma, \nu^+).$$

We observe that the degree semigroup of $\mathrm{gr}_{L,\Sigma}(R)$ is Σ. If $g \in R$ and $\nu = \nu_{L,\Sigma}(g)$, we shall call the initial form of g with respect to (L, Σ), the element

$$\mathrm{in}_{L,\Sigma}(g) = g + I(L, \Sigma, \nu^+) \in \mathrm{gr}^{\nu}_{L,\Sigma}(R).$$

If J is an ideal in R, we shall call the initial ideal of J with respect to L, Σ, the following ideal in $\mathrm{gr}_{L,\Sigma}(R)$:

$$\mathrm{in}_{L,\Sigma}(J) = (\{\mathrm{in}_{L,\Sigma}(g)\}_{g \in J}) \cdot \mathrm{gr}_{L,\Sigma}(R).$$

The following properties hold:

(1) If $\Sigma \subset \Sigma'$ then $\nu_{L,\Sigma}(g) \leq \nu_{L,\Sigma'}(g)$, for every $g \in R$.

(2) For any non-zero $g \in R$, one has

$$\nu_{L,\Sigma}(g) = \sup\{\nu \in \mathbb{Z} \mid \nu \leq \inf\{L(A, B) \mid (A, B) \in \mathscr{E}_{\mathbf{z},\mathbf{w}}(g)\}\}.$$

In particular, if $L(\mathbb{Z}_0^c \times \mathbb{Z}_0^d) \subset \Sigma$,

$$\nu_{L,\Sigma}(g) = \inf\{L(A, B) \mid (A, B) \in \mathscr{E}_{\mathbf{z},\mathbf{w}}(g)\}.$$

(3) If $f, g \in R$, then $\nu_{L,\Sigma}(fg) \geq \nu_{L,\Sigma}(f) + \nu_{L,\Sigma}(g)$ and the equality holds if and only if $L(\mathbb{Z}_0^c \times \mathbb{Z}_0^d) \subset \Sigma$.

In fact, it is clear that $\nu_{L,\Sigma}(fg) \geq \nu_{L,\Sigma}(f) + \nu_{L,\Sigma}(g)$, because

$$I(L, \Sigma, \nu).I(L, \Sigma, \nu') \subset I(L, \Sigma, \nu + \nu').$$

On the other hand, by (2) above, $L(\mathbb{Z}_0^c \times \mathbb{Z}_0^d) \subset \Sigma$ implies the equality. Let us show the converse. Assume that $L(\mathbb{Z}_0^c \times \mathbb{Z}_0^d) \not\subset \Sigma$, and let $(A, B) \in \mathbb{Z}_0^{c+d}$ be such that $L(A, B) \notin \Sigma$. Let $\nu = \nu_{L,\Sigma}(\mathbf{z}^A\mathbf{w}^B)$, set $\nu' = L(A, B) - \nu > 0$, and let $t \in \mathbb{Z}_+$ be such that $t\nu' \geq 1$ and $(t-1)\nu' < 1$. We then have $L(tA, tB) = tL(A, B) = \nu t + \nu' t \geq \nu t + 1$, so $\nu_{L,\Sigma}((\mathbf{z}^A\mathbf{w}^B)^t) \geq \nu t + 1$, which proves that the inequality is strict.

(4) If $f, g \in R$, then $\nu_{L,\Sigma}(f + g) \geq \inf(\nu_{L,\Sigma}(f), \nu_{L,\Sigma}(g))$.

(5) If $\Sigma = \mathbb{Z}_0$, then

$$\nu_{L,\mathbb{Z}_0}(g) = \inf\{[L(A, B)] \mid (A, B) \in \mathscr{E}_{\mathbf{z},\mathbf{w}}(g)\}$$

for every $f \in R$, where [ ] means the integral part.

In the case $L(\mathbb{Z}_0^c \times \mathbb{Z}_0^d) \subset \Sigma$, we shall omit any mention of Σ, and we will simply write $\nu_L(g)$, $\mathrm{gr}_L(R)$, $\mathrm{in}_L(g)$, $\mathrm{in}_L(I)$ for $\nu_{L,\Sigma}(g)$, $\mathrm{gr}_{L,\Sigma}(R)$, $\mathrm{in}_{L,\Sigma}(g)$, $In_{L,\Sigma}(I)$. In this case and if, furthermore, all l_i, l'_j are different from zero, we have the following: if for $g \in R$ we write $g = \mathrm{in}_L(g)$, then

$$\mathrm{gr}_L(R) \simeq \mathbb{C}[\mathbf{z}, \mathbf{w}],$$

which is a polynomial ring in the variables $\mathbf{z} = (z_1, \ldots, z_c)$ and $\mathbf{w} = (w_1, \ldots, w_d)$, under an isomorphism that is, in general, non-graded if one considers the natural graduation on $\mathbb{C}[\mathbf{z}, \mathbf{w}]$. More generally, if Ω is the set of the indices i such that $l_i = 0$ and Ω' is the set of j such that $l'_j = 0$, then

$$\mathrm{gr}_L(R) = \mathbb{C}\{(z_i)_{i\in\Omega}, (w_j)_{j\in\Omega'}\}[(z_i)_{i\notin\Omega}, (w_j)_{j\notin\Omega'}].$$

Example 2.1.1 Choose $\delta \in \mathbb{R}_+$ and consider the positive linear form

$$L(\mathbf{x}, \mathbf{x}') = \sum_{i=1}^{c} x_i + \frac{1}{\delta}\sum_{j=1}^{d} x'_j$$

and the semigroup $\Sigma_\delta = \mathbb{Z}_0 + (1/\delta)\mathbb{Z}_0$. Corresponding to these data, we have a filtration

$$\{I(L_\delta, \nu) = I(L_\delta, \Sigma_\delta, \nu)\}_{\nu\in\Sigma_\delta}$$

on R and a graded ring $\mathrm{gr}_{L_\delta}(R)$, which is isomorphic to $\mathbb{C}[\mathbf{z}, \mathbf{w}]$, where

$$\begin{aligned} z_i &= z_i + I(L_\delta, 1^+), \qquad 1 \le i \le c, \\ w_j &= w_j + I(L_\delta, (1/\delta)^+), \ 1 \le j \le d. \end{aligned}$$

In particular, if $\delta = a/b$, where a, b are relatively prime positive integers, and if we endow $\mathbb{C}[\mathbf{z}, \mathbf{w}]$ with the graduation assigning degree a to all z_i and degree b to all w_j, the isomorphism $\mathrm{gr}_{L_\delta}(R) \simeq \mathbb{C}[\mathbf{z}, \mathbf{w}]$ is graded.

The whole problem can be described by means of the Newton diagram in the following way: to each monomial $\mathbf{z}^A\mathbf{w}^B$, we assign the point $(|B|, |A|)$. The set of these points is called the **Newton diagram** of R. Consider the straight line l of slope $-1/\delta$ passing through $(0, \nu)$. Then:

(1) The ideal $I(L_\delta, \nu)$ is generated, as an ideal of R, by the monomials corresponding to points lying on or above l.
(2) The ideal $I(L_\delta, \nu^+)$ is generated, as an ideal of R, by the monomials corresponding to points lying above l.
(3) The $\mathbb{C}$-vector space $\mathrm{gr}^\nu_{L_\delta}(R)$ is generated in $\mathbb{C}[\mathbf{z}, \mathbf{w}]$, by the monomials which, in a similar representation for $\mathbb{C}[\mathbf{z}, \mathbf{w}]$, correspond to points lying on l.

From the geometric point of view, the filtration $\{I(L_\delta, \nu)\}_{\nu\in\Sigma_\delta}$ and the graded ring $\mathrm{gr}_{L_\delta}(R)$ are intrinsic, in a sense that we are going to make clear. Let Z be a smooth $\mathbb{C}$-space, let $x \in Z$, and let W be a smooth closed $\mathbb{C}$-subspace of Z, locally around x. Denote by $M = M_{Z,x}$ the maximal ideal of $\mathcal{O}_{Z,x}$, and by $P \subset \mathcal{O}_{Z,x}$ the ideal defining W locally around x. Write, for $\nu \in \Sigma_\delta$,

$$F^{\nu}_{Z,W,\delta,x} = \sum_{\substack{\alpha,\beta\in\mathbb{Z}_0 \\ \alpha+(\beta/\delta)\geq\nu}} P^{\alpha}M^{\beta}, \qquad F^{\nu^+}_{Z,W,\delta,x} = \sum_{\substack{\alpha,\beta\in\mathbb{Z}_0 \\ \alpha+(\beta/\delta)>\nu}} P^{\alpha}M^{\beta},$$

$$\mathrm{gr}^{\nu}_{W,\delta,x}(Z) = F^{\nu}_{Z,W,\delta,x}/F^{\nu^+}_{Z,W,\delta,x}, \quad \mathrm{gr}_{W,\delta,x}(Z) = \bigoplus_{\nu\in\Sigma_\delta} \mathrm{gr}^{\nu}_{W,\delta,x}(Z),$$

$$\nu^{W,\delta}_x(f) = \mathrm{Sup}\{\nu \in \Sigma_\delta \mid f \in F^{\nu}_{Z,W,\delta,x}\} \text{ for } f \in \mathscr{O}_{Z,x},$$

$$\mathrm{in}^{W,\delta}_x(f) = f + F^{\nu^+}_{Z,W,\delta,x}, \qquad \nu = \nu^{W,\delta}_x(f).$$

Then, if we choose a local coordinate system $(\mathbf{z}, \mathbf{w})$ on Z centered at x and associated with W, the isomorphism $\mathscr{O}_{Z,x} \simeq R$ induces an isomorphism

$$\begin{aligned} F^{\nu}_{Z,W,\delta,x} &\simeq I(L_\delta, \nu), \\ F^{\nu^+}_{Z,W,\delta,x} &\simeq I(L_\delta, \nu^+), \\ \mathrm{gr}^{\nu}_{W,\delta,x}(Z) &\simeq \mathrm{gr}^{\nu}_{L_\delta}(R), \end{aligned}$$

and a graded isomorphism

$$\mathrm{gr}_{W,\delta,x}(Z) \simeq \mathrm{gr}_{L_\delta}(R).$$

Example 2.1.2 Take $\delta = 1$ in the preceding example; then we get the $(\mathbf{z}, \mathbf{w})$-adic filtration and graduation because, for $\nu \in \mathbb{Z}_0$,

$$I(L_1, \nu) = (\mathbf{z}, \mathbf{w})^{\nu}, \quad I(L_1, \nu^+) = (\mathbf{z}, \mathbf{w})^{\nu+1}.$$

In this case, we shall use the notation ν_0 for ν_{L_1}, $\mathrm{gr}_0(R)$ for $\mathrm{gr}_{L_1}(R)$, and so on. As before, this filtration and graduation are intrinsic, corresponding to the M-adic filtration in $\mathscr{O}_{Z,x}$ and the graded ring

$$\mathrm{gr}_{\mathbf{x}}(Z) = \bigoplus_{n\in\mathbb{Z}_0} (M^n/M^{n+1}).$$

Also, for $f \in \mathscr{O}_{Z,x}$, we shall use the notation $\nu_x(f)$, $\mathrm{in}_x(f)$, and so on, to express the corresponding concepts here. As a general rule, we shall write $\bar{f} = \mathrm{in}_0(f)$, for $f \in R$.

Remark 2.1.4 With the notation of Examples 2.1.1 and 2.1.2, fix a local retraction $r : Z \to W$ around x, and denote by $r_x : \mathscr{O}_{W,x} \to \mathscr{O}_{Z,x}$ the corresponding homomorphism. Consider the rings $\mathrm{gr}_x(W)$ and $\mathrm{gr}_P(\mathscr{O}_{Z,x})$, endowed with the structure of $\mathscr{O}_{W,x}$-modules in the obvious way, and consider on $\mathscr{O}_{Z,x}$ the structure of the $\mathscr{O}_{W,x}$-module given by r_x. Then, one has an isomorphism

$$\theta'_r : \mathrm{gr}_x(Z) \to \mathrm{gr}_x(W) \otimes_{\mathscr{O}_{W,x}} \mathrm{gr}_P(\mathscr{O}_{Z,x}),$$

defined as follows. Let $(\mathbf{z}, \mathbf{w})$ be a local coordinate system on Z centered at x and associated with r; then, as a $\mathbb{C}$-linear map, θ'_r is defined by

$$\begin{aligned} \theta'_r(\bar{z}_i) &= 1 \otimes (z_i + P^2), \qquad 1 \le i \le c, \\ \theta'_r(\bar{w}_j) &= (w_j + M^2_{W,x}) \otimes 1,\ 1 \le j \le d. \end{aligned}$$

It is then clear (see the last observation in Remark 2.1.2) that θ'_r depends only on r and not on a special choice of coordinates.

On the other hand, we have two canonical homomorphisms

$$\theta_1 : \mathrm{gr}_x(W) \to \mathrm{gr}_{W,\delta,x}(Z) \quad \text{and} \quad \theta_2 : \mathrm{gr}_P(\mathscr{O}_{Z,x}) \to \mathrm{gr}_{W,\delta,x}(Z),$$

determined, respectively, by the natural maps

$$\begin{aligned} M_{W,x}/M^2_{W,x} &\to F^{1/\delta}_{Z,W,\delta,x}/F^{(1/\delta)^+}_{Z,W,\delta,x}, \\ P/P^2 &\to F^{1}_{Z,W,\delta,x}/F^{1^+}_{Z,W,\delta,x}. \end{aligned}$$

Therefore, the composition $(\theta_1 \otimes \theta_2)\theta'_r$ gives a map

$$\theta_r : \mathrm{gr}_x(Z) \to \mathrm{gr}_{W,\delta,x}(Z),$$

which is an isomorphism because it is defined by the relations

$$\begin{aligned} \theta_r(\bar{z}_i) &= \tilde{z}_i,\ \ 1 \le i \le c, \\ \theta_r(\bar{w}_j) &= \tilde{w}_j,\ 1 \le j \le d, \end{aligned}$$

where $\tilde{f} = \mathrm{in}_{W,\delta,x}(f)$.

Example 2.1.3 Consider the positive linear form L_J defined by

$$L_J(x, x') = \sum_{i=1}^{c} x_i + \sum_{j=1}^{e} x'_j, \quad e \le d,$$

that we say, by definition, is associated with the ideal

$$J = (\mathbf{z}, w_1, \ldots, w_e)R.$$

Then $\{I(L_J, \nu)\}_{\nu \in \mathbb{Z}_0}$ is the J-adic filtration of R, and the corresponding graduation is the J-adic one. In this case, we shall write ν_J, gr_J, and so on, instead of the general notation. In particular, one has

$$\mathrm{gr}_J(R) \simeq \mathbb{C}\{w_{e+1}, \ldots, w_d\}[\tilde{\mathbf{z}}, \tilde{w}_1, \ldots, \tilde{w}_e]$$

with $\tilde{z}_i = z_i + J^2$ and $\tilde{w}_j = w_j + J^2$. We observe that, if $g \in R$, then

$$\nu_J(g) = \nu_0(g) \Longleftrightarrow \mathrm{in}_0(g) \notin (w_{e+1}, \ldots, w_d) \cdot \mathbb{C}[\tilde{\mathbf{z}}, \tilde{\mathbf{w}}];$$

in particular, if $e = 0$, then

$$\nu_{(z)}(g) = \nu_0(g) \Longleftrightarrow \mathrm{in}_0(g) \notin (\mathbf{w}).\mathbb{C}[\tilde{\mathbf{z}}, \tilde{\mathbf{w}}].$$

2.2 The Weierstrass Preparation Theorem

As pointed out in the preceding section, we are led to an algebraic study of the ring $R = \mathbb{C}\{\mathbf{z}, \mathbf{w}\}$ of convergent power series in the variables $\mathbf{z} = (z_1, \ldots, z_c)$ and $\mathbf{w} = (w_1, \ldots, w_d)$, with coefficients in the complex field $\mathbb{C}$. We denote by M the maximal ideal of R. This section is entirely devoted to the development of our main tool, the Weierstrass preparation theorem. We start with some notation.

Remark 2.2.1 (Notation) In our study of the Weierstrass preparation theorem, we shall frequently deal with positive linear forms L (Remark 2.1.3)

$$L(\mathbf{x}, \mathbf{x}') = \sum_{i=1}^{c} l_i x_i + \sum_{j=1}^{d} l'_j x'_j,$$

where $l_i, l'_j \in \mathbb{R}_0$ and not all the l_i are zero. In order to consider filtrations and graduations, we shall take, for every L, the sub-semigroup $\Sigma_L = L(\mathbb{Z}_0^c \times \mathbb{Z}_0^d) + \mathbb{Z}_0$ of $\mathbb{R}_0$; in this way, such filtrations and graduations only depend on L.

Given two families of elements

$$\mathbf{F} = \{f_i\}_{1 \leq i \leq k} \subset R, \quad \mathbf{A} = \{A_i\}_{1 \leq i \leq k} \subset \mathbb{Z}_0^c,$$

we denote by $S\{\mathbf{F}, \mathbf{A}\}$ the set of positive linear forms L satisfying the following condition: For each $j = 1, \ldots, k$, there exists a non-zero complex number c_j such that $\nu_L(\mathbf{z}^{A_j}) < \nu_L(f_j - c_j \mathbf{z}^{A_j})$. For every such family $\mathbf{A}$, we shall construct a partition

$$\mathbb{Z}_0^c = \bigcup_{i=0}^{k} \Box_i$$

of $\mathbb{Z}_0^c$, which we say is *associated with* $\mathbf{A}$, as follows:

$$\Box_0 = \mathbb{Z}_0^c \setminus \bigcup_{i=1}^{k} (A_i + \mathbb{Z}_0^c),$$

$$\Box_1 = A_1 + \mathbb{Z}_0^c,$$

$$\Box_i = (A_i + \mathbb{Z}_0^c) \setminus (A_i + \mathbb{Z}_0^c) \cap \bigcup_{j=1}^{i-1} \Box_j, \qquad \text{for } i > 1.$$

Example 2.2.1 Take $c = 1$. Given a non-unit $f \in R$, different from zero, and an integer $a > 0$, we say that f is **regular** in $z_1 = z$ of order a if the power series $f(z, \mathbf{0})$ is of order a. This is equivalent to saying that f contains a term of the form γz^a, for $\gamma \in \mathbb{C} \setminus \{0\}$, and no non-zero term in only z of smaller degree. If we take the families $\mathbf{F} = \{f\}$ and $\mathbf{A} = \{a\}$, then to say that $S\{\mathbf{F}, \mathbf{A}\} \neq \varnothing$ is equivalent to saying that f is regular in z of order a.

Theorem 2.2.1 (The Weierstrass preparation theorem) *Let*

$$\mathbf{F} = \{f_i\}_{1 \leq i \leq k} \subset R \quad \textit{and} \quad \mathbf{A} = \{A_i\}_{1 \leq i \leq k} \subset \mathbb{Z}_0^c$$

be two families such that $S\{\mathbf{F}, \mathbf{A}\} \neq \varnothing$, *and consider the partition*

$$\mathbb{Z}_0^c = \bigcup_{i=0}^{k} \Box_i$$

associated with $\mathbf{A}$. *Then, for every* $g \in R$, *there exists a unique family* $\{h_i\}_{0 \leq i \leq k}$ *of elements in* R *satisfying the following conditions:*

(1) $g = h_0 + \sum_{i=1}^{k} h_i f_i$.
(2) $\mathscr{E}_{\mathbf{z}}(h_i \mathbf{z}^{A_i}) \subset \Box_i$ *for* $0 \leq i \leq k$, *with* $A_0 = 0$.
(3) $\nu_L(h_i) \geq \nu_L(g) - \nu_L(f_i)$ *for every* $L \in S\{\mathbf{F}, \mathbf{A}\}$ *and* $1 \leq i \leq k$.

The remainder of this section is devoted to the proof of this theorem, which is long and requires some auxiliary results.

Remark 2.2.2 (Reduction of the initial data) The fact that $S\{\mathbf{F}, \mathbf{A}\}$ is non-empty implies, in particular, that $(f_i)_{A_i,0} \neq 0$ for all $i = 1, \ldots, k$. We take a generic $\mathbf{m} = (m_1, \ldots, m_c) \in \mathbb{R}_+^c$ and perform the change of variables $\mathbf{z} = \mathbf{m} \cdot \mathbf{z}'$, that is,

$$z_i' = z_i / m_i, \qquad \text{for } 1 \leq i \leq c.$$

We also write

$$f_i'(\mathbf{z}', \mathbf{w}) = \frac{1}{(f_i)_{A_i,0}\mathbf{m}^{A_i}} f_i(\mathbf{m} \cdot \mathbf{z}', \mathbf{w}) \quad \text{for } 1 \le i \le k,$$
$$g'(\mathbf{z}', \mathbf{w}) = g(\mathbf{m} \cdot \mathbf{z}', \mathbf{w}).$$

Now, we have:

(1) The equality $(f_i')_{A_i,0} = 1$.
(2) The equality $S\{\{f_i'\}_{1 \le i \le k}, \mathbf{A}\} = S\{\mathbf{F}, \mathbf{A}\}$.
(3) The Weierstrass preparation theorem holds for g, f_i, A_i if and only if it holds for g', f_i', A_i.

This reduction has two purposes: to make the $(f_i)_{A_i,0}$ equal to 1 for simplicity, and to introduce the factor $\mathbf{m}$, to be determined later in a suitable way. We take, until further notice, $f_i' = f_i$ and $g' = g$.

Remark 2.2.3 We are going to define inductively a family $\{h_i{}^{(j)}\}_{0 \le i \le k, j \ge 0}$ of formal power series. As before, we set $A_0 = 0$, and define the power series $\{h_i{}^{(0)}\}_{0 \le i \le k}$ by

$$h_i{}^{(0)} = \sum_{A \in \mathbb{Z}_0^c} (h_i{}^{(0)})_A \mathbf{z}^A,$$

where

$$(h_i{}^{(0)})_A = \begin{cases} 0 & \text{if } A + A_i \notin \Box_i, \\ g_{A+A_i} & \text{if } A + A_i \in \Box_i. \end{cases}$$

Then

$$\mathscr{E}_{\mathbf{z}}(h_i{}^{(0)} \mathbf{z}^{A_i}) \subset \Box_i, \quad g = \sum_{i=0}^{k} h_i{}^{(0)} \mathbf{z}^{A_i}.$$

Now assume the $\{h_i{}^{(j)}\}_{0 \le i \le k}$ have been defined, for $j < n$, where $n > 1$ is an integer. Write $p_i = f_i - \mathbf{z}^{A_i}$, and define

$$h_i{}^{(n)} = \sum_{A \in \mathbb{Z}_0^c} (h_i{}^{(n)})_A \mathbf{z}^A$$

as follows:

$$(h_i{}^{(n)})_A = \begin{cases} 0 & \text{if } A + A_i \notin \Box_i, \\ -\displaystyle\sum_{j=1}^{k} \left(\sum_{B+C=A+A_i} (h_j{}^{(n-1)})_B (p_j)_C \right) & \text{if } A + A_i \in \Box_i. \end{cases}$$

Then

$$-\sum_{i=1}^{k} h_i{}^{(n-1)} p_i = \sum_{i=0}^{k} h_i{}^{(n)} \mathbf{z}^{A_i},$$

and

$$\mathscr{E}_{\mathbf{z}}(h_i{}^{(n)} \mathbf{z}^{A_i}) \subset \square_i \quad \text{for } 0 \le i \le k.$$

Remark 2.2.4 Assume that the sums $\sum_{n=0}^{\infty} h_i{}^{(n)}$ make sense in R for $i = 0, 1, \ldots, k$, that is, that these sums define elements $h_i = \sum_{n=0}^{\infty} h_i{}^{(n)}$ in R. We are going to prove that, in this case, the h_i satisfy the conditions in Theorem 2.2.1.

For every $N \in \mathbb{Z}_+$, write

$$S_i{}^N = \sum_{n=0}^{N} h_i{}^{(n)} \quad \text{for } 0 \le i \le k;$$

then

$$\begin{aligned} S_0{}^N + \sum_{i=1}^{k} S_i{}^N f_i &= S_0{}^N + \sum_{i=1}^{k} S_i{}^N (p_i + \mathbf{z}^{A_i}) \\ &= \sum_{n=0}^{N} \left(\sum_{i=1}^{k} h_i{}^{(n)} p_i + \sum_{j=0}^{k} h_j{}^{(n)} \mathbf{z}^{A_j} \right) \\ &= \sum_{j=0}^{k} h_j{}^{(0)} \mathbf{z}^{A_j} + \sum_{i=1}^{k} h_i{}^{(N)} p_i \\ &= g + \sum_{i=1}^{k} h_i{}^{(N)} p_i. \end{aligned}$$

Then, since the sums $\sum_{n=0}^{\infty} h_i{}^{(n)}$ make sense as formal power series, one must necessarily have $\lim_{n\to\infty} h_i{}^{(n)} = 0$ for every $i = 0, 1, \ldots, k$, so

$$\lim_{N\to\infty} \sum_{i=1}^{k} h_i{}^{(N)} p_i = 0$$

and therefore,

$$g = h_0 + \sum_{i=1}^{k} h_i f_i,$$

which proves condition (1) of Theorem 2.2.1.

Now it is not difficult to check that conditions (2) and (3) of Theorem 2.2.1 hold. In fact, by construction of the h_i, for $0 \le i \le k$, one has

$$\mathscr{E}_{\mathbf{z}}(h_i \mathbf{z}^{A_i}) = \mathscr{E}_{\mathbf{z}}(\sum_{n=0}^{\infty} h_i{}^{(n)} \mathbf{z}^{A_i}) \subset \bigcup_{n=0}^{\infty} \mathscr{E}_{\mathbf{z}}(h_i{}^{(n)} \mathbf{z}^{A_i}) \subset \Box_i,$$

which proves (2) of Theorem 2.2.1. Finally, condition (3) follows from (1), (2), and the fact that $L \in S\{\mathbf{F}, \mathbf{A}\}$ implies $\nu_L(f_i) = L(A_i, 0)$.

Now, we must now prove that the sums

$$\sum_{n=0}^{\infty} h_i{}^{(n)},$$

for $0 \le i \le k$, make sense in R. This is a bit tricky, and will be essentially performed by using the next two lemmas. We recall that a formal power series $f \in \mathbb{C}[[\mathbf{z}, \mathbf{w}]]$ is convergent if and only if there exists $l \in \mathbb{R}_+$ such that

$$\|f\|_l = \sum_{(A,B) \in \mathbb{Z}_0^c \times \mathbb{Z}_0^d} |f_{A,B}|\, l^{|A|+|B|} < \infty.$$

Lemma 2.2.1 *For all $i = 0, 1, \ldots, k$, all $n, b \in \mathbb{Z}_0$, and all $l \in \mathbb{R}_+$, one has*

$$\sum_{B \in \mathbb{Z}_0^c,\, |B|=b} \|(h_i{}^{(n)})_B\|_l \le H^n \,\|g\|_l\, l^{-|A_i|-b},$$

where

$$H = \sum_{i=1}^{k} \|f_i - \mathbf{z}^{A_i}\|_l\, l^{-|A_i|} = \sum_{i=1}^{k} \|p_i\|_l l^{-|A_i|}.$$

Proof We use induction on n. Starting with $n = 0$, one has

$$\begin{aligned}\sum_{B \in \mathbb{Z}_0^c,\, |B|=b} \|(h_i{}^{(0)})_B\|_l &= \sum_{\substack{B \in \mathbb{Z}_0^c,\, |B|=b \\ B+A_i \in \Box_i}} \|g_{B+A_i}\|_l \\ &= l^{-|A_i|-b} \sum_{\substack{B \in \mathbb{Z}_0^c,\, |B|=b \\ B+A_i \in \Box_i}} \|g_{B+A_i}\|_l\, l^{|A_i|+b} \\ &\le l^{-|A_i|-b} \|g\|_l.\end{aligned}$$

Now assume that $n > 0$ and that the lemma is true for $n-1$. We then have

$$\begin{aligned}
&\sum_{\substack{B\in\mathbb{Z}_0^c\\ |B|=b}} \|(h_i{}^{(n)})_B\|_l \\
&\le \sum_{\substack{B\in\mathbb{Z}_0^c,\,|B|=b\\ B+A_i\in\square_i}} \sum_{\substack{1\le j\le k\\ B'+B''=B+A_i}} \|(h_j{}^{(n-1)})_{B'}\|_l \,\|(p_j)_{B''}\|_l \\
&\le l^{-|A_i|-b} \sum_{\substack{1\le j\le k\\ b'+b''=b+|A_i|}} \sum_{\substack{|B'|=b'\\ |B''|=b''}} \|(h_j{}^{(n-1)})_{B'}\|_l\, l^{b'} \,\|(p_j)_{B''}\|_l\, l^{b''} \\
&= l^{-|A_i|-|b|} \sum_{\substack{1\le j\le k\\ b''\le b+|A_i|\\ |B''|=b''}} \left(\sum_{|B'|=b+|A_i|-b''} \|(h_j{}^{(n-1)})_{B'}\|_l\, l^{b+|A_i|-b''}\right) \|(p_j)_{B''}\|_l\, l^{b''} \\
&\le H^{n-1}\|g\|_l\, l^{-|A_i|-b} \sum_{\substack{1\le j\le k\\ b''\le b+|A_i|\\ |B''|=b''}} \|(p_j)_{B''}\|_l\, l^{-|A_j|-b-|A_i|+b''+b+|A_i|-b''+b''} \\
&= H^{n-1}\|g\|_l\, l^{-|A_i|-b} \sum_{\substack{1\le j\le k\\ |b''|\le b+|A_i|\\ |B''|=b''}} \|(p_j)_{B''}\|_l\, l^{b''-|A_j|} \\
&\le H^{n-1}\|g\|_l\, l^{-|A_i|-b} \sum_{1\le j\le k} \left(\sum_{B''\in\mathbb{Z}_0^c} \|(p_j)_{B''}\|_l\, l^{|B''|}\right) l^{-|A_j|} \\
&= H^{n-1}\|g\|_l\, l^{-|A_i|-b} \sum_{1\le j\le k} \|p_j\|\, l^{-|A_j|} = H^n\|g\|\, l^{-|A_i|-b}.
\end{aligned}$$

This proves the lemma. □

Lemma 2.2.2 *In the initial conditions for $g'(\mathbf{z}',\mathbf{w})$, $\{f_i'(\mathbf{z}',\mathbf{w})\}_{1\le i\le k}$, there exist $\mathbf{m}\in\mathbb{R}_+^c$ and $l\in\mathbb{R}_+$ such that:*

(1) $\|g'\|_l < \infty$ *and* $\|f_i'\|_l < \infty$, *for* $1\le i\le k$, *and*
(2) $H = \sum_{i=1}^k \|f_i' - \mathbf{z}^{A_i}\|_l\, l^{-|A_i|} < 1$.

Proof Take $l_1\in\mathbb{R}_+$ such that $\|g\|_{l_1} < \infty$ and $\|f_i\|_{l_1} < \infty$ for $1\le i\le k$. Then, for every $\mathbf{m} = (m_1,\dots,m_c)\in\mathbb{R}_+^c$, if we take $\nu\ge\sup\{1, m_1,\dots,m_c\}$ and $l = l_1/\nu$, we get $\|f_i'\|_l < \infty$, $\|g'\|_l < \infty$. Therefore, if we choose l_1, ν, for every $\mathbf{m}$, subject to the above restrictions, inequality (1) holds. Therefore, the point is to choose l_1 and $\mathbf{m}$ so that (2) also holds.

Let $p_i = f_i - (f_i)_{A_i,0}\mathbf{z}^{A_i}$, for $1 \leq i \leq k$. Then, for every $(A, B) \in \mathbb{Z}_0^c \times \mathbb{Z}_0^d$, we get

$$(f_i' - \mathbf{z}^{A_i})_{A,B} = \frac{\mathbf{m}^{A-A_i}}{(f_i)_{A_i,0}}(p_i)_{A,B}.$$

Write $p_i' = f_i' = \mathbf{z}^{A_i}$; then

$$\|p_i'\|_l\, l^{-|A_i|} = \frac{1}{|(f_i)_{A_i,0}|} \sum_{(A,B)\in\mathbb{Z}_0^c\times\mathbb{Z}_0^d} \frac{\mathbf{m}^{A-A_i}}{\nu^{|A|+|B|-|A_i|}} |(p_i)_{A,B}|\, l_1^{|A|+|B|-|A_i|}.$$

Also write

$$\varphi_{i,A,B} = \frac{\mathbf{m}^{A-A_i}}{\nu^{|A|+|B|-|A_i|}};$$

then the following assertion holds:

Remark 2.2.5 (Assertion) For every $i = 1, \ldots, k$, every $(A, B) \in \mathscr{E}_{\mathbf{z},\mathbf{w}}(p_i)$ such that $|A| \leq |A_i|$, and every positive $\varepsilon < 1$, there exist $\mathbf{m} \in \mathbb{R}_+^c$ and $\nu \in \mathbb{R}_+$ such that $\varphi_{i,A,B} < \varepsilon$.

We prove the assertion now and continue with the lemma later. Let $L \in S\{\mathbf{F}, \mathbf{A}\}$ be given by

$$L(\mathbf{x}, \mathbf{x}') = \sum_{j=1}^{c} l_j x_j + \sum_{j=1}^{d} l'_{j'} x'_{j'},$$

and set

$$L_1(\mathbf{x}) = \sum_{j=1}^{c} l_j x_j, \quad L_1'(\mathbf{x}') = \sum_{j'=1}^{d} l'_{j'} x'_{j'};$$

then L_1 is such that, for every $(A, 0) \in \mathscr{E}_{\mathbf{z},\mathbf{w}}(p_i)$,

$$L_1(A) = L(A, 0) > L(A_i, 0) = L_1(A_i).$$

Next, let

$$T = \bigcup_{i=1}^{k} \{A \in \mathbb{Z}_0^c \mid (A, 0) \in \mathscr{E}_{\mathbf{z},\mathbf{w}}(p_i),\ |A| \leq |A_i|\} = \bigcup_{i=1}^{k} T_i;$$

T is a finite set, so there exists $\alpha \in \mathbb{R}_+$ such that $L_1(A)-L_1(A_i) = L_1(A-A_i) > \alpha$ for every $i = 1, \dots, k$ and $A \in T_i$. Then we may construct another linear form

$$L_2(\mathbf{x}) = \sum_{j=1}^{c} l_j^* x_j$$

and find $\beta \in \mathbb{R}_+$ such that $l_j^* > 0$ for $j = 1, \dots, c$ and $L_2(A - A_i) > \beta$ for $1 \leq i < k$ and $A \in T_i$.

Now fix i satisfying $1 \leq i \leq k$, and choose $(A, B) \in \mathscr{E}_{\mathbf{z},\mathbf{w}}(p_i)$ such that $|A| = |A_i|$. Write

$$m_j = \nu \varepsilon^{l_j^*/\beta}$$

for $1 \leq j \leq c$, and set also $A = (a_1, \dots, a_c)$, $A_i = (a_{i_1}, \dots, a_{i_c})$. Then:

(1) If $B = 0$ then $A \in T_i$, so, for any choice of ν,

$$\varphi_{i,A,0} = \frac{\mathbf{m}^{A-A_i}}{\nu|A| - |A_i|} = \prod_{j=1}^{c} \left(\frac{m_j}{\nu}\right)^{a_j - a_{ij}} = \varepsilon^{L_2(A-A_i)/\beta} < \varepsilon,$$

since $\varepsilon < 1$.

(2) If $B \neq 0$ and we choose

$$\nu > \sup_{1 \leq i' \leq k} \{\varepsilon^{-1-L_2(A_{i'})/\beta}\}$$

we get

$$\frac{1}{\nu|B|} < \frac{1}{\nu} < \varepsilon^{1+L_2(A_i)/\beta} = \varepsilon \cdot \varepsilon^{L_2(A_i)/\beta} = \varepsilon \prod_{c}^{j=1} \varepsilon^{l_j^* a_{ij}/\beta} = \varepsilon \cdot \frac{m^{A_i}}{\nu|A_i|}.$$

Therefore

$$\varphi_{i,A,B} = \frac{1}{\nu^{|B|}} \frac{\mathbf{m}^{A-A_i}}{\nu^{|A|-|A_i|}} < \varepsilon \frac{\mathbf{m}^A}{\nu^{|A|}} = \varepsilon \prod_{c}^{j=1} \left(\frac{m_j}{\nu}\right)^{a_j} < \varepsilon$$

because $m_j = \nu \varepsilon^{l_j^*/\beta}$, $\varepsilon < 1$ and $l_j^* > 0$, so $m_j < \nu$. This proves the assertion. Note that the fact $\nu > m_j$ is compatible with the condition for ν at the beginning.

We now return to the proof of the lemma. We have

$$
\begin{aligned}
H &= \sum_{i=1}^{k} \|p_i'\|_l \, l^{-|A_i|} \\
&= \sum_{i=1}^{k} \frac{1}{|(f_i)_{A_i,0}|} \sum_{(A,B)\in\mathbb{Z}_0^c\times\mathbb{Z}_0^d, |A|\le|A_i|} \varphi_{i,A,B}|(p_i)_{A,B}|\, l_1^{|A|+|B|-|A_i|} \\
&\quad + \sum_{i=1}^{k} \frac{1}{|(f_i)_{A_i,0}|} \sum_{(A,B)\in\mathbb{Z}_0^c\times\mathbb{Z}_0^d, |A|>|A_i|} \varphi_{i,A,B}|(p_i)_{A,B}|\, l_1^{|A|+|B|-|A_i|}.
\end{aligned}
$$

In the second sum above, one has $|A| + |B| - |A_i| \geq 1$ and $\varphi_{i,A,B}|(p_i)_{A,B}|$ is bounded; therefore, if we choose l_1 small enough, this sum can be made smaller than $\frac{1}{2}$. The first sum is finite and, since $\varphi_{i,A,B} < \varepsilon$ for a suitable choice of $\mathbf{m}$ and ν, we can take ε so that this sum is again smaller than $\frac{1}{2}$. This proves that $H < 1$, concluding the proof of the lemma. □

Remark 2.2.6 We are now able to prove that, in the case of Remark 2.2.4, the sums $h_i = \sum_{n=0}^{\infty} h_i^{(n)}$, for $0 \le i \le k$, actually make sense in R. Take l as in Lemma 2.2.2 (we again assume that $g = g'$ and $f_i = f_i'$ for $1 \le i \le k$). Take $\sigma \in \mathbb{R}$ in the range $0 < \sigma < l$. We then have

$$
\begin{aligned}
\|(h_i^{(n)})\| &= \sum_{B\in\mathbb{Z}_0^c} \|(h_i^{(n)})_B\|_\sigma \sigma^{|B|} \le \sum_{b=0}^{\infty} \sum_{\substack{B\in\mathbb{Z}_0^c \\ |B|=b}} \|(h_i^{(n)})_B\|_l \sigma^{|B|} \\
&\le \frac{H^n \|g\|_l}{l^{|A_i|}} \sum_{b=0}^{\infty} \left(\frac{\sigma}{l}\right)^b = \frac{H^n \|g\|_l}{l^{|A_i|}} \cdot \frac{1}{1-\sigma/l},
\end{aligned}
$$

so $\|h_i^{(n)}\|_\sigma < \infty$. Therefore $h_i^{(n)}$ converges in the polydisk

$$
D_\sigma = \{(\mathbf{z}, \mathbf{w}) \in \mathbb{C}^{c+d} \mid |z_i| \le \sigma \text{ and } |w_j| \le \sigma \text{ for all } i, j\}.
$$

As far as the h_i are concerned, we have

$$
\|h_i\|_\sigma \le \sum_{n=0}^{\infty} \|h_i^{(n)}\| \le \frac{\|g\|_l}{l^{|A_i|}} \frac{1}{1-\sigma/l} \sum_{n=0}^{\infty} H^n = \frac{\|g\|_l}{l^{|A_i|}} \frac{1}{1-\sigma/l} \frac{1}{1-H}.
$$

Since $H < 1$, we have $\|h_i\|_\sigma < \infty$ and h_i converges in D_σ. This ends the proof of the existence part in Theorem 2.2.1.

Remark 2.2.7 We now see the uniqueness of the h_i, $0 \le i \le k$, in Theorem 2.2.1. For this, it is enough to show that, if we have power series $h_i \in R$, for $0 \le i \le k$, such that

$$0 = h_0 + \sum_{i=1}^{k} h_i f_i$$

and

$$\mathscr{E}_z(h_i \mathbf{z}^{A_i}) \subset \square_i$$

for $0 \le i \le k$, then $h_i = 0$ for $i = 0, 1, \ldots, k$.

As before, we may assume, without loss of generality, that $(f_i)_{A_i,0} = 1$, and so $f_i = p_i + \mathbf{z}^{A_i}$. Fix $L \in S\{\mathbf{F}, A\}$; we have $L(A_i, 0) < \nu_L(p_i)$ for $1 \le i \le k$. Consider the relation

$$h_0 + \sum_{i=1}^{k} h_i \mathbf{z}^{A_i} = -\sum_{i=1}^{k} h_i p_i.$$

If we take quotients with respect to the ideal $(\mathbf{w})R$ and use a superscript $*$ to denote the classes modulo this ideal, we have

$$h_0^* + \sum_{i=1}^{k} h_i^* \cdot (\mathbf{z}^*)^{A_i} = -\sum_{i=1}^{k} h_i^* p_i^*.$$

If there were an index i_0, $1 \le i_0 \le k$, such that $h_{i_0}^* \ne 0$, we would have

$$\begin{aligned}\nu_L\left(h_0^* + \sum_{i=1}^{k} h_i^* \cdot (\mathbf{z}^*)^{A_i}\right) &= \min_{0\le i\le k} \nu_L(h_i^*) + L(A_i, 0) \le \min_{1\le i\le k} \nu_L(h_i^*) + L(A_i, 0) \\ &< \min_{1\le i\le k} \nu_L(h_i^*) + \nu_L(p_i^*) \le \nu_L\left(-\sum_{i=1}^{k} h_i^* p_i^*\right),\end{aligned}$$

contradicting the assumption that $h_0 + \sum_{i=1}^{k} h_i f_i = 0$.

Therefore, $h_i \in (\mathbf{w})R$ for $1 \le i \le k$, so $h_0 \in (\mathbf{w})R$. By repeating this procedure, taking quotients with respect to the ideals $(\mathbf{w})^2 R$, $(\mathbf{w})^3 R$, and so on, we conclude that

$$h_i \in \bigcap_{n=1}^{\infty} (\mathbf{w})^n R$$

for all $i = 0, 1, \ldots, k$, so $h_i = 0$. This ends the proof of the Weierstrass preparation theorem.

Corollary 2.2.1 (The classical Weierstrass division theorem) *Assume that $c = 1$, as in Example 2.2.1, and let $f \in R$ be a non-unit different from zero and regular in z of order a. Then, for every $g \in R$, there exist $h_0, h_1 \in R$, uniquely determined, such that $g = h_0 + h_1 f$ and*

$$h_0 = \sum_{i=0}^{a-1} h_{0,i}(\mathbf{w})z^i, \quad h_{0,i}(\mathbf{w}) \in \mathbb{C}\{\mathbf{w}\}.$$

Remark 2.2.8 If we read carefully the proof of Theorem 2.2.1, we should notice several very important facts:

(1) In the proof of Lemma 2.2.2, which is the key result in proving that the h_i belong to R, the choice of l depends only on $\mathbf{F}$ and $\mathbf{A}$ and not on g. The statement of the lemma can be modified so that it involves choosing the polydisk D determined by l (using only $(\mathbf{F}, \mathbf{A})$), and then saying that the assertions hold for any g holomorphic and bounded in D.
(2) The conditions $\mathscr{E}_{\mathbf{z}}(h_i \mathbf{z}^{A_i}) \subset \square_i$, for $0 \leq i \leq k$, are not affected by a parameter change, within the $\{w_1, \ldots, w_d\}$.
(3) The conditions $\mathscr{E}_{\mathbf{z}}(h_i \mathbf{z}^{A_i}) \subset \square_i$, for $0 \leq i \leq k$, are not affected by translations within the $\mathbf{w}$-part of the polydisk, that is, by a change of variables of the type $w_j = w_j' + \alpha_j$, with $\alpha_j \in \mathbb{C}$ for $1 \leq j \leq d$, satisfying $(\alpha_1, \ldots, \alpha_d) \in D \cap (\{0\} \times \mathbb{C}^d)$.
(4) As a result of these observations, we have a kind of "local" Weierstrass preparation theorem, whose statement is the following.

Theorem 2.2.2 (The polydisk Weierstrass preparation theorem) *Let*

$$\mathbf{F} = \{f_i\}_{1 \leq i \leq k} \subset R, \ \mathbf{A} = \{A_i\}_{1 \leq i \leq k} \subset \mathbb{Z}_0^c$$

be two families such that $S\{\mathbf{F}, \mathbf{A}\} \neq \varnothing$, and consider the partition

$$\mathbb{Z}_0^c = \bigcup_{i=0}^{k} \square_i.$$

Then, there exists a polydisk D centered at $\mathbf{0}$ in $\mathbb{C}^c \times \mathbb{C}^d$ such that, for any g analytic and bounded in D, there are functions $\{h_i\}_{0 \leq i \leq k}$, holomorphic in D, such that the following conditions are satisfied:

(1) *$g = h_0 + \sum_{i=1}^{k} h_i f_i$.*
(2) *$\mathscr{E}_{\mathbf{z}}(h_i \mathbf{z}^{A_i}) \subset \square_i$, and this condition is preserved by translations within the $\mathbf{w}$-part of D.*
(3) *For any $L \in S\{\mathbf{F}, \mathbf{A}\}$ and $i = 1, \ldots, k$, one has*

$$\nu_L(h_i) \geq \nu_L(g) - \nu_L(f_i).$$

Appendix to Sect. 2.2 The Implicit Function Theorem and the Jacobian Criterion for Smoothness

Among the most widely known applications of the Weierstrass preparation theorem, two play a special role: the implicit function theorem and the jacobian criterion for smoothness. We include them here by virtue of their importance.

This time, we denote by $R = \mathbb{C}\{\mathbf{z}\}$ the ring of convergent power series in the variables $\mathbf{z} = (z_1, \ldots, z_n)$. Let I be an ideal in R different from (0) and (1), and write

$$J(I) = \{(\partial f/\partial z_1, \ldots, \partial f/\partial z_n) \mid f \in I\}.$$

We may think of $J(I)$ as "an infinite matrix" with n columns. We shall also consider finite submatrices of $J(I)$, of the form

$$\frac{\partial(f_1, \ldots, f_s)}{\partial(z_1, \ldots, z_n)} = \begin{pmatrix} \dfrac{\partial f_1}{\partial z_1} & \cdots & \dfrac{\partial f_1}{\partial z_n} \\ \vdots & & \vdots \\ \dfrac{\partial f_s}{\partial z_1} & \cdots & \dfrac{\partial f_s}{\partial z_n} \end{pmatrix}, \quad \text{for } f_1, \ldots, f_s \in I;$$

such submatrices, and $J(I)$ itself, will be generically called "jacobians".

For a finite jacobian

$$\frac{\partial(f_1, \ldots, f_s)}{\partial(z_1, \ldots, z_n)}$$

we consider a polydisk $D_{(f_1,\ldots,f_s)}$ centered at 0 in which all the f_i, and hence their derivatives, converge. If $\mathbf{m} \in D_{(f_1,\ldots,f_s)}$, we shall denote by

$$\left(\frac{\partial(f_1, \ldots, f_s)}{\partial(z_1, \ldots, z_n)}\right)_{\mathbf{z}=\mathbf{m}}$$

the numerical matrix whose elements are the evaluations at $\mathbf{m}$ of the elements in $\partial(f_1, \ldots, f_s)/\partial(x_1, \ldots, x_n)$. We denote by $J(I)_{\mathbf{z}=\mathbf{0}}$ the numerical matrix whose elements are the evaluations at $\mathbf{0}$ of the elements in $J(I)$. We denote by r the maximum of the ranks of all matrices

$$\left(\frac{\partial(f_1, \ldots, f_s)}{\partial(z_1, \ldots, z_n)}\right)_{\mathbf{z}=\mathbf{m}}, \quad \text{for } \mathbf{m} \in D_{(f_1,\ldots,f_s)};$$

thus r is the generic rank of $J(I)$, that is, the rank of $J(I)$ as a matrix with entries in R.

Proposition 2.2.1 *With the notation as above, the following conditions are equivalent:*

(1) *Up to a parameter change in R, there exists a basis of I of the form*

$$(z_1 - g_1(z_{r+1}, \dots, z_n), \ \dots, \ z_r - g_r(z_{r+1}, \dots, z_n)).$$

(2) *R/I is a regular local ring.*
(3) *The matrix $J(I)_{\mathbf{z}=\mathbf{0}}$ has rank equal to r.*
(4) *There exist a basis $(f_1, \dots, f_s)$ of I and a polydisk $D \subset D_{(f_1,\dots,f_s)}$ centered at* 0 *such that the ranks of all the matrices*

$$\left\{ \left(\frac{\partial(f_1, \dots, f_s)}{\partial(z_1, \dots, z_n)} \right)_{\mathbf{z}=\mathbf{m}} \middle| \ \mathbf{m} \in D \right\}$$

coincide. (This statement will be abbreviated by simply saying that the jacobian of I is **of constant rank** in a neighborhood of the origin. This rank must be r.)
(5) *For every basis $(f_1, \dots, f_s)$ of I there exist a polydisk $D \subset D_{(f_1,\dots,f_s)}$ centered at* 0 *such that the ranks of all matrices*

$$\left\{ \left(\frac{\partial(f_1, \dots, f_s)}{\partial(z_1, \dots, z_n)} \right)_{\mathbf{z}=\mathbf{m}} \middle| \ \mathbf{m} \in D \right\}$$

coincide. (This rank must be r.)

Proof The equivalence between (1) and (2) is standard in commutative algebra, so we do not treat it specifically. A clue for it can be found in Remarks 2.1.1 and 2.1.2.

Clearly (1) implies (3). To show the converse, let $f_1, \dots, f_r \in I$ be such that

$$\operatorname{rank} \left(\frac{\partial(f_1, \dots, f_r)}{\partial(z_1, \dots, z_n)} \right)_{\mathbf{z}=\mathbf{0}} = r;$$

then the initial forms of the f_i are linearly independent linear forms. After a permutation of the variables and after taking suitable linear combinations of $\{f_1, \dots, f_r\}$ with constant coefficients, we may assume that the initial forms of the f_i are

$$\bar{f}_1 = \bar{z}_1 + a_{1,r+1}\bar{z}_{r+1} + \cdots + a_{1,n}\bar{z}_n,$$
$$\dots\dots\dots\dots\dots\dots\dots\dots\dots\dots$$
$$\bar{f}_r = \bar{z}_r + a_{r,r+1}\bar{z}_{r+1} + \cdots + a_{r,n}\bar{z}_n.$$

We then apply the Weierstrass preparation theorem in the following framework: the two sets of variables are $(z_1, \ldots, z_r)$ and $(z_{r+1}, \ldots, z_n)$; for $\mathbf{F}$ we take $\{f_1, \ldots, f_r\}$; and for $\mathbf{A}$ we take $\{A_1, \ldots, A_r\}$ with $A_i = (0, \ldots, 1, \ldots, 0)$. Note that $S\{\mathbf{F}, \mathbf{A}\} \neq \varnothing$ because, for instance,

$$L = \sum_{i=1}^{r} x_i + 2 \sum_{j=r+1}^{n} x'_j \in S\{\mathbf{F}, \mathbf{A}\}.$$

Applying the Weierstrass preparation theorem gives

$$z_j = h_{j0} + \sum_{i=1}^{r} h_{ji} f_j \quad \text{for } 1 \leq j \leq r,$$

where the h_{j0} depend only on $\{z_{r+1}, \ldots, z_n\}$. If we write $g_j = h_{j0}$ for $1 \leq j \leq r$, we get $(z_1 - g_1, \ldots, z_r - g_r) \subset I$.

It remains to show that $(z_1 - g_1, \ldots, z_r - g_r) = I$. Take any $f \in I$ and apply again the Weierstrass preparation theorem, as before, but with $F = \{z_1 - g_1, \ldots, z_r - g_r\}$. We then get

$$f = h_0 + \sum_{i=1}^{r} h_i \, (z_i - g_i),$$

where $h_0 \in \mathbb{C}\{z_{r+1}, \ldots, z_n\}$. By the definition of r one has

$$\operatorname{rank} \begin{pmatrix} 1 & 0 & \cdots & -\dfrac{\partial g_1}{\partial z_{r+1}} & \cdots & -\dfrac{\partial g_1}{\partial z_n} \\ \vdots & \vdots & \ddots & \vdots & & \vdots \\ 0 & 1 & \cdots & -\dfrac{\partial g_r}{\partial z_{r+1}} & \cdots & -\dfrac{\partial g_r}{\partial z_n} \\ 0 & 0 & \cdots & -\dfrac{\partial h_0}{\partial z_{r+1}} & \cdots & -\dfrac{\partial h_0}{\partial z_n} \end{pmatrix} = r,$$

so $0 = \partial h_0 / \partial z_{r+1} = \cdots = \partial h_0 / \partial z_n$. Consequently $h_0 = 0$, proving our claim.

Next we prove that (1) implies (5). From the assumption, it is clear that any basis $(f_1, \ldots, f_s)$ of I must contain a subset, say $\{f_1, \ldots, f_r\}$, such that the initial forms of the $f_1, \ldots, f_r$ are linearly independent linear forms. Therefore,

$$\operatorname{rank} \left(\frac{\partial(f_1, \ldots, f_r)}{\partial(z_1, \ldots, z_n)} \right)_{\mathbf{z}=\mathbf{0}} = r,$$

so this matrix contains a non-zero r-rowed minor. By continuity, this minor is not zero in a polydisk $D \subset D_{(f_1,\dots,f_s)}$, so the conclusion is now clear.

Clearly (5) implies (4). To conclude the proof of the proposition, therefore, it is enough to show that (4) implies (3). Let r' be the common rank of all the matrices in the statement of (4). By assumption the matrix $\partial(f_1, \dots, f_s)/\partial(z_1, \dots, z_n)$ has a r'-rowed minor that is a unit in R. Therefore $\operatorname{rank} J(I)_{\mathbf{z}=\mathbf{0}} \geq r'$. Assume the inequality is strict; then there is $f \in I$ such that

$$\operatorname{rank}\left(\frac{\partial(f_1, \dots, f_s, f)}{\partial(z_1, \dots, z_n)}\right)_{\mathbf{z}=\mathbf{0}} = r' + 1.$$

Then, $(f_1, \dots, f_s)$ contains a subset of r' power series whose initial forms are linearly independent linear forms, while the figure for $(f_1, \dots, f_s, f)$ is $r' + 1$. This is not possible, since $I = (f_1, \dots, f_s) = (f_1, \dots, f_s, f)$; this shows that $\operatorname{rank} J(I)_{\mathbf{z}=\mathbf{0}} = r'$. We must now prove that $r' = r$.

By the assumption, $r' \leq r$. Assume $r' < r$; then there is $g \in I$ such that

$$\operatorname{rank}\left(\frac{\partial(f_1, \dots, f_s, g)}{\partial(z_1, \dots, z_n)}\right)_{\mathbf{z}=\mathbf{0}} = r' + 1$$

and all the $(r' + 1)$-rowed minors are non-units. By considering a non-zero $(r' + 1)$-rowed minor Δ of this matrix, we can find a polydisk $D' \subset D$ and a point $\mathbf{m} \in D'$ such that $\Delta(\mathbf{m}) \neq 0$. Then there is a second polydisk $D'' \subset D'$ centered at $\mathbf{m}$ such that $\Delta(\mathbf{m})$ is not zero at any point of D''. By translating the origin to $\mathbf{m}$ we have two matrices, with respectively s and $s + 1$ rows, of constant ranks on D'', respectively equal to r' and $r' + 1$. This is not possible, as we have just seen. This proves the proposition. □

Remark 2.2.9 (Terminology) In Proposition 2.2.1, the equivalence between (1) and any of (3), (4) and (5) is called the **implicit function theorem** for analytic functions. The equivalence of (2) with (3), (4) and (5) is called the **jacobian criterion of regularity**.

Remark 2.2.10 Let X be a $\mathbb{C}$-space, and let $x \in X$. The question of deciding whether or not x is a smooth point of X is a local one, so we may assume that X is a closed complex subspace, around x, of some smooth $\mathbb{C}$-space Z. If we take a local coordinate system $z = (z_1, \dots, z_n)$ on Z centered at x, then X is defined, locally around x, by the vanishing of the functions belonging to some ideal $I \subset \mathbb{C}\{\mathbf{z}\}$. By Remark 2.1.1, the fact that x is a smooth point of X is equivalent to the regularity of $\mathbb{C}\{\mathbf{z}\}/I$. Therefore, in this case, Proposition 2.2.1 furnishes a criterion for a point to be smooth that is local, that is, formulated in terms of local coordinates. This is called the **jacobian criterion for smoothness**.

2.3 Flatness of the Graded Rings

The aim of this section is to give the algebraic counterpart to the concept of normal flatness. For this we need, first of all, some combinatorial results.

Consider a finite set of non-zero linear forms on $\mathbb{R}^n$

$$L_i(\mathbf{x}) = \sum_{j=1}^{n} a_{ij}x_j, \quad \text{for } 1 \leq i \leq s,\ a_{ij} \in \mathbb{Z},$$

satisfying the following properties:

(1) If $\mathbf{u} = (1, \ldots, 1)$, then $L_i(\mathbf{u}) = 0$, for every $i = 1, \ldots, s$.
(2) For any $i = 1, \ldots, s$, denote by $q(i)$ the lowest index such that the coefficient $a_{iq(i)}$ of $L_i(x)$ is non-zero. Then $a_{iq(i)} < 0$.

We write

$$L_i^+ = \{\mathbf{x} \in \mathbb{R}^n \mid L_i(\mathbf{x}) > 0\},$$

and we are interested in the set

$$\Gamma = \mathbb{R}^n_+ \cap L_1^+ \cap \cdots \cap L_s^+.$$

Γ is open for the classical topology, because all the $\mathbb{R}^n_+$ and L_i^+ are. Moreover, Γ is semi-conic in the sense that, if $\mathbf{v} \in \Gamma$ and $\lambda \in \mathbb{R}_+$, then $\lambda\mathbf{v} \in \Gamma$. Here is another important property of Γ:

Lemma 2.3.1 *Let* $\mathbf{v} \in \Gamma$, *denote by* μ *the straight line joining* $\mathbf{0}$ *and* $\mathbf{u} = (1, \ldots, 1)$, *let* $H_{\mathbf{v}}$ *be the hyperplane through* $\mathbf{v}$ *perpendicular to* μ, *and write* $\mathbf{v}_1 = H_{\mathbf{v}} \cap \mu$ *(note that* $\mathbf{v}_1 \neq \mathbf{v}$*). Then the half-open interval* $(\mathbf{v}_1, \mathbf{v}]$ *is contained in* Γ.

Proof The proof is straightforward by classic college analytic geometry. Let $\mathbf{v}_1 = (\alpha_1, \ldots, \alpha_n)$, where $\alpha_i > 0$ for $1 \leq i \leq n$, and write $\alpha = \alpha_1 + \cdots + \alpha_n$; then the equation of $H_{\mathbf{v}}$ is $x_1 + \cdots + x_n - \alpha = 0$. Therefore, $\mathbf{v}_1 = (\alpha/n, \ldots, \alpha/n)$, so the equations of the interval $(\mathbf{v}_1, \mathbf{v}]$ are

$$x_i = \frac{\alpha}{n} + \left(\alpha_i - \frac{\alpha}{n}\right)\lambda, \quad \text{for } 0 < \lambda \leq 1,\ 1 \leq i \leq n.$$

For any $i = 1, \ldots, s$, one has

$$\begin{aligned}\sum_{j=1}^{n} a_{ij}\left(\frac{\alpha}{n} + \left(\alpha_j - \frac{\alpha}{n}\right)\lambda\right) &= \frac{\alpha}{n}\sum_{j=1}^{n} a_{ij} + \lambda\sum_{j=1}^{n} a_{ij}\alpha_j - \lambda\frac{\alpha}{n}\sum_{j=1}^{n} a_{ij} \\ &= \lambda\sum_{j=1}^{n} a_{ij}\alpha_j > 0,\end{aligned}$$

and, for any $j = i, \dots, n$,

$$\frac{\alpha}{n} + \left(\alpha_j - \frac{\alpha}{n}\right)\lambda = \frac{\alpha}{n}(1-\lambda) + \lambda\alpha_j > 0.$$

This proves that $(\mathbf{v}_1, \mathbf{v}] \subset \Gamma$, so the lemma follows. □

Corollary 2.3.1 *If Γ is non-empty, it contains points arbitrarily close to* $\mathbf{u}$.

Proof Let $\mathbf{v} \in \Gamma$, and use the notation of the preceding lemma; then $\mathbf{v}' = (n/\alpha)\mathbf{v} \in \Gamma$, by the semi-conic property of Γ. Consider the hyperplane $H_{\mathbf{v}'}$ and the point $\mathbf{v}'_1 = H_{\mathbf{v}'} \cap \mu$; then $\mathbf{v}'_1 = \mathbf{u}$. By the preceding lemma, $(\mathbf{u}, \mathbf{v}'] \subset \Gamma$, so the corollary is now evident. □

One of the key results is the following.

Proposition 2.3.1 *It is verified that $\Gamma \neq \varnothing$.*

Proof We use induction on n, starting with $n = 2$.
Case $n = 2$. We have

$$L_i(\mathbf{x}) = a_{i1}x_1 + a_{i2}x_2 \quad \text{for } 1 \le i \le s,$$

with $a_{i1} < 0$, $a_{i2} > 0$, and $a_{i1} + a_{i2} = 0$, for $i = 1, \dots, s$. If $\beta > 1$ is an integer we have $(1, \beta) \in \Gamma$ because, for all $i = 1, \dots, s$,

$$a_{i1} + \beta a_{i2} > a_{i1} + a_{i2} = 0.$$

Therefore $\Gamma \neq \varnothing$.
Inductive step. Assume the result proved for $n - 1$. We may assume that not all the first coefficients a_{i1} are zero for, if they were, Γ would be non-empty by the induction assumption. We reorder the $L_i(\mathbf{x})$ in such a way that $a_{11} < 0, \dots, a_{r1} < 0$, and $a_{r+1,1} = \cdots = a_{s,1} = 0$. We do not disregard the case $r = s$, so the set $\{L_{r+1}(\mathbf{x}), \dots, L_s(\mathbf{x})\}$ could be empty.

Consider the linear forms in $\mathbb{R}^{n-1}$ given by

$$L'_i(\mathbf{x}) = \sum_{j=2}^{n} a_{ij}x_j \quad \text{for } 1 \le i \le r,$$

and denote by $(L'_i)^+$ the set

$$(L'_i)^+ = \{\mathbf{x} \in \mathbb{R}^{n-1} \mid L'_i(\mathbf{x}) > 0\}.$$

Then $\mathbf{u}' = (1, \dots, 1) \in (L'_i)^+$ for any $i = 1, \dots, r$, so

$$\mathbf{u}' \in \Gamma' = \mathbb{R}^{n-1}_+ \cap (L'_1)^+ \cap \cdots \cap (L'_r)^+,$$

hence $\Gamma' \neq \varnothing$. We see that Γ' is also semi-conic, so it is clear that there is a point $\mathbf{v}' = (\alpha_2, \ldots, \alpha_n) \in \Gamma'$ such that $L'_i(\mathbf{v}') > |a_{i1}|$ for any $i = 1, \ldots, r$. If we take $\mathbf{v} = (1, \alpha_2, \ldots, \alpha_n)$, we have

$$\mathbf{v} \in \Gamma'' = \mathbb{R}^n_+ \cap L_1^+ \cap \cdots \cap L_r^+,$$

so $\Gamma'' \neq \varnothing$. In particular, if $r = s$, our proof is concluded.

Assume, therefore, that $r < s$, and denote by L_i^+, for $i = r+1, \ldots, s$, the projection of L_i^+ onto $\mathbb{R}^{n-1}$, with coordinates $x_2, \ldots, x_n$. By the induction assumption,

$$\Gamma'' = \mathbb{R}^{n-1}_+ \cap L_{r+1}^+ \cap \cdots \cap L_s^+ \neq \varnothing.$$

By Corollary 2.3.1, Γ'' contains points arbitrarily close to $\mathbf{u}' = (1, \ldots, 1)$. But $\mathbf{u}' \in \Gamma'$, which is open, so $\Gamma' \cap \Gamma'' \neq \varnothing$. If $\mathbf{v}' = (\alpha_2, \ldots, \alpha_n) \in \Gamma' \cap \Gamma''$, then, by the semi-conic property of both sets, we can find $\lambda \in \mathbb{R}_+$ such that $\lambda\mathbf{v}' \in \Gamma' \cap \Gamma''$ and $L'_i(\lambda\mathbf{v}') > |a_{i1}|$ for any $i = 1, \ldots, r$. Therefore, $\mathbf{v} = (1, \lambda\alpha_2, \ldots, \lambda\alpha_n) \in \Gamma \neq \varnothing$, and our proof is complete. □

Corollary 2.3.2 *Let $(A_1, B_1), \ldots, (A_s, B_s)$ be pairs of multi-indices in $\mathbb{Z}^n_0$ such that $|A_i| = |B_i|$ and $A_i < B_i$ for the lexicographic order, for $i = 1, \ldots, s$. Let an integer $m > 1$ be given. Then there exists a positive linear form*

$$L(\mathbf{x}) = \sum_{j=1}^n \alpha_j x_j, \quad \textit{with } \alpha_j \in \mathbb{R}_+,$$

satisfying the following conditions:

(1) *For $i = 1, \ldots, s$, $L(A_i) > L(B_i)$.*
(2) *For $i = 1, \ldots, s$, $|L(A_i) - |A_i|| < \frac{1}{m}$ and $|L(B_i) - |B_i|| < \frac{1}{m}$.*

Proof Write

$$A_i = (a_{i1}, \ldots, a_{in}), \quad B_i = (b_{i1}, \ldots, b_{in}), \quad \text{for } i = 1, \ldots, s,$$

and consider the linear forms

$$L_i(\mathbf{x}) = \sum_{j=1}^n (a_{ij} - b_{ij}) x_j, \quad \text{for } 1 \leq i \leq s.$$

By Proposition 2.3.1, there exists $\mathbf{v} = (\alpha_1, \ldots, \alpha_n) \in \mathbb{R}^n_+$ such that $L_i(\mathbf{v}) > 0$ for every $i = 1, \ldots, s$. Therefore, the linear form

$$L(\mathbf{x}) = \sum_{j=1}^n \alpha_j x_j$$

satisfies (1). Our proof will be complete if we show that $\mathbf{v}$ can be chosen so as to satisfy (2).

By Corollary 2.3.1, we may choose $\mathbf{v}$ in such a way that

$$|\alpha_j - 1| < \frac{1}{mM}, \quad \text{for } j = 1, \ldots, n,$$

where

$$M = \sup_{1 \le i \le s} \{|A_i|\} = \sup_{1 \le i \le s} \{|B_i|\}.$$

In this way one has, for any $i = 1, \ldots, s$,

$$\begin{aligned} \big|L(A_i) - |A_i|\big| &= \left|\sum_{j=1}^{n} (\alpha_j - 1) a_{ij}\right| \le \sum_{j=1}^{n} |\alpha_j - 1|\, a_{ij} \\ &< \frac{1}{mM} |A_i| \le \frac{1}{mM} M = \frac{1}{m}, \end{aligned}$$

and similarly $\big|L(B_i) - |B_i|\big| < 1/m$. □

Once these combinatorial results have been settled, we start with the algebraic part of our studies on normal flatness. In general, the circumstances will be as follows. We consider the ring $R = \mathbb{C}\{\mathbf{z}, \mathbf{w}\} = \mathbb{C}\{z_1, \ldots, z_c, w_1, \ldots, w_d\}$ with maximal ideal M, the ideal $P = (\mathbf{z})R$, an ideal $I \subset P$ of R, and we work with the following graded rings (identified with polynomial rings):

$$\begin{aligned} \mathrm{gr}_P(R) &= \mathbb{C}\{\mathbf{w}\}[\tilde{\mathbf{z}}], \quad \text{where } \tilde{z}_i = z_i + P^2 \text{ for } 1 \le i \le c, \\ \mathrm{gr}_{(P/I)}(R/I) &= \mathbb{C}\{\mathbf{w}\}[\tilde{\mathbf{z}}]/\mathrm{in}_P(I). \end{aligned}$$

For any $f \in R$, we write $\tilde{f} = \mathrm{in}_P(f)$ and $z_i^* = \tilde{z}_i + \mathrm{in}_P(I) \in \mathbb{C}\{\mathbf{w}\}[\tilde{\mathbf{z}}]/\mathrm{in}_P(I)$. We shall also write $\mathrm{gr}_P(R)_n$ and $\mathrm{gr}_{(P/I)}(R/I)_n$ for the homogeneous components of degree n of these graded rings.

The main problem we shall deal with in this section will be to find characterizations of the fact that $\mathrm{gr}_{(P/I)}(R/I)$ is a flat $\mathbb{C}\{\mathbf{w}\}$-module. In this case, flatness is equivalent to freeness. In fact, assume that $\mathrm{gr}_{(P/I)}(R/I)$ is a flat $\mathbb{C}\{\mathbf{w}\}$-module. Then, for any $n \ge 0$, the homogeneous component $\mathrm{gr}_{(P/I)}(R/I)_n$, which is a $\mathbb{C}\{\mathbf{w}\}$-module of finite type, is flat, since it is a direct summand of $\mathrm{gr}_{(P/I)}(R/I)$. Therefore, $\mathrm{gr}_{(P/I)}(R/I)_n$ is a free $\mathbb{C}\{\mathbf{w}\}$-module and hence so is $\mathrm{gr}_{(P/I)}(R/I)$. This proves that $\mathrm{gr}_{(P/I)}(R/I)$ is a flat $\mathbb{C}\{\mathbf{w}\}$-module if and only if it is free. Moreover, in this case, $\mathrm{gr}_{(P/I)}(R/I)$ admits a homogeneous free basis.

We proceed to find characterizations for the flatness of $\mathrm{gr}_{(P/I)}(R/I)$. Only the lexicographic ordering for multi-indices will be considered.

For a power series $f \in R$, we define

$$u(f) = \sup\{A \in \mathbb{Z}_0^c \mid |A| = \nu_P(f) \text{ and } f_A(0) \neq 0\},$$

where, as usual, we write

$$f = \sum_{A \in \mathbb{Z}_0^c} f_A z^A \quad \text{with } f_A \in \mathbb{C}\{\mathbf{w}\}.$$

We remark that:

(1) It is verified that $u(f) = \varnothing$ if and only if $\tilde{f} \in (\mathbf{w}) \cdot \mathrm{gr}_P(R)$.
(2) It is verified that $u(f) \neq \varnothing$ if and only if $\nu_P(f) = \nu_{\mathrm{m}}(f)$, where M is the maximal ideal of R.

We also define, for any ideal $J \subset R$,

$$u(J) = \{u(f) \mid f \in J\};$$

then $u(J) = \varnothing$ if and only if $\mathrm{in}_P(J) \subset (\mathbf{w}) \cdot \mathrm{gr}_p(R)$.

Lemma 2.3.2 *The set of monomials*

$$H = \{(\mathbf{z}^*)^A \mid A \in \mathbb{Z}_0^c \setminus u(I)\}$$

is a minimal system of generators of $\mathrm{gr}_{(P/I)}(R/I)$ *as a* $\mathbb{C}\{\mathbf{w}\}$*-module.*

Proof We first show that H is a system of generators, and later prove minimality. Since H is clearly a system of generators if $u(I)$ is empty, we assume $u(I) \neq \varnothing$. What we need to prove is that the monomials $(z^*)^A$, for $A \in u(I)$, can be written as linear combinations of elements in H. We shall write, for any $n \in \mathbb{Z}_0$,

$$u(I)_n = \{A \in u(I) \mid |A| = n\}.$$

Take $n \in \mathbb{Z}_0$ such that $u(I)_n \neq \varnothing$, and let

$$u(I)_n = \{A_1, \ldots, A_r\},$$

with $A_1 < \cdots < A_r$ for the lexicographic order. For every $i = 1, \ldots, r$, we choose $f_i' \in I$ such that $u(f_i') = A_i$, and we write

$$\tilde{f}_i' = g_i' + \sum_{j=1}^{r} \alpha_{ij}' \tilde{\mathbf{z}}^{A_j} \quad \text{with } \alpha_{ij}' \in \mathbb{C}\{\mathbf{w}\},$$

where $\alpha_{ij}' \in (\mathbf{w}) \cdot \mathbb{C}\{\mathbf{w}\}$ for $i < j$, $\alpha_{ii}'(0) \neq 0$, and $\mathscr{E}_{\mathbf{z}}(g_i') \subset \mathbb{Z}_0^c \setminus u(I)_n$. By application of the usual procedure for the triangularization of a matrix, we get a family $f_1, \ldots, f_r \in I$ whose initial forms are of the type

$$\begin{array}{ll} \tilde{f_1} = \alpha_{11}\tilde{\mathbf{z}}^{A_1} + \alpha_{12}\tilde{\mathbf{z}}^{A_2} + \cdots + \alpha_{1r}\tilde{\mathbf{z}}^{A_r} + g_1, \\ \tilde{f_2} = \qquad\qquad \alpha_{22}\tilde{\mathbf{z}}^{A_2} + \cdots + \alpha_{2r}\tilde{\mathbf{z}}^{A_r} + g_2, \\ \cdots\cdots\cdots\cdots\cdots\cdots\cdots\cdots\cdots \\ \tilde{f_r} = \qquad\qquad\qquad\qquad \alpha_{rr}\mathbf{z}^{A_r} + g_r, \end{array}$$

where $\alpha_{ij} \in \mathbb{C}\{\mathbf{w}\}$ for all i, j, $\alpha_{ij} \in (w) \cdot \mathbb{C}\{w\}$ for $i < j$, $\alpha_{ii}(0) \neq 0$, and $\mathscr{E}_{\tilde{\mathbf{z}}}(g_i) \subset \mathbb{Z}_0^c \backslash u(I)_n$. By taking classes modulo $\mathrm{in}_P(I)$, we get a system of equations in the $(\mathbf{z}^*)^{A_i}$ which, when solved, gives us the $(\mathbf{z}^*)^{A_i}$ as a linear combination of monomials with exponents in $\mathbb{Z}_0^c \setminus u(I)_n$.

The minimality of H as a system of generators for $\mathrm{gr}_{(P/I)}(R/I)$ is established in the following way. Suppose H is not minimal, and let $n \geq 0$ be an integer such that there is $B \in \mathbb{Z}_0^c \setminus u(I)$ such that $|B| = n$ and $(\mathbf{z}^*)^B$ is a linear combination of the remaining monomials of degree n in H. If we denote by $\{B_1, \ldots, B_s\}$ the multi-indices in $\mathbb{Z}_0^c \setminus u(I)$ other than B, and such that $|B_1| = \cdots = |B_s| = n$, we may write

$$(\mathbf{z}^*)^B = \beta_1 \cdot (\mathbf{z}^*)^{B_1} + \cdots + \beta_s \cdot (\mathbf{z}^*)^{B_s}, \quad \text{with } \beta_j \in \mathbb{C}\{\mathbf{w}\} \text{ for } j = 1, \ldots, s.$$

This means that

$$\tilde{\mathbf{z}}^B - \beta_1\tilde{\mathbf{z}}^{B_1} - \cdots - \beta_s\tilde{\mathbf{z}}^{B_s} \in \mathrm{in}_P(I),$$

so either B or one of the B_j belongs to $u(I)$, which is a contradiction. This proves the lemma. □

The next proposition gives the first characterization of the flatness of

$$\mathrm{gr}_{(P/I)}(R/I).$$

Proposition 2.3.2 *If the module* $\mathrm{gr}_{(P/I)}(R/I)$ *is free over* $\mathbb{C}\{\mathbf{w}\}$, H *is one of its free basis.*

Proof Freeness of $\mathrm{gr}_{(P/I)}(R/I)$ implies the same componentwise. So the homogeneous components are finite free, hence minimal systems of generators are bases. Therefore, H is a free basis, which proves our assertion. □

The following result will not be strictly needed in the future, but we give it here for the sake of completeness of our algebraic study.

Proposition 2.3.3 *The following conditions are equivalent:*

(1) *The module* $\mathrm{gr}_{(P/I)}(R/I)$ *is free over* $\mathbb{C}\{\mathbf{w}\}$.
(2) *Every system of parameters in* $\mathbb{C}\{\mathbf{w}\}$ *is a regular sequence in*

$$\mathrm{gr}_{(P/I)}(R/I)$$

as a $\mathbb{C}\{\mathbf{w}\}$*-module.*

(3) *There exists a regular system of parameters in* $\mathbb{C}\{\mathbf{w}\}$ *that is a regular sequence in* $\mathrm{gr}_{(P/I)}(R/I)$ *as a* $\mathbb{C}\{\mathbf{w}\}$*-module.*

Proof (1) ⇒ (2) To prove this, it is enough to show that, under assumption (1), $(\mathbf{w})$ is a regular sequence. Since $\mathrm{gr}_{(P/I)}(R/I)$ is free over $\mathbb{C}\{\mathbf{w}\}$, no non-zero element in $\mathbb{C}\{\mathbf{w}\}$ is a zero divisor with respect to $\mathrm{gr}_{(P/I)}(R/I)$; in particular w_1 is not one. Now, take $i < d$ and assume $\{w_1, \ldots, w_i\}$ is a regular sequence. Let $\Delta, \Delta_1, \ldots, \Delta_i \in \mathrm{gr}_{(P/I)}(R/I)$ be such that $\Delta w_{i+1} = \sum_{j=1}^{i} \Delta_j w_j$, and write

$$\Delta = \sum_{A \in \mathbb{Z}_0^c \setminus u(I)} \alpha_A \cdot (\mathbf{z}^*)^A,$$

$$\Delta_j = \sum_{A \in \mathbb{Z}_0^c \setminus u(I)} \alpha_{j,A} \cdot (\mathbf{z}^*)^A \quad \text{for } 1 \leq j \leq i,$$

where $\alpha_A, \alpha_{j,A} \in \mathbb{C}\{\mathbf{w}\}$ and only a finite number of them are non-zero. Then

$$\sum_A (\alpha_A w_{i+1})(\mathbf{z}^*)^A = \sum_A \Big(\sum_j \alpha_{j,A} w_j \Big) (\mathbf{z}^*)^A,$$

so

$$\alpha_A w_{i+1} = \sum_{j=1}^{i} \alpha_{j,A} w_j$$

for every A. This proves that $\alpha_A \in (w_1, \ldots, w_i) \cdot \mathbb{C}\{\mathbf{w}\}$, so

$$\Delta \in (w_1, \ldots, w_i) \cdot \mathrm{gr}_{(P/I)}(R/I)$$

and this w_{i+1} is not a zero divisor in

$$\mathrm{gr}_{(P/I)}(R/I)/(w_1, \ldots, w_i) \cdot \mathrm{gr}_{(P/I)}(R/I).$$

(2) ⇒ (3). Clear.

(3) ⇒ (1). We observe first that, for every regular system of parameters $(\mathbf{w}')$ of $\mathbb{C}\{\mathbf{w}\}$, saying that $(\mathbf{w}')$ is a regular sequence in $\mathrm{gr}_{(P/I)}(R/I)$ is equivalent to saying that $(\mathbf{w}')$ is one in $\mathrm{gr}_{(P/I)}(R/I)_n$ for every $n \geq 0$. Since all the $\mathrm{gr}_{(P/I)}(R/I)_n$ are finitely generated over $\mathbb{C}\{\mathbf{w}\}$, this implies in particular that, if $(w'_1, \ldots, w'_d)$ is a regular sequence in $\mathrm{gr}_{(P/I)}(R/I)$, so is any permutation of it. Using this fact, we immediately see that, if $(w'_1, \ldots, w'_d)$ is a regular sequence in $\mathrm{gr}_{(P/I)}(R/I)$, so is

$$(w'_1 - \alpha_1 w'_d, \ \ldots, \ w'_{d-1} - \alpha_{d-1} w'_d, \ w'_d)$$

for $(\alpha_1, \ldots, \alpha_{d-1}) \in \mathbb{C}^{d-1}$. Keeping this in mind, we proceed to our argument.

Take a regular system of parameters $(\mathbf{w}')$ in $\mathbb{C}\{\mathbf{w}\}$ that is a regular sequence in $\mathrm{gr}_{(P/I)}(R/I)$; we need to prove that H is a free basis. To derive a contradiction, assume that there is a non-trivial relation

$$\sum_{A\in\mathbb{Z}_0^c\setminus u(I)} \alpha_A(\mathbf{z}^*)^A = 0, \quad \text{with } \alpha_A \in \mathbb{C}\{\mathbf{w}'\};$$

then none of the α_A can be a unit, since H is a minimal system of generators of $\mathrm{gr}_{(P/I)}(R/I)$. By taking a new regular system of parameters $(\mathbf{w}'')$ in $\mathbb{C}\{\mathbf{w}'\}$ defined by relations of the type

$$\begin{aligned} w_1'' &= w_1' - \mu_1 w_d', \\ &\cdots\cdots\cdots\cdots\cdots\cdots \\ w_{d-1}'' &= w_{d-1}' - \mu_d w_d', \\ w_d'' &= w_d', \end{aligned}$$

we may assume that the non-zero α_A are regular in w_d' of order equal to the order of α_A, and that $(\mathbf{w}'')$ is still a regular sequence by the preceding observations. Therefore, we may assume, from the very beginning, that $(\mathbf{w})$ itself is a regular sequence and that all the non-zero α_A are regular in w_d.

If we denote by a the infimum of the orders of the non-zero α_A, we may write the relation $\sum_A \alpha_A \cdot (\mathbf{z}^*)^A = 0$ as

$$w_d^a \sum_A \alpha_A' \cdot (\mathbf{z}^*)^A + \sum_A \beta_A \cdot (\mathbf{z}^*)^A = 0,$$

where $\alpha_A' \in \mathbb{C}\{w_d\}$ and at least one of them is a unit, and

$$\beta_A \in (w_1, \ldots, w_{d-1}) \cdot \mathbb{C}\{\mathbf{w}\}.$$

By the condition for a regular sequence, we have

$$\sum_A \alpha_A' \cdot (\mathbf{z}^*)^A = \sum_A \gamma_A \cdot (\mathbf{z}^*)^A, \quad \text{where } \gamma_A \in (w_1, \ldots, w_{d-1}) \cdot \mathbb{C}\{\mathbf{w}\}.$$

Therefore

$$\sum_A (\alpha_A' - \gamma_A)(\mathbf{z}^*)^A = 0,$$

and at least one of the coefficients $\alpha_A' - \gamma_A$ is a unit. This contradicts the minimality of H, so our proposition is proven. □

We are almost ready for the main characterization of the flatness of $\mathrm{gr}_{(P/I)}(R/I)$, Theorem 2.3.1. We still need one more concept.

Definition 2.3.1 An **equimultiple standard basis** for I is a finite set $\{f_1, \dots, f_r\}$ of elements in I satisfying the following conditions:

(1) $I = (f_1, \dots, f_r)$.
(2) The ordinary initial forms $\bar{f_i} = \mathrm{in}_M(f_i)$, for $i = 1, \dots, r$, generate $\mathrm{in}_M(I)$ as an ideal in $\mathrm{gr}_M(R)$.
(3) For all $i = 1, \dots, r$, $\nu_P(f_i) = \nu_M(f_i)$. Equivalently, $u(f_i) \neq \varnothing$.

In general, when we deal with an equimultiple standard basis $(f_1, \dots, f_r)$ for I, we shall implicitly assume that $\nu_m(f_1) \leq \cdots \leq \nu_m(f_r)$.

Here then is the main theorem of this section:

Theorem 2.3.1 *The following conditions are equivalent:*

(1) *The module* $\mathrm{gr}_{(P/I)}(R/I)$ *is a free* $\mathbb{C}\{\mathbf{w}\}$*-module.*
(2) *The ideal I possesses an equimultiple standard basis.*

Before proving the theorem, we still need to state and prove certain technical results.

Lemma 2.3.3 *Let* Ω *be a subset of* $\mathbb{Z}_0^c$ *such that* $\Omega + \mathbb{Z}_0^c = \Omega$. *Then there exists a finite family* $A_1, \dots, A_r \in \Omega$ *such that*

$$\Omega = \bigcup_{i=1}^{r} (A_i + \mathbb{Z}_0^c).$$

This representation is called **irredundant** if the union of any proper subset of $A_1 + \mathbb{Z}_0^c, \dots, A_r + \mathbb{Z}_0^c$ is different from Ω.

Proof Consider the ring $\mathbb{C}[\mathbf{z}] = \mathbb{C}[z_1, \dots, z_c]$, and the ideal J generated by all the monomials $\mathbf{z}^A$, for $A \in \Omega$. If $A \in \mathbb{Z}_0^c$ is such that $\mathbf{z}^A \in J$, then

$$\mathbf{z}^A = f_1(\mathbf{z})\mathbf{z}^{B_1} + \cdots + f_s(\mathbf{z})\mathbf{z}^{B_s}, \quad \text{where } f_i(\mathbf{z}) \in \mathbb{C}[\mathbf{z}] \text{ and } B_i \in \Omega.$$

Since $\Omega + \mathbb{Z}_0^c = \Omega$, we then have $A \in \Omega$. This proves that $\mathbf{z}^A \in J$ if and only if $A \in \Omega$. By the noetherian property of $\mathbb{C}[\mathbf{z}]$, the ideal J is generated by a finite number of monomials $\{\mathbf{z}^{A_1}, \dots, \mathbf{z}^{A_r}\}$. This automatically implies that

$$\Omega = \bigcup_{i=1}^{r} (A_i + \mathbb{Z}_0^c).$$

□

Corollary 2.3.3 *Let $J \subset R$ be an ideal distinct from* (0) *and* (1), *then there exist* $A_1, \ldots, A_r \in \mathbb{Z}_0^c$ *such that*

$$u(J) = \bigcup_{i=1}^{r} (A_i + \mathbb{Z}_0^c).$$

The proof is an immediate consequence of the preceding lemma and the fact that $u(J) + \mathbb{Z}_0^c = u(J)$.

Lemma 2.3.4 *Let $g_1, \ldots, g_s$ be a collection of non-units in R different from zero. Then there exists $K \in \mathbb{R}_+$ such that, for any $l_1, \ldots, l_c \in \mathbb{R}$ satisfying $|l_i - 1| < 1/K$ for $1 \leq i \leq c$, the form*

$$L(\mathbf{x}, \mathbf{x}') = \sum_{i=1}^{c} l_i x_i + \sum_{j=1}^{d} x'_j$$

satisfies the following condition: for any $k = 1, \ldots, s$, the initial form $\mathrm{in}_L(g_k)$ *consists of a sum of terms (monomials with coefficients) occurring only in the ordinary initial form g_k. (For the sake of brevity, we consider all initial forms as homogeneous polynomials in R.)*

Proof Let

$$\nu = \sup_{1 \leq k \leq s} \{\nu_M(g_k)\},$$

and take, for instance, $K > 10\nu$. Fix k satisfying $1 \leq k \leq s$, and take $A \in \mathbb{Z}_0^c$, $B \in \mathbb{Z}_0^d$ such that $|A| + |B| = \nu_M(g_k)$ and $A = (a_1, \ldots, a_c)$. Then

$$\begin{aligned} |L(A, B) - \nu_M(g_k)| &= \left| \sum_{i=1}^{c} (l_i - 1) a_i \right| \leq \sum_{i=1}^{c} |l_i - 1|\, a_i \\ &< \frac{1}{10\nu} |A| \leq \frac{1}{10\nu} \nu = \frac{1}{10}. \end{aligned}$$

On the other hand, take $A' \in \mathbb{Z}_0^c$, $B' \in \mathbb{Z}_0^d$ such that $|A'| + |B'| > \nu_M(g_k)$, and write $\delta = |A'| + |B'| - \nu_M(g_k)$, $A' = (a'_1, \ldots, a'_c)$. Then

$$\begin{aligned} L(A', B') - \nu_M(g_k) &= L(A', B') + \delta - |A'| - |B'| \\ &= \sum_{i=1}^{c} (l_i - 1)\, a'_i + \delta > -\frac{1}{K} |A'| + \delta. \end{aligned}$$

But $|A'| \leq \delta + \nu_M(g_k)$, so

$$\begin{aligned}
L(A', B') - \nu_M(g_k) &> \delta - \frac{1}{K}|A'| > \delta - \frac{1}{10\nu}|A'| \\
&\geq \delta - \frac{1}{10\nu}(\delta + \nu_M(g_k)) = \delta\left(1 - \frac{1}{10\nu}\right) - \frac{\nu_M(g_k)}{10\nu} \\
&> \delta\left(1 - \frac{1}{10}\right) - \frac{\nu}{10\nu} > \frac{9}{10} - \frac{1}{10} = \frac{8}{10},
\end{aligned}$$

We have, therefore, proved that

$$\begin{aligned}
|L(A, B) - \nu_M(g_k)| < 1/10 &\quad \text{for } |A| + |B| = \nu_M(g_k), \\
L(A', B') - \nu_M(g_k) > 8/10 &\quad \text{for } |A'| + |B'| > \nu_M(g_k),
\end{aligned}$$

so the lemma follows. □

We now proceed to choose the tools we need to perform our task.

Remark 2.3.1 Let $A_1 < \cdots < A_r$ be multi-indices in $u(I)$ such that

$$u(I) = \bigcup_{k=1}^{r}(A_k + \mathbb{Z}_0^c),$$

and write

$$\begin{aligned}
\square_0 &= \mathbb{Z}_0^c \setminus u(I), \\
\square_1 &= A_1 + \mathbb{Z}_0^c, \\
\square_k &= (A_k + \mathbb{Z}_0^c) \setminus (A_k + \mathbb{Z}_0^c) \cap \left(\bigcup_{l=1}^{k-1} \square_l\right) \quad \text{for } 2 \leq k \leq r.
\end{aligned}$$

Take an integer $m > 0$; by Corollary 2.3.2 we can find a positive linear form

$$L(\mathbf{x}, \mathbf{x}') = \sum_{i=1}^{c} l_i x_i + \sum_{j=1}^{d} x_j', \quad \text{for } l_i \in \mathbb{R}_+$$

satisfying these conditions:

(1) $L(B, 0) > L(A_k, 0) > L(B', 0)$ for any $k = 1, \ldots, r$, $B, B' \in \mathbb{Z}_0^c$, $|B| = |B'| = |A_k|$, $B' > A_k > B$.
(2) For every k, B, B' as before, we have

$$\begin{aligned}
\big|L(B, 0) - |B|\big| &< 1/m, \\
\big|L(A_k, 0) - |A_k|\big| &< 1/m,
\end{aligned}$$

$$\left|L(B',0)-|B'|\right|<1/m.$$

By taking m great enough (say $m>10$), we may assume that, in addition,

(3) $L(B',0)+1>L(A_k,0)$ for every $k=1,\ldots,r$, $B'\in\mathbb{Z}_0^c$, $|B'|=|A_k|$, $B'>A_k$.

We must impose on L some further restrictions so that it satisfies an additional condition. This time the matter is a little subtler. In view of the proof of Corollary 2.3.2, the coefficients l_i are very special but as close to 1 as we wish. Therefore, we may add a suitable upper bound for $|l_i-1|$ (so that our requirements are met) and be sure we still keep (1), (2) and (3). In this line, we take $p\in\mathbb{Z}_+$, $p>1$ such that $10^p-1>|A_k|$ for every $k=1,\ldots,r$. Then, for every $A\in\mathbb{Z}_0^c$ satisfying, $|A|>|A_k|$, one has

$$\begin{aligned}(10^p-1)|A|-10^p|A_k|&\geq(10^p-1)(|A_k|+1)-10^p|A_k|\\&=10^p|A_k|+10^p-|A_k|-1-10^p|A_k|\\&=(10^p-1)-|A_k|>0,\end{aligned}$$

so

$$0<\frac{(10^p-1)|A|-10^p|A_k|}{10^{p+1}}=\frac{|A|-|A_k|}{10}-\frac{|A|}{10^{p+1}}.$$

We now impose on l_i the conditions $|l_i-1|<1/10^{p+1}$ for all $i=1,\ldots,c$. Then, for every $k=1,\ldots,r$ and every $A=(a_1,\ldots,a_c)\in\mathbb{Z}_0^c$ satisfying $|A|>|A_k|$, one has

$$\left|L(A,0)-|A|\right|=\left|\sum_{i=1}^{c}(l_i-1)a_i\right|\leq\sum_{i=1}^{c}|l_i-1|a_i<\frac{1}{10^{p+1}}|A|<\frac{|A|-|A_k|}{10}.$$

Therefore,

$$\begin{aligned}L(A,0)&>|A|-\frac{|A|-|A_k|}{10}=\frac{9|A|-|A_k|}{10}\geq\frac{9(|A_k|+1)+|A_k|}{10}\\&=|A_k|+\frac{9}{10}>|A_k|+\frac{1}{10}>L(A_k,0).\end{aligned}$$

We may, therefore, choose L satisfying not only (1)–(3), but also the following condition:

(4) For $k=1,\ldots,r$, and every $A\in\mathbb{Z}_0^c$ satisfying $|A|>|A_k|$, one has $L(A,0)>L(A_k,0)$.

Remark 2.3.2 Recall that the set

$$H=\{(\mathbf{z}^*)^B\mid B\in\Box_0\}$$

is a minimal system of generators of $\mathrm{gr}_{(P/I)}(R/I)$ as a $\mathbb{C}\{\mathbf{w}\}$-module. For $A \in u(I)$, choose an arbitrary expression for $(\mathbf{z}^*)^A$ as a homogeneous linear combination of H, say

$$(\mathbf{z}^*)^A = \sum_{\substack{B\in\square_0\\|B|=|A|}} \beta_{A,B} \cdot (\mathbf{z}^*)^B, \quad \text{where } \beta_{A,B} \in \mathbb{C}\{\mathbf{w}\}.$$

(Of course, this decomposition is generally not unique.) The expression above implies

$$\tilde{\mathbf{z}}^A - \sum_{\substack{B\in\square_0\\|B|=|A|}} \beta_{A,B}\tilde{\mathbf{z}}^B \in \mathrm{in}_P(I),$$

and we may choose an element $g^{(A)} \in I$ such that

$$\tilde{g}^{(A)} = \tilde{\mathbf{z}}^A - \sum_{\substack{B\in\square_0\\|B|=|A|}} \beta_{A,B}\tilde{\mathbf{z}}^B.$$

Again, this choice is not, in general, unique; we just make it arbitrarily.

Write $\mathbf{G} = \{g^{(A_1)}, \ldots, g^{(A_r)}\}$ and $\mathbf{A} = \{A_1, \ldots, A_r\}$. Then, for any $k = 1, \ldots, r$ and $(B, C) \in \mathscr{E}_{\mathbf{z},\mathbf{w}}(g^{(A_k)})$ satisfying $(B, C) \neq (A_k, 0)$, one has:

(1) If $|B| = |A_k|$ with $B < A_k$, condition (1) implies

$$L(B, C) \geq L(B, 0) > L(A_k, 0).$$

(2) If $|B| = |A_k|$ with $B > A_k$, then $(g^{(A_k)})_B \in (\mathbf{w}) \cdot \mathbb{C}\{\mathbf{w}\}$, so (3) gives

$$L(B, C) \geq L(B, 0) + 1 > L(A_k, 0).$$

(3) If $|B| > |A_k|$, condition (4) gives

$$L(B, C) \geq L(B, 0) > L(A_k, 0).$$

This means that $\nu_L(g^{(A_k)}) = L(A_k, 0)$ and $\nu_L(g^{(A_k)} - \mathbf{z}^{A_k}) > L(A_k, 0)$, so, with the notation of the Weierstrass preparation theorem, $S\{\mathbf{G}, \mathbf{A}\} \neq \varnothing$.

Now, for any $A \in u(I)$, the Weierstrass preparation theorem allows us to write

$$\mathbf{z}^A = h_{A,0} + \sum_{k=1}^{r} h_{A,k} g^{(A_k)},$$

where, in particular, $h_{A,0}$ is such that $\mathscr{E}_{\mathbf{z}}(h_{A,0}) \subset \square_0$. We are going to prove that $\nu_P(h_{A,0}) \geq |A|$. To this end, we need to follow the construction of the

$h_{A,0}, \ldots, h_{A,r}$ as it was done in Remark 2.2.3. Schematically, the procedure is as follows:

First step: If $A \in \square_{k_0}$ satisfies $A = A_{k_0} + A'$, we write

$$\mathbf{z}^A = \mathbf{z}^{A'}\mathbf{z}^{A_{k_0}} = h^{(0)}_{A,A_{k_0}}\mathbf{z}^{A_{k_0}}.$$

Second step: Then,

$$\mathbf{z}^A = \mathbf{z}^{A'}g^{(A_{k_0})} - z\mathbf{z}^{A'}(g^{(A_{k_0})} - z^{(A_{k_0})}) = h^{(0)}_{A,A_{k_0}}g^{(A_{k_0})} + h^{(1)}_{A,0} + h^{(1)}_{A,1}\mathbf{z}^{A_1} + \cdots + h^{(1)}_{A,r}\mathbf{z}^{A_r}$$

and so on. It is clear that orders in $\mathbf{z}$ never decrease, so, in particular, $\nu_P(h_{A,0}) \leq |A|$.

Write $f^{(A)} = \mathbf{z}^A - h_{A,0}$, for any $A \in u(I)$. We may take the $f^{(A)}$ instead of the $g^{(A)}$, from the very beginning, and we have the same properties of the $g^{(A)}$, plus $\mathscr{E}_{\mathbf{z}}(f^{(A)} - \mathbf{z}^{(A)}) \subset \square_0$. We are going to work with the family $\{f^{(A)} \mid A \in u(I)\}$. In particular, if we write $\mathbf{F} = \{f^{(A_1)}, \ldots, f^{(A_r)}\}$, we have $S\{\mathbf{F}, \mathbf{A}\} \neq \varnothing$.

Lemma 2.3.5 *With the notations of Theorem* 2.3.1, *condition* (1) *implies* (2).

Proof For brevity, write f_k instead of $f^{(A_k)}$, for $1 \leq k \leq r$. If we take $f \in I$, the Weierstrass preparation theorem allows us to write

$$f = h_0 + \sum_{k=1}^{r} h_k f_k,$$

where, in particular, $\mathscr{E}_{\mathbf{z}}(h_0) \subset \square_0$. This condition, together with the fact that $h_0 \in I$ and the freeness of $\mathrm{gr}_{(P/I)}(R/I)$, implies that $h_0 = 0$. This proves, in particular, that $I = (f_1, \ldots, f_r)$. We are going to prove that $\mathbf{F}$ is an equimultiple standard basis for I. Since $\nu_P(f_k) = \nu_m(f_k) = |A_k|$ for any $k = 1, \ldots, r$, we need only prove that

$$\mathrm{in}_m(I) = (\bar{f}_1, \ldots, \bar{f}_r).$$

For $f \in I$ we have, as before, an expression

$$f = \sum_{k=1}^{r} h_k f_k$$

directly deduced from the Weierstrass preparation theorem, in which:

(1) it is verified that $\mathscr{E}_{\mathbf{z}}(h_k\mathbf{z}^{A_k}) \subset \square_k$, for $1 \leq k \leq r$, and
(2) for any $L' \in S\{\mathbf{F}, \mathbf{A}\}$ and any $k = 1, \ldots, r$,

$$\nu_{L'}(h_k) \geq \nu_{L'}(f) - \nu_{L'}(f_k).$$

Now, coming back to the construction of L satisfying (1) to (4) of Remark 2.3.1, as was done using Corollary 2.3.2, we see that the coefficients l_i are very special ones, but as close to 1 as we wish. If we combine this with Lemma 2.3.4, we see that L can be so chosen as to satisfy (1) to (4) of Remark 2.3.1, plus the following condition:

(*) For any $k = 1, \ldots, r$, each of the initial forms $\mathrm{in}_L(h_k)$ and $\mathrm{in}_L(f_k)$ consists of a sum of terms occurring only in their respective ordinary initial forms.

Actually, this statement for $\mathrm{in}_L(f_k)$ is a consequence of Remark 2.3.1, since we proved there that these conditions imply that $\mathrm{in}_L(f_k) = \mathbf{z}^{A_k}$.

Now, one has

$$\nu_L(f) \geq \inf_{1 \leq k \leq r} \{\nu_L(h_k) + \nu_L(f_k)\} \geq \nu_L(f),$$

so

$$\nu_L(f) = \inf_{1 \leq k \leq r} \{\nu_L(h_k) + \nu_L(f_k)\}.$$

This means that, if $f \neq 0$, the terms of lowest ν_L in $\{h_1 f_1, \ldots, h_r f_r\}$ do not cancel in $\sum_{k=1}^r h_k f_k$. This implies, by condition (*) that

$$\sum_{k=1}^{r} \bar{h}_k \bar{f}_k \neq 0. \tag{2.1}$$

But we can say even more, by taking the end of the proof of Lemma 2.3.4 into consideration. Let $\mathbf{z}^A \mathbf{w}^B$ and $\mathbf{z}^{A'}$ be monomials in the ordinary initial forms of h_{k_0} and f_{k_0}, respectively, where $\bar{h}_{k_0} \bar{f}_{k_0}$ is a summand in equation (2.1) at which the infimum of the ν_M is reached. Then

$$|L(A, B) - \nu_M(h_{k_0})| < 1/10, \quad |L(A', 0) - \nu_M(f_{k_0})| < 1/10.$$

Now let $\mathbf{z}^C \mathbf{w}^D$ and $\mathbf{z}^{C'}$ be monomials in the ordinary initial forms of some h_k, f_k, respectively, for which $\nu_M(h_k f_k) > \nu_M(h_{k_0} f_{k_0})$; then

$$L(C, D) - \nu_M(h_{k_0}) > 8/10, \quad L(C', 0) - \nu_M(f_{k_0}) > 8/10.$$

This implies that

$$\inf_{1 \leq k \leq r} \{\nu_L(h_k f_k)\}$$

is reached only at the ordinary initial forms of the $h_k f_k$ where

$$\inf_{1 \leq k \leq r} \{\nu_M(h_k f_k)\}$$

is reached. Therefore the forms of lowest degree in (2) do not cancel, so $\bar{f} \in (\bar{f}_1, \ldots, \bar{f}_r)$, and our proof is complete. □

Corollary 2.3.4 *For any $f \in I$, there exists an expression*

$$f = \sum_{k=1}^{r} h_k f_k,$$

where $\bar{f}$ is the sum of the forms of lowest degree among $\{\bar{h}_1 \bar{f}_1, \ldots, \bar{h}_r \bar{f}_r\}$.

Remark 2.3.3

(1) Given a form $\varphi \in \mathbb{C}[\bar{\mathbf{z}}]$, we denote by $\bar{u}(\varphi)$ the greatest (in the lexicographic order) of the multi-indices in $\mathscr{E}_{\bar{\mathbf{z}}}(\varphi)$. For a homogeneous ideal $H \subset \mathbb{C}[\bar{\mathbf{z}}]$, we write

$$\bar{u}(H) = \{\bar{u}(\varphi) \mid \varphi \text{ is a form belonging to } H\}.$$

(2) For an ideal I, we define $\tau(I)$ as the ideal in $\mathbb{C}[\bar{\mathbf{z}}]$ generated by all the forms $\varphi(\mathbf{z}, 0)$, where $\varphi(\bar{\mathbf{z}}, \bar{\mathbf{w}}) \in \mathrm{in}_M(I)$. Then if $\mathrm{gr}_{(P/I)}(R/I)$ is a free $\mathbb{C}\{\mathbf{w}\}$-module, $\bar{u}(\tau(I)) = u(I)$. In fact, take the basis $I = (f_1, \ldots, f_r)$ of Lemma 2.3.5. Since $\mathrm{in}_M(I) = (\bar{f}_1, \ldots, \bar{f}_r)$ one has $\tau(I) = (\bar{f}_1, \ldots, \bar{f}_r)$. Then, it is clear that $\bar{u}(\tau(I)) \subset u(I)$. Conversely, let $f \in I$ be such that $u(f) = B \in \mathbb{Z}_0^c$; then

$$\bar{f} = \sum_{k=1}^{r} \varphi_k(\bar{\mathbf{z}}, \bar{\mathbf{w}}) \bar{f}_k, \quad \text{where } \varphi_k(\bar{\mathbf{z}}, \bar{\mathbf{w}}) \in \mathbb{C}[\bar{\mathbf{z}}, \bar{\mathbf{w}}],$$

and $\bar{u}$ is either zero or a form of degree $\deg(\bar{f}) - \deg(\bar{f}_k)$. Then

$$\bar{f} = \sum_{k=1}^{r} \varphi_k(\bar{\mathbf{z}}, 0) \bar{f}_k,$$

which implies $B \in \bar{u}(\tau(I))$. This proves the equality.

(3) A **P-normalized standard basis** for the ideal I is an equimultiple standard basis $(g_1, \ldots, g_s)$ of I satisfying the following properties:

   (1) For every $i = 1, \ldots, s-1$, $\nu_M(g_i) \leq \nu_M(g_{i+1})$.
   (2) For every $i = 1, \ldots, s-1$, $\mathscr{E}_{\mathbf{z}}(g_i) \cap \bar{u}((\bar{g}_1, \ldots, \bar{g}_{i-1})) = \varnothing$.

Proposition 2.3.4 *If $\mathrm{gr}_{(P/I)}(R/I)$ is a free $\mathbb{C}\{\mathbf{w}\}$-module, then I possesses a P-normalized standard basis.*

Proof As in Remark 2.3.1, we consider a family $\{A_1, \ldots, A_r\}$ of multi-indices in $\mathbb{Z}_0^c$ such that

$$u(I) = \bigcup_{k=1}^{r} (A_k + \mathbb{Z}_0^c) = \bar{u}(\tau(I)),$$

and we denote by α the supremum of the $|A_k|$. We now define in $\mathbb{Z}_0^c$ a new total order relation $<^*$ by the rule

$$A <^* B \iff |A| < |B| \text{ or} |A| = |B| \text{ and } A < B \text{ (in lexicographic order).}$$

We then consider the elements of $u(I)$, ordered by means of $<^*$ from smaller to greater; it is clear what a successor (next greater element) means in this case. We also write

$$D = \sup_{<^*}\{A \in u(I) \mid |A| \leq \alpha\}.$$

We now consider the family $\mathscr{F} = \{f^{(A)} \mid A \in u(I)\}$ and choose a finite subfamily $\mathscr{G} = \{g^{(B_1)}, \ldots, g^{(B_q)}\}$ by using the following algorithm:

(1) Set $B_1 = \inf_{<^*} u(I)$, $g^{(B_1)} = f^{(B_1)}$.
(2) Set $s = 1$.
(3) Set $C = B_s$.
(4) If $D <^* C$, go to step 8.
(5) Set A to the successor of C in $u(I)$.
(6) If $\mathscr{E}_{\mathbf{z}}(f^{(A)}) \cap \bar{u}((\bar{g}^{(B_1)}, \ldots, \bar{g}^{(B_s)}))$ is empty, set $g^{(B_{s+1})} = f^{(A)}$, set $s = s + 1$, and go to (3).
(7) Set $C = A$ and go to (4).
(8) End.

This algorithm terminates, so we get our subfamily $\mathscr{G} = \{g^{(B_1)}, \ldots, g^{(B_q)}\}$. To complete our proof we will show that $\mathscr{G}$, as an ordered family for the order $<^*$, is the P-normalized standard basis of I we are looking for.

We need only show that $\text{in}_M(I) = (\bar{g}^{(B_1)}, \ldots, \bar{g}^{(B_q)}) \cdot \text{gr}_M(R)$. For this, we must take into account the fact that, by Proposition 2.3.5,

$$\text{in}_M(I) = (\{\bar{f}^{(A)} \mid A \leq^* D\}) \cdot \text{gr}_M(R),$$

since $A_k \leq^* D$ for every $k = 1, \ldots, r$. Now our problem is to show that, when we get rid of some $f^{(A)}$ in the algorithm, the ideal generated by the initial forms is not affected. This amounts to proving the following claim.

Remark 2.3.4 (Assertion) Let $\{g^{(B_1)}, \ldots, g^{(B_s)}\}$ be elements in $\mathscr{G}$ such that $B_1 <^* \cdots <^* B_s$, and such that there is no $g^{(B)} \in \mathscr{G}$ with B strictly between two of the B_i. Let $A \in u(I)$, and assume that the following conditions hold:

(1) It is $B_s <^* A$ and $\mathscr{E}_{\mathbf{z}}(f^{(A)}) \cap \bar{u}((\bar{g}^{(B_1)}, \ldots, \bar{g}^{(B_s)})) \neq \varnothing$.
(2) For any $C \in u(I)$ satisfying $C <^* A$, is $\bar{f}^{(C)} \in (\bar{g}^{(B_1)}, \ldots, \bar{g}^{(B_s)}) \cdot \text{gr}_M(R)$.

Then $(\bar{g}^{(B_1)}, \ldots, \bar{g}^{(B_s)}) \cdot \text{gr}_M(R) = (\bar{g}^{(B_1)}, \ldots, \bar{g}^{(B_s)}, \bar{f}^{(A)}) \cdot \text{gr}_M(R)$.

To prove the assertion we first observe that, since $\bar{u}(\tau(I)) = u(I)$ and

$$\mathscr{E}_{\mathbf{z}}(f^{(A)}) \cap u(I) = \{A\}$$

the only possibility for condition (1) to be satisfied is that $A \in \bar{u}((\bar{g}^{(B_1)}, \dots, \bar{g}^{(B_s)})$. Then there exists a form $\varphi \in \mathbb{C}[\bar{\mathbf{z}}]$ such that $\bar{u}(\varphi) = A$ and

$$\varphi = \varphi_1 \bar{g}^{(B_1)} + \cdots + \varphi_s \bar{g}^{(B_s)},$$

where φ_i is either zero, or a form of degree $\deg(\varphi) - \deg(\bar{g}^{(B_i)})$. We assume that the coefficient of $\bar{\mathbf{z}}^A$ in φ is 1.

Write

$$f' = \varphi_1(\mathbf{z}) g^{(B_1)} + \cdots + \varphi_s(\mathbf{z}) g^{(B_s)}.$$

Then $\bar{f}' = \varphi$ and $u(\bar{f}') = A$. Set also

$$\tilde{f}' = \tilde{\mathbf{z}}^A + \sum_{\substack{C \in u(I) \\ |C| = |A|}} \alpha_C(\mathbf{w}) \tilde{\mathbf{z}}^C + \Psi,$$

where $\alpha_C(\mathbf{w}) \in \mathbb{C}\{\mathbf{w}\}$, $\alpha_C(\mathbf{0}) = 0$ if $C > A$, and Ψ is a form in $\tilde{\mathbf{z}}$, with coefficients in $\mathbb{C}\{\mathbf{w}\}$, such that $\mathscr{E}_{\tilde{\mathbf{z}}}(\Psi) \subset \Box_0$. Now set

$$f = f' - \sum_{\substack{C \in u(I) \\ |C| = |A|}} \alpha_C(\mathbf{w}) f^{(C)}.$$

By uniqueness of expression with respect to a free basis one has $\tilde{f} = \tilde{f}^{(A)}$, so

$$\begin{aligned} \bar{f} = \bar{f}^{(A)} = \bar{f}' - &\sum_{\substack{C \in u(I) \\ |C| = |A|,\, C < A}} \alpha_C(\mathbf{0}) \bar{f}^{(C)} \\ = \varphi - &\sum_{\substack{C \in u(I) \\ |C| = |A|,\, C < A}} \alpha_C(\mathbf{0}) \bar{f}^{(C)} \in (\bar{g}^{(B_1)}, \dots, \bar{g}^{(B_s)}) \cdot \mathrm{gr}_M(R). \end{aligned}$$

This proves the claim and the proposition. □

The following result will be useful later.

Proposition 2.3.5 *If* $\mathrm{gr}_{(P/I)}(R/I)$ *is a free* $\mathbb{C}\{\mathbf{w}\}$*-module, then*

$$\mathrm{in}_P(I) = (\bar{f}^{(A_1)}, \dots, \bar{f}^{(A_r)}) \cdot \mathrm{gr}_P(R).$$

Proof We first show that

$$\mathrm{in}_P(I) = (\{\tilde{f}^{(A)} \mid A \in u(I)\}) \cdot \mathrm{gr}_P(R). \tag{2.2}$$

In fact, take a non-zero $f \in I$, and write

$$\tilde{f} = \sum_{\substack{C \in \mathbb{Z}_0^c \\ |C| = \nu_P(f)}} \sigma_C \cdot \tilde{\mathbf{z}}^C, \quad \text{where } \sigma_C \in \mathbb{C}\{\mathbf{w}\}.$$

If we consider the element

$$f - \sum_{\substack{A \in u(I) \\ |A| = \nu_P(f)}} \sigma_A f^{(A)} \in I,$$

we see that, by the free basis condition,

$$\tilde{f} - \sum_{\substack{A \in u(I) \\ |A| = \nu_P(f)}} \sigma_A \tilde{f}^{(A)} = 0,$$

which proves (2.2).

Now consider a fixed degree n and denote by $C_1 < \cdots < C_t$ all the multi-indices in $u(I)$ such that $|C_j| = n$. There is a well-defined map

$$\sigma = \{1, \ldots, t\} \to \{1, \ldots, r\}$$

such that $C_i \in \square_{\sigma(i)}$ for $1 \le i \le t$. We write $C_i = A_{\sigma(i)} + C_i'$, for $C_i' \in \mathbb{Z}_0^c$.

We now consider the family of homogeneous elements of $\mathrm{in}_P(I)$ given by

$$\tilde{g}_j = \tilde{\mathbf{z}}^{C_j'} \tilde{f}^{(A_{\sigma(j)})} \quad \text{for } j = 1, \ldots, t,$$

and we write these elements as

$$\bar{g}_j = \gamma_{j1}\tilde{\mathbf{z}}^{C_1} + \cdots + \gamma_{jt}\tilde{\mathbf{z}}^{C_t} + \sum_{C \in \square_0, |C| = n} \gamma_{j,C}\tilde{\mathbf{z}}^C,$$

where $\gamma_{jk}, \gamma_{jB} \in \mathbb{C}\{\mathbf{w}\}$, $\gamma_{jj} = 1$, and $\gamma_{jk} \in (\mathbf{w}) \cdot \mathbb{C}\{\mathbf{w}\}$ for $k > j$, by the construction of the $f^{(A_l)}$. We may write the homogeneous polynomials above in the form

$$\begin{array}{lcl} \tilde{g}_1 & = & \gamma_{11}\tilde{\mathbf{z}}^{C_1} + \cdots + \gamma_{1t}\tilde{\mathbf{z}}^{C_t} + \tilde{g}_1', \\ \cdots & & \cdots \\ \tilde{g}_t & = & \gamma_{t1}\tilde{\mathbf{z}}^{C_1} + \cdots + \gamma_{tt}\tilde{\mathbf{z}}^{C_t} + \tilde{g}_t'; \end{array}$$

hence, in matrix form,

$$\begin{pmatrix} \tilde{g}_1 \\ \vdots \\ \tilde{g}_t \end{pmatrix} = M \begin{pmatrix} \tilde{\mathbf{z}}^{C_1} \\ \vdots \\ \tilde{\mathbf{z}}^{C_t} \end{pmatrix} + \begin{pmatrix} \tilde{g}'_1 \\ \vdots \\ \tilde{g}'_t \end{pmatrix},$$

with $M = (\gamma_{ij})$. But M is invertible in $\mathbb{C}\{\mathbf{w}\}$, so

$$M^{-1} \begin{pmatrix} \tilde{g}_1 \\ \vdots \\ \tilde{g}_t \end{pmatrix} = \begin{pmatrix} \tilde{\mathbf{z}}^{C_1} \\ \vdots \\ \tilde{\mathbf{z}}^{C_t} \end{pmatrix} + M^{-1} \begin{pmatrix} \tilde{g}'_1 \\ \vdots \\ \tilde{g}'_t \end{pmatrix}.$$

By taking classes modulo $\mathrm{in}_P(I)$, we get

$$\begin{pmatrix} 0 \\ \vdots \\ 0 \end{pmatrix} = \begin{pmatrix} (\mathbf{z}^*)^{C_1} \\ \vdots \\ (\mathbf{z}^*)^{C_t} \end{pmatrix} + M^{-1} \begin{pmatrix} g^*_1 \\ \vdots \\ g^*_t \end{pmatrix}.$$

By uniqueness of expression with respect to a free basis,

$$M^{-1} \begin{pmatrix} \tilde{g}_1 \\ \vdots \\ \tilde{g}_t \end{pmatrix} = \begin{pmatrix} \tilde{f}^{(C_1)} \\ \vdots \\ \tilde{f}^{(C_t)} \end{pmatrix}.$$

Therefore $\tilde{f}^{(C_1)}, \dots, \tilde{f}^{(C_t)} \in (\tilde{f}^{(A_1)}, \dots, \tilde{f}^{(A_r)}) \cdot \mathrm{gr}_P(R)$, and the proposition is proved. □

Proposition 2.3.6 *In the statement of Theorem* 2.3.1, (2) *implies* (3).

Proof We need to prove that H is a free basis of $\mathrm{gr}_{(P/I)}(R/I)$ as a $\mathbb{C}\{\mathbf{w}\}$-module. We keep the previous notation.

Suppose, for a contradiction, that there is a non-zero $f \in I$ whose initial form $\tilde{f}$ is such that $\mathscr{E}_{\mathbf{z}}(\tilde{f}) \subset \square_0$. By the Weierstrass preparation theorem, we may write

$$f = h_0 + \sum_{k=1}^{r} h_k f^{(A_k)},$$

where, in particular, $\mathscr{E}_{\mathbf{z}}(h_0) \subset \square_0$. The construction of the $h_0, h_1, \dots, h_r$ carried out in Remark 2.2.3 tells us that $h_0 \neq 0$ and $\tilde{h}_0 = \tilde{f}$. Therefore, we may assume that $\mathscr{E}_{\mathbf{z}}(f) \subset \square_0$.

We now choose an equimultiple standard basis $(g_1, \dots, g_q)$ of I; then $\bar{f}$ is a linear combination of $(\bar{g}_1, \dots, \bar{g}_q)$. Write

$$\bar{f} = \varphi_1 \bar{g}_1 + \cdots + \varphi_q \bar{g}_q,$$

where $\varphi_1, \dots, \varphi_q \in \mathbb{C}[\bar{\mathbf{z}}, \bar{\mathbf{w}}]$ are either zero or forms of degree

$$\deg(\varphi_1) = \deg(\bar{f}) - \deg(\bar{g}_i).$$

Consider the decomposition $\bar{f} = \varphi_1 \bar{g}_1 + \cdots + \varphi_q \bar{g}_q$ as a linear combination of monomials in $\bar{\mathbf{z}}$ with indeterminate coefficients. The existence of a solution means that a certain set of homogeneous linear equations (those corresponding to monomials in $u(I)$ plus, maybe, something else), and a set of homogeneous linear inequalities (those corresponding to monomials that actually occur in $\bar{f}$) have a solution in $\mathbb{C}[\bar{\mathbf{w}}]$. But these homogeneous linear equalities and inequalities have coefficients in $\mathbb{C}$ because $\bar{g}_1, \dots, \bar{g}_q \in \mathbb{C}[\bar{\mathbf{z}}]$, so they also have a solution in $\mathbb{C}$. By considering only homogeneous components with respect to the $\bar{\mathbf{z}}$, we see that there exist forms $\varphi_1', \dots, \varphi_q' \in \mathbb{C}[\bar{\mathbf{z}}]$ such that, if we write

$$\varphi = \varphi_1' \bar{g}_1 + \cdots + \varphi_q' \bar{g}_q$$

with $\deg \varphi = \deg_{\bar{\mathbf{z}}}(\bar{f})$, then

$$\varnothing \neq \mathscr{E}_{\bar{\mathbf{z}}}(\varphi) \subset \Box_0.$$

Since the possibly occurring terms $\varphi_i' \bar{g}_i$ of degree smaller that $\deg(\varphi)$ must cancel, we may assume that the only non-zero φ_i' are those such that $\deg(\varphi_i') = \deg(\varphi) - \deg(\bar{g}_i)$. Now take

$$h = \varphi_1' g_1 + \cdots + \varphi_q' g_q;$$

then $\bar{h} = \varphi$, so $u(h) \subset \Box_0$, a contradiction. □

To conclude this section, we study what happens if, under the assumption that $\mathrm{gr}_{(P/I)}(R/I)$ is free, we consider any basis of I, not necessarily an equimultiple standard one.

Remark 2.3.5 Assume that $\mathrm{gr}_{(P/I)}(R/I)$ is free over R/P, and consider two bases of I: the standard one $\{f^{(A_i)} \mid 1 \leq i \leq r\}$ constructed in Proposition 2.3.5, and a completely arbitrary one $(g_1, \dots, g_s)$. For concreteness, assume that

$$u(I) = \bigcup_{i=1}^{r} (A_i + \mathbb{Z}_0^c)$$

is an irredundant representation, in which case we call $A_1, \dots, A_r$ the **vertices** of $u(I)$. For a point $A \in \mathbb{Z}_0^c$, we call $|A|$ the **order** of A, and we denote by $\nu_{\mathbf{z}}$ the order in $\mathbf{z}$ of any element in R.

Let $\nu = \inf\{|A_i| \mid i = 1, \dots, r\}$, and assume that $\{A_{01}, \dots, A_{0p_0}\}$ are exactly those A_j for which $|A_j| = \nu$. In the homogeneous component $\mathbb{C}\{\mathbf{w}\}[\tilde{\mathbf{z}}]_\nu$ of degree ν of $\mathbb{C}\{\mathbf{w}\}[\tilde{\mathbf{z}}]$ we have vectors

$$\tilde{f}^{(A_{01})} = \tilde{\mathbf{z}}^{A_{01}} - \sum_{\substack{|B|=\nu \\ B\notin u(I)}} (f^{(A_{01})})_B \, \tilde{\mathbf{z}}^B,$$

$$\cdots\cdots\cdots\cdots\cdots\cdots\cdots\cdots\cdots\cdots\cdots\cdots$$

$$\tilde{f}^{(A_{0p_0})} = \tilde{\mathbf{z}}^{A_{0p_0}} - \sum_{\substack{|B|=\nu \\ B\notin u(I)}} (f^{(A_{0p_0})})_B \, \tilde{\mathbf{z}}^B,$$

and the quotient of $\mathbb{C}\{\mathbf{w}\}[\tilde{\mathbf{z}}]_\nu$ by the submodule $\langle \tilde{f}^{(A_{01})}, \dots, \tilde{f}^{(A_{0p_0})} \rangle$ spanned by those vectors is a free $\mathbb{C}\{\mathbf{w}\}$-module of rank

$$\binom{\nu + c - 1}{\nu} - p_0.$$

Now look at the $(g_1, \dots, g_s)$. By Proposition 2.3.5, $\nu_{\mathbf{z}}(g_i) \geq \nu$ for all $i = 1, \dots, s$, so there must exist some g_i of order ν. Because we are dealing with a basis, there exist p_0 power series among the g_i, say $\{g_1, \dots, g_{p_0}\}$, such that $\langle \tilde{g}_1, \dots, \tilde{g}_{p_0} \rangle = \langle \tilde{f}^{(A_1)}, \dots, \tilde{f}^{(A_{p_0})} \rangle$, so there is an invertible matrix $M_0(\mathbf{w})$ with entries in $\mathbb{C}\{\mathbf{w}\}$ such that

$$(\tilde{f}^{(A_1)}, \dots, \tilde{f}^{(A_{p_0})}) = (\tilde{g}_1, \dots, \tilde{g}_{p_0}) M_0(\mathbf{w}).$$

Write

$$(f_1, \dots, f_{p_0}) = (g_1, \dots, g_{p_0}) M_0(\mathbf{0});$$

then let $(M_0(\mathbf{0})) \neq 0$ and we have

$$(f_1, \dots, f_{p_0}) = (g_1, \dots, g_{p_0}) M_0(\mathbf{0}),$$
$$u(f_j) = A_j \quad \text{for } 1 \leq j \leq p_0.$$

We also have

$$\tilde{f}_j = \tilde{f}^{(A_j)} \quad \text{for } 1 \leq j \leq p_0.$$

Moreover, we can write

$$(g_1, \dots, g_{p_0}) = (f^{(A_1)}, \dots, f^{(A_{p_0})}) N_0(\mathbf{z}, \mathbf{w}),$$

where $N_0(\mathbf{z}, \mathbf{w})$ is a matrix with entries in R. Since expressions with respect to a free basis are unique, we get $N_0(\mathbf{0}, \mathbf{w}) = M_0(\mathbf{w})^{-1}$, so let $(N_0(\mathbf{0}, \mathbf{0})) \neq 0$ and, again, $N_0(\mathbf{z}, \mathbf{w})$ is invertible. We then have

$$(f_1, \dots, f_{p_0}) = (f^{(A_1)}, \dots, f^{(A_{p_0})}) N_0(\mathbf{z}, \mathbf{w}) M_0(\mathbf{0})$$

and $N_0(\mathbf{z}, \mathbf{w})$. $M_0(\mathbf{0})$ is again invertible. Moreover, all the remaining g_i of order ν (if any) have initial forms belonging to $\langle g_1, \dots, g_{p_0} \rangle$.

Now we go to order $\nu + \mu$, where $\mu \geq 1$, and make the following induction assumptions. We have a subset

$$\{g_1, \dots, g_{p_0}, g_{p_0+1}, \dots, g_{p_{\mu-1}}\}$$

of $(g_1, \dots, g_s)$, and we have found elements

$$\{f_1, \dots, f_{p_0}, f_{p_0+1}, \dots, f_{p_{\mu-1}}\}$$

in I satisfying the following conditions:

(1) For all $j = 1, \dots, p_{\mu-1}$, $\nu(g_j) = \nu(f_j) = \nu(f^{(A_j)})$, where $A_1, \dots, A_{p_{\mu-1}}$ are exactly those vertices A_j in $u(I)$ for which $|A_j| \leq \nu + \mu - 1$, and $u(f_j) = A_j$ for $j = 1, \dots, p_{\mu-1}$.

(2) We have $$(f^{(A_1)}, \dots, f^{(A_{p_{\mu-1}})}) = (g_1, \dots, g_{p_{\mu-1}}) \begin{pmatrix} M_0(\mathbf{w}) & & Q \\ & \ddots & \\ 0 & & M_{\mu-1}(\mathbf{w}) \end{pmatrix},$$

where $M_0(\mathbf{w}), \dots, M_{\mu-1}(\mathbf{w})$ are invertible matrices with entries in $\mathbb{C}\{\mathbf{w}\}$, the entries below the diagonal are zero and the ones above the diagonal are homogeneous polynomials in $\mathbb{C}\{\mathbf{w}\}[\tilde{\mathbf{z}}]$. We denote this matrix by $M'_{\mu-1}(\mathbf{z}, \tilde{\mathbf{w}})$.

(3) $(\tilde{f}_1, \dots, \tilde{f}_{p_{\mu-1}}) = (\tilde{g}_1, \dots, \tilde{g}_{p_{\mu-1}}) M'_{\mu-1}(\tilde{\mathbf{z}}, \mathbf{0})$ and $\tilde{f}_j = \tilde{f}^{(A_j)}$ for $1 \leq j \leq p_{\mu-1}$.

(4) $(f_1, \dots, f_{p_{\mu-1}}) = (f^{(A_j)}, \dots, f^{(A_{p_{\mu-1}})}) N'_{\mu-1}(\mathbf{z}, \mathbf{w})$, where $N'_{\mu-1}(\mathbf{z}, \mathbf{w})$ is an invertible matrix with entries in R.

(5) All the remaining g_i of order $\leq \nu + \mu - 1$, if any, satisfy

$$\tilde{g}_i \in \left\langle \tilde{\mathbf{z}}^A \tilde{g}_j \mid \nu_{\mathbf{z}}(g_j) \leq \nu_{\mathbf{z}}(g_i) \text{ for } 1 \leq j \leq \mu - 1,\ |A| = \nu_{\mathbf{z}}(g_i) - \nu_{\mathbf{z}}(g_j) \right\rangle$$

in $\mathbb{C}\{\mathbf{w}\}[\tilde{\mathbf{z}}]_{\nu_{\mathbf{z}}(g_i)}$.

We now consider the submodule $S_{\mu+\nu}$ of $\mathbb{C}\{\mathbf{w}\}[\tilde{\mathbf{z}}]_{\mu+\nu}$ spanned by the vectors

$$\{\tilde{\mathbf{z}}^A \tilde{f}^{(A_j)} \mid 1 \leq j \leq p_{\nu-1},\ |A| = \mu + \nu - |A_j|\}.$$

A basis of this submodule is given by the vectors

$$\tilde{\mathbf{z}}^{A'_{p_{\mu-1}+1}} - \sum_{\substack{|B|=\nu+\mu \\ B \notin u(I)}} \gamma_{A'_{p_{\mu-1}+1}, B} \tilde{\mathbf{z}}^B,$$

$$\tilde{\mathbf{z}}^{A'_{p'_\mu}} - \sum_{\substack{|B|=\nu+\mu \\ B\notin u(I)}} \gamma_{A'_{p'_\mu},B}\tilde{\mathbf{z}}^B,$$

that are zero modulo $\mathrm{in}_P(I)$, where $\gamma_{A'_j,B} \in \mathbb{C}\{\mathbf{w}\}$ and $A'_{p_{\mu-1}+1}, \ldots, A'_{p'_\mu}$ are exactly the points in $u(I)$ of order $\nu + \mu$ belonging to

$$\bigcup_{j=1}^{p_{\mu-1}} (A_j + \mathbb{Z}_0^c).$$

These vectors are the initial forms with respect to P of elements $h_{p_{\mu-1}+1}, \ldots, h_{p'_\mu} \in I$ respectively, so they can be written as

$$(h_{p_{\mu-1}+1}, \ldots, h_{p_\mu}) = (f^{(A_1)}, \ldots, f^{(A_{p_{\mu-1}})})L_\mu,$$

where L is a $p_{\mu-1} \times (p'_\mu - p_{\mu-1})$-matrix whose entries are homogeneous polynomials in $\mathbb{C}\{\mathbf{w}\}[\mathbf{z}]$. It could happen that $u(I)$ has no vertex of order $\nu + \mu$. If this is so, we jump to degree $\nu + \mu - 1$ without adding further f_j's.

We assume that $u(I)$ has vertices of order $\nu + \mu$ and call them $A_{p_{\mu-1}+1}, \ldots, A_{p_\mu}$. Since $\tilde{f}^{(A_{p_\mu+1})}, \ldots, \tilde{f}^{(A_{p_\mu})}$ are linearly independent modulo $S_{\nu+\mu} = \langle \tilde{h}_{p_{\mu-1}+1}, \ldots, \tilde{h}_{p'_\mu}\rangle$, and because of conditions (5) and (2) above, there must exist $g_{p_{\mu-1}+1}, \ldots, g_{p_\mu}$ such that

$$\begin{aligned}(\tilde{f}^{(A_1)}, \ldots, \tilde{f}^{(A_{p_{\mu-1}})}, \tilde{f}^{(A_{p_{\mu-1}+1})}, \ldots, \tilde{f}^{(A_{p_\mu})})\\ = (\tilde{g}_1, \ldots, \tilde{g}_{p_\mu 1}\tilde{g}_{p_{\mu-1}+1}, \ldots, \tilde{g}_{p_\mu})M'_\mu(\mathbf{z}, \mathbf{w}),\end{aligned}$$

where the matrix $M'_\mu(\mathbf{z}, \mathbf{w})$ is as in (2). If we let

$$(f_1, \ldots, f_{p_{\mu-1}}, f_{p_{\mu-1}+1}, \ldots, f_{p_\mu}) = (g_1, \ldots, g_{p_\mu})M'_\mu(\mathbf{z}, \mathbf{0}),$$

we have constructed $(f_{p_{\mu-1}+1}, \ldots, f_{p_\mu})$ with similar conditions as before. This process must have an end since the number of vertices of $u(I)$ is finite.

Corollary 2.3.5 *Assume that* $\mathrm{gr}_{(P/I)}(R/I)$ *is a free* (R/I)*-module and let* $(g_1, \ldots, g_s)$ *be any basis of* I*. Then there exists an equimultiple standard basis* $(f_1, \ldots, f_r)$ *of* I *whose elements are linear combinations of the* g_i *with coefficients in* $\mathbb{C}[\mathbf{z}]$ *and such that* $\{u(f_1), \ldots, u(f_r)\}$ *is the set of vertices of* $u(I)$.

2.4 Normal Flatness and Hilbert–Samuel Functions

In the preceding section we dealt with the flatness of a graded ring at a point; we now change the point. This section is devoted to explaining what we mean by "changing the point" and what problems arise when we do so.

Let X be a $\mathbb{C}$-space, let $x \in X$ be a point, and let W be a closed $\mathbb{C}$-subspace of X containing x.

Definition 2.4.1 We say that X is **normally flat** along W at x if x is a smooth point of W and, if P denotes the prime ideal of $\mathcal{O}_{X,x}$ defining the germ of W at x, then $\mathrm{gr}_P(\mathcal{O}_{X,x})$ is a flat $(\mathcal{O}_{X,x}/P)$-module.

Remark 2.4.1 Consider a local $\mathbb{C}$-element of X centered at x; that is, we take a smooth $\mathbb{C}$-space Z containing X as a closed subspace around x. By assuming that x is a smooth point of W, we can take a local coordinate system $(\mathbf{z}, \mathbf{w}) = (z_1, \dots, z_c, w_1, \dots, w_d)$ on Z centered at x and associated with W, that is, such that W is defined, locally around x, by the ideal $P = (\mathbf{z}) \cdot \mathbb{C}\{\mathbf{z}, \mathbf{w}\}$. We write $R = \mathbb{C}\{\mathbf{z}, \mathbf{w}\}$ and $M = (\mathbf{z}, \mathbf{w})R$, and denote by $I \subset P$ the ideal defining X in Z locally around x. We also keep all the notation of the preceding section. In this case, the algebraic translation of Definition 2.4.1 is the following: X is **normally flat** along W at x if $\mathrm{gr}_{(P/I)}(R/I)$ is a free $\mathbb{C}\{\mathbf{w}\}$-module. With this interpretation, all the algebraic machinery of Sect. 2.3 applies here without change.

Here is the first major result of this section.

Theorem 2.4.1 *Let X be a $\mathbb{C}$-space, and let W be a closed $\mathbb{C}$-subspace of X. There exists a closed $\mathbb{C}$-subspace $W' \subset W$, with $W \setminus W'$ dense in W, such that the following conditions are equivalent for $x \in W$:*

(1) *X is normally flat along W at x.*
(2) $x \notin W'$.

Proof We assume that W has smooth points, for otherwise there is nothing to prove. We choose a smooth point $x_0 \in W$, and take a local $\mathbb{C}$-element of X centered at x_0, that is, a smooth $\mathbb{C}$-space Z containing X as a closed subspace around x_0, and a local coordinate system $(\mathbf{z}, \mathbf{w})$ on Z centered at x_0 and associated with W. By passing to germs, we have the ring $R = \mathbb{C}\{\mathbf{z}, \mathbf{w}\}$, and the germs of W and X will be defined by respective ideals $P = (\mathbf{z}) \cdot R$ and $I \subset P$. Choose a basis $(f_1, \dots, f_r)$ of I and a polydisk $\Delta \subset \mathbb{C}^{c+d}$ centered at $\mathbf{0}$ such that all the f_i converge on Δ; then X is defined within Δ by the vanishing of the functions that correspond to $f_1, \dots, f_r$. The theorem will be proved if we show the following:

Remark 2.4.2 (Assertion) There is a finite set of elements $g_1(\mathbf{w}), \dots, g_p(\mathbf{w}) \in \mathbb{C}\{\mathbf{w}\}$, convergent on the polydisk Δ' which is the projection of Δ onto $\mathbb{C}^d$, and such that, if W'_Δ is the subset of $W \cap \Delta$ defined by the vanishing of the corresponding functions, then, for every $y \in W \cap \Delta$ the following conditions are equivalent:

(a) The space X is normally flat along W at x.
(b) The point $y \notin W'_\Delta$.

To prove this assertion, it is clearly enough to show the following:

Remark 2.4.3 (Assertion) For every point $x \in W \cap \Delta$ such that X is normally flat along W at x, there exists a finite set of elements $v_1(\mathbf{w}), \dots v_s(\mathbf{w}) \in \mathbb{C}\{\mathbf{w}\}$, convergent on Δ', such that, if we denote by W'' the set of points in $W \cap \Delta$ at which the product of the functions corresponding to the $v_i(\mathbf{w})$ vanishes, then the point with coordinates $\mathbf{w}$ does not belong to W'' and X is normally flat along W at any point $y \in (W \cap \Delta) \setminus W''$.

To prove the second assertion, let $\boldsymbol{\alpha} = (\alpha_1, \dots, \alpha_d) \in \mathbb{C}^d$ be such that $x = (\mathbf{0}, \boldsymbol{\alpha})$, consider new indeterminates $\mathbf{w}' = (w'_1, \dots, w'_d)$ and write $w_i = w'_i + \alpha_i$ for $1 \le i \le d$, or $\mathbf{w} = \mathbf{w}' + \boldsymbol{\alpha}$ for short. Then $\mathcal{O}_{Z,x}$ and the germ of X at x are defined, respectively, by the ring $R' = \mathbb{C}\{\mathbf{z}, \mathbf{w}'\}$ and the ideal $I' = (f_i(\mathbf{z}, \mathbf{w}'+\boldsymbol{\alpha}), \dots, f_r(\mathbf{z}, \mathbf{w}'+\boldsymbol{\alpha}))$. Write $f'_i(\mathbf{z}, \mathbf{w}') = f_i(\mathbf{z}, \mathbf{w}'+\boldsymbol{\alpha})$ for $1 \le i \le r$ and $P' = (\mathbf{z}) \cdot R'$. By assumption, X is normally flat along W at x, so $\mathrm{gr}_{(P/I')}(R'/I')$ is (R'/P')-free. By Corollary 2.3.5 there exists an equimultiple standard basis $\{g'_j(\mathbf{z}, \mathbf{w}')\}_{j=1,\dots,s}$ of I' such that

$$g'_j(\mathbf{z}, \mathbf{w}') = \sum_{i=1}^{r} h_{ij} f'_i(\mathbf{z}, \mathbf{w}'),$$

with $h_{ji} \in \mathbb{C}[\mathbf{z}]$. Moreover, if $u(g'_j) = A_j$, then $\{A_1, \dots, A_s\}$ is the set of vertices of $u(I')$. This implies in particular that, if we write as usual

$$g'_j(\mathbf{z}, \mathbf{w}') = \sum_{A \in \mathbb{Z}_0^c} (g'_j)_A \mathbf{z}^A, \quad \text{with } (g'_j)_A \in \mathbb{C}\{\mathbf{w}'\},$$

then the $(g'_j)_A$ converge on the polydisk Δ'. Write $v'_j(\mathbf{w}') = (g'_j)_{A_j}$ and $v_j(\mathbf{w}) = v'_j(\mathbf{w} - \boldsymbol{\alpha})$ for $j = 1, \dots, s$, and take W'' as in the second assertion. Then $x \notin W''$, since the $v'_j(\mathbf{w}')$ are units. It remains to show that X is normally flat along W at any point $y \in (W \cap \Delta) \setminus W''$.

Take such a y, write $y = (\mathbf{0}, \boldsymbol{\beta})$, consider the variable $\mathbf{w}''$ and write $\mathbf{w}' = \mathbf{w}'' + \boldsymbol{\beta}$; then $\mathbf{w} = \mathbf{w}'' + \boldsymbol{\alpha} + \boldsymbol{\beta}$ and all the $v'_j(\mathbf{w}'' + \boldsymbol{\beta})$, for $j = 1, \dots, s$, are units in $\mathbb{C}\{\mathbf{w}''\}$. By dividing each g'_j by $v'_j(\mathbf{w}')$—which we can do within $(W \cap \Delta) \setminus W''$—and taking suitable linear combinations with coefficients of the form $\gamma \mathbf{z}^A$, for $\gamma \in \mathbb{C}\{\mathbf{w}'\}$ convergent on $(W \cap \Delta) \setminus W''$, we may assume that

$$g'_j = \mathbf{z}^{A_j} + \sum_{\substack{|B|=|A_j| \\ B \notin u(I')}} (g'_j)_B \mathbf{z}^B.$$

Consider $g''_j(\mathbf{z}, \mathbf{w}'') = g'_j(\mathbf{z}, \mathbf{w}''+\boldsymbol{\beta})$ for $1 \leq j \leq s$; then

$$g''_j = \mathbf{z}^{A_j} + \sum_{\substack{|B|=|A_j| \\ B\notin u(I')}} (g''_j)_B \mathbf{z}^B,$$

where $(g''_j)_B(\mathbf{w}'') = (g'_j)_B(\mathbf{w}'' + \boldsymbol{\beta})$.

Fix a degree, say ν, in $\mathbb{C}\{\mathbf{w}''\}[\mathbf{z}]$. We have a set of vectors in $\mathbb{C}\{\mathbf{w}''\}[\mathbf{z}]_\nu$ of the form

$$\left\{\mathbf{z}^A + \sum_{\substack{|B|=\nu \\ B\notin u(I')}} \Lambda_{A,B} \cdot \mathbf{z}^B \;\middle|\; A \in u(I'),\ |A| = \nu\right\},$$

which are obtained by suitable linear combinations of the g''_j with coefficients of the form $\gamma \cdot \mathbf{z}^A$ with $\gamma \in \mathbb{C}\{\mathbf{w}''\}$ convergent on $(W \cap \Delta) \setminus W''$. Then everything will be proved if we show that there are no non-trivial relations (modulo $\mathrm{in}_{\mathbf{z}}(f_1(\mathbf{z}, \mathbf{w}''+\boldsymbol{\alpha}+\boldsymbol{\beta}), \ldots, f_r(\mathbf{z}, \mathbf{w}''+\boldsymbol{\alpha}+\boldsymbol{\beta})) = \mathrm{in}_{\mathbf{z}}(I'')$ among the $\mathbf{z}^B$, $|B| = \nu$, $B \notin u(I')$.

Assume to the contrary that

$$\Lambda = \sum_{\substack{|B|=\nu \\ B\notin u(I')}} \Lambda_B \tilde{\mathbf{z}}^B \in \mathrm{in}_{\mathbf{z}}(I'')$$

is such a non-trivial relation. This means that Λ is the initial form of some power series of the type

$$\sum_{i=1}^{r} h_i(\mathbf{z}, \mathbf{w}'') f_i(\mathbf{z}, \mathbf{w}''+\boldsymbol{\alpha}+\boldsymbol{\beta}) = \sum_{i=1}^{r} h_i(\mathbf{z}, \mathbf{w}'') f_i(\mathbf{z}, \mathbf{w}''+\boldsymbol{\beta}). \tag{2.3}$$

Suppose, for a moment, that the h_i have indeterminate coefficients (of the monomials $\mathbf{z}^{A'}$) up to degree ν. Let S be the subring of $\mathbb{C}\{\mathbf{w}''\}$ consisting of the Taylor expansions, at y, of functions defined in $(W \cap \Delta) \setminus W''$. Then the existence of a relation such as (2.3) means that a certain system of linear homogeneous equalities and inequalities, corresponding to monomials $\mathbf{z}^{A'}$ that occur in the initial form of (2.3) or that are canceled in degrees $\leq \nu$ in the indeterminate coefficients of the h_i, **with coefficients in** S, has an actual solution in $\mathbb{C}\{\mathbf{w}''\}$ that is non-trivial. But, if this is so, it has a solution in S, which means the existence of a non-trivial relation in $\mathbb{C}\{\mathbf{w}'\}[\mathbf{z}]$ among the $\mathbf{z}^B$, $|B| = \nu$, $B \notin u(I')$. This is impossible. □

Corollary 2.4.1 *With the notation of Definition* 2.4.1, *if X is normally flat along W at x then the rank of every homogeneous component of* $\mathrm{gr}_P(\mathcal{O}_{X,x})$, *as a free $(\mathcal{O}_{X,x}/P)$-module, is locally constant.*

We now go deeper into the geometric meaning of normal flatness.

Remark 2.4.4 Consider:

$$W \hookrightarrow Y \hookrightarrow X \hookrightarrow Z,$$

where Z is a smooth $\mathbb{C}$-space and W, Y, X are closed $\mathbb{C}$-subspaces of Z with W, Y smooth. Then we have a diagram

$$\begin{array}{ccccc} T_Y(W) & \longrightarrow & T_Z(W) & \longrightarrow & T_{Z,Y}|_W \\ \Vert & & \uparrow & & \uparrow \\ T_Y(W) & \xrightarrow{\alpha} & C_X(W) & \xrightarrow{\beta} & C_{X,Y}|_W \end{array} \tag{2.4}$$

whose horizontal lines are the fundamental tangent-normal sequences (see Definition 1.6.5 in Chap. 1) associated with $W \subset Y \subset Z$ and $W \subset Y \subset X$, and whose vertical lines correspond to the inclusion $X \subset Z$. The top line is an exact sequence of vector bundles. Take a retraction $r : Z \to Y$ of the inclusion and consider its restriction to X and the morphism

$$\varphi_r : C_X(W) \to T_Y(W) \times C_{X,Y}|_W,$$

which is the product of the tangent map $dr : C_X(W) \to T_Y(W)$ and β. This φ_r corresponds to a morphism of $\mathcal{O}_W$-algebras

$$\tan_X(W) \to \tan_Y(W) \otimes_{\mathcal{O}_W} i_W^*(\mathrm{gr}_Y(X)),$$

where $i_W : W \to Y$ is the inclusion. In other words, if we use the isomorphism

$$\tan_X(W) \simeq \mathrm{gr}_{\Delta W}(W \times X)$$

of Example 1.6.3 in Chap. 1, this morphism of $\mathcal{O}_W$-algebras is written as

$$\Psi_r : \mathrm{gr}_{\Delta W}(W \times X) \to \mathrm{gr}_{\Delta W}(W \times Y) \otimes_{\mathcal{O}_Y} \mathrm{gr}_Y(X). \tag{2.5}$$

For example, if $X = Z$ by restriction to a local $\mathbb{C}$-element $\mathbf{u}$ of Z having a local coordinate system $(\mathbf{w}, \mathbf{y}, \mathbf{z})$, where $\mathbf{w} = (w_1, \ldots, w_d)$ is a coordinate system on $\mathbf{u} \cap W$ and $(\mathbf{w}, \mathbf{y}) = (w_1, \ldots, w_c, y_1, \ldots, y_s)$ is a coordinate system on $\mathbf{u} \cap y$, then Ψ_r is the morphism

$$\Psi_r : \mathcal{O}_W[\bar{\mathbf{y}}, \bar{\mathbf{z}}] \to \mathcal{O}_W[\bar{\mathbf{y}}] \otimes_{\mathcal{O}_Y} \mathcal{O}_Y[\bar{\mathbf{z}}]$$

given by

$$\Psi_r(\bar{y}_i) = \bar{y}_i \otimes 1, \quad \Psi_r(\bar{z}_j) = 1 \otimes \bar{z}_j,$$

which is clearly an isomorphism.

Now, we shall say that $(\boldsymbol{\alpha}, \boldsymbol{\beta})$ **is an exact sequence** if the following conditions hold:

(1) The cone $C_X(W)$ is invariant by the action of $T_Y(W)$ on $T_Z(W)$ by translation. In other words, $T_Y(W)$ acts by translation on $C_X(W)$.
(2) The isomorphism $T_Z(W)/T_Y(W) \to T_{Z,Y}|_W$ given by the exactness of the top row in (1) induces an isomorphism $C_X(W)/T_Y(W) \to C_{X,Y}(W)$.

If we fix a retraction $r : Z \to Y$ as above, these two conditions are equivalent to Ψ_r in (2.5) being an isomorphism. Moreover, if we consider the diagram induced by the inclusion $X \subset Z$,

$$\begin{array}{ccccccc}
0 \longrightarrow \ker b & \longrightarrow & \mathrm{gr}_{\Delta W}(W\times Z) & \xrightarrow{\;b\;} & \mathrm{gr}_{\Delta W}(W\times X) & \longrightarrow & 0 \\
 & & \uparrow & & \uparrow & & \\
\mathrm{gr}_{\Delta W}(W\times Y)\otimes_{\mathcal{O}_Y}\mathrm{in}_Y(X) & \longrightarrow & \mathrm{gr}_{\Delta W}(W\times Y)\otimes_{\mathcal{O}_Y}\mathrm{gr}_Y(X) & \longrightarrow & \mathrm{gr}_{\Delta W}(W\times Y)\otimes_{\mathcal{O}_{\mathscr{Y}}}\mathrm{gr}_Y(X) & \longrightarrow & 0
\end{array} \tag{2.6}$$

the condition that Ψ_r is an isomorphism is equivalent to saying that the image of $\mathrm{gr}_{\Delta W}(W \times Y) \otimes_{\mathcal{O}_Y} \mathrm{in}_Y(X)$ in $\mathrm{gr}_{\Delta W}(X \times Z)$ generates $\ker b$.

In particular, if we assume that $W = \{x\}$ is a single point, (2.4) becomes the following diagram of vector spaces and cones:

$$\begin{array}{ccccccccc}
0 & \longrightarrow & T_{Y,x} & \longrightarrow & T_{Z,x} & \longrightarrow & C_{Z,Y,x} & \longrightarrow & 0 \\
 & & \| & & \uparrow & & \uparrow & & \\
 & & T_{Y,x} & \xrightarrow{\;\alpha'\;} & C_{X,x} & \xrightarrow{\;\beta'\;} & C_{X,Y,x} & &
\end{array} \tag{2.7}$$

In this case, the exactness if (α', β') means the following:

(1) The vector space $T_{Y,x}$ acts on $C_{X,x}$ by translation.
(2) The isomorphism of vector spaces $T_{Z,x}/T_{Y,x} \simeq C_{Z,Y,x}$ induces an isomorphism $C_{X,x}/T_{Y,x} \to C_{X,Y,x}$.

On the other hand, fix a retraction $r : Z \to Y$; then the top row of (2.7) splits by dr and we have an isomorphism

$$T_{Z,x} \simeq T_{Y,x} \times C_{Z,Y,x}.$$

This isomorphism induces another isomorphism

$$C_{X,x} \to T_{Y,x} \times C_{X,Y,x} \tag{2.8}$$

if and only if (α', β') is exact. Set $R = \mathcal{O}_{Z,x}$, let M be the maximal ideal of R, let P and I, respectively, be the ideals of Y and X in R, and take a local coordinate system $(\mathbf{z}, \mathbf{w})$ associated with r. Then (2.8) translates to an isomorphism

$$\Psi_r : \mathbb{C}[\bar{\mathbf{z}}, \bar{\mathbf{w}}]/\mathrm{in}_M(I) \to \mathrm{gr}_{(P/I)}(R/I) \otimes_{\mathbb{C}\{\mathbf{w}\}} \mathbb{C}[\bar{\mathbf{w}}].$$

Also, the exactness of (α', β') means that the image of $\mathrm{in}_P(I) \otimes_{\mathbb{C}\{\mathbf{w}\}} \mathbb{C}[\bar{\mathbf{w}}]$ in $\mathbb{C}[\bar{\mathbf{z}}, \bar{\mathbf{w}}]$ generates $\mathrm{in}_M(I)$.

Definition 2.4.2 As before, consider $W \hookrightarrow Y \hookrightarrow X \hookrightarrow Z$, and take a point $x \in W$. We say that X is **W-normally flat along Y at x** if the sequence (α, β) is exact in a neighborhood of x. We say that X is **normally flat along Y at x** if (α', β') is an exact sequence. Finally, we shall say that X is **normally flat** along Y if and only in X is normally flat along Y at x, for any $x \in Y$.

We now prove that this definition is only a new version of the old concept of normal flatness given before.

Proposition 2.4.1 *With the notation of Definition* 2.4.2, *the following conditions are equivalent:*

(1) *The space X is normally flat along Y at x.*
(2) *The $\mathcal{O}_{Y,x}$-module $\mathrm{gr}_Y(X)_x$ is flat.*
(3) *The $\mathcal{O}_{Y,x}$-module $\mathscr{P}_X^{(\nu)}(Y)_x$ is flat for every $\nu \geq 0$. (See Example* 1.6.3 *in Chap.* 1.*)*
(4) *The $\mathcal{O}_{Y,x}$-module $\tan_Y(X)_x$ is flat.*

Proof (1) $\Longleftrightarrow$ (2) X is normally flat along Y at x if and only if $\mathrm{in}_P(I)$ generates $\mathrm{in}_M(I)$ in $\mathrm{gr}_M(R)$. This is equivalent to the existence of a basis $I = (f_1, \ldots, f_s) \cdot \mathcal{O}_{Z,x}$ satisfying these conditions:

(1) for $1 \leq i \leq s$, $\nu_P(f_i) = \nu_M(f_i)$ and
(2) it is $\mathrm{in}_M(I) = (\bar{f}_1, \ldots, \bar{f}_s) \cdot \mathrm{gr}_M(R)$;

in other words, it is equivalent to the existence of an equimultiple standard basis for I. By Theorem 2.3.1 this is equivalent to the flatness of $\mathrm{gr}_Y(X)_x$ as a $\mathcal{O}_{Y,x}$-module.
(3) $\Longleftrightarrow$ (4). Since

$$\tan_Y(X) = \bigoplus_{\nu \geq 0} \ker(\tau_\nu)$$

(see Example 1.6.3 in Chap. 1), the sequence

$$0 \longrightarrow \tan_X^\nu(Y)_x \longrightarrow \mathscr{P}_X^{(\nu)}(Y)_x \longrightarrow \mathscr{P}_X^{(\nu-1)}(Y)_x \longrightarrow 0$$

is exact, then, for $\nu = 0$,

$$\mathscr{P}_X^{(0)}(Y)_x = \tan_X^0(Y)_x = \mathscr{O}_{Y,x},$$

and inductively, the flatness of $\tan_X^\nu(Y)_x$ for every $\nu \geq 0$ is equivalent to the flatness of $\mathscr{P}_X^\nu(Y)_x$.
(2) $\iff$ (4) If we let $W = Y$ in (2) we get a morphism

$$\Psi_r : \mathrm{gr}_{\Delta Y}(Y \times X) \to \mathrm{gr}_{\Delta Y}(Y \times Y) \otimes_{\mathscr{O}_Y} \mathrm{gr}_X(X),$$

which is an isomorphism of $\mathscr{O}_Y$-algebras, since the flatness of $\mathrm{gr}_{\Delta Y}(Y \times Y)$ over $\mathscr{O}_Y$ makes the bottom row of (3) exact. By taking into account Example 1.6.3 of Chap. 1, we can write this isomorphism as

$$\Psi_r : \tan_X(Y) \simeq \tan_Y \otimes_{\mathscr{O}_Y} \mathrm{gr}_Y(X).$$

If we consider a local coordinate system $(\mathbf{z}, \mathbf{y})$ centered at x and associated with the retraction, this Ψ_r becomes, at x,

$$\Psi_r : \tan_X(Y)_x \simeq \mathscr{O}_{Y,x}[\mathbf{y}] \otimes_{\mathscr{O}_{Y,x}} \mathrm{gr}_Y(X)_x,$$

and, since $\mathscr{O}_{Y,x}[\mathbf{y}]$ is $\mathscr{O}_{Y,x}$-free, $\tan_X(Y)_x$ is flat if and only if $\mathrm{gr}_Y(X)_x$ is flat. □

Definition 2.4.3 Let X be a $\mathbb{C}$-space and $x \in X$ a point. The **Hilbert–Samuel function** $H_{X,x}$ of X at x is the map $H_{X,x} : \mathbb{Z}_0 \to \mathbb{Z}_0$ given by

$$H_{X,x}(d) = \sum_{i=0}^{d} \mathrm{rank}_{\mathbb{C}}\, \mathrm{gr}_x^i(X).$$

In other words, if M is the maximal ideal of $\mathscr{O}_{X,x}$,

$$H_{X,x}(d) = \mathrm{rank}_{\mathbb{C}}(\mathscr{O}_{X,x}/M^{d+1}) = \mathrm{rank}_{\mathbb{C}}(\mathscr{P}_X^{(d)}(x)).$$

Proposition 2.4.2 *Let X be a $\mathbb{C}$-space, and Y a smooth closed connected $\mathbb{C}$-subspace. Then X is normally flat along Y if and only if $H_{X,x}$ is independent of the point $x \in Y$.*

Proof Normal flatness is equivalent to the flatness of $\mathrm{gr}_Y(X)_x$ for every $x \in Y$. The result then follows from Theorem 2.4.1 and the fact that Y is smooth and connected. □

Remark 2.4.5 In the hypersurface case, the Hilbert–Samuel function is completely determined by the multiplicity of the point. Therefore, by Proposition 2.4.2, normal flatness is equivalent to equimultiplicity along Y.

We conclude this section with the concept of Samuel stratification.

Remark 2.4.6 As an immediate consequence of Theorem 2.4.1 and the characterization of normal flatness, one has the following result: If X is a $\mathbb{C}$-space and Y a reduced closed $\mathbb{C}$-subspace of X, there exists an open dense subset Y' of Y such that X is normally flat along Y at any point of Y'. This result is refined in the following two lemmas:

Lemma 2.4.1 *Let X be a $\mathbb{C}$-space and Y a reduced closed $\mathbb{C}$-subspace of X. Then, for every point $y_0 \in Y$, there exists a neighborhood U_0 of y_0 in Y such that the set*

$$\mathscr{H}_0 = \{H : \mathbb{Z}_0 \to \mathbb{Z}_0 \mid H = H_{X,y} \textit{ for some } y \in U_0\}$$

is finite.

Proof We work by induction on $\dim Y$. If $\dim Y = 0$, the result is clear. Assume $\dim Y > 0$ and the result true in smaller dimensions. Let U be the dense open subset of Y (see Theorem 2.4.1) such that X is normally flat along Y at every point in U. If $y_0 \in U$, there exists an open subset $U_0 \subset U$ such that $\mathscr{H}_0$ has a single element, namely H_{X,y_0}. So assume $y_0 \notin U$. Then, since the complement of y in Y is a closed $\mathbb{C}$-subspace $Y_1 \subset Y$, one has $y_0 \in Y_1$ and $\dim Y_1 < \dim Y$. Then there exists a neighborhood U_0 of y_0 in Y that has a non-empty intersection with only a finite number of connected components of U, and such that (by the induction assumption) there is only a finite number of Hilbert–Samuel functions within $U_0 \cap Y_1$. This proves our lemma. □

For the next lemma, we consider the set of functions $\mathbb{Z}_0 \to \mathbb{Z}_0$ as an ordered set, by taking the product ordering, namely

$$H \geq H' \iff H(d) \geq H'(d) \text{ for all } d \in \mathbb{Z}_0.$$

Lemma 2.4.2 *Let X be a $\mathbb{C}$-space and $H : \mathbb{Z}_0 \to Z_0$ a given function. Then the set*

$$C_H = \{x \in X \mid H_{X,x} \geq H\}$$

is a closed $\mathbb{C}$-subspace of X.

This result is usually known as the **upper semicontinuity of the Hilbert–Samuel function**.

Proof The assertion that C_H is closed is of a global nature, whereas the fact that it has analytic equations is local. We start with the latter.

If we apply Lemma 2.4.1 to $Y = Y_{\text{red}}$ at $x \in X$, we find a neighborhood U of x in X such that

$$\{H_{X,y} \mid y \in U\} = \{H_1, \ldots, H_s\}.$$

Assume that $H_i \geq H$ for $1 \leq i \leq s'$ and $H_i \not\geq H$ for $s'+1 \leq i \leq s$. Then, for $s'+1 \leq i \leq s$, there exists d_i such that $H_i(d_i) < H(d_i)$. Then

$$C_H \cap U = \{y \in U \mid H_{X,x}(d_i) \geq H(d_i) \text{ for all } i\}$$

and, since $H_{X,x}(d) = \operatorname{rank}_{\mathbb{C}} \mathscr{P}_X^{(\nu)}(y)$ and $\mathscr{P}_X^{(\nu)}$ is coherent, $C_H \cap U$ has a finite number of analytic equations.

To prove that C_H is closed we follow a similar pattern. If $x \notin C_H$ then $H_{X,x} \not\geq H$, so there exists $d \in \mathbb{Z}_0$ such that $H_{X,x}(d) < H(d)$. By the coherence of $\mathscr{P}_X^{\nu}$ this holds for a neighborhood of x, which is, therefore, disjoint from C_H. □

Definition 2.4.4 Let X be a $\mathbb{C}$-space. A **$\mathbb{C}$-stratification** of X is a decomposition of X into a disjoint union

$$X = \bigcup_{\alpha \in \Lambda} X_\alpha$$

with the following properties:

(1) For all $\alpha \in \Lambda$, the closure $\bar{X}_\alpha$ of X_α in X and the frontier $\bar{X}_\alpha \setminus X_\alpha$ are closed $\mathbb{C}$-subspaces of X.
(2) The family $\{X_\alpha\}_{\alpha \in \Lambda}$ is a locally finite family.

The X_α are called the **strata** of the stratification.

Proposition 2.4.3 *Let X be a $\mathbb{C}$-space. There exists a stratification*

$$X = \bigcup_{\alpha \in \Lambda} S_\alpha$$

such that, for every $x, y \in X$, one has $H_{X,x} = H_{X,y}$ if and only if x, y belong to the same stratum.

Proof We take as the index set Λ the set

$$\Lambda = \{H : \mathbb{Z}_0 \to \mathbb{Z}_0 \mid H = H_{X,x} \text{ for some } x \in X\},$$

and define the strata as

$$S_H = \{x \in X \mid H_{X,x} = H\} \quad \text{for } H \in \Lambda.$$

We need to check that the family $\{S_H\}_{H \in \Lambda}$ satisfies the conditions in the definition.

It is clear that $X = \bigcup_{H \in \Lambda} S_H$ and that the S_H are disjoint. Moreover, if $x \in S_H$ then $H_{X,x} \geq H$, so $S_H \subset C_H$. If $H_{X,x} \not\geq H$, by Lemma 2.4.2, there exists a neighborhood of x in X disjoint from S_H. On the other hand, if $x \in S_H$, there is an open neighborhood U of x with only a finite number of Hilbert–Samuel functions $\{H, H_1, \ldots, H_n\}$, where $H_i \geq H$ for $1 \leq i \leq s'$, $H_i \not\geq H$ for $s'+1 \leq i \leq n$, and

$$S_H \cap U = \{x \in C_H \mid H_{X,x} \neq H_i \text{ for } s'+1 \leq i \leq n\} = \left(C_H \setminus \bigcup_{i=s'+1}^{n} C_{H_i}\right) \cap U.$$

Then $\bar{S}_H \setminus S_H$ and $\bar{S}_H$ are closed $\mathbb{C}$-subspaces of X and the local finiteness of $\{S_H\}_{H\in\Lambda}$ is a consequence of the existence of a finite number of Hilbert–Samuel functions in a neighborhood of any point. □

Definition 2.4.5 The stratification in the preceding proposition is called the **Samuel stratification** of X and each S_H is called the **Samuel stratum** corresponding to the function H.

We observe that, in the Samuel stratification, $\bar{S}_H \setminus S_H$ is not necessarily a union of strata, as can be seen in the following example. Let X be the surface in $\mathbb{C}^3$ given by the equation $z_1^3 - z_2^3 z_3^2 = 0$, and denote by l_i the z_i-axis, for $i = 1, 2, 3$. The locus of singular points is $\mathrm{Sing}(X) = l_2 \cup l_3$, and the set of regular points is $\mathrm{Reg}(X) = X \setminus \mathrm{Sing}(X)$. The Samuel stratification of X is

$$X = \mathrm{Reg}(X) \cup l_3 \cup (l_2 \setminus \{\mathbf{0}\}).$$

In this case,

$$\overline{(l_2 \setminus \{\mathbf{0}\})} \setminus (l_2 \setminus \{\mathbf{0}\}) = \{\mathbf{0}\},$$

which is not a union of strata.

2.5 Normal Flatness and Blow-Ups

The aim of this section is to analyze the behavior of the Hilbert–Samuel functions under monoidal transformations. We show that, for an appropriate choice of the centers of the blow-ups, the Hilbert–Samuel function cannot increase (Bennett's theorem [8]), or decrease indefinitely (finiteness theorem).

Remark 2.5.1 (1) Let X be a $\mathbb{C}$-space and $x \in X$ a point. Since the tangent cone $C_{X,x}$ to X at x is also a $\mathbb{C}$-space, it makes sense to consider the Hilbert–Samuel function $H_{C_{X,x},x}$ of $C_{X,x}$ at x. We clearly have

$$H_{C_{X,x},x} = H_{X,x}.$$

(2) Let $H : \mathbb{Z}_0 \to \mathbb{Z}_0$ be a function, and extend to the negative integers with the value 0. Set

$$H^{(-1)}(d) = H(d) - H(d-1) \quad \text{for } d \in \mathbb{Z}_0,$$

$$H^{(+1)}(d) = \sum_{i=0}^{d} H(i),$$

and, inductively, for $e \in \mathbb{Z}$,

$$H^{(e\pm1)} = (H^{(e)})^{(\pm1)}.$$

This defines a collection $\{H^{(m)} \mid m \in \mathbb{Z}\}$ of functions that will be useful later on.

(3) Let $(A, \mathfrak{m})$ be a noetherian local ring, and let $H_A : \mathbb{Z}_0 \to \mathbb{Z}_0$ be the ordinary Hilbert–Samuel function of A with respect to $\mathfrak{m}$, that is, the function defined by

$$H_A(d) = \operatorname{rank}_{(A/\mathfrak{m})}(A/\mathfrak{m}^{d+1}) \quad \text{for } d \in \mathbb{Z}_0.$$

The following properties of H_A are well-known from elementary commutative algebra (see [107]):

(1) The function H_A becomes a polynomial for all values of the variable greater than a given number. This is called the **Hilbert–Samuel polynomial** of A.
(2) It is verified that $H_A^{(-1)}(1) = \operatorname{rank}_{(A/\mathfrak{m})}(\mathfrak{m}/\mathfrak{m}^2) = H_A(1) - 1$.
(3) The degree of the Hilbert–Samuel polynomial of A is equal to the dimension of A. If $\dim A = n$, the leading coefficient of this polynomial is of the form $m/n!$, where m is a positive integer, called the **multiplicity** of the local ring $(A, \mathfrak{m})$.
(4) For any positive integer e, one has $H_A^{(e)} = H_{A_e}$, where

$$A_e = A[[t_1, \ldots, t_e]]$$

and $\{t_1, \ldots, t_e\}$ are variables.

Remark 2.5.2 Now consider a $\mathbb{C}$-space X and a point $x \in X$. By letting $A = \mathcal{O}_{X,x}$ and $\mathfrak{m} = \mathfrak{m}_{X,x}$, the above concepts apply word for word, so we get the **Hilbert–Samuel polynomial** of X at x, the **multiplicity** of X at x, and so on. Particularly interesting is the so-called **Zariski tangent space** to X at x, which is the $\mathbb{C}$-vector space $\mathfrak{m}_{X,x}/\mathfrak{m}_{X,x}^2$. We have

$$\operatorname{rank}_{\mathbb{C}}(\mathfrak{m}_{X,x}/\mathfrak{m}_{X,x}^2) = H_{X,x}(1) - 1.$$

Property (4) above becomes

$$H_{X,x}^{(e)} = H_{X\times\mathbb{C}^e,\,(x,\mathbf{0})} \quad \text{for all } e \in \mathbb{Z}_+.$$

Lemma 2.5.1 *Let $(A, \mathfrak{m})$ be a noetherian local ring, take $z \in \mathfrak{m}$, and write $B = A/zA$. Then*

$$H_B^{(1)} \geq H_A.$$

Proof Write $k = A/\mathfrak{m}$ and, for any integer $d \geq 0$, put $A(d) = A/\mathfrak{m}^{d+1}$ and $B(d) = (A/zA)/(\mathfrak{m}/zA)^{d+1} = A/(\mathfrak{m}^{d+1} + zA)$. Choose an integer $d \geq 0$; by considering the exact sequences of finite-dimensional k-vector spaces

$$0 \longrightarrow z^{i+1}A(d) \longrightarrow z^i A(d) \longrightarrow z^i A(d)/z^{i+1}A(d) \longrightarrow 0$$

we see that

$$H_A(d) = \mathrm{rank}_k\, A(d) = \sum_{i=0}^{d} \mathrm{rank}_k \left(z^i A(d)/z^{i+1}A(d)\right).$$

Multiplication by z^i gives a map $A \to A$ sending $\mathfrak{m}^{d-i+1}$ into $\mathfrak{m}^{d+1}$, so we have two surjective maps

$$A(d-i) \longrightarrow z^i A(d), \qquad zA(d-i) \longrightarrow z^{i+1}A(d),$$

which give rise to a surjective map

$$B(d-i) = A(d-i)/zA(d-i) \longrightarrow z^i A(d)/z^{i+1}A(d).$$

We therefore have

$$\begin{aligned} H_A(d) &= \sum_{i=0}^{d} \mathrm{rank}_k \left(z^i A(d)/z^{i+1}A(d)\right) \\ &\leq \sum_{i=0}^{d} \mathrm{rank}_k(B(d-i)) = \sum_{i=0}^{d} H_B(d-i) = H_B^{(1)}(d). \end{aligned}$$

This proves the lemma. □

The first key result of this section is due to Bennett [8].

Theorem 2.5.1 *Let X be a $\mathbb{C}$-space, Y a smooth closed $\mathbb{C}$-subspace of X, and $x \in Y$ a point. Let $p : X' \to X$ be the blow-up of X with center Y* (see Definition 1.3.2 in Chap. 1). *If X is normally flat along Y at x, then*

$$H_{X',x'} \leq H_{X,x}$$

for every point $x' \in p^{-1}(x)$.

Proof The theorem is a consequence of the following three facts, which we prove below:

(1) $H_{X,x} = H^{(r)}_{C_{X,Y,x}}$ where r is the dimension of Y at x.
(2) $H_{C_{X,Y,x}} \geq H^{(1)}_{p^{-1}(x),x'}$ for any $x' \in p^{-1}(x)$.
(3) $H_{p^{-1}(x),x'} \geq H_{X',x'}$ for any $x' \in p^{-1}(x)$.

We first prove (1). The assumption of normal flatness implies the existence of an isomorphism

$$C_{X,x} \simeq T_{Y,x} \times C_{X,Y,x}$$

and, since $T_{Y,x}$ is a r-dimensional complex vector space, we have

$$H_{X,x} = H_{C_{X,x},x} = H^{(r)}_{C_{X,Y,x}}.$$

Next we prove (3). Let $x' \in p^{-1}(x)$. The question being local, we may take a local $\mathbb{C}$-element of X at x, that is, a smooth $\mathbb{C}$-space Z such that, locally around x, X is a closed $\mathbb{C}$-subspace of Z. We may further choose a local coordinate system $(\mathbf{z}, \mathbf{y}) = (z_1, \ldots, z_s, y_1, \ldots, y_r)$, associated with Y, such that $(z_1, z_2/z_1, \ldots, z_s/z_1, \mathbf{y})$ is a local coordinate system on the monoidal transform Z' of Z with center Y, centered at x' (see Sect. 1.7 in Chap. 1). Now, locally around x', $p^{-1}(x)$ is defined by the ideal $(\mathbf{z}, \mathbf{y}) \cdot \mathscr{O}_{X'x'} = (z_1, \mathbf{y}) \cdot \mathscr{O}_{X',x'}$, which is generated by $r+1$ elements. Then relation (3) is a consequence of Lemma 2.5.1 applied $r+1$ times.

Finally we prove (2). Pick a local coordinate system as above. Then

$$C_{X,Y,x} = \operatorname{Specan}(\mathbb{C}[\mathbf{z}]/I)$$

(see Sect. 1.5 in Chap. 1), where I is a homogeneous ideal. The fiber $p^{-1}(x)$ is the projective space associated with $C_{X,Y,x}$. If we write $u_i = z_i/z_1$ for $i = 2, \ldots, s$, and denote by I' the ideal of $\mathbb{C}[u_2, \ldots, u_s]$ generated by

$$\{f/z_1^d \mid f \in I_d\},$$

where I_d is the homogeneous component of degree of I, then the algebra of functions on the corresponding affine open neighborhood of x' in $p^{-1}(x)$ is just $\mathbb{C}[u_2, \ldots, u_s]/I'$. Set $G = \mathbb{C}[\mathbf{z}]$, $A = \mathbb{C}[u_2, \ldots, u_s]$, denote by G_m the homogeneous component of degree m of G, and write

$$A(m) = A/(u_2, \ldots, u_s)^{m+1}A.$$

Then we have a map $\lambda_m : I_m \to A(m)$ defined by $\lambda_m(f) = f/z_1^m$, which is injective and whose image is contained in $I' \cdot A(m)$. In this case,

$$\begin{aligned}
H_{p^{-1}(x),x'}(m) &= \operatorname{rank}_{\mathbb{C}}[A(m)/I'A(m)] \\
&\leq \operatorname{rank}_{\mathbb{C}}[A(m)/\lambda_m(I_m)] = \operatorname{rank}_{\mathbb{C}} A(m) - \operatorname{rank}_{\mathbb{C}} \lambda_m(I_m) \\
&= \operatorname{rank}_{\mathbb{C}} G_m - \operatorname{rank}_{\mathbb{C}} I_m = \operatorname{rank}_{\mathbb{C}}(G/I)_m = H^{(-1)}_{C_{X,Y,x}}.
\end{aligned}$$

Therefore, $H^{(1)}_{p^{-1}(x),x'} \leq H_{C_{X,Y,x}}$, and this ends our proof. □

The second key result of this section is due to Hilbert:

Theorem 2.5.2 (Local finiteness theorem) *Let*

$$\cdots \longrightarrow X_2 \xrightarrow{p_2} X_1 \xrightarrow{p_1} X_0$$

be an infinite sequence of blow-ups of $\mathbb{C}$*-spaces, and let*

$$\cdots \longrightarrow x_2 \xrightarrow{p_2} x_1 \xrightarrow{p_1} x_0$$

be a sequence of points $x_i \in X_i$ for all $i \geq 0$, such that the center Y_{i-1} of each p_i is a smooth closed $\mathbb{C}$-subspace of X_{i-1} containing x_{i-1}, and such that X_{i-1} is normally flat along Y_{i-1} at x_{i-1}. Then there exists an index $q \geq 0$ such that, for all $i \geq q$, the tangent cone C_{X_i,x_i} is (non-canonically) isomorphic to C_{X_q,x_q}.

Proof Since our statement is clearly local, we may restrict the diagram in the statement of the theorem to a suitable neighborhood of x_0 in X_0. Let, then, Z_0 be a smooth ambient space for X_0 locally around x_0 (so that X_0 is locally a closed $\mathbb{C}$-subspace of Z_0), and consider the successive blow-ups $\pi_i : Z_i \to Z_{i-1}$ with centers Y_{i-1}, for each $i \in \mathbb{Z}_+$.

Now, each Z_i can be considered, in a natural way, as a smooth ambient space for X_i, so we have a diagram of blow-ups

$$\begin{array}{ccccccc}
\cdots \longrightarrow & Z_2 & \xrightarrow{\pi_2} & Z_1 & \xrightarrow{\pi_1} & Z_0 \\
 & \uparrow & & \uparrow & & \uparrow \\
\cdots \longrightarrow & X_2 & \xrightarrow{p_2} & X_1 & \xrightarrow{p_1} & X_0
\end{array}$$

On the other hand, by Theorem 2.5.1, for every $r \in \mathbb{Z}_0$, one has a non-increasing sequence

$$H_{X_0,x_0}(r) \geq H_{X_1,x_1}(r) \geq \cdots \geq H_{X_i,x_i}(r) \geq \cdots$$

of non-negative integers. Such a sequence must be stationary. Therefore, given a fixed $m \in \mathbb{Z}_+$, there exists an integer $i(m) \in \mathbb{Z}_+$ such that, for each $i \geq i(m)$ and each r satisfying $0 \leq r \leq m$, we have

$$H_{X_i,x_i}(r) = H_{X_{i(m)},x_{i(m)}}(r).$$

This amounts to the following. Form the rings $\mathrm{gr}_{x_j}(Z_j)$, for $j \geq 0$, and take in each one the homogeneous ideal

$$I_j = \mathrm{in}_{x_j}(Z_j, X_j)$$

that defines the tangent cone to X_j at x_j. Then, for every $m \in \mathbb{Z}_+$, there exists an integer $i(m)$ such that, if $I_{j,k}$ denotes the homogeneous component of degree k of I_j, then

$$\operatorname{rank}_{\mathbb{C}} I_{j,k} = \operatorname{rank}_{\mathbb{C}} I_{i(m),k}$$

for $i \geq i(m)$ and $0 \leq k \leq m$. Now, it is easy to see that we can construct a family $\{f_j\}_{j \in \mathbb{Z}_+}$ of (non-canonical) isomorphisms

$$f_j : \operatorname{gr}_{x_j}(Z_j) \xrightarrow{\sim} \mathbb{C}[z_1, \ldots, z_n],$$

all homogeneous of degree zero, such that

$$f_i(I_i, k) = f_{i(m)}(I_{i(m),k}),$$

for $i \geq i(m)$ and every $0 \leq k \leq m$.

Now, the subset

$$I = \bigcup_{m \in \mathbb{Z}_0} f_{i(m)}\left(I_{i(m),m}\right)$$

of $\mathbb{C}[z_1, \ldots, z_n]$ is a homogeneous ideal and, by the Hilbert basis theorem, finitely generated. Hence there exists $m_0 \in \mathbb{Z}_+$ such that

$$I = \left(\bigcup_{0 \leq m \leq m_0} f_{i(m)}(I_{i(m),m})\right) \cdot \mathbb{C}[z_1, \ldots, z_n],$$

so there is an index q such that $f_i(I_{i,m}) = I_n$ for $i \geq q$ and $m \leq m_0$. Therefore $I = f_q(I_q) = f_i(I_i)$ for $i \geq q$, and hence

$$C_{X_i,x_i} \simeq C_{X_q,x_q},$$

as desired. □

Corollary 2.5.1 *In the conditions of Theorem 2.5.2, there exists an index $q \geq 0$ such that*

$$H_{X_i,x_i} = H_{X_q,x_q}$$

for all $i \geq q$.

Chapter 3
Maximal Contact

3.1 Plane Curve Singularities

As an example of our methods we show how to study singularities of plane curves.

Remark 3.1.1 Let X be a reduced complex-analytic curve in $\mathbb{C}^2$ passing through the origin. Set $Z_0 = \mathbb{C}^2$, $X_0 = X$, $x = x_0 = \mathbf{0} \in X$. We are going to compute the strict transform of X_0 by the quadratic transformation of Z_0 with center at x_0. For every coordinate system (z_0, z_1) in $\mathbb{C}^2$ with origin x, X is defined by an equation $f = 0$ locally around x, where $f \in \mathbb{C}\{z_0, z_1\}$. We write

$$f = \sum_{i \geq \nu} f^{(i)},$$

where $\nu = \nu_x(f) = \nu_x(X)$ is the multiplicity of X at x, the $f^{(i)}$ are either zero or homogeneous polynomials of degree i, and $f^{(\nu)} \neq 0$. (When the multiplicity of X at x is ν, we also say that x is a **ν-fold point** in X_t.) It is easily seen that there exists a coordinate system (z_0, z_1) in $\mathbb{C}^2$ such that $z_0 = 0$ is not tangent to X at x, that is, such that

$$f^{(\nu)}(z_0, z_1) = (z_1 - \alpha_1 z_0)^{\nu_1} \cdots (z_1 - \alpha_s z_0)^{\nu_s}, \tag{3.1}$$

where the α_i are distinct complex numbers, all the ν_i are positive integers, and the sum of the ν_i is ν. In this case we say that the z_1-axis is **transversal to X at** x.

Denote by $\pi_1 : Z_1 \to Z_0$ the quadratic transformation of Z_0 with center x. Let X_1 be the strict transform of X_0, and write $p_1 = \pi_1 |_{X_1}$. By (3.1), $p_1^{-1}(x)$ is contained in the chart $Z_{1,0}$ of Z_1 with coordinates $(z_0, z_1/z_0)$ and, in this chart, X_1 is locally defined by an equation of the type

$$f_{10} = (w_1 - \alpha_1)^{\nu_1} \cdots (w_1 - \alpha_s)^{\nu_s} + w_0 g(w_0, w_1) = 0,$$

J. M. Aroca et al., *Complex Analytic Desingularization*,
https://doi.org/10.1007/978-4-431-49822-3_3

$$w_0 = z_0, \qquad w_1 = z_1/z_0,$$

where $g(w_0, w_1) \in \mathbb{C}\{w_0, w_1\}$. Therefore

$$p_1^{-1}(x) = \{(0, \alpha_1), \dots, (0, \alpha_s)\}$$

and, if we write $x_{1,i} = (0, \alpha_i)$ for $i = 1, \dots, s$, we have

$$\nu_{x_{1,i}}(f_{10}) = \nu_{x_{1,i}}(X_1) = \min\{\nu_i, \nu_{x_{1,i}}(w_0 g)\}.$$

Hence

$$\nu_{x_{1,i}}(X_1) \leq \nu_i.$$

In this case we have the following facts:

(1) For all $i = 1, \dots, s$, $\nu_{x_{1,i}}(X_1) \leq \nu_x(X_0)$.
(2) If there is an index i such that $\nu_{x_{1,i}}(X_1) = \nu_x(X_0)$, then $s = 1$ and $p_1^{-1}(x)$ consists of only one point, which we denote simply by x_1.

So long as we are in case (2), we apply again a quadratic transformation with center x_1, and so on. In this way, we get a diagram

$$\begin{array}{ccccccccc}
Z_0 & \xleftarrow{\pi_1} & Z_1 & \longleftarrow & \cdots \longleftarrow & Z_{j-1} & \xleftarrow{\pi_j} & Z_j & \longleftarrow \cdots \\
\cup\uparrow & & \cup\uparrow & & & \cup\uparrow & & \cup\uparrow & \\
X_0 & \xleftarrow{p_1} & X_1 & \longleftarrow & \cdots \longleftarrow & X_{j-1} & \xleftarrow{p_j} & X_j & \longleftarrow \cdots
\end{array}$$

and a sequence of points

$$x_0 \xleftarrow{p_1} x_1 \longleftarrow \cdots \longleftarrow x_{j-1} \xleftarrow{p_j} x_j \longleftarrow \cdots$$

with $x_i \in X_i$ for each i. We shall prove later that, if $\nu > 1$, such a sequence of quadratic transformations is necessarily finite: there exists an integer t such that no point of $p_t^{-1}(x_{t-1})$ is a ν-fold point of X_t.

Definition 3.1.1 The points $x_0, x_1, \dots, x_{t-1}$ are called the infinitely near ν-fold points to x in X.

Remark 3.1.2 To compute the number of infinitely near ν-fold points to x in X, we will introduce a numerical invariant of the curve at x.

After restricting to a sufficiently small neighborhood of x in $\mathbb{C}^2$, we may write X as a union of irreducible components

$$X = X^{(1)} \cup X^{(2)} \cup \cdots \cup X^{(e)}.$$

For instance, if X is defined by the equation

$$(z_1^2 - z_0^2)(z_1^2 - z_0^3) + z_0^6 = 0$$

locally around $\mathbf{0}$, we can write, in a small ball around $\mathbf{0}$,

$$X = X^{(1)} \cup X^{(2)} \cup X^{(3)},$$

where the whole of X and the irreducible components are as in Fig. 3.1 on the next page.

Definition 3.1.2 Let W be a smooth curve passing through x. The **contact exponent of** W with X at x is the number

$$\delta_x(W, X) = \min_{1 \le i \le e} \frac{(W, X^{(i)})_x}{\nu_x(X^{(i)})},$$

where $(\ ,\ )_x$ denotes the intersection number at x. The **first characteristic exponent** of X at x is the number

$$\delta_x(X) = \sup \delta_x(W, X),$$

where the supremum is taken over all smooth curves W passing through x. (Clearly, $\delta_x(X) = \infty$ if and only if X is smooth at x.) A smooth curve has **maximal contact** with X at x if

$$\delta_x(W, X) = \delta_x(X).$$

Our aim now is to prove the following.

Theorem 3.1.1 *The number of infinitely near ν-fold points to x in X is equal to the integral part of $\delta_x(X)$.*

The proof of this theorem is a consequence of four lemmas, which we are going to state and prove immediately. The first gives a procedure to compute $\delta_x(X)$. The second sets up a criterion for a smooth curve W to have maximal contact with X at x. The third proves the stability of the maximal contact by quadratic transformations. The last lemma proves the existence, for every curve X and every point $x \in X$, of a smooth curve having maximal contact with X at x.

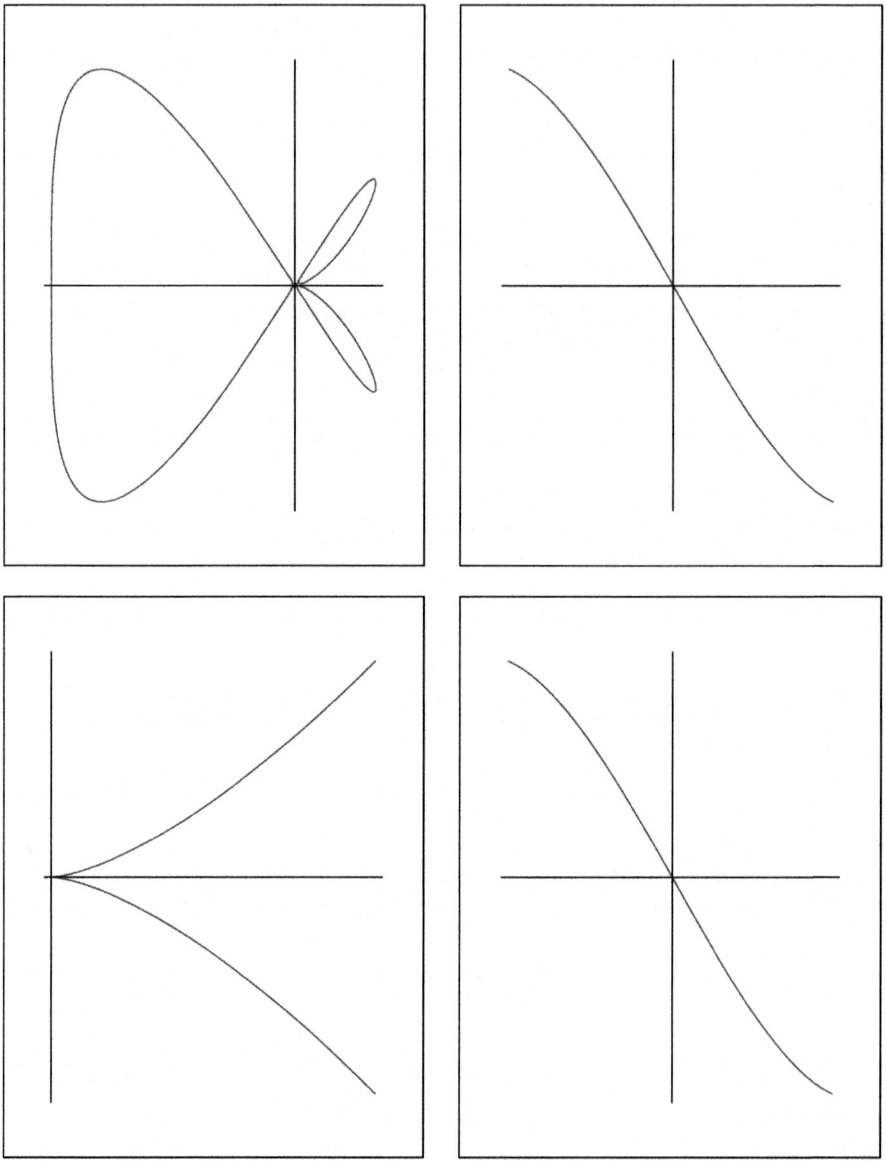

Fig. 3.1 Clockwise from top left: The curve X and the components $X^{(1)}$, $X^{(2)}$, and $X^{(3)}$

Remark 3.1.3 Let W be a fixed smooth curve passing through x, and choose a local coordinate system (z_0, z_1) in $\mathbb{C}^2$, centered at x, such that the axis $z_0 = 0$ is transversal to X at x, and W is defined by the equation $z_1 = 0$. Let $f = 0$ be the equation of X at x, with Taylor expansion

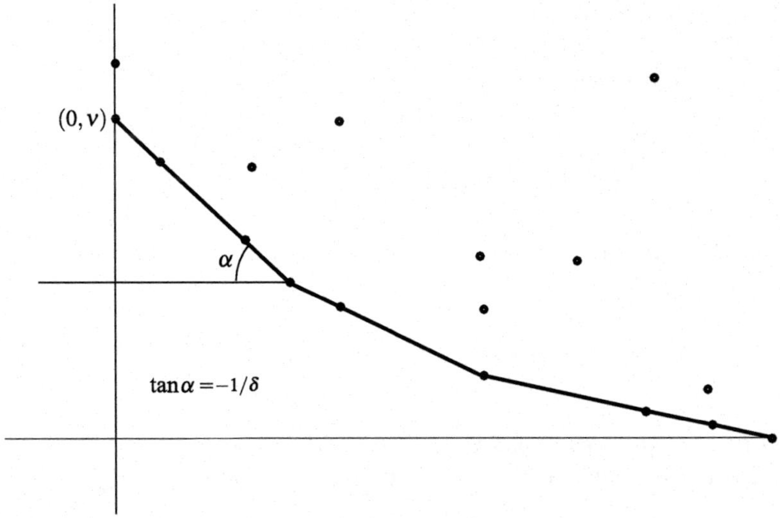

Fig. 3.2 First segment of the Newton polygon

$$f = \sum_{(i,j)\in Z_0^2} a_{ij} z_0^i z_1^j, \quad \text{with } a_{ij} \in \mathbb{C},\ \nu = \nu_x(f).$$

Write

$$\Delta(f) = \bigcup_{a_{ij}\neq 0} (i, j) + \mathbb{R}_0^2$$

and denote by $\mathbf{N}(f)$ the convex hull of $\Delta(f)$ in $\mathbb{R}_0^2$. The boundary of $\mathbf{N}(f)$ consists of two half-lines, one vertical and one horizontal, plus a finite number of line segments. The segments, plus the horizontal half line, form the **Newton polygon** of **f** with respect to the coordinate system (z_0, z_1). We consider the first segment of the Newton polygon the one having one end point on the vertical half line; see Fig. 3.2.

If $f = f_1 f_2 \cdots f_e$ is the factorization of f in $\mathbb{C}\{z_0, z_1\}$, then

$$\mathbf{N}(f) = \mathbf{N}(f_1) + \mathbf{N}(f_2) + \cdots + \mathbf{N}(f_e) \tag{3.2}$$

in the sense of addition of vectors in $\mathbb{R}^2$.

Lemma 3.1.1 *With the notation above, the slope of the first segment of the Newton polygon of f with respect to (z_0, z_1) is $-1/\delta_x(W, X)$.*

Proof Assume first that f is irreducible; then the Newton polygon of f consists of only one line segment of slope, say, $-1/\delta$. One can easily see that

$$\mathrm{rank}_{\mathbb{C}}(\mathbb{C}\{z_0, z_1\}/(z_1, f)\mathbb{C}\{z_0, z_1\}) = \nu\delta,$$

and this rank is, by definition, the intersection number $(W, X)_x$, so the result follows.

If f is reducible, and $f = f_1 f_2 \cdots f_e$ is the factorization of f in $\mathbb{C}\{z_0, z_1\}$, denote by $-1/\delta_i$ the slope of the first segment of the Newton polygon of f_i. Then, by (3.2) of Remark 3.1.3, the slope of the first segment of the Newton polygon of f is equal to $-1/\delta$ with $\delta = \min\{\delta_i : 1 \leq i \leq e\}$, and the lemma follows from the irreducible case. □

Lemma 3.1.2 *With the notation of Remark* 3.1.3, *let L be the first segment of the Newton polygon of f, and let*

$$F(z_0, z_1) = \sum_{(i,j)\in L} a_{ij} z_0^i z_1^j;$$

we call F the ***characteristic polynomial*** *of f with respect to* (z_0, z_1). *Then W does not have maximal contact with X at x if and only if*

$$F(z_0, z_1) = c(z_1 - d z_0^{\delta})^{\nu} \quad \textit{with } c, d \in \mathbb{C} \setminus \{0\},\ \delta, \nu \in \mathbb{N}. \tag{3.3}$$

If this is the case, we must have $\delta = \delta_x(W, X)$.

Proof Assume that $F(z_0, z_1)$ does have an expression of the form (3.3), and let V be the curve defined by the equation $z_1 - d z_0^{\delta} = 0$; then V is smooth containing x. Write $z_1' = z_1 - d z_0^{\delta}$; then the Newton polygon of f with respect to the coordinates (z_0, z_1') lies strictly above the line of slope $-1/\delta$ passing through $(0, \nu)$ (apart from the point $(0, \nu)$, of course). Therefore, the slope of the first segment of this new polygon is equal to $-1/\delta_1$ with $\delta_1 > \delta$, so

$$\delta_x(V, X) = \delta_1 > \delta = \delta_x(W, X),$$

and W does not have maximal contact with X at x.

Conversely, assume that there exists a smooth curve V through x such that $\delta_x(V, X) > \delta_x(W, X)$, and let $h = 0$ be an equation of V. Then the linear part of h must depend on z_1 for, otherwise, V would not be tangent to X at x, in contradiction with $\delta_x(V, X) > \delta_x(W, X) \geq 1$. By an argument converse to the one in the preceding paragraph, the lemma follows. □

Lemma 3.1.3 *With the hypothesis as above, and with the notation of Remark* 3.1.1, *let* $x_1 \in X_1$ *be an infinitely near* ν*-fold point to x in X. Suppose W has maximal contact with X at x, and let* W_1 *be the strict transform of W by* π_1. *Then:*

(1) $x_1 \in W_1$.
(2) W_1 *has maximal contact with* X_1 *at* x_1.
(3) $\delta_x(X_1) = \delta_{x_1}(W_1, X_1) = \delta_x(W, X) - 1 = \delta_x(X) - 1$.

Proof Choose a local coordinate system (z_0, z_1) in $Z_0 = \mathbb{C}^2$, centered at x, such that $z_0 = 0$ is transversal to X at x, and such that $(z_0, z_1/z_0)$ is a local coordinate system for Z_1, centered at x_1 (for this, it is enough to choose $z_1 = 0$ as the tangent to X at x). If, as above,

$$f = \sum_{(i,j)\in\mathbb{Z}_0^2} a_{ij} z_0^i z_1^i \quad \text{with } a_{ij} \in \mathbb{C},\ \nu = \nu_x(f),$$

then

$$f_{10} = \sum_{(i,j)\in\mathbb{Z}_0^2} a_{ij} z_0^{i+j-\nu} (z_1/z_0)^j = 0$$

is an equation for X_1, locally at x_1. Since x_1 is an infinitely near ν-fold point to x in X, we have $\nu_{x_1}(f_{10}) = \nu$, which implies $a_{ij} = 0$ for all $i + j - \nu + j < \nu$, and the first segment of the Newton polygon of f with respect to (z_0, z_1) has slope $-1/\delta$, with $\delta \geq 2$. If W has maximal contact with X at x, then $\delta_x(W, X) \geq 2$, which means that W is tangent to X at x. Then, by choosing (z_0, z_1) in such a way that $z_1 = 0$ defines W, again $(z_0, z_1/z_0)$ is a local coordinate system of Z_1, centered at x_1; this proves assertion (1). Further, the characteristic polynomial F_{10} of f_{10} with respect to $(z_0, z_1/z_0)$ is

$$F_{10}(z_0, z_1/z_0) = F(z_0, z_1)/z_0.$$

Then assertion (2) follows by applying Lemma 3.1.2 to F and F_1. Assertion (3) easily follows by looking at the change of the Newton polygon, by the transformation from (z_0, z_1) to $(z_0, z_1/z_0)$. □

Lemma 3.1.4 *Given a complex-analytic plane curve X and a point $x \in X$, there always exists a smooth curve W passing through x and having maximal contact with X at x.*

Proof If $\delta_x(X) = 1$, any W has maximal contact with X at x. Assume that $\delta_x(X) > 1$. Start with any W_0 such that $\delta_x(W_0, X) > 1$. Choose a local coordinate system $(z_0, z_1^{(0)})$ on $\mathbb{C}^2$, centered at x, with $z_0 = 0$ transversal to X at x, and such that W_0 is defined by the equation $z_1^{(0)} = 0$. If $\delta_x(W_0, X) < \delta_x(X)$, then, by Lemma 3.1.2, the characteristic polynomial $F(z_0, z_1^{(0)})$ of $f \in \mathbb{C}\{z_0, z_1^{(0)}\}$ such that $f = 0$ is a local equation for X is of the form

$$F(z_0, z_1^{(0)}) = c(z_1^{(0)} - dz_0^{\delta})^{\nu}.$$

Let W_1 be defined by $z_1 = z_1^{(0)} - dz_0^{\delta} = 0$. If $\delta_x(W_1, X) < \delta_x(X)$ we repeat the same process. This can be continued indefinitely only if $\delta_x(X) = \infty$, because the sequence $\{\delta_x(W_i, X)\}$ tends to infinity. If $\delta_x(X) = \infty$, X is smooth and $W = X$ has maximal contact with X at x. Otherwise, we arrive at a W_n having maximal contact with X at x. □

3.2 Maximal Contact

Our objective is to generalize to higher dimensions the results of the last section. Let Z be a smooth complex analytic space of dimension $c+d$, let $x \in Z$ be a point, let W be a smooth d-dimensional closed complex subspace of Z containing x, and $I = I_{Z,W,x}$ the ideal at x of W in Z. With the notation of Example 2.1.1 in Chap. 2, define, for a real number $\delta \geq 1$,

$$\mathrm{in}_x^{W,\delta}(X, Z) = \left(\{\mathrm{in}_x^{W,\delta}(f) \mid f \in I\}\right) \cdot \mathrm{gr}_{W,\delta,x}(Z),$$

$$\mathrm{gr}_{W,\delta,x}(X) = \mathrm{gr}_{W,\delta,x}(Z)/\mathrm{in}_x^{W,\delta}(X, Z).$$

Recall from Remark 2.1.4 in Chap. 2 that, for any local retraction $r : Z \to W$ around x, there exists an isomorphism

$$\theta_r : \mathrm{gr}_x(Z) \to \mathrm{gr}_{W,\delta,x}(Z).$$

In this chapter, we will not use very general retractions, but rather some of a special type to be described in this Section.

Definition 3.2.1 For the initial condition of this chapter, we shall say that the retraction $r : Z \to W$ is **transversal** to X at x if the tangent map $(dr)_x : T_{Z,x} \to T_{W,x}$ induces a flat morphism $(dr)_x : C_{X,x} \to T_{W,x}$. This condition amounts to saying that $\mathrm{gr}_x(X)$ is $\mathrm{gr}_x(W)$-flat via the natural morphism

$$\mathrm{gr}_x(W) \xrightarrow{r^*} \mathrm{gr}_x(Z) \longrightarrow \mathrm{gr}_x(X).$$

Lemma 3.2.1 *For a retraction $r : Z \to W$, the following conditions are equivalent:*

(1) *The retraction r is transversal to X at x.*
(2) *The initial forms of the elements in a minimal base of $I_{Z,r^{-1}(x),x}$ form a regular sequence of $\mathrm{gr}_x(X)$ as a $\mathrm{gr}_x(Z)$-module.*

Proof If we choose a local coordinate system $(\mathbf{z}, \mathbf{w})$ on Z, centered at x, and associated with r, and take the usual notation, saying that r is transversal to X at x is equivalent to saying that the natural homomorphism

$$\mathbb{C}[\bar{\mathbf{w}}] \to \mathbb{C}[\bar{\mathbf{z}}, \bar{\mathbf{w}}]/\mathrm{in}_M(I)$$

is flat, where $M = (\mathbf{z}, \mathbf{w}) \cdot \mathbb{C}\{\mathbf{z}, \mathbf{w}\}$ and I is the ideal of $\mathbb{C}\{\mathbf{z}, \mathbf{w}\}$ defining the germ of X at x. But the flatness of this natural homomorphism is equivalent to the condition that $(\bar{\mathbf{w}})$ is a regular sequence in the $\mathbb{C}[\bar{\mathbf{w}}]$-module $\mathbb{C}[\bar{\mathbf{z}}, \bar{\mathbf{w}}]/\mathrm{in}_M(I)$. (Proving this is easy: one could modify slightly the proof of Proposition 2.3.3, in Chap. 2.) This condition is, in turn, the same as (2), since any minimal base of $I_{Z,r^{-1}(x),x}$ can be written as $(\mathbf{w})$ for a suitable choice of the local coordinate system $(\mathbf{z}, \mathbf{w})$ associated with r. □

Example 3.2.1 Let $Z = \mathbb{C}^2$ and let X be the subspace of Z defined around $x = \mathbf{0} = (0, 0)$ by the equation $y^2 - x^2 + x^3 = 0$. Consider a smooth curve W through the origin and a retraction $r : Z \to W$; we look for the conditions of transversality to X at x. Let $g(x, y) = 0$ be an equation of W; then the initial form of g should be of the type $ax + by$, with $(a, b) \neq (0, 0)$. Moreover, if r is the projection parallel to the line $cx + dy = 0$ then $ad - bc \neq 0$, and we have a new local coordinate system, this time associated with r, given by

$$z = g(x, y), \quad w = cx + dy.$$

Transversality means that the morphism $\mathbb{C}[w] \to \mathbb{C}[x, y]/(y^2 - x^2)$ taking w to $cx + dy + (y^2 - x^2)$ is flat, or equivalently, that $cx + dy$ is not a zero divisor in $\mathbb{C}[x, y]/(y^2 - x^2)$. This amounts to saying that $cx + dy$ is not of the form $a(y \pm x)$. Geometrically, transversality means, in this case, that the direction of the projection is not tangent to X:

$$T_{r^{-1}(\mathbf{0}),\mathbf{0}} \not\subset C_{X,\mathbf{0}}.$$

In some cases, we shall say that the curve $r^{-1}(\mathbf{0})$ is transversal to X at x.

This result is valid for plane curves, but not in general. In the general case, we can assert, at most, that $T_{r^{-1}(x),x} \subset C_{X,x}$ implies $\mathrm{in}_x(X) \subset (\mathbf{w}) \cdot \mathrm{gr}_x(X)$, which in turn implies that r is not transversal to X at x. The converse implications are false, in general.

Lemma 3.2.2 *If r is transversal to X at x then*

$$C_{X,x} \cap T_{r^{-1}(x),x} = C_{X \cap r^{-1}(x),x}.$$

Proof We have

$$C_{X,x} \cap T_{r^{-1}(x),x} = \mathrm{Specan}\left(\mathrm{gr}_x(X)/\mathrm{in}_x(I_{Z,r^{-1}(x),x}) \cdot \mathrm{gr}_x(X)\right).$$

But, by taking coordinates as before,

$$\mathrm{gr}_x(X)/\mathrm{in}_x(I_{Z,r^{-1}(x),x}) \cdot \mathrm{gr}_x(X) = \mathrm{gr}_x(X)/(\mathbf{w}) \cdot \mathrm{gr}_x(X).$$

Therefore, it is enough to prove that, as a consequence of the transversality of r, the kernel of the canonical epimorphism

$$\mathrm{gr}_x(X) \to \mathrm{gr}_x(X \cap r^{-1}(x)) = \mathrm{gr}_{\mathfrak{m}_{X,x}/(\mathbf{w})\cdot\mathcal{O}_{X,x}}(\mathcal{O}_{X,x}/(\mathbf{w}) \cdot \mathcal{O}_{X,x})$$

is generated by $(\bar{\mathbf{w}})$. Since $(\bar{\mathbf{w}})$ is a regular sequence, $\bar{w}_1$ is not a zero divisor in $\mathrm{gr}_x(X)$, so

$$\mathrm{gr}_x(X)/w_1 \cdot \mathrm{gr}_x(X) \simeq \mathrm{gr}_{\mathfrak{m}_{X,x}/(\bar{w}^1)\cdot\mathcal{O}_{X,x}}(\mathcal{O}_{X,x}/(\bar{w}_1) \cdot \mathcal{O}_{X,x}).$$

This process can be inductively continued with $\bar{w}_2, \ldots, \bar{w}_d$, so that we get the desired result. □

We observe, finally, that, since the transversality condition depends only on $I_{Z,r^{-1}(x),x}$, then, for any projection $\mathbf{r}$ (see Remark 2.1.2 in Chap. 2), if $r_1, r_2 \in \mathbf{r}$ then r_1 is transversal to X at x if and only if r_2 is too. Therefore, it makes sense to speak about a projection being transversal to X at x.

Now, we denote by $T_{X,x}$ the vector subspace of $T_{Z,x}$ given by

$$T_{X,x} = \{\mathbf{v} \in T_{Z,x} \mid \mathbf{v} + C_{X,x} = C_{X,x}\};$$

then $T_{X,x}$ is independent of the embedding $X \to Z$, it coincides with the tangent space to X at x if x is a smooth point of X, and it is called the strict tangent space to X at x. Moreover, from Sect. 2.4 in Chap. 2 it easily follows that $C_{X,x}$ is normally flat along $T_{X,x}$. Therefore, if W is such that $T_{W,x} \subset T_{X,x}$, every retraction $r : Z \to W$ is transversal to X at x.

We now define a stronger condition for retractions than transversality.

Definition 3.2.2 We say that $r : Z \to W$ is **perpendicular** to X at x if the following conditions are satisfied:

(1) The retraction r is transversal to X at x.
(2) $T_{X\cap r^{-1}(x),x} = \{\mathbf{0}\}$.

If X and W are hypersurfaces and we pick a local coordinate system $(z, \mathbf{w}) = (z_1, w_1, \ldots, w_d)$ associated with r, then X is locally defined by a unique equation $f(z, \mathbf{w}) = 0$. In this case, r is transversal to X at x if and only if it is regular in z of order $\nu = \nu(f)$, that is, if f contains the monomial z^ν with a non-zero coefficient (equivalently, $\bar{f} \notin (\bar{\mathbf{w}}) \cdot \mathbb{C}[\bar{z}, \bar{\mathbf{w}}]$). In this case, $r^{-1}(x) \cap X = \{x\}$ (ν-fold) and so $T_{X\cap r^{-1}(x),x} = \{\mathbf{0}\}$. In other words, for hypersurfaces, transversality and perpendicularity are equivalent.

More generally, it can be easily seen that, if $T_{W,x} = T_{X,x}$, perpendicularity holds. An example of a retraction that is transversal but not perpendicular is the following. Let X be the cylinder in $\mathbb{C}^3$ given by the equation $z_1^2 - z_2^3 = 0$ (coordinates (z_1, z_2, w)), let W be given by $z_1 = z_2 = 0$, and let $r : \mathbb{C}^3 \to W$ be such that $r(z_1, z_2, w) = w$. Then r is transversal, but not perpendicular, to X at x.

If r is transversal to X at x, then

$$T_{X,x} \cap T_{r^{-1}(x),x} \subset T_{X\cap r^{-1}(x),x}.$$

To see this, remember that, if $\mathbf{v} \in T_{X,x}$, then $\mathbf{v}+C_{X,x} = C_{X,x}$. Since r is transversal,

$$C_{X\cap r^{-1}(x),x} = C_{X,x} \cap T_{r^{-1}(x),x},$$

so if, in addition, $\mathbf{v} \in T_{r^{-1}(x),x}$, one has

$$\mathbf{v} + C_{X \cap r^{-1}(x),x} = C_{X \cap r^{-1}(x),x},$$

hence $\mathbf{v} \in T_{X \cap r^{-1}(x),x}$.

However, it is not true in general that $T_{X,x} \cap T_{r^{-1}(x),x} = T_{X \cap r^{-1}(x),x}$, as is easily seen for the surface of $\mathbb{C}^3$ given by $z_1^2 - z_2^2 w + w^3 = 0$.

It is now clear that, if r is perpendicular to X at x, then

$$T_{X,x} \cap T_{r^{-1}(x),x} = \{\mathbf{0}\}.$$

As in the earlier case of transversality, it is easily seen that, if a local projection is such that one of its retractions is perpendicular to X at x, all of them are so. Therefore, it makes sense to speak about a local projection perpendicular to X at x.

Definition 3.2.3

(1) The **weighted tangent cone** to X at x in the direction of W with weight δ is

$$C_{X,x}^{W,\delta} = \mathrm{Specan}(\mathrm{gr}_{W,\delta,x}(X)),$$

that is, the complex subspace of $\mathbb{C}^{c+d}$ given by the ideal

$$\mathrm{in}_x^{W,\delta}(X, Z).$$

(2) The **contact exponent** $\delta_x(X, W, r)$ of W and X at x with respect to r is the supremum of the set of real numbers δ' with the following property: for all real δ satisfying $1 \le \delta \le \delta'$, the map θ_r induces an isomorphism

$$C_{X,x} \simeq C_{X,x}^{W,\delta}.$$

(3) The **contact exponent** of W and X at x is

$$\delta_x(X, W) = \inf\{\delta_x(X, W, r)\},$$

where the infimum is taken over all retractions $Z \to W$ at x.

(4) The **first characteristic exponent** of X at x is

$$\delta_x(X) = \sup\{\delta_x(X, W)\},$$

where the supremum is taken over all smooth curves W through x.

(5) If $\dim T_{X,x} \le 1$ and W is a smooth curve through x, we say that W has **numerical maximal contact** with X at x if there exists a retraction r such that

$$\delta_x(X) = \delta_x(X, W, r).$$

Remark 3.2.1 Assume that $r : Z \to W$ is transversal to X at x, and choose a local coordinate system $(\mathbf{z}, \mathbf{w})$ associated with r at x (see Remark 2.1.2 in Chap. 2). Then, for all $f \in I_{Z,X,x}$,

$$\mathrm{in}_x(f) \notin \mathbb{C}[\bar{\mathbf{w}}],$$

and there are two possibilities:

(1) The ideal $\mathrm{in}_x(Z, X)$ cannot be generated by elements in $\mathbb{C}[\bar{\mathbf{z}}]$, that is, $T_{W,x} \not\subset T_{X,x}$. Then $\delta_x(X, W, r) = 1$.
(2) If $T_{W,x} \subset T_{X,x}$, then $\delta_x(X, W, r)$ is the least upper bound of the $\delta \in \mathbb{R}$, $\delta \geq 1$, such that $\mathrm{in}_x^{W,\delta}(Z, X)$ is generated by elements of the image of the canonical injection

$$\theta_r : \mathrm{gr}_P(\mathcal{O}_{Z,x}) \to \mathrm{gr}_{W,x,\delta}(Z),$$

that is, by polynomials in $\tilde{\mathbf{z}}$.

As a consequence we observe that, if r is transversal, $\delta_x(X, W, r)$ does not depend on r, and it coincides with $\delta_x(X, W)$; moreover, $\delta_x(X, W) > 1$ if and only if $T_{W,x} \subset T_{X,x}$. Then we have for any $t > 1$ a canonical homomorphism

$$\theta_r : \mathrm{gr}_x(W) \to \mathrm{gr}_{W,t,x}(Z),$$

which gives a morphism

$$\pi_t : C_{X,x}^{W,t} \to T_{W,x},$$

and we have

$$\delta_x(X, W) = \sup\{\delta \in \mathbb{R} \mid \text{ either } \delta = 1 \text{ or } \pi_t \text{ is flat for } 1 \leq t \leq \delta\}.$$

Definition 3.2.4 Let $\mathbb{D}$ be an open disk in $\mathbb{C}$ around 0. A **test-map** of W at x is a holomorphic map $h : \mathbb{D} \to W$ such that $h(0) = x$. Given a test-map $h : \mathbb{D} \to W$, and the retraction $r : Z \to W$, we shall write

$$Z_h = Z \times_W \mathbb{D}, \quad X_h = X \times_W \mathbb{D}, \quad x_h = (x, 0).$$

It is obvious that $x_h \in X_h \subset Z_h$, and that h induces a smooth curve on Z_h, which we denote by $\mathbb{D}_h$.

Definition 3.2.5 We say that r has **maximal contact** with X at x (or that W has maximal contact with X at x with respect to r) if the following conditions hold:

(1) The retraction r is perpendicular to X at x.
(2) For every test-map h of W at x, $\mathbb{D}_h$ has numerical maximal contact with X_h at x_h (in particular, $\dim T_{X_h,x_h} \leq 1$).

Remark 3.2.2 Assume for the moment that both X and W are hypersurfaces in Z, locally around x. Take on Z a local coordinate system $(z, \mathbf{w})$, where $\mathbf{w} = (w_1, \ldots, w_d)$, associated with the retraction r. Then we may locally identify Z with $\mathbb{C}^{d+1}$, W with the hyperplane $z = 0$ of C^{d+1}, x with the origin, and r with the projection $(z, \mathbf{w})$ on $(\mathbf{w})$.

Assume that

$$I_{Z,X,x} = (f) \cdot \mathcal{O}_{Z,x},$$

with

$$f = \sum_{A \in Z_0, B \in Z_0^d} f_{AB} z^A \mathbf{w}^B \quad \text{for } f_{AB} \in \mathbb{C}.$$

Put $\mathrm{gr}_x(Z) = \mathbb{C}[\bar{z}, \bar{\mathbf{w}}]$, where $\bar{z} = z \bmod \mathfrak{m}_{Z,x}^2$ and $\bar{w}_i = w_i \bmod \mathfrak{m}_{Z,x}^2$ for $i = 1, \ldots, d$. Then, if $\nu_x(f) = \nu$, the tangent cone $C_{X,x}$ is defined by the equation

$$f = \sum_{\substack{A \in Z_0,\ B \in Z_0^d \\ A + |B| = \nu}} f_{AB} \bar{z}^A \bar{\mathbf{w}}^B = 0, \tag{3.4}$$

the Zariski tangent space $T_{W,x}$ is given by the equation $\bar{z} = 0$, and the map $(dr)_x : C_{X,x} \to T_{W,x}$ becomes

$$(dr)_x(\bar{z}, \bar{\mathbf{w}}) = (\bar{\mathbf{w}}).$$

The flatness of $(dr)_x$ means that $(\bar{w}_1, \ldots, \bar{w}_d)$ is a regular sequence of the $\mathbb{C}[\bar{z}, \bar{\mathbf{w}}]$-module $\mathbb{C}[\bar{z}, \bar{\mathbf{w}}]/(f)$, which is equivalent to the fact that the coefficient of z^ν in (3.4) is not zero, that is, that f is regular in z of order ν as we have seen in Sect. 2.2 of Chap. 2.

Given a test-map $h : \mathbb{D} \to W$, a coordinate function t on $\mathbb{D}$, and the local coordinate system $(z, \mathbf{w})$, we may consider on Z_h, around x_h, the local coordinate system (z, t) induced by z and t. In these coordinates, $\mathbb{D}_h$ has the equation $z = 0$, and the equation of X_h is $f(z, h(t)) = 0$ (here x_h is the origin). Then the conditions of Definition 3.2.5 are equivalent in the hypersurface case to the following:

(1) The function f is regular in z of order ν.
(2) For every smooth curve Γ in Z_h, one has

$$\delta_{x_h}(X_h, \Gamma) \leq \delta_{x_h}(X_h, \mathbb{D}_h).$$

Theorem 3.2.3 (Existence of maximal contact, hypersurface case) *Let Z be a smooth complex-analytic space, let $x \in Z$, and let X be a hypersurface in Z, locally around x. Then there exist a smooth hypersurface W in Z, locally around x, and a local retraction $r : Z \to W$ around x, such that W has maximal contact with X at x with respect to r.*

Proof If ν is the multiplicity of X at x, there exist a local coordinate system $(z', \mathbf{w})$ on Z centered at x, with $\mathbf{w} = (w_1, \ldots, w_d)$, and an equation $f' = 0$ for X in this coordinate system that is regular in z' of order ν. By the classical Weierstrass preparation theorem, we may assume that this equation has the form

$$f'(z', \mathbf{w}) = z'^{\nu} + \varphi'_1(\mathbf{w})z'^{\nu-1} + \cdots + \varphi'_\nu(\mathbf{w}) = 0,$$

with $\varphi'_i(\mathbf{w}) \in \mathbb{C}\{\mathbf{w}\}$ for $i = 1, \ldots, \nu$, and $\nu_x(f') = \nu$.

If we write

$$z = z' + \frac{1}{\nu}\varphi'_1(\mathbf{w})$$

(this is called a **Tschirnhaus transformation**), the local equation of X with respect to this new coordinate system is

$$f(z, \mathbf{w}) = z^{\nu} + \varphi_2(\mathbf{w})z^{\nu-2} + \cdots + \varphi_\nu(\mathbf{w}) = 0$$

with $\varphi_i(\mathbf{w}) \in \mathbb{C}\{\mathbf{w}\}$ and, again, $\nu_x(f) = \nu$.

Let W be the hypersurface of equation $z = 0$, and let $r : Z \to W$ the retraction defined by $r(z, \mathbf{w}) = \mathbf{w}$; then the coordinate system $(z, \mathbf{w})$ is associated with r, and we are going to prove that r has maximal contact with X at x. For this, we observe that, for every test-map $h : \mathbb{D} \to W$, the curve X_h in Z_h has the equation

$$g(z, t) = z^{\nu} + \varphi_2(h(t))z^{\nu-2} + \cdots + \varphi_\nu(h(t)) = 0$$

and satisfies

$$\delta_{x_h}(X_h, \Gamma) \leq \delta_{x_h}(X_h, \mathbb{D}_h),$$

for every smooth curve Γ through x_h. □

3.3 Tangential Intersections

Definition 3.3.1 Let H be a closed complex subspace of Z which is a hypersurface locally around x. We shall say that H is **suited to** X at x with respect to r if the following conditions hold:

(1) There exists a neighborhood U of x in Z such that $X|_U \subset H|_U$.
(2) The retraction r is transversal to H at x.
(3) For every test-map h of W at x,

$$\delta_{x_h}(X_h, \mathbb{D}_h) \leq \delta_{x_h}(H_h, \mathbb{D}_h).$$

Now, let H be a hypersurface in Z, locally around x, containing X. We say that H is **totally suited to** X at x with respect to the projection $\mathbf{r}$ (or with respect to every retraction belonging to $\mathbf{r}$) if, for every smooth complex-analytic space V and for every holomorphic map $g : V \to W$ such that $g^{-1}(x)$ is non-empty, H_g is suited to X_g at every point $y \in g^{-1}(x)$ with respect to every retraction belonging to $\mathbf{r}_g$, where $\mathbf{r}_g$ is the projection containing $r \times \mathrm{id}_V$ (this r_g does not depend on the representative r). Since every test-map of V at every point $y \in g^{-1}(x)$ gives rise to a test-map of W at x, the hypersurface H is totally suited to X at x with respect to $\mathbf{r}$ if and only if the following two conditions hold:

(1) The retraction $\mathbf{r}$ is transversal to H at x, i.e., every local retraction belonging to $\mathbf{r}$ is transversal to H at x.
(2) For every test-map h of W at x, and for every retraction s belonging to $\mathbf{r}_h$, one has

$$\delta_{x_h}(X_h, \mathbb{D}_h) \leq \delta_{x_h}(H_h, \mathbb{D}_h),$$

where $\mathbb{D}_h \hookrightarrow Z_h$ is the embedding associated with s.

Definition 3.3.2 Let $\{H^{(i)}\}_{i=1,\dots,m}$ be a family of hypersurfaces, locally around x, in Z. We say X is a **tangential intersection** of this family at x if the following conditions are satisfied:

(1) In some neighborhood of x in Z, one has $X = \bigcap_{i=1}^m H^{(i)}$.
(2) $C_{X,x} = \bigcap_{i=1}^m C_{H^{(i)},x}$.

In this case, we shall write, for brevity, that $X = \bigcap_{i=1}^m H^{(i)}$ is a tangential intersection at x. If, moreover, each $H^{(i)}$ is suited to X at x with respect to r, we say that $X = \bigcap_{i=1}^m H^{(i)}$ is a **tangential intersection suited to** X at x with respect to r. We define analogously the notion of a **tangential intersection totally suited to** X at x with respect to $\mathbf{r}$.

Example 3.3.1 Consider, in $\mathbb{C}^3$, the surfaces $H^{(1)}$, $H^{(2)}$, $H^{(3)}$, defined by the equations

$$f_1 = z_1 z_2 - w^3,$$
$$f_2 = z_1^3 - w^4,$$
$$f_3 = z_2 w^4 - z_1^2 w^3 = z_1^2 f_1 - z_2 f_2.$$

Then $X = H^{(1)} \cap H^{(2)}$ is not a tangential intersection at $\mathbf{0}$, but $X = H^{(1)} \cap H^{(2)} \cap H^{(3)}$ is.

Remark 3.3.1 (Stability of transversality by base change) Let $\mathscr{U}$ be a complex-analytic space, let $u \in \mathscr{U}$, and let Y be a closed complex subspace of $\mathscr{U} \times \mathbb{C}$ containing $(u, 0)$ and containing $\{u\} \times \mathbb{C}$ around $(u, 0)$. Let J be the ideal of Y in $\mathscr{U} \times \mathbb{C}$ around $(u, 0)$, and let (t) be a coordinate function on $\mathbb{C}$ around the origin.

We shall denote by Y^t the subspace of $\mathscr{U} \times \mathbb{C}$ defined, locally around $(u, 0)$, by the ideal

$$J^* = \bigcup_{\nu=1}^{\infty} (J : t^\nu).$$

Lemma 3.3.1 *Let* $\mathbf{z} = (z_1, \ldots, z_c)$, $\mathbf{w}' = (w_1, \ldots, w_d)$, *and* (t) *be local coordinate systems on* $\mathbb{C}^c$, $\mathbb{C}^d$, *and* $\mathbb{C}$, *centered at the origins. Let* $q : \mathbb{C}^c \times \mathbb{C}^d \times \mathbb{C} \to \mathbb{C}^d \times \mathbb{C}$ *be the projection, and let* X *be a closed complex subspace of* $\mathbb{C}^c \times \mathbb{C}^d \times \mathbb{C}$ *containing* $(\mathbf{0}, \mathbf{0}, 0)$ *and such that, locally around this point, it contains* $\mathbf{0} \times \mathbf{0} \times \mathbb{C}$. *Then if* X_0^t *denotes the fiber at* $t = 0$ *of the map* $X^t \to \mathbb{C}$ *induced by* $q : X^t \to \mathbb{C}^d \times \mathbb{C}$, *and if* $q_0 : X_0^t \to \mathbb{C}^d \times \mathbf{0}$ *is flat at the origin, we have:*

(1) *The map* $q : X^t \to \mathbb{C}^d \times \mathbb{C}$ *is flat at the origin* $(\mathbf{0}, 0)$.
(2) *For every smooth complex-analytic space* V, *and for every holomorphic map* $h : V \to \mathbb{C}^d$, *if we write*

$$X' = X \times_{\mathbb{C}^d \times \mathbb{C}} (V \times \mathbb{C}),$$

then

$$(X')^t = X^t \times_{\mathbb{C}^d \times \mathbb{C}} (V \times \mathbb{C}).$$

Proof Property (1) is evident; we prove (2). We look first at the difference between the ideals of X, X^t, X', and $(X')^t$. If we choose a local coordinate system $\mathbf{u}$ on V, and if in this system the equations of h are

$$w_i = w_i(\mathbf{u}), \quad \text{for } 1 \leq i \leq d,$$

then $J^* = \{g(\mathbf{z}, \mathbf{w}, t) \mid t^\nu \cdot g(\mathbf{z}, \mathbf{w}, t) \in J \text{ for some } \nu \geq 1\}$, the ideal of X' is $J' = \{g(\mathbf{z}, \mathbf{w}(\mathbf{u}), t) \mid g(\mathbf{z}, \mathbf{w}, t) \in J\} \cdot \mathbb{C}\{\mathbf{z}, \mathbf{u}, t\}$, and

$$J'^* = \{h(\mathbf{z}, \mathbf{u}, t) \mid t^\nu \cdot h(\mathbf{z}, \mathbf{u}, t) \in J' \text{ for some } \nu \geq 1\}.$$

The ideal of $X^t \times_{\mathbb{C}^d \times \mathbb{C}} (V \times \mathbb{C})$ is $J^* = \{g(\mathbf{z}, \mathbf{w}(\mathbf{u}), t) \mid g(\mathbf{z}, \mathbf{w}, t) \in J^*\}$. We need to compare J'^* and J^*. It is clear that $J^* \subset J'^*$, and it is also clear that $h(\mathbf{z}, \mathbf{u}, t) \in J'^*$ implies the existence of $\nu \geq 1$ such that $t^\nu \cdot h(\mathbf{z}, \mathbf{u}, t) \in J^{*\prime}$. Thus the only difference between $(X')^t$ and $X^{t'} \times_{\mathbb{C}^d \times \mathbb{C}} (V \times \mathbb{C})$ is the t-torsion. But $X^{t'}$ is flat over $V \times \mathbb{C}$; therefore t is not a zero-divisor in $\mathbb{C}\{\mathbf{z}, \mathbf{u}, t\}/J^{*\prime}$, and $J^{*\prime} = J'^*$. □

Lemma 3.3.2 *Using the initial conditions of this section, if* V *is a smooth* $\mathbb{C}$*-space,* $h : V \to W$ *is a holomorphic map, and* x' *is an arbitrary point in* $h^{-1}(x)$, *then the square*

$$\begin{array}{ccc} C_{X,x} & \xleftarrow{\alpha_3} & C_{X_h,x'} \\ {\scriptstyle \alpha_1}\downarrow & & \downarrow{\scriptstyle \alpha_4} \\ T_{W,x} & \xleftarrow{\alpha_2} & T_{V,x'} \end{array}$$

is cartesian, where α_1 *and* α_2 *are the differentials of* r *and* h*, respectively, and* α_3 *and* α_4 *are the morphisms associated with the projections* $X_h = X \times_W V \to X$ *and* $X_h \to V$.

Proof By means of the local coordinate system $(\mathbf{z}, \mathbf{w})$, we may locally identify Z with $\mathbb{C}^c \times \mathbb{C}^d$. Let J be the ideal in $\mathbb{C}\{\mathbf{z}, \mathbf{w}, t\}$ defined by

$$J = (f(t\mathbf{z}, t\mathbf{w}))_{f \in I_{Z,X,x}} \mathbb{C}\{\mathbf{z}, \mathbf{w}, t\},$$

where t is a new variable. Let X_0 be a subspace of $\mathbb{C}^c \times \mathbb{C}^d \times \mathbb{C}$ that represents the germ defined by the ideal J. Then $X_0^t = C_{X,x}$, and the assumption of transversality for r becomes the assumption made for X in Lemma 3.3.1. But then conclusion (2) of Lemma 3.3.1, at $t = 0$, is the conclusion of the present lemma. □

Remark 3.3.2 Since W and Z are smooth, the square of Lemma 3.3.2 for W and Z becomes

$$\begin{array}{ccc} T_{Z,x} & \xleftarrow{(dq_1)_{x'}} & T_{Z_h,x'} \\ {\scriptstyle (dr)_x}\downarrow & & \downarrow{\scriptstyle (dq_2)_{x'}} \\ T_{W,x} & \xleftarrow{(dh)_{x'}} & T_{V,x'} \end{array}$$

where q_1 and q_2 are the projections $Z_h \to Z$ and $Z_h \to V$, respectively.

If H is locally, around x, a hypersurface of Z, and $f \in \mathcal{O}_{Z,x}$ is an equation of H in $\mathcal{O}_{Z,x}$, then $\mathrm{in}_x(f)$ is a polynomial function on $T_{Z,x}$. If we consider the function

$$\mathrm{in}_x(f) \cdot (dq_1)_{x'}$$

on $T_{Z_h,x'}$, we have two possibilities:

(1) If $\nu_x(f) < \nu_{x'}(f \cdot q_1)$, then $\mathrm{in}_x(f) \cdot (dq_1)_{x'}$ is obviously zero.
(2) If $\nu_x(f) = \nu_{x'}(f \cdot q_1)$, then $\mathrm{in}_x(f) \cdot (dq_1)_{x'} = \mathrm{in}_x(f \cdot q_1)$.

Corollary 3.3.1 *With the notation and hypothesis of Lemma* 3.3.2*, if*

$$X = \bigcap_{i=1}^{m} H^{(i)}$$

is a tangential intersection at x, then

$$X_h = \bigcap_{i=1}^{m} H_h^{(i)}$$

is a tangential intersection at x'.

Proof Assume that $I_{Z,H^{(i)},x} = (f_i)\mathscr{O}_{Z,x}$ for $i = 1, \dots, m$, and let A be the set of indices $i = 1, \dots, m$ such that

$$\text{in}_x(f_i) \cdot (dq_1)_{x'} \neq 0.$$

By Lemma 3.3.2 and the definition of a tangential intersection, we have

$$\begin{aligned} C_{X_h,x'} &= (dq_1)_{x'}^{-1}(C_{X,x}) = (dq_1)_{x'}^{-1}\left(\bigcap_{i=1}^{m} C_{H^{(i)},x}\right) \\ &= \bigcap_{i=1}^{m}(dq_1)_{x'}^{-1}(C_{H^{(i)},x}) = \bigcap_{i\in A} C_{H_h^{(i)},x'} = \bigcap_{i=1}^{m} C_{H_h^{(i)},x'}. \end{aligned}$$

This last equality holds because $C_{H_h^{(i)},x'} \supset C_{X_h,x'}$, for all i. □

Lemma 3.3.3 *Let $X = \bigcap_{i=1}^{m} H^{(i)}$ be a tangential intersection at x, such that r is transversal to each $H^{(i)}$ at x for $i = 1, \dots, m$. Then, if δ is any real number such that*

$$1 \leq \delta \leq \min_{1\leq i\leq m} \{\delta_x(H^{(i)}, W)\},$$

we have

$$C_{X,x}^{W,\delta} = \bigcap_{i=1}^{m} C_{H^{(i)},x}^{W,\delta}.$$

We shall in fact prove a slightly more general lemma, from which we shall also be able to study the relation between maximal contact and numerical maximal contact in the sequel.

Lemma 3.3.4 *Let S be a subsemigroup of $(\mathbb{R}_0^+)$ containing $\mathbb{Z}_0$, and let $X = \bigcap_{i=1}^{m} H^{(i)}$ be a tangential intersection at x such that r is transversal to each $H^{(i)}$ at x, for $1 \leq i \leq m$. Write $I_{Z,H^{(i)},x} = (f_i)\mathscr{O}_{Z,x}$ for $i = 1, \dots, m$ and let $\varepsilon \in \mathbb{R}_0$ be such that*

$$1 \leq \varepsilon \leq \min_{1\leq i\leq m} \{\delta_x(H^{(i)}, W)\}.$$

Define

$$L_{\varepsilon}(\mathbf{x}, \mathbf{x}') = \sum_{i=1}^{c} x_i + \frac{1}{\varepsilon}\sum_{i=1}^{d} x_i'.$$

Then:

(1) *For every* $g \in I_{Z,X,x}$ *there is a family* $\{h_i\}_{1 \le i \le m}$ *of elements of such that:*

(a) $\nu_{L_\varepsilon,S}(h_i) \ge \nu_{L_\varepsilon,S}(g) - \nu_{L_\varepsilon,S}(f_i)$ *for* $i = 1, \ldots, m$, *and*
(b) $\nu_x(g - \sum_{i=1} h_i f_i) > \nu_x(g)$.

(2) $\mathrm{in}_{L_\varepsilon,S}(I_{Z,X,x}) = (\mathrm{in}_{L_\varepsilon,S}(f_1), \ldots, \mathrm{in}_{L_\varepsilon,S}(f_m)) \cdot \mathrm{gr}_{L_\varepsilon,S}(\mathbb{C}\{\mathbf{z}, \mathbf{w}\})$.

Proof If $\varepsilon = 1$ the result is trivially verified. Assume $\delta > 1$.

Fix $g \in I_{Z,X,x}$. Set $\nu = \nu_x(g)$, $\mu = \nu_{L_\varepsilon,S}(g)$, and $\nu_i = \nu_x(f_i)$ for $1 \le i \le m$. From the inequality $\varepsilon > 1$ and from the transversality of r to each $H^{(i)}$ it follows that, for each $1 \le i \le m$, we have $\bar{f_i} = \mathrm{in}_x(f_i) \in \mathbb{C}[\bar{\mathbf{z}}]$ and $\nu_i = \nu_x(f_i) = \nu_{L_\varepsilon,S}(f_i)$.

Also, we can always write

$$\bar{g} = \mathrm{in}_x(g) = \sum_{\alpha=1}^{t} \varphi_\alpha(\bar{\mathbf{w}}).\psi_\alpha(\bar{\mathbf{z}}),$$

where $\varphi_\alpha(\bar{\mathbf{w}})$ and $\psi_\alpha(\bar{\mathbf{z}})$ are homogeneous polynomials $1 \le \alpha \le t$ such that:

(i) $\nu_{L_\varepsilon,S}(\psi_\alpha) + \nu_{L_\varepsilon,S}(\varphi_\alpha) \ge \mu$, $\deg(\psi_\alpha) + \deg(\varphi_\alpha) = \nu$, and
(ii) the $\{\varphi_\alpha\}_{1 \le \alpha \le t}$ are $\mathbb{C}$-linearly independent.

X being a tangential intersection of the $H^{(i)}$, for $1 \le i \le m$, by (i), we have

$$\psi_\alpha(\bar{\mathbf{z}}) \in (\bar{f_1}, \ldots, \bar{f_m})\mathbb{C}[\bar{\mathbf{z}}] \quad \text{for } 1 \le \alpha \le t;$$

hence

$$\psi_\alpha(\bar{\mathbf{z}}) = \sum_{i=1}^{m} \psi_{\alpha,i}(\bar{\mathbf{z}}).\bar{f_i}(\bar{\mathbf{z}}),$$

where the $\psi_{\alpha,i}$ are homogeneous polynomials (or zero) and

$$\deg(\psi_{\alpha,i}) = \deg(\psi_\alpha) - \nu_i \quad \text{for } 1 \le \alpha \le t \text{ and } 1 \le i \le m.$$

Now write

$$h_i = \sum_{\alpha=1}^{t} \varphi_\alpha(\mathbf{w}).\psi_{\alpha,i}(\mathbf{z}) \in \mathbb{C}[\mathbf{z}, \mathbf{w}] \subset \mathbb{C}\{\mathbf{z}, \mathbf{w}\} \quad \text{for } 1 \le i \le m.$$

This family $\{h_i\}_{1\leq i\leq m}$ satisfies conclusion (1) of the lemma. Indeed,

$$\begin{aligned}\nu_{L_\varepsilon,S}(h_i) &\geq \min_{1\leq\alpha\leq t}(\nu_{L_\varepsilon,S}(\varphi_\alpha)+\nu_{L_\varepsilon,S}(\varphi_{\alpha,i}))\\ &= \min_{1\leq\alpha\leq t}(\nu_{L_\varepsilon,S}(\varphi_\alpha)+\nu_{L_\varepsilon,S}(\psi_\alpha)-\nu_i)\geq\mu-\nu_i\end{aligned}$$

by property (i) above. On the other hand,

$$\nu_x(h_i)\geq\nu-\nu_i$$

by the same property, so that

$$\mathrm{in}_x\left(\sum_{i=1}^m h_i f_i\right)=g,$$

which proves the second part of conclusion (1).

To prove conclusion (2), observe first that, for any $g\in\mathcal{O}_{Z,x}$, we have

$$\begin{aligned}\varepsilon(\nu_{L_\varepsilon,S}(g)+1) &\geq \varepsilon(\nu_{L_\varepsilon\mathbb{Z}_0}(g)+1)\\ &= \min\left\{\varepsilon\,|A|+\varepsilon\left(|B|/\varepsilon\right)+\varepsilon \;\middle|\; (AB)\in\mathscr{E}_{\mathbf{z},\mathbf{w}}(g)\right\}\\ &> \min\left\{\varepsilon|A|+|B| \;\middle|\; (AB)\in\mathscr{E}_{\mathbf{z},\mathbf{w}}(g)\right\}\geq\nu_x(g)\\ &\geq \min\left\{|A|+1/\varepsilon|B| \;\middle|\; (AB)\in\mathscr{E}_{\mathbf{z},\mathbf{w}}(g)\right\}\\ &\geq \nu_{L_\varepsilon,\mathbb{R}_0}(g)\geq\nu_{L_\varepsilon,S}(g).\end{aligned}$$

Then, if $g\in I_{Z,X,x}$ by finite iterated application of the lemma's conclusion (1) and by adding the results, we can find a family of functions $\{h_i^*\}_{1\leq i\leq m}$ such that $\nu_{L_\varepsilon,S}(h_i^*)\geq\mu-\nu_i$ for $1\leq i\leq m$ and that

$$\nu_{L_\varepsilon,S}(g-\sum_{i=1}^m h_i^* f_i)>\nu_{L_\varepsilon,S}(g)=\mu.$$

In this case

$$\mu=\min_{1\leq i\leq m}\{\nu_{L_\varepsilon,S}(h_i^*)+\nu_{L_\varepsilon,S}(f_i)\},$$

and if we define Δ as the set of indices $i=1,\ldots,m$ such that $\mu=\nu_{L_\varepsilon,S}(h_i^*)+\nu_{L_\varepsilon,S}(f_i)$, we obtain

$$\mathrm{in}_{L_\varepsilon,S}(g)=\sum_{i\in\Delta}\mathrm{in}_{L_\varepsilon,S}(h_i^*)\cdot\mathrm{in}_{L_\varepsilon,S}(f_i),$$

which proves the lemma. □

Proposition 3.3.1 *Let $X = \bigcap_{i=1}^m H^{(i)}$ be a tangential intersection suited to X at x with respect to r. Then, for every test-map h of W at x, we have*

$$\delta_{x_h}(X_h, \mathbb{D}_h) = \min_{1 \le t \le m} \{\delta_{x_h}(H_h^{(i)}, \mathbb{D}_h)\}.$$

Proof By Corollary 3.3.1, for every test-map h of W at x, $X_h = \bigcap_{i=1}^m H_h^{(i)}$ is a tangential intersection at x_h. Also, from the definition of transversality, it is obvious that the retraction $r_h : Z_h \to \mathbb{D}_h$ induced by r is transversal to $H_h^{(i)}$ at x_h, for $1 \le i \le m$. Therefore we are in a position to apply Lemma 3.3.3.

Let $\delta \in \mathbb{R}_0$ be such that $1 \le \delta \le \min_{1 \le i \le m}\{\delta_{x_h}(H_h^{(i)}, \mathbb{D}_h)\}$. Then

$$\delta_{x_h}(C_{X_h, x_h}, \mathbb{D}_h) = \delta_{x_h}\left(\bigcap_{i=1}^m C_{H_h^{(i)}, x_h}, \mathbb{D}_h\right).$$

Furthermore, if $\delta < \min_{1 \le i \le m}\{\delta_{x_h}(H_h^{(i)}, \mathbb{D}_h)\}$, it is clear that the ideal of $C_{H_h^{(i)}, x_h}$ is generated by a polynomial in $\mathbb{C}[\bar{\mathbf{z}}]$. From this we conclude that, for all $\delta \in \mathbb{R}_0$ satisfying

$$1 \le \delta < \min_{1 \le i \le m} \{\delta_{x_h}(H_h^{(i)}, \mathbb{D}_h)\},$$

one has

$$\delta \le \delta_{x_h}(X_h, \mathbb{D}_h),$$

hence

$$\delta_{x_h}(X_h, \mathbb{D}_h) \ge \min_{1 \le i \le m} \{\delta_{x_h}(H_h^{(i)}, \mathbb{D}_h)\}. \tag{3.5}$$

(This last relation is trivially true if $\min_{1 \le i \le m}\{\delta_{x_h}(H_h^{(i)}, \mathbb{D}_h)\} = 1$.) From this and from the fact that each $H^{(i)}$, for $i = 1, \ldots, m$, is suited to X at x with respect to r, we obtain the proposition. □

Corollary 3.3.2 *With the notation and hypothesis of Proposition* 3.3.1, *we have*

$$\delta_x(X, W) = \min_{1 \le i \le m} \{\delta_x(H^{(i)}, W)\}.$$

Proof Observe first that there exists $\varepsilon \in \mathbb{R}_+$ such that, for every $\boldsymbol{\xi} = (\xi_1, \ldots, \xi_d) \in \mathbb{C}^d$ with $\|\boldsymbol{\xi}\| = \varepsilon$, the map $h_{\boldsymbol{\xi}} : \mathbb{D} \to W$ defined by $h_{\boldsymbol{\xi}}(t) = t\boldsymbol{\xi}$ has a sense and is, of course, a test-map (here $t\boldsymbol{\xi}$ means the point of coordinates $(t\xi_1, \ldots, t\xi_d)$ in the local coordinate system induced on W by $(\mathbf{w})$).

Then, X_h is locally defined by the ideal

$$(f(\mathbf{z}, t\boldsymbol{\xi}))_{f \in I_{Z,X,x}} \mathbb{C}\{\mathbf{z}, t\}$$

and $\mathbb{D}_h$ is defined by the ideal $(\mathbf{z})\mathbb{C}\{\mathbf{z}, t\}$. It can easily be seen, by looking at the Newton polygon, that we always have

$$\delta_{x_{h_\xi}}(X_{h_\xi}, \mathbb{D}_{h_\xi}) \geq \delta_x(X, W)$$

and that there is at least one element $\boldsymbol{\xi} \in \mathbb{C}^d$ with $\|\boldsymbol{\xi}\| = \varepsilon$ such that the equality holds, that is,

$$\delta_x(X, W) = \min_{\boldsymbol{\xi} \in \mathbb{C}^d,\ \|\boldsymbol{\xi}\| = \varepsilon} \{\delta_{x_{h_\xi}}(X_{h_\xi}, \mathbb{D}_{h_\xi})\}.$$

The corollary follows from this and from Proposition 3.3.1. □

Remark 3.3.3 (Idealistic exponents) We now introduce the concept of an idealistic exponent as a generalization of the concept of a rational number, and we use it to generalize the concept of the first characteristic exponent, as we announced at the beginning of Sect. 3.2.

Definition 3.3.3 (Zariski[132]) Let A be a commutative ring and let I, J be two ideals of A. We say that J is **integral** over I if each $\xi \in J$ satisfies a relation of the type

$$\xi^d + a_1\xi^{d-1} + \cdots + a_{d-1}\xi + a_d = 0,$$

where $a_i \in I^i$ for $i = 1, \ldots, d$.

Example 3.3.2 Let $A = \mathbb{C}\{x_1, x_2\}$ be the ring of the convergent power series in two independent variables over $\mathbb{C}$. Put $I = (x_1^2, x_2^2)$ and $J = (x_1^2, x_1x_2, x_2^2)$; then J is integral over I because $(x_1x_2)^2 - x_1^2x_2^2 = 0$ and $x_1^2x_2^2 \in I^2$.

Lemma 3.3.5 *Let $(A, \mathfrak{m})$ be a noetherian normal local ring, and let I, J be ideals of A. If I is invertible and J is integral over I, then $J \subset I$.*

Proof Put $I = tA$; if $t = 0$ then $J = (0)$ because A has no non-zero nilpotent elements. Assume that $t \neq 0$. For every $\xi \in J$, we have a relation

$$\xi^d + a_1\xi^{d-1} + \cdots + a_d = 0,$$

where $a_i \in t^iA$ for $i = 1, \ldots, d$. Write $a_i = a_1't^i$ for $i = 1, \ldots, d$; in the total ring of fractions of A we may write

$$(\xi/t)^d + a_1'(\xi/t)^{d-1} + \cdots + a_d' = 0,$$

which is an equation of integral dependence for ξ/t over A. Since A is normal, ξ/t lies in A, hence $\xi \in tA$ and so $J \subset I$. □

Lemma 3.3.6 *Let $(A, \mathfrak{m})$ be a noetherian local ring, and let $I \supset J$ be ideals of A, where I is invertible. If I is integral over J, then $I = J$.*

Proof Suppose $I = tA$, and consider $J' = J \cdot I \subset A$. Then A is integral over J', so J' is not contained in $\mathfrak{m}$. This proves that $t \in J$. □

Lemma 3.3.7 *Let A be an integral domain, and let I, J be ideals of A. Suppose that I is finitely generated. If $a \in A$ is such that*

$$aI \subset J \cdot I,$$

then a depends integrally on J.

Proof Let $b_1, \dots, b_s$ a system of generators of I. There exists a matrix $(a_{ij})_{1 \le i,j \le s}$ with coefficients $a_{ij} \in J$ and such that

$$ab_i = \sum_{j=1}^{s} a_{ij} b_j \quad \text{for } 1 \le i \le s$$

implies $\big((a_{ij}) - aI_s\big)(b_1, \dots, b_s)' = \mathbf{0}$, where $(\ )'$ means the transpose and I_s is the $s \times s$ identity matrix. Multiplying by the adjoint of $(a_{ij}) - aI_s$ we see that

$$\det\big((a_{ij}) - aI_s\big)(b_1, \dots, b_s)' = \mathbf{0},$$

hence $\det((a_{ij}) - aI_s) = 0$, and this is a relation of integral dependence for a over J. □

Definition 3.3.4

(1) Let W be a complex-analytic space, and let $\mathbf{J}, \mathbf{I}$ be coherent ideals on W. We say that $\mathbf{I}$ is **integral over J** if, for every point $x \in W$, the stalk $\mathbf{I}_x$ is integral over the stalk $\mathbf{J}_x$.
(2) We say that $\mathbf{J}$ and $\mathbf{I}$ are **integrally equivalent** if $\mathbf{J}$ is integral over $\mathbf{I}$ and $\mathbf{I}$ is integral over $\mathbf{J}$.

Note 3.3.1 Let W be a complex-analytic space, and let $\mathbf{I}$ be a coherent ideal on W. For each complex-analytic map $h : \mathbb{D} \to W$, where $\mathbb{D}$ is a disk in $\mathbb{C}$ around 0, we may construct the ideal $\mathbf{I} \cdot \mathscr{O}_{\mathbb{D}}$, the pull-back of $\mathbf{I}$ by h. The stalk of this ideal at 0 is an ideal of the ring $\mathbb{C}\{t\}$ of convergent power series in one variable over $\mathbb{C}$; therefore it makes sense to speak of the order of this ideal, which we denote by $\nu_0(\mathbf{I} \cdot \mathscr{O}_{\mathbb{D}})$.

The key result of this section in the following.

Theorem 3.3.1 *Let W be a smooth complex-analytic space, and let $\mathbf{J}$ and $\mathbf{I}$ be coherent ideals on W. The following conditions are equivalent:*

(1) *The ideals $\mathbf{J}$ and $\mathbf{I}$ are integrally equivalent.*
(2) *For every morphism $\varphi : V \to W$ of complex-analytic spaces such that V is normal and $\mathbf{I} \cdot \mathscr{O}_V$ is invertible, one has $\mathbf{J} \cdot \mathscr{O}_V = \mathbf{I} \cdot \mathscr{O}_V$.*
(2′) *For every proper surjective morphism $\varphi : V \to W$ of complex-analytic spaces such that V is normal and $\mathbf{I} \cdot \mathscr{O}_V$ is invertible, one has $\mathbf{J} \cdot \mathscr{O}_V = \mathbf{I} \cdot \mathscr{O}_V$.*

(3) *(Valuational criterion) For every morphism* $h : \mathbb{D} \to W$ *of C-spaces, where* $\mathbb{D}$ *is a disk around the origin in* $\mathbb{C}$*, we have*

$$\nu_0(\mathbf{J} \cdot \mathcal{O}_{\mathbb{D}}) = \nu_0(\mathbf{I} \cdot \mathcal{O}_{\mathbb{D}}).$$

Proof The scheme of the proof will be (1) $\Rightarrow$ (2) $\Rightarrow$ (3) $\Rightarrow$ (2′) $\Rightarrow$ (1). That (2) implies (3) is clear.

To show that (1) $\Rightarrow$ (2), take any point $x \in V$, and consider the stalks $(\mathbf{J} \cdot \mathcal{O}_V)_x$ and $(\mathbf{I} \cdot \mathcal{O}_V)_x$ as ideals in $\mathcal{O}_{V,x}$. By Lemma 3.3.5, $(\mathbf{I} \cdot \mathcal{O}_V)_x \subset (\mathbf{J} \cdot \mathcal{O}_V)_x$. By Lemma 3.3.6, $(\mathbf{J} \cdot \mathcal{O}_V)_x = (\mathbf{I} \cdot \mathcal{O}_V)_x$ for every $x \in V$, and this proves (2).

Next we prove that (3) $\Rightarrow$ (2′). Since the statement is symmetric, it is enough to prove that $\mathbf{I} \cdot \mathcal{O}_V \subset \mathbf{J} \cdot \mathcal{O}_V$. Since φ is surjective, it is enough to show that, for all $x \in W$, for all $f \in \mathbf{I}_x$ and for all $x' \in \varphi^{-1}(x)$, we have $\varphi^*(f) \cdot \mathcal{O}_{V,x'} \in (\mathbf{J} \cdot \mathcal{O}_V)_{x'}$. If $(\mathbf{J} \cdot \mathcal{O}_V)_x = t \cdot \mathcal{O}_{V,x'}$, we can construct $g = f/t$. If g is holomorphic, the result follows; let's show that g is necessarily holomorphic. If g were not holomorphic, there would exist a morphism $h' : \mathbb{D} \to V$ such that $\nu_0(g\mathcal{O}_{\mathbb{D}}) < 0$, where $g\mathcal{O}_{\mathbb{D}}$ is taken via h'. By constructing $\varphi h : \mathbb{D} \to W$, we would have a contradiction with (3).

Finally, we show that (2′) implies (1). By symmetry, we prove only that $\mathbf{J}$ is integral over $\mathbf{I}$.

Let $\varphi : V \to W$ be the blow-up of $\mathbf{I}$ in W, and let $\Psi : V' \to V$ be the normalization. $\mathbf{I} \cdot \mathcal{O}_{V'}$ is invertible; then, by (2′),

$$\mathbf{I} \cdot \mathcal{O}_{V'} = \mathbf{J} \cdot \mathcal{O}_{V'}.$$

Now take $x \in W$, $f \in \mathbf{J}_x$, and a neighborhood U of x in W such that:

(1) $\mathbf{I}|_U = (g_1, \ldots, g_s) \cdot \mathcal{O}_W|_U$;
(2) The set $\varphi^{-1}(U)$ is covered by the sets

$$U_i = \text{Specan}\, \mathcal{O}_W|_U\left(\frac{g_1}{g_i}, \ldots, \frac{\hat{g}_i}{g_i}, \ldots, \frac{g_s}{g_i}\right), \quad \text{for } 1 \le i \le s;$$

(3) If we denote again by f a representative of f over U, there exists $\lambda \in \mathcal{O}_V(U)$ such that $\lambda \cdot f \in \mathbf{J} \cdot \mathcal{O}_V(\varphi^{-1}(U))$.

Then, for each $1 \le i \le s$, we can choose n_i such that $f \cdot g_i^{n_i} \in \mathbf{J}_x \cdot \mathbf{I}_x^{n_i}$ for all i. This implies $f \cdot \mathbf{I}_x^N \subset \mathbf{J}_x \cdot \mathbf{I}_x^N$ for some N. Since I_x^N is finitely generated and $\mathcal{O}_{W,x}$ is an integral domain, f is integral over $\mathbf{J}_x$ (Lemma 3.3.7). □

Remark 3.3.4 Let W be a smooth complex-analytic space. On the set of pairs $(\mathbf{J}, b)$, where $\mathbf{J}$ is a coherent ideal on W and b is a positive integer, we can define an equivalence relation by putting $(\mathbf{J}_1, b_1) \sim (\mathbf{J}_2, b_2)$ if and only if $\mathbf{J}_1^{b_2}$ and $\mathbf{J}_2^{b_1}$ are integrally equivalent.

Let $\mathbb{D}$ be a disk around the origin in $\mathbb{C}$, and write, for every complex-analytic map $h : \mathbb{D} \to W$,

$$\nu_h(\mathbf{J}, b) = \nu_0(\mathbf{J} \cdot \mathcal{O}_{\mathbb{D}})/b.$$

Then, by Theorem 3.3.1, $(\mathbf{J}_1, b_1) \sim (\mathbf{J}_2, b_2)$ if and only if $\nu_h(\mathbf{J}_1, b_1) = \nu_h(\mathbf{J}_2, b_2)$ for all $h : \mathbb{D} \to W$.

Definition 3.3.5 The $\sim$-equivalence class of a pair $(\mathbf{J}, b)$ will be written as $((\mathbf{J}, b))$ and will be called an **idealistic exponent** on W.

If $\mathscr{E} = ((\mathbf{J}, b))$, then, for every complex-analytic map $h : \mathbb{D} \to W$, the number $\nu_h(\mathscr{E}) = \nu_h(\mathbf{J}, b)$ does not depend on the representative. It is called the **order** of $\mathscr{E}$ relative to h.

Remark 3.3.5 Idealistic exponents generalize rational numbers in the following way. Let W be a disk in $\mathbb{C}$ around the origin, and consider the set of all idealistic exponents $((\mathbf{J}, b))$ such that

$$\mathbf{J}|_{W\setminus\{0\}} = \mathscr{O}_W|_{W\setminus\{0\}}.$$

Then the correspondence that assigns to $\mathscr{E} = ((\mathbf{J}, b))$ the rational number $\nu_{\mathrm{id}_W}(\mathscr{E})$ is one-to-one over $\mathbb{Q}$.

Definition 3.3.6 Let $\mathscr{E} = ((\mathbf{J}, b))$ be an idealistic exponent on W, and let $w \in W$. We denote by $\nu_w(\mathscr{E})$ the number $\nu_w(\mathbf{J})/b$, where

$$\nu_w(\mathbf{J}) = \min\{\nu_w(f) \mid f \in \mathbf{J}_w\};$$

this number is called the **order** of $\mathscr{E}$ at $w \in W$. Moreover, we denote by $\mathrm{Sing}(\mathscr{E})$ the set

$$\mathrm{Sing}(\mathscr{E}) = \{w \in W \mid \nu_w(\mathscr{E}) \geq 1\}.$$

Theorem 3.3.2 *Let Z be a smooth complex-analytic space, and let $x \in Z$. Suppose X and W are complex subspaces of Z around x, with W smooth, and let $r : Z \to W$ be a local retraction, around x, transversal to X at x. Suppose moreover that $X = \bigcap_{i=1}^m H^{(i)}$ is a tangential intersection suited to X at x with respect to r. Then there exist a neighborhood U of x in W, a coherent ideal $\mathbf{J}$ of $\mathscr{O}_W|_U$, and a positive integer b such that, for every test-map h of W at x, one has*

$$\delta_{x_h}(X_h, \mathbb{D}_h) = \nu_h((\mathbf{J}, b)).$$

Proof For each $i = 1, \dots, m$, set $I_{Z,H^{(i)},x} = (f_i)\mathbb{C}\{\mathbf{z}, \mathbf{w}\}$, and write

$$f_i = \sum_{A \in \mathbb{Z}_0^c} (f_i)_A \mathbf{z}^A,$$

where $(f_i)_A \in \mathbb{C}\{\mathbf{w}\}$ for each i. Then, let U be a neighborhood of x in W in which all the $(f_i)_A$ are holomorphic; set $b = \nu_1! \times \cdots \times \nu_m!$, where $\nu_i = \nu_x(f_i)$; and finally let $\mathbf{J}$ be the ideal in $\mathscr{O}_W|_U$ generated by the elements $(f_i)_A^{b/(\nu_i - |A|)}$, for $1 \leq i \leq m$

and A ranging over the set

$$\{A \in \mathbb{Z}_0^c \mid 0 \leq |A| \leq \nu_i - 1\}.$$

We shall prove that U, $\mathbf{J}$ and b have the required properties.

By Proposition 3.3.1 we have, for every test-map h of W at x,

$$\delta_{x_h}(X_h, \mathbb{D}_h) = \min_{1 \leq i \leq m} \{\delta_{x_h}(H_h^{(i)}, \mathbb{D}_h)\}. \tag{3.6}$$

On the other hand, consider a coordinate function (t) on $\mathbb{D}$ around 0, and let $(\mathbf{z}, t)$ be the coordinate system on Z_h induced by $(\mathbf{z}, \mathbf{w}, t)$. Construct, for every $i = 1, \ldots, m$, the Newton polygon for $H_h^{(i)}$, by taking the exponent of t on the horizontal axis and the total exponent of $\mathbf{z}$ (that is, the sum of the components of each exponent) on the vertical axis. In this way we always get the point $(0, \nu_i)$ and, by considering the first segment of the polygon, we see that

$$-\frac{1}{\delta_{x_h}(H_h^{(i)}, \mathbb{D}_h)} = \min_{0 \leq |A| \leq \nu_i - 1} \left\{ -\frac{\nu_i - |A|}{\nu_0((f_i)_A(h(t)))} \right\} \quad \text{for } 1 \leq i \leq m,$$

and thus

$$\delta_{x_h}(H_h^{(i)}, \mathbb{D}_h) = \min_{0 \leq |A| \leq \nu_i - 1} \left\{ -\frac{1}{\nu_i - |A|} \nu_0((f_i)_A(h(t))) \right\} \quad \text{for } 1 \leq i \leq m. \tag{3.7}$$

Finally, we obviously have

$$\frac{1}{\nu_i - |A|} \nu_0\big((f_i)_A(h(t))\big) = \frac{1}{b} \nu_0\big((f_i)_A(h(t))\big)^{b/(\nu_i - |A|)}. \tag{3.8}$$

The theorem follows from (3.6), (3.7), (3.8) above and the definition of $\mathbf{J}$. □

Definition 3.3.7 The idealistic exponent $((\mathbf{J}, b))$, where $\mathbf{J}$ and b satisfy the condition of Theorem 3.3.2, is called the **contact exponent** of r with X at x.

Remark 3.3.6 (The existence of suited tangential intersections) We now translate the result of Sect. 2.3 of Chap. 2 into the language of suited tangential intersections.

Let f_i, for $1 \leq i \leq k$, be non-units in $\mathcal{O}_{Z,x} = \mathbb{C}\{\mathbf{z}, \mathbf{w}\}$. Take initial forms $\bar{f}_i = \text{in}_x(f_i) \in \mathbb{C}[\bar{\mathbf{z}}, \bar{\mathbf{w}}]$, and write them in the form

$$\bar{f}_i = \varphi_i(\bar{\mathbf{z}}) + \psi_i(\bar{\mathbf{z}}, \bar{\mathbf{w}}),$$

with $\varphi_i(\bar{\mathbf{z}}) = \bar{f}_i(\bar{\mathbf{z}}, 0)$ and, in consequence, $\psi_i(\bar{\mathbf{z}}, \bar{\mathbf{w}}) \in (\bar{\mathbf{w}}) \cdot \mathbb{C}[\bar{\mathbf{z}}, \bar{\mathbf{w}}]$. We know that the retraction associated with the coordinate system is transversal to the hypersurface defined (locally at x) by $f_i = 0$ if $\varphi_i(\bar{\mathbf{z}}) \neq 0$, and that, with the notation introduced in Chap. 2, Remark 2.3.3, we have

$$\bar{u}(I) = \bar{u}(\varphi_1, \ldots, \varphi_k),$$

where I is the ideal generated by the f_i.

Proposition 3.3.2 *Let P be the ideal of a smooth subspace W of Z containing x, let $\{f_i\}_{1\le i\le m}$ be a P-normalized standard basis of $I_{Z,X,x}$ with respect to $(\mathbf{z}, \mathbf{w})$ and, for each $i = 1, \ldots, m$, let $H^{(i)}$ be the hypersurface in Z defined, locally around x, by $(f_i)\mathcal{O}_{Z,x}$. Then, for every $i = 1, \ldots, m$, the hypersurface $H^{(i)}$ is suited to X at x with respect to r, so that $X = \bigcap_{i=1}^{m} H^{(i)}$ is a tangential intersection suited to X at x with respect to r.*

Proof Since r is transversal to each $H^{(i)}$ at x, all we need to show is that, for every test-map $h : \mathbb{D} \to W$, we have

$$\delta_{x_h}(X_h, \mathbb{D}_h) \le \delta_{x_h}(H_h^{(i)}, \mathbb{D}_h) \quad \text{for } 1 \le i \le m.$$

We observe first that the conditions of Chap. 2, Definition 2.3.1 and Remark 2.3.3 (3), are not affected under a base extension by h. Therefore, in order to prove the proposition, we may assume that W is a curve, in which case we must prove that

$$\delta_x(X, W) \le \delta_x(H^{(i)}, W) \quad \text{for } 1 \le i \le m.$$

If $\delta_x(X, W) = 1$, the result is clear, so we assume that $\delta_x(X, W) > 1$. To prove the desired inequality, it is enough to show that, for δ in the interval $\big(1, \delta_x(X, W)\big)$, we have

$$\delta \le \delta_x(H^{(i)}, W) \quad \text{for } 1 \le i \le m.$$

Assume, to the contrary, that there exists $\delta \in \big(1, \delta_x(X, W)\big)$ and an integer j such that $\delta > \delta_x(H^{(j)}, W)$. We may take j to be minimal. Since $\delta < \delta_x(X, W)$, there is a family $\{g_i\}_{1\le i\le q}$ of elements in $I_{Z,X,x}$ satisfying these properties:

(1) For $1 \le i \le q$, is $\nu_x^{W,\delta}(g_i) = \nu_x(g_i)$.
(2) For $1 \le i \le q$, is $\tilde{g}_i = \mathrm{in}_x^{W,\delta}(g_i) \in \mathbb{C}[\tilde{\mathbf{z}}]$.
(3) The family $(\tilde{g}_1, \ldots, \tilde{g}_q)$ is a basis of $\mathrm{in}_x^{W,\delta}(Z, X)$.

On the other hand, since $(\bar{f}_1, \ldots, \bar{f}_m)\mathbb{C}[\bar{\mathbf{z}}, \bar{\mathbf{w}}] = \mathrm{in}_x(Z, X)$, and since $\delta < \delta_x(X, W)$, we have

$$\bar{g}_i = \tilde{g}_i(\bar{\mathbf{z}}) \in (\bar{f}_1, \ldots, \bar{f}_m)\mathbb{C}[\bar{\mathbf{z}}, \bar{\mathbf{w}}] \quad \text{for } 1 \le i \le q;$$

by condition (2) above and the definition of the φ_i, taken with $\bar{\mathbf{w}} = \mathbf{0}$, we also have

$$\tilde{g}_i(\bar{\mathbf{z}}) \in (\varphi_1, \ldots, \varphi_m)\mathbb{C}[\bar{\mathbf{z}}] \quad \text{for } 1 \le i \le q.$$

Condition (3) gives

$$\tilde{f}_i(\tilde{\mathbf{z}}, \tilde{\mathbf{w}}) \in (\tilde{g}_1, \ldots, \tilde{g}_q)\mathbb{C}[\tilde{\mathbf{z}}, \tilde{\mathbf{w}}] \quad \text{for } 1 \le i \le m.$$

Putting the last three equations together we get

$$\tilde{f}_i(\bar{\mathbf{z}}, \bar{\mathbf{w}}) \in (\varphi_1, \ldots, \varphi_m)\mathbb{C}[\bar{\mathbf{z}}, \bar{\mathbf{w}}] \quad \text{for } 1 \le i \le m.$$

But $\delta > \delta_x(H^{(j)}, W)$ yields

$$\nu_x^{W,\delta}(f_j) < \nu_x(f_j);$$

therefore

$$\deg_{\bar{\mathbf{z}}}(\tilde{f}_j(\bar{\mathbf{z}}, \bar{\mathbf{w}})) \le \nu_x^{W,\delta}(f_j) < \nu_x(f_j) \le \deg(\varphi_i) \quad \text{for any } i \ge j,$$

where $\deg_{\bar{\mathbf{z}}}(\tilde{f}_j(\bar{\mathbf{z}}, \bar{\mathbf{w}}))$ means the total degree, in $\bar{\mathbf{z}}$, of $\tilde{f}_j(\bar{\mathbf{z}}, \bar{\mathbf{w}})$. Then

$$\tilde{f}_j(\bar{\mathbf{z}}, \bar{\mathbf{w}}) \in (\varphi_1, \ldots, \varphi_{j-1})\mathbb{C}[\bar{\mathbf{z}}, \bar{\mathbf{w}}],$$

so that

$$\mathscr{E}_x(f_j) \cap \bar{u}(\varphi_1, \ldots, \varphi_{j-1}) \supset \mathscr{E}_{\bar{\mathbf{z}}}(\tilde{f}_j(\bar{\mathbf{z}}, \bar{\mathbf{w}})) \cap \bar{u}(\varphi_1, \ldots, \varphi_{j-1})$$

is non-empty, contradicting the assumption of normalization for the f_i. □

We return to the conditions we had throughout this section.

Lemma 3.3.8 *With the notation and hypothesis of Proposition* 3.3.2, *each hypersurface* $H^{(i)}$, *for* $i = 1, \ldots, m$, *is totally suited to* X *at* x *with respect to the projection* $\mathbf{r}$ *to which* r *belongs.*

Proof As in the proof of Proposition 3.3.2, the conditions of Chap. 2, Definition 2.3.1 and Remark 2.3.3(3) are not affected by a basis extension $g : V \to W$. Therefore, what we need to prove is that these conditions are also not affected by a change of retraction inside the projection.

If $r' : Z \to W'$ is a local retraction, around x, belonging to the same projection $\mathbf{r}$ as r, we can find $\sigma(\mathbf{w}) = (\sigma_1(\mathbf{w}), \ldots, \sigma_c(\mathbf{w}))$, with $\sigma_i \in \mathbb{C}\{\mathbf{w}\}$, such that W' is defined by $\mathbf{z} - \sigma(\mathbf{w}) = \mathbf{0}$. Let $\boldsymbol{\xi} = \mathbf{z} - \sigma(\mathbf{w})$; then $(\boldsymbol{\xi}, \mathbf{w})$ is a local coordinate system of Z at x, associated with r'. Then the only problem is in normalization, but by this change of variables, the φ_i are not affected except for the replacement of $\mathbf{z}$ by $\boldsymbol{\xi}$. Moreover, for every $f \in \mathscr{O}_{Z,x}$,

$$\mathscr{E}_{\boldsymbol{\xi}}(f) \subset \{A \in \mathbb{Z}_0^c \mid A + B \in \mathscr{E}_{\mathbf{z}}(f) \text{ for some } B \in \mathbb{Z}_0^c\},$$

so that the property of being normalized is not affected. □

Theorem 3.3.3 *Let* Z *be a smooth complex-analytic space, and let* $x \in Z$. *Suppose* X *and* W *are closed complex subspaces of* Z *around* x, *with* W *smooth, and let* $r : Z \to W$ *be a local retraction, around* x, *transversal to* X *at* x. *Then there* *exists a family* $\{H^{(i)}\}_{1\le i\le m}$ *of hypersurfaces in* Z*, locally around* x*, such that* $X = \bigcap_{i=1}^{m} H^{(i)}$ *is a tangential intersection at* x*, and each* $H^{(i)}$ *is totally suited to* X *at* x *with respect to the projection to which* r *belongs. Moreover, if* Γ *is a smooth subspace of* Z*, locally around* x*, containing* $W \cap X$*, and such that* X *is normally flat along* W *at* x*, then the family* $\{H^{(i)}\}_{1\le i\le m}$ *can be chosen in such a way that* $H^{(i)}$ *is equimultiple along* Γ *around* x*, for* $1 \le i \le m$*.*

Proof This is a geometric translation of the results in Sect. 2.3 of Chap. 2, with the addition of Lemma 3.3.8. To incorporate the condition of equimultiplicity along Γ is a simple exercise. □

3.4 Existence of Maximal Contact

The aim of this section is to prove an existence theorem for maximal contact. We need some preliminary technical lemmas. Throughout this section, Z will be a smooth complex-analytic space, $x \in Z$ a point, X and W are closed complex subspaces of Z containing x, with W smooth, and $\mathbf{r} : Z \to W$ will be a local projection, around x, perpendicular to X at x.

Let $r : Z \to W$ be any local retraction around x, and let $(\mathbf{z}, \mathbf{w})$, with $\mathbf{z} = (z_1, \dots, z_c)$ and $\mathbf{w} = (w_1, \dots, w_d)$, be a local coordinate system on Z centered at x associated with r.

For each integer $\alpha = 1, \dots, c$:

(1) Set $\mathbf{z}(\alpha) = (z_1, \dots, z_\alpha)$.
(2) Let Z_α be a complex subspace of Z representing the germ defined by $(\mathbf{z}(\alpha))\mathcal{O}_{Z,x}$. We shall write also $Z_0 = Z$, and observe that $Z_c = W$.
(3) Define $r_\alpha : Z_{\alpha-1} \to Z_\alpha$ as the local retraction, around x, defined by

$$r_\alpha(z_\alpha, \dots, z_c, \mathbf{w}) = (z_{\alpha+1}, \dots, z_c, \mathbf{w}),$$
$$r_c(z_c, \mathbf{w}) = (\mathbf{w}).$$

(4) Define $r''_\alpha : Z \to Z_\alpha$ as $r''_\alpha = r_\alpha \circ \dots \circ r_1$ and define $r'_\alpha : Z_\alpha \to W$ as $r'_\alpha = r_c \circ \dots \circ r_{\alpha+1}$; then $r = r'_\alpha \cdot r''_\alpha$.
(5) We shall often write any $g \in \mathcal{O}_{Z,x}$ in the form

$$g = \sum_{A \in \mathbb{Z}_0^\alpha} g_A \cdot \mathbf{z}(\alpha)^A,$$

where $g_A \in \mathbb{C}\{z_{\alpha+1}, \dots, z_c, \mathbf{w}\}$.

Lemma 3.4.1 *Let* H *be a hypersurface in* Z *containing* x*, defined around* x *by the ideal* $(f) \cdot \mathcal{O}_{Z,x}$*, and such that* r *is a transversal to* H *at* x*. Let* α *be an integer, with* $1 \le \alpha \le c$*, such that there exists an element* $A \in \mathbb{Z}_0^\alpha$ *satisfying the following conditions:*

(a) $\mathbf{0} \leq |A| < \nu = \nu_x(f)$.
(b) $\nu = \nu_x(f_A) + |A|$.
(c) *If H_A is the hypersurface in Z_α defined, at x, by $(f_A)\mathcal{O}_{Z_\alpha,x}$, r'_α has maximal contact with H_A at x.*

Then, if Γ is a smooth curve in Z contained in Z_α, containing x, and such that r'_α induces an isomorphism $\Gamma \to r'_\alpha(\Gamma)$ at x, we have

$$\delta_x(H, \Gamma) \leq \delta_x(H, r'_\alpha(\Gamma)).$$

Proof Since $r'_\alpha(\Gamma)$ is a smooth curve contained in W, we can always choose the local coordinate system on Z, centered at x, and associated with r, in such a way that $r'_\alpha(\Gamma)$ is defined in Z, around x, by the ideal $(\mathbf{z}, w_2, \ldots, w_d)\mathcal{O}_{Z,x}$. In this coordinate system Γ is defined in Z, around x, by an ideal of the type

$$(\mathbf{z}(\alpha),\ z_{\alpha+1}-\sigma_{\alpha+1}(w_1),\ \ldots,\ z_c-\sigma_c(w_1),\ w_2,\ \ldots,\ w_d)\,\mathcal{O}_{Z,x},$$

where $\sigma_j(w_1) \in \mathbb{C}\{w_1\}$ for $\alpha + 1 \leq j \leq c$.

Set $\delta = \delta_x(H, \Gamma)$. If $\delta = 1$, the assertion of the lemma is trivially true; let's assume, then, that $\delta > 1$. For brevity, we shall write

$$\mu = \min_{\alpha+1\leq j\leq c} \{\nu_x(\sigma_j(w_1))\},$$

$$\mathbf{z}' = (z_1, \ldots, z_\alpha, z_{\alpha+1} - \sigma_{\alpha+1}(w_1), \ldots, z_c - \sigma_c(w_1)),$$

$$\mathbf{w}' = (w_2, \ldots, w_d).$$

Then

$$f = \sum_{E,F,c} f_{E,F,c}\mathbf{z}'^E \mathbf{w}'^F w_1^c, \quad \text{with } |E| + |F| + \frac{c}{\delta} \geq \nu,$$

and we have the following facts:

(1) If $(E, F, c) \in \mathcal{E}_{\mathbf{z}',\mathbf{w}',w_1}(f) \subset \mathbb{Z}_0^c \times \mathbb{Z}_0^{d-1} \times \mathbb{Z}_0$, then $|E| + |F| + c/\delta \geq \nu$.
(2) There exists $(E, \mathbf{0}, 0) \in \mathcal{E}_{\mathbf{z}',\mathbf{w}',w_1}(f)$ such that $|E| = \nu$.

Now we must consider two cases:

Case 1. $\mu \geq \delta$. If we change the coordinate system from $(\mathbf{z}', \mathbf{w}', w_1)$ to $(\mathbf{z}, \mathbf{w}, w_1)$, to compute $\delta_x(H, r'_\alpha(\Gamma))$, facts (1) and (2) imply that one has $|E| + |F| + (1/\delta)c \geq \nu$ whenever $(E, F, c) \in \mathcal{E}_{\mathbf{z},\mathbf{w}',w_1}(f)$, and that there exists $(E, \mathbf{0}, 0) \in \mathcal{E}_{\mathbf{z},\mathbf{w}',w_1}(f)$ such that $|E| = \nu$. Hence

$$\delta_x(H, r'_\alpha(\Gamma)) \geq \delta,$$

and the lemma holds.

Case 2. $\mu < \delta$. We shall see that we cannot have this possibility, under the assumptions of the lemma.

Identify, around x, the image $r'_\alpha(\Gamma)$ with the unit disk $\mathbb{D}$ in $\mathbb{C}$ and let $h : \mathbb{D} \to W$ be the embedding corresponding to this identification. Then, by assumption (c) of the lemma, we have

$$\delta_{x_h}((H_A)_h, \Gamma_h) \leq \delta_{x_h}((H_A)_h, \mathbb{D}_h), \tag{3.9}$$

where, in this case, the basis extension by h is the section by $\mathbf{w}' = \mathbf{0}$.

Since, by assumption (c), r'_α is transversal to H_A at x, we have

$$f_A \notin (\mathbf{w}) \cdot \mathbb{C}[z_{\alpha+1}, \ldots, z_c, \mathbf{w}];$$

hence there exists $B \in \mathbb{Z}_0^{c-\alpha}$ such that $|B| = \nu - |A| \neq 0$, and $((A, B), \mathbf{0}) \in \mathscr{E}_{\mathbf{z},\mathbf{w}}(f)$. From this it follows that

$$\delta = \delta_x(H, \Gamma) \leq \delta_{x_h}((H_A)_h, \Gamma_h).$$

Because $\mu < \delta$, and recalling the definition of μ and the identification of $\mathbb{D}$ with $r'_\alpha(\Gamma)$, we deduce from the preceding inequality that

$$\begin{aligned}\delta_{x_h}((H_A)_h, \Gamma_h) \geq \delta > \mu &= \delta_{x_h}(\mathbb{D}_h, \Gamma_h)\\ &\geq \min\{\delta_{x_h}((H_A)_h, \mathbb{D}_h), \delta_{x_h}(\mathbb{D}_h, \Gamma_h)\}.\end{aligned}$$

Thus

$$\delta_{x_h}((H_A)_h, \mathbb{D}_h) = \delta_{x_h}(\mathbb{D}_h, \Gamma_h) = \mu < \delta_{x_h}((H_A)_h, \Gamma_h),$$

which contradicts inequality in (3.9) of the hypothesis, and so proves the lemma. □

Lemma 3.4.2 *Let Z be a smooth $\mathbb{C}$-space, let X, W be closed subspaces of Z, and take $x \in X \cap W$. Let $\mathbf{r} : Z \to W$ be a local projection around x, perpendicular to X at x, and let $r_1 : Z \to W_1$ and $r_2 : Z \to W_2$ be local retractions belonging to* $\mathbf{r}$. *If $\delta_x(X, W_2) > \min\{\delta_x(X, W_2), \delta_x(W_1, W_2)\}$, then $\delta_x(X, W_2) = \delta_x(W_1, W_2)$.*

Proof Since r_1 and r_2 belong to the same projection, we may choose two local coordinate systems on Z centered at x, say

$$(\mathbf{z}, \mathbf{w}) = (z_1, \ldots, z_c, w_1, \ldots, w_d) \quad \text{and} \quad (\mathbf{z}', \mathbf{w}') = (z'_1, \ldots, z'_c, w'_1, \ldots, w'_d),$$

associated with r_1 and r_2, and such that

$$\begin{aligned} z'_j &= z_j - \sigma_j(\mathbf{w}), && \text{with } \sigma_j(\mathbf{w}) \in \mathbb{C}\{\mathbf{w}\} \text{ for } 1 \leq j \leq c,\\ w'_i &= w'_i(\mathbf{w}), && \text{with } w'_i(\mathbf{w}) \in \mathbb{C}\{\mathbf{w}\} \text{ for } 1 \leq j \leq d,\end{aligned}$$

where the second relation is a coordinate change within W. Then the local equations of W_1 on $(\mathbf{z}', \mathbf{w}')$ are

$$z'_j + \sigma_j(\mathbf{w}) = 0 \quad \text{for } 1 \leq j \leq c,$$

where $w_i = w_i(\mathbf{w}')$ by inverting $w'_i = w'_i(\mathbf{w})$. Hence

$$\delta_x(W_1, W_2) = \min_{1 \leq j \leq c} \{\nu_x(\sigma_j(\mathbf{w}))\}. \tag{3.10}$$

Let $X = \bigcap_{i=1}^m H^{(i)}$ be a tangential intersection totally suited to X at x, with respect to $\mathbf{r}$, and let $(f_i)\mathcal{O}_{Z,x}$ be the ideal of H^i at $\mathcal{O}_{Z,x'}$. Then, by Corollary 3.3.2,

$$\begin{aligned} \delta_x(X, W_1) &= \min_{1 \leq i \leq m} \{\delta_x(H^{(i)}, W_1)\}, \\ \delta_x(X, W_2) &= \min_{1 \leq i \leq m} \{\delta_x(H^{(i)}, W_2)\}. \end{aligned} \tag{3.11}$$

Let $g \in \mathcal{O}_{Z,x}$, and set

$$\begin{aligned} \mu &= \min\{\delta_x(X, W_2), \delta_x(W_1, W_2)\}, \\ \bar{g} &= \mathrm{in}_x(g), \\ \tilde{g} &= \mathrm{in}_x^{W_2\mu}(g), \\ \hat{g} &= \mathrm{in}_x^{W_1\mu}(g), \\ \sigma_j^*(\mathbf{w}) &= \begin{cases} 0 & \text{if } \nu_x(\sigma_j(\mathbf{w})) > \mu, \\ \bar{\sigma}_j(\mathbf{w}) & \text{if } \nu_x(\sigma_j(\mathbf{w})) = \mu, \end{cases} \\ \boldsymbol{\sigma}^*(\mathbf{w}) &= (\sigma_i^*(\mathbf{w}), \ldots, \sigma_c^*(\mathbf{w})). \end{aligned}$$

By assumption,

$$\delta_x(X, W_1) > \mu,$$

and by (3.11) we have, for $1 \leq i \leq m$,

$$\begin{aligned} \bar{f}_i &\in \mathbb{C}[\bar{\mathbf{z}}], \\ \hat{f}_i(\hat{\mathbf{z}}) &= f_i(\hat{\mathbf{z}}), \\ \tilde{f}_i(\tilde{\mathbf{z}}', \tilde{\mathbf{w}}') &= \bar{f}_i(\tilde{\mathbf{z}}' + \boldsymbol{\sigma}^*(\tilde{\mathbf{w}})). \end{aligned}$$

Now, for a contradiction, assume that the lemma is false. There are two possibilities:

Case 1. $\delta_x(W_1, W_2) > \delta_x(X, W_2)$, or, equivalently, $\delta_x(W_1, W_2) > \mu$. Then, by (3.10), we have $\nu_x(\sigma_i(\mathbf{w})) > \mu$ for $1 \leq i \leq c$, hence $\boldsymbol{\sigma}^*(\mathbf{w}) = \mathbf{0}$, hence

$\tilde{f}_i(\tilde{\mathbf{z}}', \tilde{\mathbf{w}}) = \tilde{f}_i(\tilde{\mathbf{z}}')$ for $1 \le i \le m$. It follows that $\delta_x(H^{(i)}, W_2) > \mu$ for $1 \le i \le m$, hence $\delta_x(X, W_2) > \mu$, in contradiction with the definition of μ.

Case 2. $\delta_x(X, W_2) > \delta_x(W_1, W_2)$, or, equivalently, $\delta_x(X, W_2) > \mu$. Then, by (3.11), we have $\delta_x(H^{(i)}, W_2) > \mu$ for $1 \le i \le m$, hence $\tilde{f}_i \in \mathbb{C}[\tilde{\mathbf{z}}]$ for $1 \le i \le m$. The expression of $\tilde{f}(\tilde{\mathbf{z}}', \tilde{\mathbf{w}})$ then yields

$$\tilde{f}_i(\tilde{\mathbf{z}}' + \boldsymbol{\sigma}^*(\tilde{\mathbf{w}})) = \tilde{f}_i(\tilde{\mathbf{z}}') \quad \text{for } 1 \le i \le m.$$

On the other hand, since $\delta_x(X, W_2) > \mu$, we have

$$\mu = \delta_x(W_1, W_2) = \min_{1 \le i \le c} (\nu_x(\sigma_i(\mathbf{w}))),$$

so there exists at least one $j = 1, \ldots, c$ such that

$$\sigma_j^*(\mathbf{w}) \neq 0.$$

Then, for some $\mathbf{a} \in \mathbb{C}^d$, we have

$$\mathbf{v} := \boldsymbol{\sigma}^*(\mathbf{a}) \neq 0,$$

and $\tilde{f}_i(\tilde{\mathbf{z}} + \mathbf{v}) = \tilde{f}_i(\tilde{\mathbf{z}})$ implies $\mathbf{v} \in T_{\mathbf{r}^{-1}(x) \cap X, x}$, in contradiction with the assumption of perpendicularity of $\mathbf{r}$. □

Lemma 3.4.3 *Let* $\mathbf{r}$ *be a local retraction belonging to* $\mathbf{r}$ *perpendicular to* X *at* x *and let* $X = \bigcap_{i=1}^m H^{(i)}$ *be a tangential intersection totally suited to* X *at* x *with respect to* $\mathbf{r}$. *Let* $(f_i)\mathcal{O}_{Z,x}$ *be the ideal of* $H^{(i)}$ *in* $\mathcal{O}_{Z,x}$, *for* $i = 1, \ldots, m$, *and assume that, for each* $\alpha = 1, \ldots, c$, *there exists a pair* $(i(\alpha), A(\alpha))$, *where* $1 \le i(\alpha) \le m$ *and* $A(\alpha) \in \mathbb{Z}_0^{\alpha-1}$, *satisfying the following conditions:*

(1) $0 \le |A(\alpha)| < \nu_{i(\alpha)} = \nu_x(f_{i(\alpha)})$.

(2) $\nu_x(f_{i(\alpha), A(\alpha)}) = \nu_{i(\alpha)} - |A(\alpha)|$, *where we have written*

$$f_{i(\alpha)} = \sum_{A \in \mathbb{Z}_0^{\alpha-1}} f_{i(\alpha), A}\mathbf{z}(\alpha-1)^A \quad \textit{with } f_{i(\alpha), A} \in \mathbb{C}\{z_\alpha, z_{\alpha+1}, \ldots, z_c, \mathbf{w}\}.$$

(3) *If* H_α *is the hypersurface in* $Z_{\alpha-1}$ *defined, around* x, *by the ideal*

$$(f_{i(\alpha), A(\alpha)})\mathcal{O}_{Z_{\alpha-1}, x}$$

then r_α *has maximal contact with* H_α *at* x.

Proof Let $h : \mathbb{D} \to W$ be a test-map at x. It is clear that all the assumptions of the lemma remain true after basis extension by h. Therefore, in order to prove the lemma, it is enough to consider the case $\dim_x(W) = 1$ and to prove that

$$\delta_x(X) = \delta_x(X, W).$$

Assume to the contrary that $\delta_x(X) \neq \delta_x(X, W)$. Then there exists a smooth curve Γ in Z containing x and such that

$$\delta_x(X, \Gamma) > \delta_x(X, W).$$

Since, by assumption, r is perpendicular to X at x, and since $\delta_x(X, \Gamma) > 1$, we see that $T_{\Gamma,x} \subset T_{X,x}$ implies

$$T_{\Gamma,x} \cap T_{r^{-1}(x),x} \subset T_{X,x} \cap T_{r^{-1}(x)} = \{0\}.$$

Then

$$T_{\Gamma,x} \cap T_{r^{-1}(x),x} = \{0\},$$

so r induces, locally, an isomorphism $r : \Gamma \to W$. Therefore, there exist functions $\sigma_i(w) \in \mathbb{C}\{w\}$, for $1 \leq i \leq c$, such that the ideal of Γ at x is

$$(z_1-\sigma_1(w), \ \ldots, \ z_c-\sigma_c(w))\mathcal{O}_{Z,x};$$

hence

$$\delta_x(\Gamma, W) = \min_{1\leq i\leq c} \{\nu_x(\sigma_i(w))\}.$$

We had above that $\delta_x(X, \Gamma) > \delta_x(X, W)$; since we also have $\delta_x(X, W) = \delta_x(\Gamma, W)$, we conclude that

$$\delta_x(X, \Gamma) > \delta_x(\Gamma, W) = \min_{1\leq i\leq c} \{\nu_x(\sigma_i(w))\}.$$

We denote this number by δ.

Observe that we may assume that

$$\sigma_i(w) = \lambda_i w^\delta, \text{ with } \lambda_i \in \mathbb{C} \text{ for } 1 \leq i \leq c;$$

in fact, if $\sigma_i(w) = \sum_j \lambda_{ij} w^{\delta+j}$, the curve Γ' defined in Z around x by the ideal

$$(z_1-\lambda_{10}w^\delta, \ \ldots, \ z_c-\lambda_{c0}w^\delta)\mathcal{O}_{Z,x}$$

also satisfies

$$\delta_x(X, \Gamma') > \delta,$$

so we may use Γ' instead of Γ.

Let α be the least index such that $\lambda_\alpha \neq 0$. If $\Gamma_\beta = r''_\beta(\Gamma)$, for all β, $1 \leq \beta \leq c$, we have $\Gamma_{\alpha-1} = \Gamma$ and $\Gamma_\alpha \neq \Gamma$. Since we have the diagram

$$\begin{array}{ccccc} Z & \xrightarrow{r''_{\alpha-1}} & Z_{\alpha-1} & \xrightarrow{r_\alpha} & Z_\alpha \\ \uparrow & & \uparrow & & \uparrow \\ \Gamma & \xrightarrow{r''_{\alpha-1}} & \Gamma_{\alpha-1} & \xrightarrow{r_\alpha} & \Gamma_\alpha \end{array}$$

we are in a position to apply Lemma 3.4.2 to $(Z, r''_\alpha, Z_\alpha, H^{i(\alpha)})$ instead of (Z, r, W, H) with the curve $\Gamma \subset Z_{\alpha-1}$, thus obtaining

$$\delta_x(H^{(i(\alpha))}, \Gamma) \leq \delta_x(H^{(i(\alpha))}, \Gamma_\alpha).$$

Since r_α has maximal contact with H_α at x, and because the intersection is totally suited, we have

$$\delta_x(H_\alpha, \Gamma_\alpha) \geq \delta_x(H^{i(\alpha)}, \Gamma_\alpha) \geq \delta_x(H^{i(\alpha)}, \Gamma) \geq \delta_x(X, \Gamma) > \delta = \delta_x(W, \Gamma_\alpha)$$

But

$$\delta_x(H, W) \geq \delta_x(H_\alpha, \Gamma_\alpha) > \delta = \delta_x(W, \Gamma_\alpha)$$

implies $\delta_x(H_\alpha, \Gamma_\alpha) = \delta_x(W, \Gamma_\alpha) = \delta$, in contradiction with the preceding set of inequalities. □

Theorem 3.4.1 *Let Z be a smooth complex-analytic space, take $x \in Z$, and let X, W be closed complex subspaces of Z containing x, with W smooth. Let* $\mathbf{r}$ *be a local projection, around x, perpendicular to X at x. Then there exists a local retraction r belonging to* $\mathbf{r}$ *that has maximal contact with X at x.*

Proof We will show the existence of a local retraction r belonging to $\mathbf{r}$ and satisfying the assumption of Lemma 3.4.1. An application of that lemma then proves the theorem. For this, we shall proceed by induction in the following way.

Let $X = \bigcap_{i=1}^m H^{(i)}$ be a tangential intersection totally suited to X at x with respect to $\mathbf{r}$, and let $(f_i)\mathcal{O}_{Z,x}$ be the ideal of $H^{(i)}$ in $\mathcal{O}_{Z,x}$, for $i = 1, \ldots, m$. We shall take an integer α in the range $1 \leq \alpha \leq c$, and we shall assume that there is a local retraction r' belonging to $\mathbf{r}$ for which, and for all β with $1 \leq \beta < \alpha$, the assumptions of Lemma 3.4.1 hold. Our goal will be to prove that, under this induction hypothesis, one can find a local retraction r belonging to $\mathbf{r}$ for which, and for all $\beta = 1, \ldots, \alpha$, the assumptions of Lemma 3.4.1 hold.

Let $(\mathbf{z}', \mathbf{w})$ be a local coordinate system on Z centered at x and associated with r'. Observe first that, for every

$$\sigma(z'_{\alpha+1}, \ldots, z'_c, \mathbf{w}) \in \mathbb{C}\{z'_{\alpha+1}, \ldots, z'_c, \mathbf{w}\},$$

and for every family $\{\xi_j\}_{\alpha<j\le c}$, of complex numbers ξ_j, the local retraction r associated with the local coordinate system on Z, centered at x, defined by

$$\begin{aligned} z_i &= z'_i \quad \text{for } 1 \le i < \alpha, \\ z_\alpha &= z'_\alpha + \sigma(z'_{\alpha+1}, \ldots, z'_c, \mathbf{w}), \\ z_j &= z'_j + \xi_j z'_\alpha, \quad \text{for } \alpha < j \le c, \end{aligned}$$

belongs to $\mathbf{r}$, and the assumptions of Lemma 3.4.1 are satisfied for r and for every β in the range $1 \le \beta < \alpha$. We shall show that σ and the ξ_j can be chosen in such a way that these assumptions are also satisfied for $\beta = \alpha$.

For $1 \le i \le m$, write $\bar{f}_i$ as

$$\bar{f}_i(\bar{\mathbf{z}}', \bar{\mathbf{w}}) = \varphi'_i(\bar{\mathbf{z}}') + \psi'_i(\bar{\mathbf{z}}', \bar{\mathbf{w}}),$$

where $\psi'_i(\bar{\mathbf{z}}', \bar{\mathbf{w}}) \in (\bar{\mathbf{w}})\mathbb{C}[\bar{\mathbf{z}}', \bar{\mathbf{w}}]$. Since r is perpendicular to X at x, there must exist an i in the range $1 \le i \le m$ such that z'_α occurs in $\varphi'_i(\mathbf{z}')$ in some monomial with non-zero coefficient; we denote such an arbitrary i by $i(\alpha)$. If we change to the new coordinate system, we can write

$$\bar{f}_i(\bar{\mathbf{z}}, \bar{\mathbf{w}}) = \varphi_i(\bar{\mathbf{z}}) + \psi_i(\bar{\mathbf{z}}, \bar{\mathbf{w}}) \quad \text{for } 1 \le i \le m,$$

where $\psi_i(\bar{\mathbf{z}}, \bar{\mathbf{w}}) \in (\bar{\mathbf{w}})\mathbb{C}[\bar{\mathbf{z}}, \bar{\mathbf{w}}]$, and, by a suitable choice of the $\xi_j \in \mathbb{C}$ for $\alpha < j \le c$, one can ensure that $\varphi_{i(\alpha)}(\bar{\mathbf{z}})$ includes a non-zero term in $\bar{\mathbf{z}}^A_{\alpha-1}\bar{z}^a_\alpha$, with $a > 0$ and $A \in \mathbb{Z}_0^{\alpha-1}$. If we write $A(\alpha) = A$, it is clear that, for arbitrary $\sigma(z'_{\alpha+1}, \ldots, z'_c, \mathbf{w})$, $i(\alpha)$, and $A(\alpha)$ verify conditions (a) and (b) in Lemma 3.4.1.

On the other hand, by the Weierstrass preparation theorem, one can write

$$f_{i(\alpha),A(\alpha)} = u \cdot g,$$

where u is a unit and $g = z^a_\alpha + g_1 z^{a-1}_\alpha + \cdots + g_a$, with $g_i \in \mathbb{C}\{z_{\alpha+1}, \ldots, z_c, \mathbf{w}\}$ for each i. Then if we take σ so that $g_1 = 0$, which is always possible, condition (c) in Lemma 3.4.1 is also verified (compare the proof of Theorem 3.2.3). □

3.5 The Continuity of Perpendicularity

The aim of this section is to prove the following result. Let Z be a smooth complex-analytic space, let X be a closed subspace of Z, and let W be a smooth subspace of Z. Take $x \in X \bigcap W$, and let S be the Samuel stratum of X containing x and $r : Z \to W$ a local retraction around x. With this hypothesis we shall see that, if r is perpendicular to X at x, there exists a neighborhood U of x in $S \bigcap W$ such that r is perpendicular to X at every point of U.

Proposition 3.5.1 (Strict tangent space) *Let X be a complex-analytic space, $a : E \to X$ a complex vector bundle, $b : C \to X$ a cone fiber subspace of E. If b is flat, there exists a vector fiber subspace T of E, over X, such that*

$$T(x) = \{\mathbf{v} \in E(x) \mid C(x) + \mathbf{v} = C(x)\}$$

for all $x \in X$.

Proof Denote by G the graded $\mathcal{O}_X$-algebra of linear LF type such that $E = \operatorname{Specan} G$. Let $\mathbf{J}$ be the ideal of C in G, set $\bar{G} = G/\mathbf{J}$, and let $\mathbf{J}_d$, G_d, and $\bar{G}_d$ be the homogeneous arts of degree d of $\mathbf{J}$, G, and $\bar{G}$. All these homogeneous parts are $\mathcal{O}_X$-modules; furthermore, since $b : C \to X$ is flat and E is a vector bundle they are locally free $\mathcal{O}_X$-modules.

We shall write, for every $k \in \mathbb{Z}_+$,

$$F^{(k)} = \mathbf{Hom}_{\mathcal{O}_X}\left(\bigoplus_{i=0}^{k} J_i, \bigoplus_{i=0}^{k} \bar{G}_i\right).$$

Since J and the $\bar{G}_i$ are locally free, by taking the residual fiber at an arbitrary point $x \in X$ we have

$$F^{(k)}(x) = \operatorname{Hom}\left(\bigoplus_{i=0}^{k} J_i(x), \bigoplus_{i=0}^{k} \bar{G}_i(x)\right). \tag{3.12}$$

On the other hand, if

$$D = \mathbf{Hom}_{\mathcal{O}_X}(G_1, \mathcal{O}_X)$$

is the $\mathcal{O}_X$-module dual to G_1, then D can be extended in a natural way to $\operatorname{Der}_{\mathcal{O}_X}(G)$, the sheaf of $\mathcal{O}_X$-derivations of G. This gives, for every $k \in \mathbb{Z}_+$, a natural morphism $\varphi^k : D \to F^{(k)}$, and, by passage to the dual, another morphism

$$\psi^k : F^{(k)^*} = \mathbf{Hom}_{\mathcal{O}_X}(F^{(k)}, \mathcal{O}_X) \to G_1.$$

Associated with these morphisms, we can construct two sequences of coherent $\mathcal{O}_X$-modules,

$$\{\ker \varphi^k\}_{1 \le k \le \infty} \quad \text{and} \quad \{\operatorname{im} \psi^k\}_{1 \le k \le \infty},$$

which are monotone, non-decreasing, and locally stationary. In fact, for all $k \geq 2$, the inclusion

$$\bigoplus_{i=0}^{k-1} J_i \to \bigoplus_{i=0}^{k} J_i$$

and the projection

$$\bigoplus_{i=0}^{k} \bar{G}_i \to \bigoplus_{i=0}^{k-1} \bar{G}_i$$

induce a morphism $\mu^{k-1} : F^{(k)} \to F^{(k-1)}$ such that $\varphi^{k-1} = \mu^{k-1}\varphi^k$. This means that the sequence

$$\{\ker \varphi^k\}_{1\leq k\leq\infty}$$

is monotone and non-decreasing, and, by duality, so is the sequence

$$\{\operatorname{im} \psi^k\}_{1\leq k\leq\infty}.$$

Both sequences are locally stationary because their terms are coherent.

Now, if we write

$$L = \bigcup_{k=1}^{\infty} \operatorname{im} \psi^k,$$

L is a coherent $\mathcal{O}_X$-module (not locally free in general), contained in G_i. Then $L\mathcal{O}_E$ defines a vector fiber subspace T of E that, as we shall see, satisfies the condition of the proposition.

To show this, observe that, if we denote by $L\langle x\rangle$, for every $x \in X$, the image of $L(x)$ under the homomorphism $L(x) \to G_1(x)$ associated with the inclusion $L \to G_1$, then $T(x)$ is defined in $E(x)$ by $L\langle x\rangle \cdot G(x)$ and, in addition,

$$L\langle x\rangle = \bigcup_{k=1}^{\infty} \operatorname{im} \psi^k(x).$$

Then, by taking into account the behavior at the fibers of the $F^{(k)}$ given by (3.12), it is enough to take $X = \{x\}$ in order to prove that T satisfies the desired conditions.

Then $G = \mathbb{C}[t_1, \ldots, t_c]$, with the natural graduation, where the t_i are independent variables. Moreover,

$$D = \sum_{i=1}^{c} \mathbb{C}\frac{\partial}{\partial t_i},$$

and we have $J = (h_1, \ldots, h_m)G$, where the h_i are homogeneous, for $1 \leq i \leq m$. Then, if

$$k_0 = \sup_{1\leq i\leq m} \{\deg(h_i)\},$$

the definitions of φ^k and ψ^k imply that $\ker\varphi^k = \ker\varphi^{k_0}$, for all $k \geq k_0$, and that $\operatorname{im}\psi^k = \operatorname{im}\psi^{k_0}$, for all $k \geq k_0$. Moreover, for any $\partial \in D$,

$$\partial \in \ker\varphi^{k_0} \Leftrightarrow \partial(g) \in J \text{ for all } g \in J. \tag{3.13}$$

Combining this with the definitions of L, φ^k, and ψ^k, we see that we may always choose the variables $\{t_1, \ldots, t_c\}$ in such a way that

$$L = \sum_{j=1}^{e} \mathbb{C} \cdot t_j \quad \text{and} \quad \ker\varphi_0^k = \sum_{i=e+1}^{c} \mathbb{C} \cdot \frac{\partial}{\partial t_i} \text{ for } 1 \leq e \leq c. \tag{3.14}$$

By (3.13), for every $g \in J$ and for every $\boldsymbol{\alpha} = (\alpha_{e+1}, \ldots, \alpha_c) \in \mathbb{Z}_0^{c-e}$, we have

$$\frac{\partial^{|\boldsymbol{\alpha}|}}{\partial \mathbf{t}^{\boldsymbol{\alpha}}} g \in J.$$

Therefore J is generated by elements in $\mathbb{C}[t_1, \ldots, t_e]$, and consequently T—which, by (3.14), is defined by the equations $t_1 = \cdots = t_e = 0$—leaves C invariant by translations.

With these arguments we have shown that, if T' is the maximum subspace of E that leaves C invariant by translations, then $T \subset T'$. We now prove the opposite inclusion.

Observe that we may always choose the variables t_i in such a way that (3.14) holds, and that there exists an integer e' in the range $1 \leq e' \leq e$, such that T' is defined by

$$L' = \sum_{i=1}^{e'} \mathbb{C} \cdot t_i.$$

Since T' leaves C invariant by translations, J has a basis in $\mathbb{C}[t_1, \ldots, t_{e'}]$. By (3.14), we then have

$$\sum_{i=e'+1}^{c} \mathbb{C} \frac{\partial}{\partial t_i} \subset \ker\varphi^{k_0}.$$

Therefore $e' \geq e$, hence $T' = T$. □

Remark 3.5.1 Consider this assumptions: Z is a smooth complex-analytic space, X a closed subspace of Z, W a smooth subspace of Z, and x a point in $X \cap W$. Let S be the Samuel stratum of X containing x, and take a local retraction $r : Z \to W$, around x. The results in Sect. 2.4 of Chap. 2 then imply the following facts:

(1) The $\mathcal{O}_s$-module $\tan_S(X)$ is a locally free $\mathcal{O}_s$-module.

(2) For every $s \in S$, the natural homomorphism

$$\mathrm{gr}\,\mathscr{P}_X(S)(s) \to \mathrm{gr}(\mathscr{P}_X)(s)$$

is an isomorphism.
(3) If $b : T_Z(S) \to S$ is the tangent vector bundle to Z along S, then $T_Z(S) = \mathrm{Specan}(\tan_S(Z))$.

Lemma 3.5.1 *With the assumptions of Remark* 3.5.1, *there exists a cone fiber subspace $a : C_X(S) \to S$ of $T_Z(S)$ such that, for every $s \in S$, the fiber $a^{-1}(s)$ is the tangent cone $C_{X,S}$ to X at S.*

Proof Let K be the kernel of the natural homomorphism

$$dr : \tan_S(Z) \to \tan_S(X).$$

By Remark 3.5.1, K defines a cone fiber subspace $C_X(S)$ of $T_Z(S)$, which is the desired cone fiber subspace. □

Definition 3.5.1 The cone fiber space $(C_X(S), a)$ of Lemma 3.5.1 will be called the **tangent cone fiber space** to X along S.

Theorem 3.5.1 *With the assumptions of Remark* 3.5.1, *the map $a : C_X(S) \to S$ is flat.*

The theorem is the geometrical translation of the following result:

Lemma 3.5.2 *Let A, B be local rings, let $d : A \to B$ be a local morphism and let I be an ideal of A, such that A and B are separated respectively for the I and the IB topologies. Then d is flat if an only if the following conditions are verified:*

(1) *The module B/IB is A/I flat.*
(2) *The canonical homomorphism*

$$(B/IB) \bigotimes_{A/I} \mathrm{gr}_I(A) \to \mathrm{gr}_I(B)$$

is an isomorphism.

Proof If d is flat, B/IB is A/I flat and, for all n:

$$(B/IB) \bigotimes_{A/I} (I^n/I^{n+1}) \to IB^n/IB^{n+1}$$

is an isomorphism.

Now by induction, conditions (1) and (2) imply that:

$$\forall n \geq 1 B/IB^n \text{is flat over } A.$$

Let us prove that for any A-module of finite type N and P and for any injective homomorphism $u : N \to P$, $1 \otimes u : B \bigotimes_A N \to B \bigotimes_A P$ is injective. Being both $B \bigotimes_A N$, $B \bigotimes_A P$, B-modules of finite type separated for the IB-adic topology it suffices to prove that the completed morphism:

$$\widehat{1 \otimes u} : \widehat{B \bigotimes_A N} \to \widehat{B \bigotimes_A P}$$

is injective, and that happens because, being $B/I^n B$ A-flat,

$$u_n : (B/I^n B) \bigotimes_A N \longrightarrow (B/I^n B) \bigotimes_A P$$

is injective and then $\widehat{1 \otimes u} = \underleftarrow{lim}\, u_n$ is also. □

Remark 3.5.2 If S is not a Samuel stratum, $C_X(S)$ as above does not exist, in general. For instance, if X is defined by $z_i^2 - z_3 z_2^2 = 0$ in $\mathbb{C}^3$ and Y is the z_2-axis, and if U is any neighborhood of $\mathbf{0}$ in Y, there is no cone fiber subspace $C_X(U)$ of T_z such that $C_X(U)(y) = C_{X,y}$ for all $y \in U$. In fact, $C_{X,y}$ is the z_1z_2-plane for all $y \in U - \{0\}$, while $C_{X,\mathbf{0}}$ is the double z_2z_3-plane.

Even when it exists, this cone fiber subspace may not be flat. For instance, if X' is another subspace defined by $z_3^2 - z_1 z_2^2 = 0$ in $\mathbb{C}^3$, there exists a closed subspace $C_{X \cup X'}(Y)$ of T_Z such that $C_{X \cup X'}(Y)(y) = C_{X \cup X',y}$ for all $y \in Y$. But then the map a induces $C_{X \cup X'}(Y) \to Y$, which is not flat at 0.

Theorem 3.5.2 *If r is transversal to X at x, there exists a neighborhood U of x in $S \cap W$ such that r is transversal to X a every point of U.*

Proof We can construct the diagram

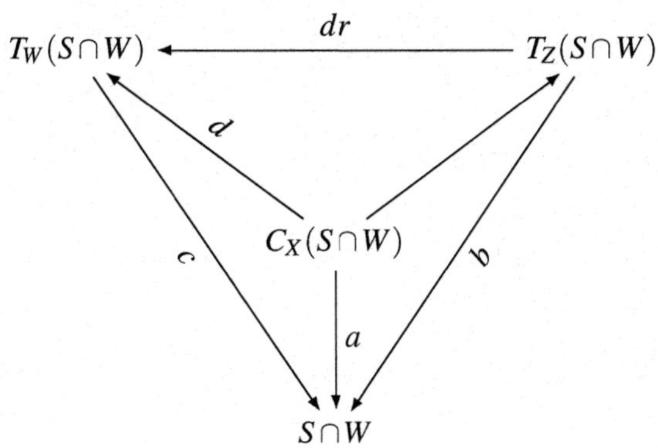

where dr is the morphism induced by r, the morphisms a and c are flat, and d is the restriction of dr to $C_X(S \cap W)$. One has, for every $s \in S \cap W$,

$$C_X(S \cap W)(s) = C_{X,s}$$

and, by the construction of d and $C_X(S \cap W)$, d induces on the fibers the canonical morphism

$$d_s : C_{X,s} \to T_{W,s}$$

for every $s \in S \cap W$.

By the definition of transversality, proving the theorem is equivalent to proving that, if d_x is flat, there exists a neighborhood U of x in $S \cap W$ such that d_s is flat for every $s \in U$. To prove this last point observe that, since the flatness is an open condition invariant under scalar multiplication, in order to show that a morphism between cones is flat, it is enough to show that it is flat at the vertex of the initial cone of the morphism.

We will now apply Lemma 3.5.2 in the following conditions: A and B are the local rings of $T_W(S \cap W)$ and $C_X(S \cap W)$, respectively, at the vertex of the fiber of $C_X(S \cap W)$ at x, A_0 is the local ring of $S \cap W$ at x, M is the maximal ideal of A_0, and $I = MA$. Then for the homomorphism $A \to B$ induced by d_x (localization at x of the restriction of dr to $C_X(S \cap W)$), condition (1) of the lemma follows from the flatness of a. On the other hand, because c and d_x are flat, we have

$$\operatorname{gr}_I(A) \approx (A/I) \bigotimes_{A_0} \operatorname{gr}_M(A_0),$$

$$\operatorname{gr}_I(B) \approx (B/IB) \bigotimes_{A_0} \operatorname{gr}_M(A_0),$$

$$\operatorname{gr}_I(B) \approx (B/IB) \bigotimes_{A/I} \operatorname{gr}_I(A).$$

This implies condition (2) of Lemma 3.5.2. Applying the lemma, we conclude that d is flat at the vertex of the fiber at x of $C_x(S \cap W)$, hence it is flat in the neighborhood of this vertex. In particular, c is flat at the vertices of the fibers of $C_x(S \cap W)$ at the points of a neighborhood U of x in $S \cap W$: then the morphism d_s induced by d is flat, for every point $s \in U$, as we wished to show. □

Lemma 3.5.3 *If r is transversal to X at every point of a neighborhood U of x in $S \cap W$, there exists a cone fiber subspace $e : C_{X|W}(U) \to U$ of $T_Z(U)$ such that e is flat and*

$$C_{X|W}(U)(s) = C_{X \cap r^{-1}(s)s}$$

for every $s \in U$.

Proof Consider the morphism

$$d : C_X(S \cap W) \to T_W(S \cap W)$$

as in the proof of Theorem 3.5.2. Let U_0 be the zero section on U of $T_W(S \cap W)$, write $C_{X|W}(U) = d^{-1}(U_0)$, and let $e : C_{X|W}(U) \to U$ the morphism induced by $a : C_X(S \cap W) \to S \cap W$. It is clear that $(C_{X|W}(U), e)$ is a cone fiber subspace of $T_Z(U)$ and that e is flat.

Consider the diagram

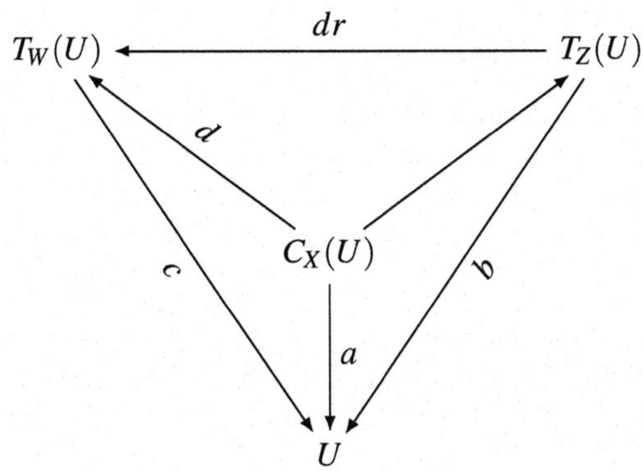

obtained from the diagram in the proof of Theorem 3.5.2 by restricting to U.

Then, by putting $E = dr^{-1}(U_0)$ and by taking the restriction $b' : E \to U$ of b, we obtain a commutative diagram

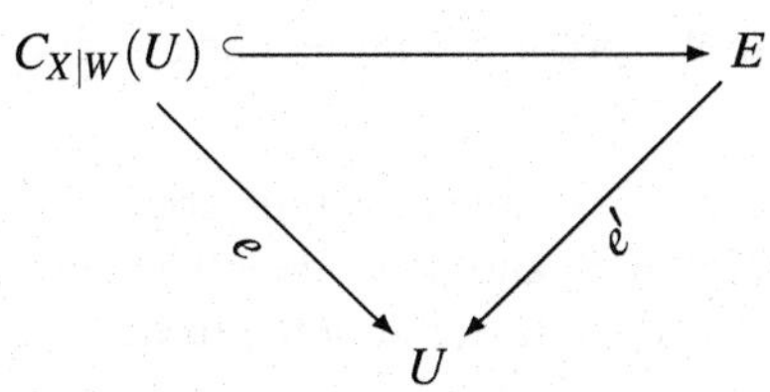

It is obvious from the construction of $C_{X|W}$ that

$$C_{X|W}(U)(s) = C_X(U)(s) \cap E(s),$$

for all $s \in U$. At the same time, from the construction of E and $C_x(U)$ we get

$$C_X(U)(s) = C_{X,s} \cap E(s) = C_{X,s} \cap T_{r^{-1}(s),s},$$

for all $s \in U$, and, from the transversality of r to X at every point $s \in U$,

$$C_{X \cap r^{-1}(s),s} = C_{X,s} \cap T_{r^{-1}(s),s}.$$

These three display relations together imply the conclusion of the lemma. □

Corollary 3.5.1 *Under the notation and hypothesis of Lemma* 3.5.3 *there exists a vector fiber subspace*

$$e' : T_{X|U}(U) \to U$$

of $T_Z(U)$ satisfying the following conditions:

(1) *The vector fiber $T_{X|W}(U)$ is a closed analytic subspace of $C_{X|W}(U)$.*
(2) *For every $s \in U$,*

$$T_{X|W}(U)(s) = \{\mathbf{v} \in T_{r^{-1}(s),s} \mid C_{X\cap r^{-1}(s),s} + \mathbf{v} = C_{X\cap r^{-1}(s),s}\}.$$

Proof Apply Lemma 3.5.3 and Proposition 3.5.1. □

Theorem 3.5.3 (Continuity of the perpendicularity) *Let Z be a smooth complex analytic space, X a closed subspace of Z, W a smooth closed subspace of Z, x a point in $X \cap W$, S the Samuel stratum on X containing x, and $r : Z \to W$ a local retraction around x. If r is perpendicular to X at x, there exists a neighborhood U of x in $S \cap W$ such that r is perpendicular to X at every point of U.*

Proof The retraction r is perpendicular to X at $s \in U$ if and only if it is transversal and $T_{X|U}(U)(s) = (\mathbf{0})$. The theorem follows from the continuity of transversality and Lemma 3.5.3. □

3.6 Two Technical Lemmas

This section is devoted to the proof of two technical Lemmas 3.6.1 and 3.6.2 that will be needed later.

Remark 3.6.1 Let X and Y be complex-analytic spaces, $f : X \to Y$ a morphism of analytic spaces, x a point in X, and $y = f(x)$. Then the morphism $f_x : \mathcal{O}_{Y,y} \to \mathcal{O}_{X,x}$ induced by f can always be factored in the form

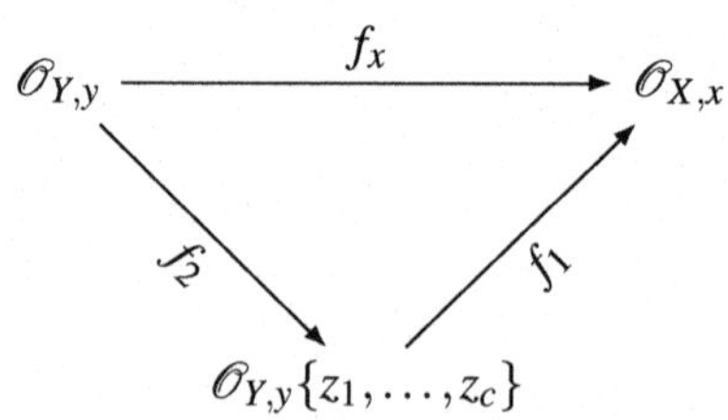

where f_2 is the inclusion, f_1 is an epimorphism, and $z_1, \ldots, z_c$ are independent variables. This implies that, locally around x, f can be factored as

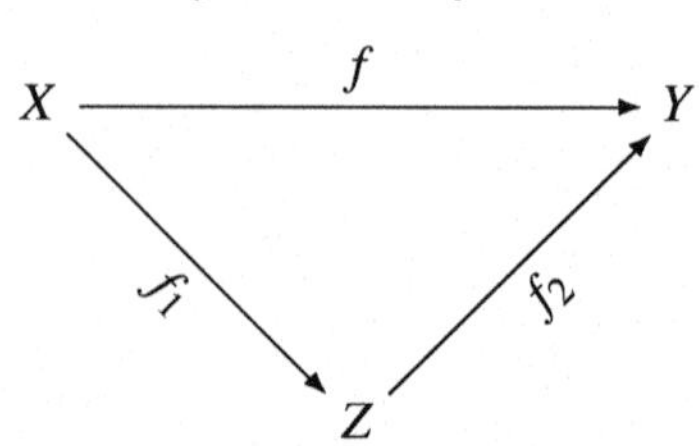

where $Z = Y \times \mathbb{C}^c$, f_1 is a closed embedding, and f_2 is the projection.

Lemma 3.6.1 *Let $l : X \to Y$ and $h : Y' \to Y$ be morphisms of complex-analytic spaces. Set $X' = X \times_Y Y'$. Let $y \in Y$, take $x \in l^{-1}(y)$, $y' \in h^{-1}(y)$, and set $x' = (x, y')$. Assume $(dl)_x : C_{X,x} \to C_{Y,y}$ is flat. Then:*

(1) $C_{X',x'} = C_{X,x} \times_{C_{Y,y}} C_{Y',y'}$.
(2) *If $l' : X' \to Y'$ is the projection, $(dl)_{x'} : C_{X',x'} \to C_{Y',y'}$ is flat.*

Proof Observe first that (2) is an immediate consequence of (1) and of the flatness of $(dl)_x$. Also, in view of Remark 3.6.1, the local character of the lemma, and the fact that the composition of pull-backs is again a pull-back, it is enough to prove the lemma in the following special cases:

Case 1: The morphism h is a closed embedding.
Case 2: $Y' = Y \times \mathbb{C}^s$ and $h : Y' \to Y$ is the first projection.

In both cases we shall assume that $l : X \to Y$ is factored in the form of Remark 3.6.1, as

$$X \xrightarrow{\ l_1\ } Z = Y \times \mathbb{C}^c \xrightarrow{\ l_2\ } Y,$$

and we shall denote by $(\mathbf{z}) = (z_1, \dots, z_c)$ the result of the lifting to Z, around $(y, \mathbf{0})$, via the second projection, of a local coordinate system on $\mathbb{C}^c$ centered at $\mathbf{0}$. Then, x can be identified with $(y, \mathbf{0})$ via l_1.

We denote by R the ring $\mathcal{O}_{Y,y}$, by N its maximal ideal, and we write $I = I_{Z,X,x}$; then $\mathcal{O}_{Z,x} = R\{\mathbf{z}\}$ and $M = N.R\{\mathbf{z}\} + (\mathbf{z})R\{\mathbf{z}\}$ is its maximal ideal.
Case 1: The morphism h is a closed embedding. Here $X' = X \times_Y Y' = l^{-1}(Y')$; we can identify y' with y and x' with x, and conclusion (1) of the lemma is equivalent to

$$C_{X',x} = (dl)_x^{-1}(C_{Y',y'}) \tag{3.15}$$

in $C_{X,x}$ or, equivalently, in $C_{Z,x}$. If we write $J = I_{Y,Y',y'}$, condition (3.15) is obviously equivalent to

$$\mathrm{in}_M(I + J \cdot R\{\mathbf{z}\}) = \mathrm{in}_N(I) + \mathrm{in}_M(J) \cdot \mathrm{gr}_M(R\{\mathbf{z}\}). \tag{3.16}$$

Proving this is equivalent to showing that, for every $g \in I + J \cdot R\{\mathbf{z}\}$, there exist $g_1 \in I$ and $g_2 \in J \cdot R\{\mathbf{z}\}$ such that $g = g_1 + g_2$ and $\mu(g_1, g_2) = \min\{\nu_M(g_1), \nu_M(g_2)\} \geq \nu_M(g)$. We shall prove now this assertion.

Take any $g \in I + J \cdot R\{\mathbf{z}\}$ and any decomposition of g as $g = g_1 + g_2$ with $g_1 \in I$ and $g_2 \in J \cdot R\{\mathbf{z}\}$. If $\mu(g_1, g_2) \geq \nu_M(g)$, the lemma is already proved; assume, then, that $\mu(g_1, g_2) < \nu_M(g)$. In this case

$$\begin{aligned} &\mu(g_1, g_2) = \nu_M(g_1) = \nu_M(g_2), \\ &\mathrm{in}_M(g_1) = -\mathrm{in}_M(g_2) \in \mathrm{in}_M(I) \cap \mathrm{in}_N(J) \cdot \mathrm{gr}_M(R\{\mathbf{z}\}). \end{aligned} \tag{3.17}$$

Further,

$$\mathrm{gr}_M(R\{\mathbf{z}\}) = \mathrm{gr}_N(R)[\bar{\mathbf{z}}],$$

with $\bar{z}_i = \mathrm{in}_M(z_i)$ for $1 \le i \le c$. Since $(dl)_x$ is assumed flat, so is the natural homomorphism

$$\mathrm{gr}_N(R) \to \mathrm{gr}_M(R\{\mathbf{z}\})/\mathrm{in}_M(I).$$

Then, since $\mathrm{in}_N(J)$ is a homogeneous ideal in $\mathrm{gr}_N(R)$, we have

$$\mathrm{in}_M(I) \cap \mathrm{in}_N(J)\mathrm{gr}_M(R\{\mathbf{z}\}) = \mathrm{in}_M(I) \cdot \mathrm{in}_N(J).$$

Therefore, by (3.17), $\mathrm{in}_M(g_1)$ lies in $\mathrm{in}_M(I) \cdot \mathrm{in}_N(J)$ and is homogeneous of degree $\mu(g_1, g_2)$; hence

$$\mathrm{in}_M(g_1) = \sum_\beta \mathrm{in}_M(f_\beta) \cdot \mathrm{in}_N(b_\beta), \tag{3.18}$$

with $f_\beta \in I$, $b_\beta \in J$, and $\nu_M(f_\beta) + \nu_N(b_\beta) = \mu(g_1, g_2)$, for every β.

Now write

$$k = \sum_\beta f_\beta b_\beta \in I \cdot J \subset I \cap J \cdot R\{\mathbf{z}\}; \tag{3.19}$$

by (3.17) and (3.18) we have

$$\mathrm{in}_M(k) = \mathrm{in}_M(g_1) = -\mathrm{in}_M(g_2). \tag{3.20}$$

Setting $g_1' = g_1 - k$ and $g_2' = g_2 + k$, we then have $g = g_1' + g_2'$, $g_1' \in I$, $g_2' \in J \cdot R\{\mathbf{z}\}$ (by (3.19)), and $\mu(g_1', g_2') > \mu(g_1, g_2)$ (by (3.20)).

Since $\nu_M(g)$ is finite, after a finite number of repetitions of the above procedure, we obtain our assertion, and the lemma is proved in this case.

Case 2: $Y' = Y \times \mathbb{C}^s$ and $h : Y' \to Y$ is the first projection. Denote by $\mathbf{t} = (t_1, \dots, t_s)$ the result of the lifting to Y', around $y' = (y, \mathbf{0})$, of a local coordinate system on $\mathbb{C}^s$ centered at $\mathbf{0}$; then $\mathscr{O}_{Y',y'} = R\{\mathbf{t}\}$. Put $Z' = Y \times \mathbb{C}^s \times \mathbb{C}^c$; then $\mathscr{O}_{Z',(y,\mathbf{0},\mathbf{0})} = R\{\mathbf{z}, \mathbf{t}\}$, and we can embed X' in Z' and identify $(y, \mathbf{0}, \mathbf{0})$ with x'. As a result, X' is defined around x' by the ideal $I \cdot R\{\mathbf{z}, \mathbf{t}\}$.

Denote by M' the maximal ideal of $R\{\mathbf{z}, \mathbf{t}\}$. If we express any $g \in I \cdot R\{\mathbf{z}, \mathbf{t}\}$ in the form

$$g = \sum_{A \in \mathbb{Z}_0^s} g_A \mathbf{t}^A \quad \text{with } g_A \in R\{\mathbf{z}\},$$

we conclude that $g_A \in I$ for all A. Since we can always find a subset $\Delta \subset \mathbb{Z}_0^s$ such that

$$\mathrm{in}_{M'}(g) = \sum_{A\in\Delta} \mathrm{in}_M(g_A)\mathbf{t}^A,$$

conclusion (3.15) of the lemma is proved in this case as well. □

Corollary 3.6.1 *Let $l : X \to Y$ be a morphism of complex-analytic spaces. Take $x \in X$ and $y = l(x)$. Assume that $(dl)_x : C_{X,x} \to C_{Y,y}$ is flat, and let*

$$X \xrightarrow{\ l_1\ } Z = Y \times \mathbb{C}^c \xrightarrow{\ l_2\ } Y,$$

be a factorization of l locally around x into a closed embedding and a projection. If $I = I_{Z,X,x}$, then the ideal of $C_{l^{-1}(y),x}$ in $\mathrm{gr}_x(\mathcal{O}_{l^{-1}(y),x})$ is

$$\mathrm{in}_x(I) \cdot \mathrm{gr}_x(\mathcal{O}_{l^{-1}(y),x}).$$

Proof Apply Lemma 3.6.1, taking $Y' = \{y\}$. □

Lemma 3.6.2 *Under the notation and hypothesis of Lemma 3.6.1, let K be an ideal in R and let $(f_1, \ldots, f_m)$ be a basis of I such that the following conditions are satisfied:*

(a) *The map $(dl)_x : C_{X,x} \to C_{Y,y}$ is flat.*
(b) *The set $\{\mathrm{in}_M(f_i)\}_{1\le i\le m}$ is a basis of $\mathrm{in}_M(I)$.*
(c) *If $P = (\mathbf{z})\cdot R\{\mathbf{z}\}$, $L = K\cdot R\{\mathbf{z}\} + P$, $P' = (\mathbf{z})\cdot(R\{\mathbf{z}\}/N\cdot R\{\mathbf{z}\})$ and $f_i' = f_i$ (mod N), then $\nu_M(f_i) = \nu_L(f_i) = \nu_{P'}(f_i') = \nu_i$, for $1 \le i \le m$.*

Then, for every $g \in I$, there exist elements $h_1, \ldots, h_m$ in $R\{\mathbf{z}\}$ satisfying these conditions:

(1) *If we write g as*

$$g = \sum_{A\in\mathbb{Z}_0^c} g_A\mathbf{z}^A, \quad \text{with } g_A \in R,$$

then for every $i = 1, \ldots, m$ the element h_i belongs to the ideal of $R\{\mathbf{z}\}$ generated by all the g_A satisfying the relation

$$\nu_L(g_A) + |A| = \nu_L(g).$$

(2) *For $1 \le i \le m$ is $\nu_L(h_i) \ge \nu_L(g) - \nu_i$.*
(3) *$\nu_L\big(g - \sum_{i=1}^m h_i f_i\big) > \nu_L(g)$.*

Proof For every $g \in R\{\mathbf{z}\}$, set $\bar{g} = \mathrm{in}_M(g)$; in particular, if $g \in N$, then $\mathrm{in}_N(g) = \mathrm{in}_M(g) = \bar{g}$. Set also

$$H = \mathrm{gr}_N(R), \quad G = \mathrm{gr}_M(R\{\mathbf{z}\}) = H[\bar{\mathbf{z}}],$$

and let

$$\delta : H[\bar{\mathbf{z}}] \to \mathbb{C}[\bar{\mathbf{z}}]$$

be the natural homomorphism; by (a) one can apply Corollary 3.6.1, and by (b) and (c) we can assert that

$$\delta(\mathrm{in}_M(I)) = (\delta(\bar{f}_1), \ldots, \delta(\bar{f}_m)) \cdot \mathbb{C}[\bar{\mathbf{z}}]. \tag{3.21}$$

In the rest of this proof we shall use the following additional notation. For any homogeneous polynomial $\phi(\bar{\mathbf{z}}) = \sum_{|A|=n} \phi_A \bar{\mathbf{z}}^A$, with $\phi_A \in \mathbb{C}$, set $\mathrm{lex}_{\bar{\mathbf{z}}}(\phi(\bar{\mathbf{z}})) = \sup\{A \mid \phi_A \neq 0\}$. Set

$$E = \{\mathrm{lex}_{\bar{\mathbf{z}}}(\xi) \mid \xi \text{ is a homogeneous polynomial in } \delta(\mathrm{in}_M(I))\}.$$

For $g \in R\{\mathbf{z}\}$ and $\mu, \nu \in \mathbb{Z}_0$, set

$$\Delta^{\nu}(g) = \{A \in \mathbb{Z}_0^c \mid \nu_N(g_A) + |A| = \nu\},$$

$$\Delta_L^{\mu}(g) = \{A \in \mathbb{Z}_0^c \mid \nu_L(g_A) + |A| = \mu\}.$$

For $A, B \in \mathbb{Z}_0^c$, we write $A \preceq B$ if and only if $(|A|, A) \leq (|B|, B)$ in lexicographical order; obviously $\preceq$ is an order relation on $\mathbb{Z}_0^c$.

It is clear that, for every $g \in R\{\mathbf{z}\}$, the relation $\Delta^{\nu}(g) \neq \varnothing$ implies that $\nu_N(g) \leq \nu$; similarly, $\Delta_L^{\mu}(g) \neq \varnothing$ implies $\nu_L(g) \leq \mu$. It is also clear that, for every $A \in \mathbb{Z}_0^c$, the set

$$\{B \in \mathbb{Z}_0^c \mid B \preceq A\}$$

is finite.

The remainder of the proof of the lemma will be divided into four steps.

Step 1 *If $g \in I$ and $\nu_L(g) = \mu$, there exist elements $h_1, \ldots, h_m$ in $R\{\mathbf{z}\}$ satisfying the following conditions:*

(1.1) $h_i \in (g_A)_{A \in \Delta_L^{\mu}(g)} \cdot R\{\mathbf{z}\}$.
(1.2) *For $i = 1, \ldots, m$ is $\nu_L(h_i) \geq \nu_L(g) - \nu_i$.*
(1.3) $\Delta_L^{\mu}(g') \cap E = \varnothing$, *where $g' = g - \sum_{i=1}^m h_i f_i$.*

Proof Note that condition (1.3) can be replaced by

(1.4) $\max_{\preceq}(\Delta_L^{\mu}(g') \cap E) \preceq \max_{\preceq}(\Delta_L^{\mu}(g) \cap E)$, where $g' = g - \sum_{i=1}^m h_i f_i$.

This is because, by applying (1.1), (1.2), and (1.4) a finite number of times, each time replacing g by g', we eventually obtain (1.3).

We now turn to the construction of the h_i. Set

$$A = \max_{\preceq}(\Delta_L^{\mu}(g) \cap E);$$

then $A = \operatorname{lex}_{\bar{\mathbf{z}}}(\xi)$ with $\xi \in \delta(\operatorname{in}_M(I))$. We may assume, in addition, that ξ is homogeneous of degree $|A|$ and that the coefficient of $\bar{\mathbf{z}}^A$ in ξ is 1. Then, by Lemma 3.3.1, we have

$$\xi = \sum_{i=1}^{m} a_i(\bar{\mathbf{z}})\delta(\bar{f_i}),$$

where the $a_i(\bar{\mathbf{z}}) \in \mathbb{C}[\bar{\mathbf{z}}]$ are either zero or homogeneous of degree $|A| - \nu_i$.

Next set $h_i(\mathbf{z}) = g_A \cdot a_i(\mathbf{z})$ for $1 \leq i \leq m$. Condition (1.1) is satisfied by construction. For condition (1.2), we see that

$$\nu_L(h_i) \geq \nu_L(g_A) + \nu_L(a_i(\mathbf{z})) = \nu_L(g) - |A| + |A| - \nu_i = \nu_L(g) - \nu_i.$$

We turn to the verification of (1.4). Let $B \in \Delta_L^{\mu}(g') \cap E$. Suppose first that $B \in \Delta_L^{\mu}(g) \cap E$; then $B \preceq A$, by the construction of A, and, since no term in $\mathbf{z}^A$ occurs in g', we have $B \neq A$, hence $B \prec A$.

If, on the contrary, $B \notin \Delta_L^{\mu}(g) \cap E$, we have

$$B \in \Delta_L^{\mu}\left(g_A \cdot \sum_{i=1}^{m} a_i(\mathbf{z}) f_i\right) \cap E,$$

hence $g'_B = g_A \cdot \lambda$, with $\lambda \in R\{\mathbf{z}\}$. By construction, $B < A$ (lexicographically), so $g'_B = g_A \cdot \lambda$ implies

$$\mu = \nu_L(g'_B) + |B| \geq \nu_L(g_A) + \nu_L(\lambda) + |B| = \mu - |A| + \nu_L(\lambda) + |B|,$$

hence $|A| \geq |B| + \nu_L(\lambda)$, and $|B| \leq |A|$. Therefore $B \preceq A$ in this case as well, which concludes the proof of step 1. □

Step 2 *If $g \in I$ is such that $\nu_M(g) = \nu$, $\nu_L(g) = \mu$ and $\Delta_L^{\mu}(g) \cap E = \varnothing$, then there exist elements $k_1, \ldots, k_m$ in $R\{\mathbf{z}\}$ such that:*

(2.1) *For $i = 1, \ldots, m$ we have $\nu_L(k_i) > \nu_L(g) - \nu_i$, and*
(2.2) *$\Delta^{\nu}(g') \cap E = \varnothing$, with $g' = g - \sum_{i=1}^{m} k_i f_i$.*

Proof The procedure here is analogous to that in the previous step, using $\Delta^{\nu}(g)$ instead of $\Delta_L^{\mu}(g)$. In the computation of $\nu_L(k_i)$, we must observe that, since by assumption $\nu_L(g) = \mu$ and $\Delta_L^{\mu}(g) \cap E = \varnothing$, one has

$$\nu_L(k_i) \geq \nu_L(g_A) + \nu_L(a_i(\mathbf{z})) > \nu_L(g) - |A| + |A| - \nu_i = \nu_L(g) - \nu_i,$$

so (2.1) holds. The proof of (2.2) is analogous to that of the corresponding statement in step 1. □

Step 3 *If $g \in I$ and $\nu_M(g) \geq \nu$, then $\Delta^{\nu}(g) \cap E = \varnothing$ implies $\nu_M(g) > \nu$.*

Proof Assume, for a contradiction, that $\Delta^\nu(g) \cap E = \varnothing$ yet $\nu_M(g) = \nu$. The assumptions imply $\delta(\bar{g}) = 0$, otherwise $\mathrm{lex}_{\mathbf{z}}(\delta(\bar{g})) \in \Delta^\nu(g) \cap E$. The assumptions also imply that

$$\bar{g} = \sum_{A \in \Delta^\nu(g)} \bar{g}_A \bar{\mathbf{z}}^A.$$

Put $\alpha = \min\{\nu_N(g_A) \mid A \in \Delta^\nu(g)\}$; obviously $\alpha \geq 1$. Moreover,

$$\bar{g} \in \mathrm{in}_N(N)^\alpha \cdot G \cap \mathrm{in}_M(I).$$

From assumption (a) of the lemma,

$$\mathrm{in}_N(N)^\alpha \cdot G \cap \mathrm{in}_M(I) = \mathrm{in}_N(N)^\alpha \cdot \mathrm{in}_M(I);$$

therefore

$$\bar{g} = \sum_\gamma \eta_\gamma \zeta_\gamma$$

with $\eta_\gamma \in \mathrm{in}_N(N)^\alpha$ and $\zeta_\gamma \in \mathrm{in}_M(I)$, of degree $\nu - \alpha$. We may always assume that, in this expression, the η_γ are $\mathbb{C}$-linearly independent; therefore, by taking into account the construction of α, we have $\delta(\zeta_\gamma) \neq 0$ for some γ and

$$\mathrm{lex}_{\bar{\mathbf{z}}}(\delta(\zeta_\gamma)) \in \Delta^\nu(g) \cap E,$$

which contradicts the assumption that $\Delta^\nu(g)$ and E are disjoint. □

Step 4 *If $g \in I$ and $\nu_L(g) \geq \mu$, the relation*

$$\Delta^\mu_L(g) \cap E = \varnothing$$

implies $\nu_L(g) > \mu$.

Proof Assume, to the contrary, that $\Delta^\mu_L(g) \cap E = \varnothing$ yet $\nu_L(g) = \mu$. Together with the results from steps 2 and 3, this implies the existence of elements $k_1, \ldots, k_m \in R\{\mathbf{z}\}$ such that

$$\nu_L(k_i) > \nu_L(g) - \nu_i = \mu - \nu_i \quad \text{for } i = 1, \ldots, m \tag{3.22}$$

and $\nu_M(g') > \nu_M(g)$, where $g' = g - \sum_{i=1}^m k_i f_i$. From (3.22) we get

$$\nu_L(g') = \nu_L(g) = \nu, \quad \Delta^\mu_L(g') = \Delta^\mu_L(g), \quad \mathrm{in}_L(g') = \mathrm{in}_L(g). \tag{3.23}$$

Applying the construction of the k_i a suitable number of times, and taking into account that $\nu_M(g') > \nu_M(g)$, we arrive at a contradiction with (3.23). This proves step 4. □

Lemma 3.6.2 follows immediately from steps 1–4. □

Corollary 3.6.2 *With the notation and hypothesis of Lemma* 3.6.2, *one has:*

(1) *The map* $R/K \to \mathrm{gr}_L(R\{\mathbf{z}\})/\mathrm{in}_L(I)$ *is flat.*
(2) *The ideal* $\mathrm{in}_L(I)$ *is generated by* $\{F_i\}_{1\leq i\leq m}$, *where* $F_i = \mathrm{in}_L(f_i)$ *for* $i = 1, \ldots, m$.

Proof This will be proved if we show that, for any integer $r \geq 0$,

$$[\mathfrak{N}^r \cdot \mathrm{gr}_L(R\{\mathbf{z}\})] \cap \mathrm{in}_L(I) \subset \mathfrak{N}^r \cdot (F_1, \ldots, F_m) \cdot \mathrm{gr}_L(R\{\mathbf{z}\}),$$

where $\mathfrak{N} = N/K$. Let $g \in I$ be such that

$$\mathrm{in}_L(g) \in \mathfrak{N}^r \cdot \mathrm{gr}_L(R\{\mathbf{z}\});$$

then

$$\mathrm{in}_L(g_A) \in \mathfrak{N}^r \quad \text{for all } A \in \Delta_L^{\nu_L(g)}(g).$$

By Lemma 3.6.2, there is a set of indices $\Delta \subset \{1, \ldots, m\}$ such that

$$\mathrm{in}_L(g) = \sum_{i\in\Delta} \mathrm{in}_L(h_i) \cdot F_i,$$

with $h_i \in R\{\mathbf{z}\}$ and $\mathrm{in}_L(h_i) \in \mathfrak{N}^r$ for $i \in \Delta$. Therefore

$$\mathrm{in}_L(g) \in \mathfrak{N}^r \cdot (F_1, \ldots, F_m) \cdot \mathrm{gr}_L(R\{\mathbf{z}\}),$$

which concludes our proof. □

3.7 Characteristic Cones: The Test of Maximal Contact

Unless the contrary is explicitly stated, throughout this section: Z will be a smooth complex-analytic space, X a closed subspace of Z, W a smooth closed subspace of Z, x a point in $X \cap W$, and $r : Z \to W$ a local retraction around x. Also, when necessary, we shall use a local coordinate system $(\mathbf{z}, \mathbf{w}) = (z_1, \ldots, z_c, w_1, \ldots, w_d)$ on Z, centered at x, and associated with r.

First we introduce the concept of a characteristic cone, and use the result of Sect. 3.6 to obtain the local equation of such a cone. Then we use characteristic cones to characterize the contact idealistic exponent of r with X at x and the maximal contact of r with X at x, and to compare maximal contact and numerical maximal contact.

Remark 3.7.1 We restrict ourselves, throughout this section, to work locally around x, so we shall speak "globally" about everything. Let $\mathscr{E} = ((\mathbf{J}, b))$ be an idealistic exponent on W and let $g : W' \to W$ be a proper injective morphism of analytic spaces satisfying these conditions:

(1) The space W' is normal and irreducible.
(2) There is a holomorphic function k on W' such that

$$\mathbf{J}\mathscr{O}'_W = (k^b)\mathscr{O}'_W$$

(The existence of such that morphism will be proved in Sect. 3.8.)

If we change the basis by g, we write

$$Z' = Z \times_W W', \quad X' = X \times_W W'.$$

We denote by k' the pull-back of k under the projection $Z' \to W'$; similarly, for $1 \leq i \leq c$, we denote by z'_i the pull-back of z_i under the projection $Z' \to Z$.

Then we can define a subspace K' of Z' by the ideal

$$\mathbf{I}_{Z',K'} = (\mathbf{z}', k') \cdot \mathscr{O}_{Z'} = \mathbf{I}_{Z',W'} + (k') \cdot \mathscr{O}_{Z'};$$

then K' is a hypersurface in W', defined in W' by the ideal $\mathbf{I}_{W',K'} = (k) \cdot \mathscr{O}_{W'}$, and is determined by g and $\mathscr{E}$ because of the normality of W'. The idealistic exponent $(((k) \cdot \mathscr{O}_{W'}, 1))$ is denoted $g^*(\mathscr{E})$.

Lemma 3.7.1 *With the notation and hypothesis of Remark* 3.7.1, *let* $x' \in g^{-1}(x)$. *Then:*

(1) $(\mathbf{z}')$ *is a regular sequence in* $\mathscr{O}_{Z',x'}$.
(2) *If* k' *is not identically zero around* x' *(that is, if* $k \neq 0$ *in* $\mathscr{O}_{W',x'}$*), then* $(\mathbf{z}', k')$ *is a regular sequence in* $\mathscr{O}_{Z',x'}$.

Proof Assertion (1) is trivial by construction; we prove (2). We have $k'|_{W'} = k$, that is, k is the image of the class of k' under the natural isomorphism

$$\mathscr{O}_{Z',x'}/(\mathbf{z}') \cdot \mathscr{O}_{Z',x'} \to \mathscr{O}_{W',x'}.$$

Since $k \neq 0$ in $\mathscr{O}_{W',x'}$, and since W' is reduced and irreducible, k is not a zero divisor in $\mathscr{O}_{W',x'}$. Hence $(\mathbf{z}', k')$ is a regular sequence in $\mathscr{O}_{Z',x'}$. □

Remark 3.7.2 Write $\overline{\overline{z}}'_i = z'_i \mod I^2_{Z',K',x'}$ for $1 \leq i \leq c$, and $\overline{\overline{k}}' = k' \mod I^2_{Z',K',x'}$; also set

$$\mathrm{gr}_{K'}(Z') = \bigoplus_{i \geq 0} I^i_{Z',K'}/I^{i+1}_{Z',K'},$$

$$\mathrm{gr}_{K'}(W') = \bigoplus_{i\geq 0} I^i_{W',K'}.$$

Then, as a consequence of Lemma 3.7.1, we have:

Corollary 3.7.1 *If $k \neq 0$ in $\mathcal{O}_{W',x'}$, then*

$$\mathrm{gr}_{K'}(Z')_{x'} = \mathcal{O}_{K',x'}[\bar{\bar{\mathbf{z}}}', \bar{\bar{k}}'],$$

a polynomial algebra in $c+1$ indeterminates. If $k = 0$ in $\mathcal{O}_{W',x'}$, then

$$\mathrm{gr}_{K'}(Z')_{x'} = \mathcal{O}_{k',x'}[\bar{\bar{\mathbf{z}}}'],$$

a polynomial algebra in c indeterminates.

And, writing $\bar{\bar{k}} = \mathrm{in}_{K'}(k)$:

Corollary 3.7.2 *If $k \neq 0$ in $\mathcal{O}_{W',x'}$, then*

$$\mathrm{gr}_{K'}(W')_{x'} = \mathcal{O}_{K',x'}[\bar{\bar{k}}].$$

If $k = 0$ in $\mathcal{O}_{w',x'}$, then

$$\mathrm{gr}_{K'}(W')_{x'} = \mathcal{O}_{K',x'}.$$

Then, if we define

$$E_{\mathcal{E},g} = \mathrm{Specan}\, \mathrm{gr}_{K'}(Z'),$$
$$N_{\mathcal{E},g} = \mathrm{Specan}\, \mathrm{gr}_{K'}(W'),$$

we see that $E_{\mathcal{E},g} \to K'$ and $N_{\mathcal{E},g} \to K'$ are complex vector bundles over K', that $E_{\mathcal{E},g}$ has dimension c or $c+1$ on each connected component of K', that $N_{\mathcal{E},g}$ has dimension zero or one on each connected component of K', that $\mathrm{gr}_{K'}(Z')$ and $\mathrm{gr}_{K'}(W')$ are the sheaves of holomorphic functions that are polynomials on the fibers of $E_{\mathcal{E},g}$ and $N_{\mathcal{E},g}$, respectively, and that the embedding $W \hookrightarrow Z$ allows us to consider $N_{\mathcal{E},g}$ as a vector sub-bundle of $E_{\mathcal{E},g} \to K'$.

Now consider $\mathrm{gr}_{K'}(X') = \bigoplus_{i\geq 0}(I^i_{X',K'}/I^{i+1}_{X',K'})$, which is a finitely presented $\mathcal{O}_{K'}$-algebra; we may construct

$$C_{\mathcal{E},g} = \mathrm{Specan}(\mathrm{gr}_{K'}(X')).$$

Then $C_{\mathcal{E},g} \to K'$ is a cone fiber subspace of $E_{\mathcal{E},g} \to K'$, defined, as a subspace of $E_{\mathcal{E},g}$, by the homogeneous ideal $\mathrm{in}_{K'}(Z', X')$.

Definition 3.7.1 The cone fiber subspace $C_{\mathscr{E},g} \to K'$ of $E_{\mathscr{E},g} \to K'$ will be called the **characteristic cone** of (X, r) with respect to $(\mathscr{E}, g)$. When the context is clear, we shall write E_g, N_g, C_g for $E_{\mathscr{E},g}, N_{\mathscr{E},g}, C_{\mathscr{E},g}$.

Remark 3.7.3 Assume r transversal to X at x, and let $X = \bigcap_{i=1}^m H^{(i)}$ be a tangential intersection at x such that r is transversal to each $H^{(i)}$ at x, for $1 \leq i \leq m$. Let $I_{Z,H^{(i)},x} = (f_i)\mathscr{O}_{Z,x}$ and $\nu_i = \nu_x(f_i)$. Write

$$f_i = \sum_{A \in \mathbb{Z}_0^c} (f_i)_A \mathbf{z}^A \quad \text{for } 1 \leq i \leq m,$$

where, as usual, $(f_i)_A \in \mathscr{O}_{W,x}$ for each $A \in \mathbb{Z}_0^c$, and $(f_i)_A$ is viewed as an element of $\mathscr{O}_{Z,x}$ by the retraction r.

Let $\mathscr{E} = ((\mathbf{J}, b))$ be the idealistic exponent on W defined by

$$b = \prod_{i=1}^m (\nu_i!), \quad \mathbf{J} = \left(\{(f_i)_A^{b/(\nu_i - |A|)}\}_{1 \leq i \leq m,\, |A| < \nu_i}\right) \cdot \mathscr{O}_W,$$

and let $g : W' \to W$ be a morphism of analytic spaces such as in Remark 3.7.1. Note that, if the $H^{(i)}$ are suited to X at x with respect to r then $\mathscr{E}$ is the contact exponent of r with X at x (Theorem 3.3.2).

Recall the notation of Remark 3.7.1, and set also

$$H^{(i)'} = H^{(i)} \times_W W',$$

and denote by f_i' the pull-back of f_i via the projection $Z' \to Z$, for $i = 1, \dots, m$. Set

$$\mathscr{L} = (\mathbf{z}', k')\mathscr{O}_{Z'} = I_{Z',K'}.$$

Then

$$E_g = \operatorname{Specan} \operatorname{gr}_{\mathscr{L}}(\mathscr{O}_{Z'}).$$

Lemma 3.7.2 *With the notation and hypothesis of Remark* 3.7.3, *let* $x' \in g^{-1}(x)$, *and set* $L = (\mathbf{z}', k')\mathscr{O}_{Z',x'}$. *Then the ideal of* C_g *in* $\operatorname{gr}_L(\mathscr{O}_{Z',x'})$ *is*

$$(F_i, \dots, F_m) \cdot \operatorname{gr}_L(\mathscr{O}_{Z',x'}),$$

where $F_i = \operatorname{in}_L(f_i)$ *for* $1 \leq i \leq m$.

Proof We first show that we are in a position to apply Lemma 3.6.2 and Corollary 3.6.2.

(a) By assumption, r is transversal to X at x, so

$$(dr)_x : C_{X,x} \to T_{W,x}$$

is flat. Therefore, if $r' : X' \to W'$ is the map defined by $r' = r \times \mathrm{id}_{W'}$,

$$(dr)_{x'} : C_{X',x'} \to C_{W',x'}$$

is flat.

(b) $X = \bigcap_{i=1}^{m} H^{(i)}$ is a tangential intersection at x, so $X' = \bigcap_{i=1}^{m} H^{(i)'}$ is a tangential intersection at x'. Hence $\{\mathrm{in}_{x'}(f_1'), \ldots, \mathrm{in}_{x'}(f_m')\}$ is a basis of the ideal $\mathrm{in}_{x'}(Z', X')$.

(c) Write $P' = (\mathbf{z}') \cdot (\mathcal{O}_{Z',x'}/\mathfrak{m}_{W',x'}\mathcal{O}_{Z',x'})$, where $\mathfrak{m}_{W',x'}$ is the maximal ideal of $\mathcal{O}_{W,x'}$. If

$$f_i'' = f_i' \mod \mathfrak{m}_{W',x'} \quad \text{for } 1 \le i \le m,$$

then $\nu_{P'}(f_i'') = \nu_x(f_i')$, since r' is transversal to X' at x'. On the other hand,

$$\left(\{(f_i)_A^{b/(\nu_i-|A|)}\}_{1\le i\le m,\,|A|<\nu_i}\right) \cdot \mathcal{O}_{W'} = (k^b)\mathcal{O}_{W'}$$

by the construction of $\mathcal{E}$ and because $\mathbf{J}\mathcal{O}_{W'} = (k^b)\mathcal{O}_{W'}$. This means that there is a family $\{(\varepsilon_i)_A\}_{1\le i\le m,\,|A|<\nu_i}\}$ of holomorphic functions on W' such that

$$\left(\{(\varepsilon_i)_A\}_{1\le i\le m,\,|A|<\nu_i}\right)\mathcal{O}_W = \mathcal{O}_{W'}$$

and such that

$$(\varepsilon_i)_A k^{\nu_i-|A|} = (f_i)_A g \quad \text{for } 1 \le i \le m|A| < \nu_i.$$

As a consequence of this and of the transversality of r' to X' at x', we obtain

$$\nu_L(f_i') = \nu_x(f_i').$$

In view of (a), (b), and (c), we may apply Lemma 3.6.2 and Corollary 3.6.2, which yields our lemma. □

Remark 3.7.4 We turn to the construction of the initial forms of the f_i' in the filtration defined by L in $\mathcal{O}_{Z',x'}$. Set $(\varepsilon_i)_A = (f_i)_A \circ g$ for $|A| = \nu_i$, and

$$(\overline{\overline{\varepsilon}}_i)_A = (\varepsilon_i)_A \mod (I_{W',K',x'}) \in \mathcal{O}_{K',x'}.$$

Then, in $\mathrm{gr}_L(\mathcal{O}_{Z',x'}) = \mathrm{gr}_{K'}(Z')(x')$, we have

$$\mathrm{in}_L(f_i') = \sum_{|A|\le\nu_i} (\overline{\overline{\varepsilon}}_i)_A \cdot \overline{\overline{k'}}^{\nu_i-|A|} \overline{\overline{\mathbf{z}}}'^A.$$

Definition 3.7.2 We say that a morphism $g : Y' \to Y$ of analytic spaces is **complete** if, for each $y \in Y$, there exist a neighborhood N of y in Y and a compact subset $E \subset Y'$ such that, for every analytic morphism $h : \mathbb{D} \to Y$ with $h(0) \in N$, there exist analytic morphisms $h' : \mathbb{D} \to Y'$ and $\varphi : \mathbb{D} \to \mathbb{D}$ with φ non-constant and $\varphi(0) = 0$, such that $h'(0) \in E$ and $gh' = h\varphi$.

Remark 3.7.5 With the conditions as above, one can easily check that:

(1) We can always assume that φ is of the type $\varphi(t) = at^e$, with $a \in \mathbb{R}_+$ and $e \in \mathbb{Z}_+$, where t is the parameter for $\mathbb{D}$.
(2) If g is complete then, for every $y \in Y$, there exists a compact subset $E \subset Y'$ such that $g(E)$ is a neighborhood of y.
(3) If g is proper and surjective, g is complete.

Definition 3.7.3 Let Y be a smooth complex-analytic space, and $\mathscr{E}_1, \mathscr{E}_2$ idealistic exponents on Y. We shall say that $\mathscr{E}_1$ **dominates** $\mathscr{E}_2$ at $y \in Y$, and we shall write $\mathscr{E}_1 >_y \mathscr{E}_2$, if, for every test-map h of Y at y, one has $\nu_h(\mathscr{E}_1) \geq \nu_h(\mathscr{E}_2)$. With the notation and hypothesis of this section, if $\mathscr{E}$ is the contact exponent of r with X at x, and if $\mathscr{E} >_x \mathscr{E}_2$, we say that $\mathscr{E}_2$ is a **subcontact exponent** of r with X at x.

Also, for any idealistic exponent $\mathscr{E}$, and for $g : W' \to W$ as in Remark 3.7.1, we define

$$T^0_{x'}(\mathscr{E}, X, r, g) = \{\mathbf{v} \in N_{\mathscr{E},g}(x') \mid \mathbf{v} + C_{\mathscr{E},g}(x') = C_{\mathscr{E},g}(x')\}.$$

For the rest of this section the assumptions will be the usual ones, assuming, in addition, that r is transversal to X at x, and that we have a tangential intersection $X = \bigcap_{i=1}^m H^{(i)}$, suited to X at x with respect to r. Then this contact exponent $\mathscr{E}$ of r with X at x can be constructed as usual.

Proposition 3.7.1 *Let $\mathscr{E} = ((\mathbf{J}, b))$ be the contact exponent of r with X at x, and let $g : W' \to W$ be a morphism of analytic spaces as in Remark* 3.7.1. *Then for every $x' \in g^{-1}(x)$, $T^0_{x'}(\mathscr{E}, X, r, g) = (0)$.*

Proof If $k = 0$ in $\mathscr{O}_{W',x'}$, then $N_g(x') = (0)$; therefore $T^0_{x'}(\mathscr{E}, X, r, g) = (0)$ and the proposition is proved. Assume then that $k \neq 0$ in $\mathscr{O}_{W',x'}$.

Consider first the case where $W' = \mathbb{D}$, so that g is a test-map at x. Our assumptions then imply that

$$\delta := \delta_{x'}(X', \mathbb{D}) = \min_{1 \leq i \leq m} \delta_{x'}(H^{(i)'}, \mathbb{D}) = \nu_g(\mathscr{E}) = \nu_0(k \cdot \mathscr{O}_D); \tag{3.24}$$

we shall prove that this contradicts the fact that

$$T^0_{x'}(\mathscr{E}, X, r, g) \neq (0).$$

This latter inequality is equivalent to the fact that the kernel of the homomorphism,

$$q : \mathrm{gr}_{K'}(Z')(x') \to \mathrm{gr}_{K'}(X')(x'),$$

has a basis contained in $\mathbb{C}[\overline{\overline{\mathbf{z}}}']$ (recall from Corollary 3.7.1 that $\mathrm{gr}_{K'}(Z')(x') = \mathbb{C}[\overline{\mathbf{z}}', \overline{k}']$). Now consider the homomorphism

$$\varphi : \mathbb{C}[\overline{\overline{\mathbf{z}}}', \overline{\overline{k}}'] \to \mathbb{C}[\tilde{\mathbf{z}}', \tilde{w}'] = \mathrm{gr}_{W',x',\delta}(Z')$$

defined by

$$\varphi(\overline{\overline{k}}') = (\tilde{w}')^{\delta}, \quad \varphi(\overline{\overline{z}}'_i) = \tilde{z}'_i \text{ for } 1 \leq i \leq c;$$

the image under φ of the kernel of q generates the kernel of the natural homomorphism

$$\mathrm{gr}_{W',x',\delta}(Z') \to \mathrm{gr}_{W',x',\delta}(X') = \mathrm{gr}_{W',x',\delta}(Z')/\mathrm{in}_{x'}^{W',\delta'}(Z', X').$$

Therefore, the kernel of this latter homomorphism has a basis contained in $\mathbb{C}[\tilde{\mathbf{z}}']$, hence $\delta_{X'}(X', \mathbb{D}) > \delta$, contradicting (3.24).

Now consider a general W'. Since $k \neq 0$ in $\mathscr{O}_{W',x'}$, we cannot have k identically zero around x' in Z'. Hence $K' \neq W'$ around x', and it is possible to find a test-map $h' : \mathbb{D} \to W'$ at x' such that $h'(\mathbb{D}') \not\subset K'$ around x'. Write $\tilde{g} = h' \circ g$ and consider $(\mathbb{D}, \tilde{g})$ instead of (W', g). We see that $(\mathbb{D}, \tilde{g})$ also satisfies conditions (a) and (b) in the statement of the proposition. It follows from the case $W' = \mathbb{D}$, already proved, that

$$T^0_{x'}(\mathscr{E}, X, r, \tilde{g}) = (0) \quad \text{for all } x' \in g^{-1}(x). \tag{3.25}$$

Assume, for a contradiction, that $T^0_{x'}(\mathscr{E}, X, r, g) \neq (0)$. Then the kernel of

$$\mathrm{gr}_{K'}(Z')(x') \to \mathrm{gr}_{K'}(X')(x')$$

has a basis contained in $\mathbb{C}[\overline{\overline{\mathbf{z}}}']$. By Lemma 3.7.2, this property is not affected by a change of basis via h', so

$$T^0_{x'}(\mathscr{E}, X, r, \tilde{g}) \neq (0),$$

contradicting (3.25). □

Proposition 3.7.2 *Under the notation and hypothesis of Proposition* 3.7.1, *assume that* $\tilde{\mathscr{E}} = ((\mathbf{H}, c))$ *is a subcontact exponent of* r *with* X *at* x *(see Definition* 3.7.3*) and let* $g : W' \to W$ *be a proper surjective morphism satisfying these properties:*

(1) *W' is normal and irreducible.*
(2) *There exists a holomorphic function k on W' such that*

$$\mathbf{H} \cdot \mathscr{O}'_W = (k^c) \cdot \mathscr{O}'_W.$$

(3) *The morphism g is complete.*

Assume, moreover, that

$$T^0_{x'}(\mathscr{E}, X, r, g) = (0) \quad \textit{for every } x' \in g^{-1}(x).$$

Then $\mathscr{E}$ is the contact exponent of r with X at x.

Proof Let $\tilde{\mathscr{E}} = ((\mathbf{J}, b))$ be the contact exponent of r with X at x. We have $\mathscr{E} >_x \tilde{\mathscr{E}}$. Therefore, in order to prove the proposition, it is enough to show that

$$\nu_h(\tilde{\mathscr{E}}) \geq \nu_h(\mathscr{E})$$

for every test-map h of W at x.

Recall from Remark 3.7.5 that, since g is complete, there exist for every test-map h as above a holomorphic map $h' : \mathbb{D} \to W'$ and a map $\varphi : \mathbb{D} \to \mathbb{D}$ of the form given by $\varphi(t) = at^e$, with $a \in \mathbb{R}_+$ and $e \in \mathbb{Z}_t$, and such that $h\varphi = gh'$. Moreover, it is clear that, for every ideal $\mathscr{L}$ on W, one has $\nu_{h\varphi}(\mathscr{L}) = e \cdot \nu_h(\mathscr{L})$. Therefore

$$\begin{aligned} \nu_h(\tilde{\mathscr{E}}) \geq \nu_h(\mathscr{E}) &\iff \nu_{h\varphi}(\tilde{\mathscr{E}}) \geq \nu_{h\varphi}(\mathscr{E}) \\ &\iff \nu_{gh'}(\tilde{\mathscr{E}}) \geq \nu_{gh'}(\mathscr{E}) \\ &\iff \nu_{h'}(g^*(\tilde{\mathscr{E}})) \geq \frac{\nu_{h'}(\mathbf{J} \cdot \mathscr{O}'_W)}{b}. \end{aligned}$$

It follows that, if $x' = h'(0)$, then $g^*(\tilde{\mathscr{E}})$ is dominated, at x', by $((\mathbf{J} \cdot \mathscr{O}_{W'}, b))$. It also follows that the proposition will be proved if we show that

$$\nu_{h'}(g^*(\tilde{\mathscr{E}})) \geq \nu_{h'}(\mathbf{J} \cdot \mathscr{O}_{W'})/b \tag{3.26}$$

for every $x' \in g^{-1}(x)$ and for every test-map h' of W' at x'.

Now, by condition (2) in the statement of the proposition, we have

$$g^*(\tilde{\mathscr{E}}) = ((k) \cdot \mathscr{O}_{W'}, 1).$$

Write $\mathbf{J}' = \mathbf{J} \cdot \mathscr{O}_{W'}$. We consider two cases:

Case 1: The composition $k \circ h' = 0$. Then

$$\nu_{h'}(g^*(\tilde{\mathscr{E}})) = \nu_0(k, \mathscr{O}_{\mathbb{D}}) = \infty,$$

and (3.26) is true.

Case 2: The composition $k \circ h' \neq 0$; in particular, $k \neq 0$ in $\mathscr{O}_{W',x'}$. As already observed, $g^*(\tilde{\mathscr{E}})$ is dominated at $h'(0)$ by $((\mathbf{J} \cdot \mathscr{O}_{W'}, b))$; together with the normality of W', this implies that

$$\mathbf{J}'_{x'} \subset (k^b) \cdot \mathscr{O}_{W',x'}.$$

If the two sides are equal, the proposition is proved. Assume, for a contradiction, that they are different. Then there exists an ideal P of $\mathcal{O}_{W',x'}$ such that $P \subset \mathfrak{m}_{W',x'}$ and $\mathbf{J}'_{x'} = (k^b) \cdot P$. By Lemma 3.7.2, the kernel of the homomorphism

$$\mathrm{gr}_{K'}(Z')_{x'} \to \mathrm{gr}_{K'}(X')_{x'}$$

has $(F_1, \ldots, F_m)$ for a basis, where

$$F_i = \sum_{|A| \leq \nu_i} (\bar{\varepsilon}_i)_A \overline{\overline{k}}'^{\nu_i - |A|} \overline{\overline{\mathbf{z}}}'^A \quad \text{for } 1 \leq i \leq m,$$

with

$$(\varepsilon_i)_A^{b/(\nu_i - |A|)} \in P^b \quad \text{for } 1 \leq i \leq m \text{ and } |A| < \nu_i.$$

Therefore

$$(\varepsilon_i)_A \in \mathfrak{m}_{W',x'} \quad \text{for } 1 \leq i \leq m \text{ and } |A| < \nu_i.$$

On the other hand, the kernel of

$$\mathrm{gr}_{K'}(Z')(x') \to \mathrm{gr}_{K'}(X')(x')$$

has a basis consisting of the images of the F_i under the natural homomorphism

$$\mathrm{gr}_{K'}(Z')_{x'} \to \mathrm{gr}_{K'}(Z')(x').$$

Since the $(\varepsilon_i)_A$ lie in $\mathfrak{m}_{W',x'}$, such a basis is contained in $\mathbb{C}[\overline{\overline{\mathbf{z}}}']$. Therefore

$$T^0_{x'}(\tilde{\mathcal{E}}, X, r, g) \neq 0,$$

contradicting the assumptions of the proposition. □

Proposition 3.7.3 *With the notation and hypothesis of Proposition* 3.7.1, *suppose moreover that r has maximal contact with X at x. Then*

$$T_{x'}(\mathcal{E}, X, r, g) = (0)$$

for every $x' \in g^{-1}(x)$.

Proof Let $x' \in g^{-1}(x)$. If $k = 0$ in $\mathcal{O}_{W',x'}$, we have

$$T_{x'}(\mathcal{E}, X, r, g) = (0),$$

since r is perpendicular to X at x; therefore the proposition holds in this case. Assume, then, that $k \neq 0$ in $\mathcal{O}_{W',x'}$.

Consider first the case where $W' = \mathbb{D}$ and g is a test-map at x; then $E_g = C_{Z',K'}$ and $C_g = C_{X',K'}$. Since the maximal contact assumption is preserved under basis extension via g, the proposition follows.

In the general case, since we have assumed that $k \neq 0$ in $\mathscr{O}_{W',x'}$, there exists a test-map h of W' at x' such that $h(\mathbb{D}) \not\subset K'$ around x', so $k \circ h \neq 0$ around zero in $\mathbb{D}$. One can easily check that $gh : \mathbb{D} \to W$ satisfies conditions (1) and (2) in Proposition 3.7.1 with respect to $\mathscr{E}$. Therefore we may construct E_{gh} and C_{gh}, obtaining

$$E_{gh} = E_g \times_{W'} \mathbb{D} \quad \text{and} \quad C_{gh} = C_g \times_{W'} \mathbb{D}.$$

It follows that, setting $x'' = (x', 0)$, we have

$$E_{gh}(x'') = E_g(x') \quad \text{and} \quad C_{gh}(x'') = C_g(x'),$$

hence

$$T_{x''}(\mathscr{E}, X, r, gh) = T_{x'}(\mathscr{E}, X, r, g).$$

Thus, since the proposition is true when $W' = \mathbb{D}$ and g is a test-map, it is true also in the general case. □

Proposition 3.7.4 *With the notation and hypothesis of Proposition* 3.7.2, *and with the same assumptions, suppose in addition that r is perpendicular to X at x, and that $T_{x'}(\mathscr{E}, X, r, g) = (0)$ for every $x' \in g^{-1}(x)$. Then r has maximal contact with X at x'.*

Proof By Proposition 3.7.2, $\mathscr{E}$ is the contact exponent of r with X at x. Take $x' \in g^{-1}(x)$ and let $h : \mathbb{D} \to W'$ be a test-map at x. Consider first the case $kh = 0$: then, setting $x'' = (x', 0)$, one has

$$\delta_{x''}(X_h, \mathbb{D}_h) = \nu_0(k \cdot \mathscr{O}_D) = \infty,$$

so $\mathbb{D}_h$ has maximal contact with X_h at x''.

If, on the contrary, $kh \neq 0$, the completeness of g implies the existence of $h' : \mathbb{D} \to W'$ and $\varphi : \mathbb{D} \to \mathbb{D}$ with $\varphi(t) = at^e$, $a \in \mathbb{R}_+$, $e \in \mathbb{Z}_+$, and $gh' = h\varphi$. Now $\mathbb{D}_h$ has maximal contact with X_h at x_h if and only if $\mathbb{D}_{gh'}$ has maximal contact with $X_{gh'}$ at $x_{gh'}$. Then, since conditions (1) and (2) of Proposition 3.7.2 are translated to $(\mathbb{D}, gh')$ by a similar argument to the one in the preceding proposition, Proposition 3.7.4 follows. □

Remark 3.7.6 (Maximal contact and numerical maximal contact) Let Z be a smooth complex analytic space, X a closed subspace of X, W a smooth curve in Z and x a point in $W \cap X$. Suppose also that $\dim T_{X,x} \leq 1$. Our objective is to analyze the relation between maximal contact and numerical maximal contact (Definition 3.2.3) as an example of an application of the techniques developed earlier. It is clear that if there exists a local retraction $r : Z \to W$ around x that has maximal contact with X at x, then W has numerical maximal contact with X at x. Our aim is to prove the converse:

Theorem 3.7.1 *If W has numerical maximal contact with X at x, every local retraction $r : Z \to W$ perpendicular to X at x has maximal contact with X at x.*

Proof Let $(\mathbf{z}, \mathbf{w}) = (z_1, \dots, z_c, w)$ be a local coordinate system on Z centered at x associated with r. Assume that

$$\delta = \delta_x(X, W) \in \mathbb{Z}.$$

Denote by K the subspace of W defined around x by $(\mathbf{z}, w^\delta)\mathcal{O}_{X,x}$; the normal cone to Z along K is

$$C_{Z,K} = \operatorname{Specan} \operatorname{gr}_K(Z),$$

where in this case

$$\operatorname{gr}_K(Z)_x = \mathcal{O}_{K,x}[\overline{\overline{\mathbf{z}}}, \overline{\overline{k}}],$$

with

$$\overline{\overline{k}} = w^\delta \bmod (I^2_{Z,K,x}),$$
$$\overline{\overline{z}}_i = z_i \bmod (I^2_{Z,K,x}) \quad \text{for } 1 \leq i \leq c,$$

and the fiber at x is

$$\operatorname{gr}_K(Z)(x) = \mathbb{C}[\overline{\overline{\mathbf{z}}}, \overline{\overline{k}}].$$

The normal cone to X along K is defined around x by the ideal $\operatorname{in}_K(X)$ generated for the initial forms of the elements of $I_{Z,X,x}$ in the filtration of $\mathbb{C}\{\mathbf{z}, \mathbf{w}\}$ defined by the ideal $(\mathbf{z}, w^\delta)\mathcal{O}_{Z,x}$.

By applying Lemma 3.3.3 to $S = \mathbb{Z}_0$, we know that if $X = \bigcap_{i=1}^m H^{(i)}$ is a tangential intersection totally united to X at x with respect to r we have

$$C_{X,K} = \bigcap_{i=1}^m C_{H^{(i)},K}.$$

By a basis change,

$$C_{X,K}(x) = \bigcap_{i=1}^m C_{H^{(i)},K}(x).$$

Then, if $\dim T_{X,x} = 1$ and if r is perpendicular to X at x, we prove now that the following conditions are equivalent:

(i) W does not have numerical maximal contact with X at x.
(ii) $\delta = \delta_x(X, W) \in \mathbb{Z}$, and

$$T = \{\mathbf{v} \in C_{Z,K}(x) \mid \mathbf{v} + C_{X,K}(x) = C_{X,K}(x)\} \neq \{0\},$$

where K is as above.

Suppose (i) is satisfied. There exists a smooth curve W' through x in Z such that

$$\delta_x(X, W') > \delta_x(X, W);$$

hence W' has local equations at x of the form

$$z_i = \sigma_i(w), \quad \text{with } \sigma_i(w) \in \mathbb{C}\{w\} \text{ for } 1 \leq i \leq c,$$

and r induces a local retraction $r' : Z \to W'$, which belongs to the same projection that r. Then, by Lemma 3.3.2,

$$\delta = \min_{1 \leq i \leq c} \{\nu_0(\sigma_i(w))\};$$

hence δ is an integer. Moreover, we have a cartesian square

$$\begin{array}{ccc} C_{Z,K}(x) & \xleftarrow{\ \varphi\ } & C^{W,\delta}_{Z,x} \\ \uparrow & & \uparrow \\ C_{X,K}(x) & \longleftarrow & C^{W,\delta}_{X,x} \end{array}$$

where φ corresponds to the map

$$\mathbb{C}[\bar{\bar{\mathbf{z}}}, \bar{\bar{k}}] \to \mathbb{C}[\tilde{\mathbf{z}}, \tilde{w}]$$

sending $\bar{\bar{z}}_i$ to $\tilde{z}_i$ and $\bar{\bar{k}}$ to $\tilde{w}^\delta$.

Now, if we change coordinates from $(\mathbf{z}, w)$ to $(\mathbf{z}', w')$, with $z'_i = z_i - \sigma_i(w)$ and $w' = w'$, a basis of the ideal of $C^{W',\delta}_{X,x}$ is contained in $\mathbb{C}[\tilde{\mathbf{z}}']$; then a basis of the ideal of $C_{X,K}(x)$ is contained in $\mathbb{C}[\bar{\bar{\mathbf{z}}}']$ and $T \neq 0$. Hence (ii) holds.

Conversely, suppose that (ii) holds. We have a cartesian square

$$\begin{array}{ccc} C_{X,K}(x) & \hookrightarrow & C_{Z,K}(x) \\ \uparrow & & \uparrow \\ C = C_{r^{-1}(x)\cap X,x} & \longrightarrow & T_{r^{-1}(x),x} \end{array}$$

Then $T \cap T_{r^{-1}(x),x} = \{0\}$; otherwise, there would be a non-zero $\mathbf{v}$ in $T_{r^{-1}(x),x}$ such that $\mathbf{v} + C_{X,K}(x) = C_{X,K}(x)$ and, since the preceding diagram is cartesian, there would be a non-zero $\mathbf{v} \in T_{r^{-1}(x),x}$, such that $\mathbf{v} + C = C$, contradicting the perpendicularity of r to X at x. As a consequence,

$$C_{Z,K}(x) \simeq T \times T_{r^{-1}(x),x}.$$

If $p : C_{Z,K}(x) \to T_{r^{-1}(x),x}$ is the second projection, we have

$$C_{X,K}(x) = p^{-1}(C),$$

since $C_{X,K}(x)$ is invariant by T. Moreover, p is defined by equations

$$\overline{\overline{z}}_i \circ p = \overline{\overline{z}}_i + \varepsilon_i \overline{\overline{k}}, \quad \text{with } \varepsilon_i \in \mathbb{C} \text{ for } 1 \le i \le c;$$

therefore if we consider the curve W' in Z defined locally around x by

$$z_i + \varepsilon_i w^\delta = 0 \quad \text{for } 1 \le i \le c,$$

we get

$$\delta_x(X, W') > \delta_x(X, W),$$

which proves (i). This shows the equivalence between (i) and (ii).

Then, if $h : \mathbb{D} \to X$ is a map whose ramification index at x is e, and if $r : Z \to X$ is a local retraction around x perpendicular to X at x, by reasoning over the Newton diagrams corresponding to the hypersurface of a tangential intersection it is easy to prove that

$$\delta_{x_h}(X_h, W_h) = e\delta_x(X, W).$$

Observe also that

$$C_{Z_h,K_h}(x_h) = C_{Z,K}(x),$$
$$C_{X_h,K_h}(x_h) = C_{X,K}(x);$$

then $e > 1$ implies $\dim T_{X_h,x_h} = 1$, and by applying the equivalence above we get

$$\delta_{x_h}(X_h) = \delta_x(X).$$

If $e = 1$ this assertion is trivial. The theorem follows. □

3.8 Continuity for the Maximal Contact

Throughout this section, we fix a smooth complex-analytic space Z, a closed subspace X of Z, a smooth closed subspace W of Z, a point $x \in X \cap W$, and a local retraction $r : Z \to W$ around x. Our aim is to prove the following result:

Theorem 3.8.1 (Continuity for the maximal contact) *Let S be the Samuel stratum of X containing x, and let $\mathscr{E}$ be the contact exponent of r with X at x. If r has maximal contact with X at x, there exists a neighborhood U of x in Z satisfying these conditions:*

(1) *$S|_U \subset W|_U$ and $S|_U \subset \mathrm{Sing}(\mathscr{E})|_U$.*
(2) *$\mathscr{E}$ is the contact exponent of r with X at every point of $S \cap U$.*
(3) *r has maximal contact with X at every point of $S \cap U$.*

We first prove the existence of U satisfying (1). Then we prove that, in the initial conditions (without assuming that r has maximal contact with X at x), if $\mathscr{E} = ((\mathbf{J}, b))$ is any idealistic exponent on W, there exists a morphism $g : W' \to W$ such that g is complete, W' is normal and irreducible and there exists a holomorphic function k on W' such that $\mathbf{J} \cdot \mathscr{O}_{W'} = (k^b) \cdot \mathscr{O}_{W'}$. Finally, we prove that conditions (2) and (3) are satisfied.

Remark 3.8.1 Let $\mathscr{E} = ((\mathbf{J}, b))$ be an idealistic exponent on W, and V a smooth complex-analytic space. Set $Z' = Z \times V$, $X' = X \times V$, $W' = W \times V$, $r' = r \times \mathrm{id}_V$, and

$$\mathscr{E}' = ((\mathbf{J} \cdot \mathscr{O}_{W'}, b)).$$

Fix $\mathbf{v} \in V$, and write $x' = (x, \mathbf{v})$. We can easily check the following facts.

(1) $C_{X',x'} = C_{X,x} \times T_{V,\mathbf{v}'}$, $T_{W',x'} = T_{W,x} \times T_{V,\mathbf{v}}$, and $(dr')_{x'} = (dr)_x \times \mathrm{id}_{T_{V,\mathbf{v}}}$.
(2) The retraction r is transversal or perpendicular to X at x if and only if r' is, respectively, transversal or perpendicular to X' at x'.
(3) The space $X = \bigcap_{i=1}^m H^{(i)}$ is a tangential intersection at x if and only if $X' = \bigcap_{i=1}^m H^{(i)\prime}$ is a tangential intersection at x', where $H^{(i)\prime} = H^{(i)} \times V$. Furthermore, if r is transversal to X at x, a hypersurface H, locally around x, is suited or totally suited to X at x with respect to r if and only if $H' = H \times V$ is, respectively, suited or totally suited to X' at x' with respect to r'.
(4) The exponent $\mathscr{E}'$ is independent of the representative of $\mathscr{E}$ we use to construct it. Furthermore $\mathscr{E}$ is the contact exponent of r with X at x if and only if $\mathscr{E}'$ is the contact exponent of r' with X' at x'.
(5) The retraction r has maximal contact with X at x if and only if r' has maximal contact with X' at x'.
(6) If V is connected and S is the Samuel stratum of X containing x, then $S' = S \times V$ is the Samuel stratum of X' containing x'.

Lemma 3.8.1 *Assume that r is transversal to X at x. Let $\mathscr{E}$ be the contact exponent of r with X at x, and let Γ be a smooth curve in Z defined locally at x, in Z. Then, locally around x,*

$$\Gamma \subset \mathrm{Sing}(\mathscr{E}) \Longleftrightarrow \Gamma \subset S \cap W.$$

Proof Suppose $\Gamma \subset \mathrm{Sing}(\mathscr{E})$. We restrict ourselves to work locally around x, so we consider a local coordinate system

$$(\mathbf{z}, \mathbf{w}) = (z_1, \dots, z_c, w_1, \dots, w_d)$$

on Z centered at x and associated with r, consisting "globally" of holomorphic functions. Now, we may assume that Γ is defined in Z by the ideal $(\mathbf{z}, w_2, \dots, w_d) \cdot \mathscr{O}_Z$, because $\Gamma \subset \mathrm{Sing}(\mathscr{E})$ implies $\Gamma \subset W$.

Let $X = \bigcap_{i=1}^m H^{(i)}$ be a tangential intersection, suited to X at x with respect to r, and let $(f_i) \cdot \mathscr{O}_Z$ be the ideal of $H^{(i)}$ in z, for $1 \leq i \leq m$. It is always possible to consider the f_i as "globally" holomorphic. Write $\nu_i = \nu_x(f_i)$ for $1 \leq i \leq m$: since $\mathscr{E}$ is the contact exponent of r with X at x, we obtain from Theorem 3.3.2:

$$\begin{aligned} \Gamma \subset \mathrm{Sing}(\mathscr{E}) &\Longrightarrow \nu_\Gamma((f_i)_A^{b/(\nu_i - |A|)}) \geq b \text{ for } 1 \leq i \leq m \text{ and } |A| < \nu_i \\ &\Longrightarrow \nu_\Gamma((f_i)_A) \geq \nu_i - |A| \text{ for } 1 \leq i \leq m \text{ and } |A| < \nu_i \\ &\Longrightarrow \nu_\Gamma(f_i) \geq \nu_i = \nu_X(f_i) \text{ for } 1 \leq i \leq m \\ &\Longrightarrow \nu_\Gamma(f_i) = \nu_X(f_i) \text{ for } 1 \leq i \leq m. \end{aligned}$$

Therefore the $H^{(i)}$ are equimultiple along Γ, hence X is normally flat along Γ and Γ is contained in $S \cap W$.

Conversely, suppose that $\Gamma \subset S \cap W$. Then X is normally flat along Γ and, by Theorem 3.3.3, one can find a tangential intersection $X = \bigcap_{i=1}^m H^{(i)}$ totally suited to X at x with respect to $\mathbf{r}$ such that $\nu_\Gamma(H^{(i)}) = \nu_x(H^{(i)})$ for $1 \leq i \leq m$. Then $\nu_\Gamma(\mathscr{E}) \geq 1$, so $\Gamma \subset \mathrm{Sing}(\mathscr{E})$. □

Lemma 3.8.2 *With the notation and hypothesis of Lemma* 3.8.1, *let Γ be, locally at x, an irreducible curve in W containing x. Then, locally around x,*

$$\Gamma \subset \mathrm{Sing}(\mathscr{E}) \Longleftrightarrow \Gamma \subset S \cap W.$$

Proof By the local character of the lemma, we may assume that the normalization of Γ is $n : \mathbb{D} \to \Gamma$. Then, using Remark 3.8.1 with $V = \mathbb{D}$ and $\mathbf{v} = 0$, and denoting by Γ^* the smooth curve in W' defined by the image of the composition of maps

$$\mathbb{D} \xrightarrow{\quad \Delta \quad} \mathbb{D} \times \mathbb{D} \xrightarrow{\quad n \times \mathrm{id}_{\mathbb{D}} \quad} W',$$

(where Δ is the diagonal morphism), we conclude that Γ^* is smooth in x' and, by the above lemma, $\Gamma^* \subset \mathrm{Sing}(\mathscr{E}')$ if and only if $\Gamma^* \subset S' \cap W'$.

From item (6) of Remark 3.8.1 it follows that $\mathrm{Sing}(\mathscr{E}') = \mathrm{Sing}(\mathscr{E}) \times \mathbb{D}$ and $S' \cap W' = (S \cap W) \times \mathbb{D}$; therefore, by projecting onto the first components, we conclude that $\Gamma \subset S \cap W$ if and only if $\Gamma \subset \mathrm{Sing}(\mathscr{E})$. □

Corollary 3.8.1 *If r is transversal to X at x and $\mathscr{E}$ is the contact exponent of r with X at x, there exists a neighborhood U of x in Z such that*

$$\mathrm{Sing}(\mathscr{E}|_U) = (S \cap W)|_U.$$

Proof This follows immediately from the preceding lemma. □

Lemma 3.8.3 *If r has maximal contact with X at x and Γ is a smooth curve in Z at x, such that r induces an isomorphism $\Gamma \to r(\Gamma)$, locally around x, then the relation $\Gamma \subset S$ implies $\Gamma \subset W$.*

Proof Since $r(\Gamma)$ is, at x, a smooth curve in W, we can pick a local coordinate system $(\mathbf{z}, \mathbf{w}) = (z_1, \ldots, z_c, w_1, \ldots, w_d)$ on Z, centered at x and associated with r, such that the ideal of $r(\Gamma)$ in $\mathscr{O}_{Z,x}$ is $(\mathbf{z}, w_2, \ldots, w_d) \cdot \mathscr{O}_{Z,x}$. Then the ideal of Γ is of the type $(\mathbf{z}', w_2, \ldots, w_d) \cdot \mathscr{O}_{Z,x}$, with

$$z_i' = z_i - \sigma_i(\mathbf{w}), \quad \text{with } \sigma_i(\mathbf{w}) \in \mathbb{C}\{\mathbf{w}\}, \text{ for } 1 \leq i \leq c.$$

Since $\Gamma \subset S$, X is normally flat along Γ, and one can find by Theorem 3.3.3 a tangential intersection $X = \bigcap_{i=1}^m H^{(i)}$ totally suited to X at x with respect to r and such that

$$\nu_\Gamma(f_i) = \nu_x(f_i) \quad \text{for } 1 \leq i \leq m,$$

where $I_{Z,H^{(i)},x} = (f_i) \cdot \mathscr{O}_{Z,x}$.

Now identify $\mathbb{D}$ with $r(\Gamma)$ by means of a suitable localization, and denote by $h : \mathbb{D} \to W$ the inclusion. Then we have a test-map h of W at x, and

$$\delta_{x_h}(H_h^{(i)}, \Gamma_h) = \infty \quad \text{for } 1 \leq i \leq m,$$

because, in this case, changing the basis via h is equivalent to cutting by

$$w_2 = \cdots = w_d = 0.$$

Therefore

$$\delta_{x_h}(X_h, \Gamma_h) = \infty,$$

because $X = \bigcap_{i=1}^m H^{(i)}$ is totally suited. Since r has maximal contact with X at x and h is a test-map, we have

$$\delta_{x_h}(H_h, \mathbb{D}_h) \geq \delta_{x_h}(X_h, \Gamma_h),$$

hence

$$\delta_{x_h}(X_h, \mathbb{D}_h) = \infty.$$

Lemma 3.4.2 then gives

$$\delta_{x_h}(X_h, \mathbb{D}_h) = \delta_{x_h}(\Gamma_h, \mathbb{D}_h) = \infty,$$

which implies that $\mathbb{D}_h = \Gamma_h$. By projecting Z_h to Z we obtain $\Gamma \subset W$, as desired. □

Lemma 3.8.4 *If r has maximal contact with X at x, there exists a neighborhood of x in Z such that $S|_U \subset W|_U$.*

Proof It is easy to show, by an argument similar to that used in proving Lemma 3.8.2, that, for every irreducible curve Γ, locally around x in Z, the relation $\Gamma \subset S$ implies $\Gamma \subset W$. The lemma follows. □

Lemma 3.8.5 *Let W be an arbitrary smooth complex-analytic space, and let $\mathscr{E} = ((\mathbf{J}, b))$ be an idealistic exponent on W. There exists a morphism $g : W' \to W$ of analytic spaces satisfying these conditions:*

(1) *W' is normal and irreducible.*
(2) *There is a holomorphic function k on W' such that*

$$\mathbf{J} \cdot \mathscr{O}_W = (k^b) \cdot \mathscr{O}_W.$$

(3) *The morphism g is complete.*

Proof Since W is smooth, we can express W as a disjoint union $W = W_1 \cup W_2$, in such a way that $\mathbf{J}|_{W_1}$ is locally nowhere zero and $\mathbf{J}|_{W_2}$ is zero everywhere. Denote by $h_1' : W_1' \to W_1$ the blow-up of $\mathbf{J}|_{W_1}$, by $h_2' : V_2' \to W_2$ the identity, and by V' the disjoint union $V' = V_1' \cup V_2'$. Then, there exists a unique morphism $h' : V' \to W$ of analytic spaces such that

$$h'|_{V_1'} = h_1' \quad \text{and} \quad h'|_{V_2'} = h_2'.$$

By the construction of h', $\mathbf{J} \cdot \mathscr{O}_{V'}$ is locally principal, hence there exist an open covering $\{M_\alpha\}_{\alpha \in \Lambda}$ of V' and a family of functions $\{d_\alpha\}_{\alpha \in \Lambda}$ such that, for every $\alpha \in \Lambda$, the map d_α is holomorphic on M_α, and $\mathbf{J} \cdot (\mathscr{O}_V'|_{M_\alpha}) = (d_\alpha)(\mathscr{O}_V'|_{M_\alpha})$.

Now consider, for each $\alpha \in \Lambda$, the subspace M_α' of $M_\alpha \times \mathbb{C}$ defined by the ideal

$$(t^b - d_\alpha) \cdot (\mathscr{O}_{V' \times \mathbb{C}} \mid M_\alpha \times \mathbb{C}),$$

where t is a parameter on $\mathbb{C}$. Denote by $f_\alpha : M_\alpha' \to M_\alpha$ the first projection. Then, for every $\alpha \in \Lambda$, there exists a holomorphic function k_α on M_α' such that $d_\alpha f_\alpha = k_\alpha$, and therefore

$$\mathbf{J} \cdot (\mathscr{O}_{V' \times \mathbb{C}}|_{M_\alpha \times \mathbb{C}}) = (k_\alpha^b) \cdot (\mathscr{O}_{V' \times \mathbb{C}}|_{M_\alpha \times \mathbb{C}}).$$

For $\alpha \in \Lambda$, denote by M''_α the normalization of $(M'_\alpha)_{\mathrm{red}}$. Further, denote by W' the disjoint union of the family $\{M''_\alpha\}_{\alpha\in\Lambda}$, and by $g : W' \to W$ the morphism obtained by composing together h', $\{f'_\alpha\}_{\alpha\in\Lambda}$ and the morphism of normalization.

Under these conditions, W' is normal by construction and, if k is the holomorphic function on W' such that, for each $\alpha \in \Lambda$, $k|_{M''_\alpha}$ is the pull-back of k_α under the normalization morphism, one has

$$\mathbf{J} \cdot \mathscr{O}_{W'} = (k^b) \cdot \mathscr{O}_{W'}.$$

Finally, we shall see that g is complete. Let $y \in W$ and let N be a compact neighborhood of y in W: by construction, h' is proper, hence $h'^{-1}(N)$ is a compact subset of V', hence there exists a finite subset $\Delta \subset \Lambda$ such that

$$h'^{-1}(N) \subset \bigcup_{\alpha\in\Delta} M_\alpha,$$

and, for each $\alpha \in \Delta$, one can find a compact set L_α in such a way that $L_\alpha \subset M_\alpha$ and

$$h'^{-1}(N) \subset \bigcup_{\alpha\in\Delta} L_\alpha.$$

On the other hand, the morphisms $\lambda_\alpha : M''_\alpha \to M_\alpha$ that arise from the composition of the normalization morphisms and the f_α are proper; hence

$$L = \bigcup_{\alpha\in\Delta} \lambda_\alpha^{-1}(L_\alpha)$$

is a compact subset of W'. It can be immediately checked that, since h' and $\{\lambda_\alpha\}_{\alpha\in\Delta}$ are proper and surjective (hence complete), L and N satisfy the conditions in the definition of a complete morphism. □

Proof of Theorem 3.8.1 As a consequence of Theorem 3.5.3, Corollary 3.8.1, and Lemma 3.8.4, one can find, under the assumptions of Theorem 3.8.1, a neighborhood U_1 of x in Z such that, if $\mathscr{E}$ is the contact exponent of r with X at x, the following conditions hold:

$$S|_{U_1} \subset W|_{U_1}; \tag{3.27}$$

r is perpendicular to X at every point of $W \cap U_1$; $\mathscr{E} = ((\mathbf{J}, b))$ is a subcontact exponent of r with X at every point of $S \cap U_1$ (by an argument similar to that in Lemma 3.8.4); and

$$S|_{U_1} = \mathrm{Sing}(\mathscr{E})|_{U_1}. \tag{3.28}$$

On the other hand, by Lemma 3.8.5, there exists an analytic morphism $g : W' \to W$ such that W' is normal, g is complete, and there is a holomorphic function k on W' satisfying $\mathbf{J} \cdot \mathscr{O}_{W'} = (k^b) \cdot \mathscr{O}_{W'}$.

Therefore we can construct the characteristic cone of (X, r) with respect to $(\mathscr{E}, g)$, $C_g \to K'$, and, for each $x' \in g^{-1}(x)$, the stalk at x' of C_g is flat. By generic flatness, for each $x' \in g^{-1}(x)$ in K', there exists a neighborhood $U_{x'}$ of x' such that $C_g(U_{x'}) \to U_{x'}$ is flat.

Now write

$$U' = \bigcup_{x' \in g^{-1}(x)} U_{x'}.$$

U' is an open set in Z', containing $g^{-1}(x)$. Since g is complete, we can apply part (3.28) of Remark 3.7.5 to find an open neighborhood U of x, contained in U_1 and such that $g^{-1}(U) \subset U'$. Then, it follows from Corollary 3.8.1 that

$$C_g(g^{-1}(S \cap U)) \to g^{-1}(S \cap U)$$

is flat; hence, by Proposition 3.5.1, there exists a vector fiber subspace $T(\mathscr{E}, X, r, g)$ of $E_g(g^{-1}(S \cap U))$ such that

$$T(\mathscr{E}, X, r, g)_y = T_y(\mathscr{E}, X, r, g) \quad \text{for every } y \in g^{-1}(S \cap U).$$

In view of this, the theorem follows from (3.27) and (3.28) above, together with Propositions 3.7.2, 3.7.3, and 3.7.4. □

3.9 Contact Stability Theorems

Throughout this section, Z will represent a smooth complex-analytic space and X a closed complex subspace of Z. Consider the commutative diagram

$$\begin{array}{ccc} X' & \longrightarrow & Z' \\ {\scriptstyle p}\downarrow & & \downarrow{\scriptstyle \pi} \\ E \subset X & \longrightarrow & Z \end{array} \tag{3.29}$$

where E is a smooth closed complex subspace of X, p and π are the blow-ups of X and Z, respectively, with center E, and X' is canonically identified with the strict transform of X by π in Z.

Remark 3.9.1 If X is normally flat along E at x, we also have the exact sequence

$$0 \to T_{E,x} \to C_{X,x} \to C_{X,E,x} \to 0. \tag{3.30}$$

On the other hand, from Sect. 1.7 of Chap. 1 we have

$$\mathbb{P}(C_{Z,X,x}) = \pi^{-1}(x), \tag{3.31}$$

and

$$\mathbb{P}(C_{X,E,x}) = p^{-1}(x),$$

from which, and if X is normally flat along E at x, (3.30) implies

$$\mathbb{P}(C_{X,x}/T_{E,x}) = p^{-1}(x). \tag{3.32}$$

As a consequence of (3.31), we can establish a one-to-one correspondence between the points of $\pi^{-1}(x)$ and the complex line through $\mathbf{0}$ of $N_{Z,X,x}$; we denote by $L_{\pi,x'}$ the image of each $x' \in \pi^{-1}(x)$ under this correspondence. If $x' \in \pi^{-1}(x)$ is such that $L_{\pi,x'} \in \mathbb{P}(T_{X,x}/T_{E,x})$, we denote by $M_{\pi,x'}$ the inverse image of $L_{\pi,x'}$ under the natural homomorphism

$$T_{X,x} \to T_{Z,x}/T_{E,x}.$$

Observe that, under these conditions, the vectors of $M_{\pi,x'}$ leave the cone $C_{X,x}$ invariant by translation, and therefore the tangent cone to $C_{X,x}$ at the points of $M_{\pi,x'}$ can be canonically identified with $C_{Z,x}$, by identifying previously the tangent space to $T_{Z,x}$ at each of its points with $T_{Z,x}$ itself.

Taking these identifications into account, one observes that, by (3.32), and assuming that X is normally flat along E at x, the differential of the projection

$$\{C_{X,x} \setminus T_{E,x}\} \to p^{-1}(x)$$

induces, at the points of $M_{\pi,x'} \setminus T_{E,x}$, a canonical morphism

$$C_{X,x} \to C_{p^{-1}(x),x'},$$

and therefore a morphism, also canonical,

$$C_{X,x}/M_{\pi,x'} \to C_{p^{-1}(x),x'}.$$

Lemma 3.9.1 (Tangency lemma) *Assume that, in* (1) *of Remark* 3.9.1, *X is normally flat along E at $x \in E$, and let $x' \in p^{-1}(x)$. If $H_{X',x'} = H_{X,x}$, where $H_{X,x}$ is the Samuel function of X at x, we have:*

(i) *The complex line $L_{\pi,x'}$ corresponding to x' is contained in $T_{X,x}/T_{E,x}$.*
(ii) *The canonical morphism*

$$C_{X,x}/M_{\pi,x'} \to C_{p^{-1}(x),x'}$$

is an isomorphism. Moreover, the intersection $p^{-1}(x) = X' \cap \pi^{-1}(x)$ is transversal at x' and

$$C_{X',x'} \cap T_{\pi^{-1}(x'),x'} = C_{p^{-1}(x),x'}.$$

Proof Under the conditions of the lemma, and for every $x' \in p^{-1}(x)$, Bennett's theorem (Theorem 2.5.1 in Chap. 2) gives

$$H_{X,x}(m) = H^{(+r)}_{C_{X,E}(x),x}(m) \geq H^{(+r+1)}_{p^{-1}(x),x'}(m) \geq H_{X',x'}(m), \tag{3.33}$$

for every $m \in \mathbb{Z}_0$, where $r = \dim_x E$. The fact that $H_{X,x} = H_{X',x'}$ implies the equality of all members of (3.33); in particular it implies that

$$H_{C_{X,E}(x),x} = H^{(+1)}_{p^{-1}(x),x'}. \tag{3.34}$$

Now let $L_{\pi,x'}$ be the complex line in $N_{Z,E}(x)$ corresponding to x', and let ξ an arbitrary point of $L_{\pi,x'} \setminus \{0\}$. Since $x' \in p^{-1}(x)$, one has

$$L_{\pi,x'} \subset C_{X,E}(x)$$

and, in addition, the germ of $C_{X,E}(x)$ at ξ is isomorphic to the germ of $p^{-1}(x) \times C$ at $(x', 0)$. Therefore

$$H_{C_{X,E}(x),\xi} = H^{(+1)}_{p^{-1}(x),x'};$$

from this and from (3.34), taking into account that x is identified with $\mathbf{0}$ in $C_{X,E}(x)$, it follows that $C_{X,E}(x)$ has the same Samuel function at each one of the points of $L_{\pi,x'}$ and, hence, $C_{X,E,x}$ is normally flat along $L_{\pi,x'}$. It follows that $C_{X,E}(x)$, which is the tangent cone to $C_{X,E}(x)$ at 0, is invariant under translation by $L_{\pi,x'}$; hence

$$L_{\pi,x'} \subset T_{X,x}/T_{E,x},$$

which proves statement (i).

Let ξ be an arbitrary point of $M_{\pi,x'} \setminus T_{E,x}$. From (3.29) and from (3.32) of Remark 3.9.1 it follows that the germ of $C_{X,x}$ at ξ is isomorphic to the germ of $p^{-1}(x) \times M_{\pi,x'}$ at $x' \times \xi$. Therefore, the canonical morphism

$$C_{X,x}/M_{\pi,x'} \to C_{p^{-1}(x),x'}$$

defined in Remark 3.9.1 is an isomorphism.

The validity of statement (ii) is now clear if we observe that, by the equality in (3.30), one has

$$H^{(+r+1)}_{p^{-1}(x),x} = H_{X',x'},$$

so that the intersection $p^{-1}(x) = X' \cap \pi^{-1}(x)$ is transversal at x'. □

Remark 3.9.2 Assume that, in the conditions of (3.29) in Remark 3.9.1, W is a smooth complex subspace of Z containing E, let $P = I_{Z,E,x}$ be ideal defining E at x in Z, and let q be the blow-up of W with center E; then we have a diagram

$$\begin{array}{ccccc} W' & \hookrightarrow & Z' & \hookleftarrow & X' \\ \downarrow q & & \downarrow \pi & & \downarrow p \\ E \subset W & \hookrightarrow & Z & \hookleftarrow & X \supset E \end{array} \tag{3.35}$$

Take $x \in W \cap X$ and $x' \in p^{-1}(x) \cap W'$, and let $(\mathbf{z}, \mathbf{w})$ denote a local coordinate system on Z, centered at x, satisfying these conditions:

(a) $\mathbf{z} = (z_1, \ldots, z_c)$ and $\mathbf{w} = (\mathbf{u}, \mathbf{t}) = (u_1, \ldots, u_s, t_1, \ldots, t_d)$.
(b) $I_{Z,W,x} = (\mathbf{z})\mathcal{O}_{Z,x'}$ and $I_{Z,E,x} = (\mathbf{z}, \mathbf{u})\mathcal{O}_{Z,x}$.
(c) If $\mathbf{z}' = (z_1/u_1, \ldots, z_c/u_1)$ and $\mathbf{w}' = (\mathbf{u}, \mathbf{t}) = (u_1, u_2/u_1, \ldots, u_s/u_1, \mathbf{t})$, then $(\mathbf{z}', \mathbf{w}')$ is a local coordinate system on Z' centered at x'.

Under these conditions, we shall denote by $r : Z \to W$ the local retraction with which is associated the coordinate system $(\mathbf{z}, \mathbf{w})$.

Lemma 3.9.2 *Using the conditions of* (3.33), *Remark* 3.9.1, *and* (3.33), *Remark* 3.9.2, *we make the following assumptions:*

(i) *The retraction r is transversal to X at x.*
(ii) *The space X is normally flat along E at x.*
(iii) $H_{X',x'} = H_{X,x}$.
(iv) *The family $(f_1, \ldots, f_m)$ is a P-normalized standard base of $\mathbf{I}_{Z,X,x}$ with respect to $(\mathbf{z}, \mathbf{w})$ (being $P = I_{Z,E,x}$).*
(v) *For $1 \leq i \leq m$ is $\nu_i = \nu_x(f_i)$ and $f_i' = u_1^{\nu_i} f_i$.*

Then:

(1) *For $1 \leq i \leq m$ is $f_i' \in \mathcal{O}_{Z',x'}$ and $\nu_{x'}(f_i') = \nu_i$.*
(2) *The family $(f_1', \ldots, f_m')$ is a normalized standard base of $I_{Z',X',x'}$ with respect to $(\mathbf{z}', \mathbf{w}')$.*

Proof With the usual notation $\bar{g} = \mathrm{in}_x(g) \in \mathrm{gr}_x(Z)$ for $g \in \mathcal{O}_{Z,x}$, one has

$$\mathrm{gr}_x(Z) = \mathbb{C}[\bar{\mathbf{z}}, \bar{\mathbf{w}}]$$

and, obviously, $M_{\pi,x'}$ is defined in $\mathrm{gr}_x(Z)$ by the ideal $(\bar{\mathbf{z}}, \bar{u}_2, \ldots, \bar{u}_s) \cdot \mathrm{gr}_x(Z)$. By conditions (ii), (iii) and conclusion (3.33) of Lemma 3.9.1, $\mathrm{in}_x(Z, X)$ is generated by elements of $\mathbb{C}[\bar{\mathbf{z}}, \bar{u}_2, \ldots, \bar{u}_s]$. Therefore, by (i) and (iv), we can assert that, for every $i = 1, \ldots, m$, f_i is normalized by $(f_1, \ldots, f_{i-1})$ with respect to $(\mathbf{z}, \mathbf{w})$, that $(f_1, \ldots, f_m)$ is a base of $\mathrm{in}_x(Z, X)$, and that

$$\bar{f}_i \in \mathbb{C}[\bar{\mathbf{z}}, \bar{u}_2, \ldots, \bar{u}_s].$$

Since $(f_1, \ldots, f_m)$ is E-normalized, we have $f_i \in (\mathbf{z}, \mathbf{u})^{\nu_i} \cdot \mathcal{O}_{Z,x}$ for $i = 1, \ldots, m$. Write

$$f_i = \sum_{A \in Z_0^c} (f_i)_A \mathbf{z}^A \quad \text{with } (f_i)_A \in \mathbb{C}\{\mathbf{w}\}.$$

Since $\bar{f}_i \in \mathbb{C}[\bar{\mathbf{z}}, \bar{\mathbf{u}}_2, \ldots, \bar{\mathbf{u}}_s]$, we have

$$(f_i)_A \in (u_2, \ldots, u_s)^{\nu_i - |A|} \mathbb{C}\{\mathbf{w}\} + u_1 \cdot (\mathbf{u}, \mathbf{t})^{\nu_i - |A| - 1} (\mathbf{u}, \mathbf{t}) \cdot \mathbb{C}\{\mathbf{w}\}.$$

From this, writing

$$f_i' = \sum_{A \in Z_0^c} (f_i')_A (\mathbf{z}')^A \quad \text{with } (f_i')_A \in \mathbb{C}\{\mathbf{w}'\},$$

we deduce that

$$(f_i')_A \in (u_1', \mathbf{t}')\mathbb{C}\{\mathbf{w}\}$$

for all i and A.

Now denote by $\hat{f}_i'$ the initial form in $\mathrm{gr}_{x'}(\pi^{-1}(x))$ of $f_i' (\mathrm{mod}\ I_{Z,\pi^{-1}(x),x'})$, for $1 \leq i \leq m$. Since, by (i) and (iv), $\bar{f}_i \notin (\mathbf{u}, \mathbf{t})\mathrm{gr}_x(Z)$ for $1 \leq i \leq m$, we conclude that $\deg(\hat{f}_i') = \nu_i$ for $1 \leq i \leq m$. Taking into account the isomorphism in conclusion (2) of Lemma 3.9.1, one has

$$\mathrm{in}_{x'}(p^{-1}(x), \pi^{-1}(x)) = (\hat{f}_1', \ldots, \hat{f}_m')\mathrm{gr}_{x'}(\pi^{-1}(x)),$$

and, since $p^{-1}(x) = X' \cap \pi^{-1}(x)$ is a transversal intersection (Lemma 3.9.1), there exists a homogeneous base $(g_1, \ldots, g_m)$ of $\mathrm{in}_{x'}(Z', X')$ such that each $\hat{f}_i'$ is the image of g_i under the natural homomorphism

$$\mathrm{gr}_{x'}(Z') \to \mathrm{gr}_{x'}(\pi^{-1}(x)).$$

As a consequence, if we set $\bar{\bar{g}} = \mathrm{in}_{x'}(g)$ for $g \in \mathcal{O}_{Z',x'}$ (so that $\mathrm{gr}_{x'}(Z') = \mathbb{C}[\bar{\bar{\mathbf{z}}}', \bar{\bar{\mathbf{w}}}']$), and if we denote by $\varphi_i = \varphi_i(\bar{\mathbf{z}}) \in \mathbb{C}[\bar{\mathbf{z}}]$ the function such that $\bar{f}_i - \varphi_i \in (\bar{\mathbf{u}}, \bar{\mathbf{t}}) \cdot \mathrm{gr}_x(Z)$, we have

(a) $g_i - \varphi_i(\bar{\bar{\mathbf{z}}}') \in (\bar{\bar{\mathbf{u}}}', \bar{\bar{\mathbf{t}}}') \cdot \mathrm{gr}_{x'}(Z')$ for $1 \leq i \leq m$.

Now, by (iv) (see the proof of Lemma 3.6.2),

$$\mathcal{E}_{\bar{\mathbf{z}}}(f_i) \cap \mathrm{lex}_{\bar{\mathbf{z}}}((\varphi_1, \ldots, \varphi_{i-1}) \cdot \mathbb{C}[\bar{\mathbf{z}}]) = \varnothing \quad \text{for } 1 \leq i \leq m;$$

hence, by comparing the Taylor expansions of f_i and f_i', we get

(b) $\mathscr{E}_{\mathbf{z}'}(f_i') \cap \operatorname{lex}_{\overline{\overline{\mathbf{z}}}'}((\varphi_1(\overline{\overline{\mathbf{z}}}'), \ldots, \varphi_{i-1}(\overline{\overline{\mathbf{z}}}')) \cdot \mathbb{C}[\overline{\overline{\mathbf{z}}}']) = \varnothing$ for $1 \leq i \leq m$.

Then, if we assume that $\nu_{x'}(f_i') < \nu_i$ for some i, and taking into account that

$$\operatorname{in}_{x'}(Z', X') = (g_1, \ldots, g_m) \cdot \operatorname{gr}_{x'}(Z'),$$

we have

$$\overline{\overline{f}}_i' \in (g_1, \ldots, g_{i-1}) \cdot \operatorname{gr}_{x'}(Z');$$

this, together with (a), leads to a contradiction with (b). This proves that $\nu_{x'}(f_i') \geq \nu_i$ for $i = 1, \ldots, m$.

On the other hand, $\varphi_i \neq 0$ for $i = 1, \ldots, m$; then, by the definition of the $\{g_i\}$ and by (1), one has $\nu_{x'}(f_i') = \nu_i$ for $i = 1, \ldots, m$, which proves conclusion (1) of the lemma.

To prove (2), it is enough to observe that, for every $i = 1, \ldots, m$, $\hat{f}_i'$ is the image of $\overline{\overline{f}}_i'$ under the natural homomorphism $\operatorname{gr}_{x'}(Z') \to \operatorname{gr}_{x'}(\pi^{-1}(x))$. Then, by conclusion (2) of Lemma 3.9.1, one has

$$(\overline{\overline{f}}_1', \ldots, \overline{\overline{f}}_m') \cdot \operatorname{gr}_{x'}(Z') = \operatorname{in}_{x'}(Z', X'),$$

and, by (b), f_i' is normalized by $(f_1', \ldots, f_{i-1}')$ with respect to $(\mathbf{z}', \mathbf{w})$, for every $i = 1, \ldots, m$, which proves the lemma. □

Remark 3.9.3 Assume that, in the case of the diagram (1) of Remark 3.9.2, we have a retraction $r : Z \to W$. In general, it is not possible to define a retraction $r' : Z' \to W'$ compatible with r, that is, such that the diagram

$$\begin{array}{ccc} Z' & \xrightarrow{\ r'\ } & W' \\ \pi \downarrow & & \downarrow q \\ Z & \xrightarrow{\ r\ } & W \end{array}$$

commutes. But it is possible to construct such an r' by a "suitable reduction" of Z', in the following way.

Let $F = r^{-1}(E)$; then F is smooth in a neighborhood of E, and the intersection $F \cap W = E$ is transversal in Z. As a consequence, if F' is the strict transform of F by, π, we have $F' \cap W' = \varnothing$. Put $Z'' = Z' - F'$; then Z'' is clearly a neighborhood of W' in Z'. As we shall see, there exists a unique retraction $r' : Z'' \to W'$ such that $q.r' = r.(\pi|_{Z''})$.

Uniqueness is evident. Existence is easily seen by considering the diagram

$$\begin{array}{ccccc} Z'_1 & \xrightarrow{\pi'} & Z_1 & & W' \\ {\scriptstyle \pi'_1}\downarrow & & \downarrow{\scriptstyle \pi_1} & & \downarrow{\scriptstyle q} \\ Z' & \xrightarrow{\pi} & Z & \xrightarrow{r} & W \end{array}$$

where π_1, π'_1 and π' are, respectively, the blow-ups of Z, Z' and Z_1 with centers F, F' and $\pi_1^{-1}(E)$. Then there exists a unique morphism $r_1 : Z_1 \to W'$ that makes the diagram commutative. Since π'_1 induces an isomorphism in Z'', $r_1 \cdot \pi' \cdot (\pi_1'^{-1}|_{Z''})$ gives on Z'' the desired retraction.

Definition 3.9.1 With the notation and hypothesis of Remark 3.9.3, we shall call r' the **retraction transform** of r by the blow-up π, with center E.

Remark 3.9.4 In the conditions of the diagram (1) of Remark 3.9.2 and of Lemma 3.9.2, assume that $\mathscr{E} = ((\mathbf{J}, b))$ is an idealistic exponent on W and that E (which is a smooth closed complex subspace of W) is contained in $\mathrm{Sing}(\mathscr{E})$. Under these conditions, if P is the ideal of E in $\mathscr{O}_W$, we have $\mathbf{J} \subset P^b$ and, therefore, $\mathbf{J}\cdot\mathscr{O}_{W'} \subset P^b\cdot\mathscr{O}_{W'}$. Since q is the blow-up of W with center E, $P\mathscr{O}_{W'}$ is invertible as a $\mathscr{O}_{W'}$-module; hence there exists an ideal $\mathbf{J}'$ in $\mathscr{O}_{W'}$ such that $\mathbf{J}'P^b\cdot\mathscr{O}_{W'} = \mathbf{J}\cdot\mathscr{O}_{W'}$. Then, if $q^*(\mathscr{E})$ is the idealistic exponent defined on W' by $((\mathbf{J}', b))$, the value of $q^*(\mathscr{E})$ is independent of the representative $((\mathbf{J}, b))$ of $\mathscr{E}$ on W as we have chosen to define it.

Definition 3.9.2 With the notation and hypothesis of Remark 3.9.4, we shall call $q^*(\mathscr{E})$ the **transform** of $\mathscr{E}$ by the blow-up q, with center E.

Remark 3.9.5 Note that the transform of a retraction or an idealistic exponent by a blow-up depends, not only on the morphism π and q, but also on the center E. For instance, the identity mapping u can be considered as a blow-up with empty center, or as a blow-up with center an arbitrary hypersurface, and $u^*(\mathscr{E})$ can vary, depending on the choice of the center.

Theorem 3.9.1 (Stability Theorem 1) *Consider a smooth complex-analytic space Z, a closed complex subspace X of Z, a smooth closed complex subspace E of X, and a smooth closed complex subspace W of Z containing E. Let π, p, q be, respectively, the blow-ups of Z, X, W with center E; we have a commutative diagram*

$$\begin{array}{ccccc} W' & \hookrightarrow & Z' & \hookleftarrow & X' \\ {\scriptstyle q}\downarrow & & \downarrow{\scriptstyle \pi} & & \downarrow{\scriptstyle p} \\ E \subset W & \hookrightarrow & Z & \hookleftarrow & X \end{array}$$

Let $r : Z \to W$ be a retraction and $\mathscr{E}$ an idealistic exponent on W; take $x \in E$ and $x' \in p^{-1}(x) \cap W'$. Assume that the following conditions are verified:

(a) *X is normally flat along E at x.*
(b) *$H_{X',x'} = H_{X,x}$.*
(c) *r is transversal to X at x.*
(d) *$\mathscr{E}$ is the contact exponent of r with X at x.*

Then:

(1) *If r' is the transform of r by π, r' is transversal to X' at x'.*
(2) *If $\mathscr{E}'$ is the transform of $\mathscr{E}$ by the blow-up with center E, $\mathscr{E}'$ is the contact exponent of r' with X' at x'.*
(3) *$x' \in \mathrm{Sing}(\mathscr{E}')$.*

Proof We can construct a local coordinate system on Z centered at x, associated with r, and with the properties (a), (b), and (c) of Remark 3.9.2. Using the notation from the proof of Lemma 3.9.2, we may write

$$\mathrm{gr}_x(Z) = \mathbb{C}[\bar{\mathbf{z}}, \bar{\mathbf{w}}], \quad \mathrm{gr}_{x'}(Z') = \mathbb{C}[\bar{\bar{\mathbf{z}}}', \bar{\bar{\mathbf{w}}}'].$$

Clearly the differential of the mapping

$$T_{Z,x} \setminus T_{E,x} \to \mathbb{P}(T_{Z,x}/T_{E,x}) = \pi^{-1}(x),$$

at an arbitrary point of $M_{\pi,x'} \setminus T_{E,x}$ maps $\bar{\bar{u}}_i'$ onto $\bar{u}_i$, for $2 \le i \le s$, and $\bar{\bar{z}}_j'$ onto $\bar{z}_j$, for $1 \le j \le c$. Also, this differential induces the isomorphism of conclusion (2) of Lemma 3.9.1. Moreover, condition (c) of our theorem implies that $(\bar{\bar{u}}_2', \dots, \bar{\bar{u}}_s')$ is a regular sequence for $\mathrm{gr}_{x'}(p^{-1}(x))$.

On the other hand, by conditions (a) and (b), we may apply Lemma 3.9.1, and since $p^{-1}(x) = X' \cap \pi^{-1}(x)$ is a transversal intersection, we have

$$\mathrm{gr}_{x'}(p^{-1}(x)) = \mathrm{gr}_{x'}(X')/(\bar{\bar{u}}_1', \bar{\bar{\mathbf{t}}}')\mathrm{gr}_{x'}(X')$$

and $(\bar{\bar{u}}_1', \bar{\bar{\mathbf{t}}}')$ is a regular sequence for $\mathrm{gr}_{x'}(X')$. Therefore, $(\bar{\bar{u}}_1', \bar{\bar{u}}_2', \dots, \bar{\bar{u}}_s', \bar{\bar{\mathbf{t}}}')$ is a regular sequence for $\mathrm{gr}_{x'}(X')$. From this we deduce that $\mathrm{gr}_{x'}(X')$ is a flat $\mathbb{C}[\bar{\bar{u}}', \bar{\bar{\mathbf{t}}}']$-module. Now, since $\mathbb{C}[\bar{\bar{u}}', \bar{\bar{\mathbf{t}}}' = \mathrm{gr}_{x'}(W')$, conclusion (1) is proved.

We now prove (2) and (3). Let S be the Samuel stratum of X containing x; from (c), (d) and Corollary 3.8.1 we deduce that $\mathrm{Sing}(\mathscr{E})$ and $(S \cap W)_{\mathrm{red}}$ coincide, locally around x. Moreover, by condition (a), E is contained in S, locally around x; therefore, by localizing, we may assume that $\mathrm{Sing}(\mathscr{E}) = (S \cap W)_{\mathrm{red}}$ and that $E \subset \mathrm{Sing}(\mathscr{E})$.

Let $(f_1, \dots, f_m)$ be a E-normalized standard basis of $I_{Z,X,x}$ with respect to $(\mathbf{z}, \mathbf{w})$. By using the notation in the proof of Lemma 3.9.2, and by taking into account Theorem 3.3.2, we may write that $\mathscr{E} = ((\mathbf{J}, b))$ with $b = \prod_{i=1}^m \nu_i!$, that

$$\mathbf{J}_x = \left(\{(f_i)_A^{b/(\nu_i - |A|)}\}_{1 \le i \le m \; |A| < \nu_i}\right) \cdot \mathcal{O}_{W,x},$$

and that

$$q^*(\mathcal{E}) = \mathcal{E}' = ((\mathbf{J}', b)),$$

with

$$\mathbf{J}'_{x'} = \left(\{(f_i)_A^{b/(\nu_i - |A|)}/u_1^b\}_{1 \le i \le m \; |A| < \nu_i}\right) \cdot \mathcal{O}_{W',x'},$$

or, equivalently,

$$\mathbf{J}'_{x'} = \left(\{(f'_i)_A^{b/(\nu_i - |A|)}\}_{1 \le i \le m \; |A| < \nu_i}\right) \cdot \mathcal{O}_{W',x'} \tag{3.36}$$

Now, by Lemma 3.9.2, $(f'_1, \ldots, f'_m)$ is a normalized standard basis of $I_{Z',W',x'}$, $\nu_{x'}(f'_i) = \nu_i$, and $(\mathbf{z}', \mathbf{w}')$ is a local coordinate system associated with r', which is transversal to X' at x'. So (1) implies that $q^*(\mathcal{E}) = \mathcal{E}'$ is the contact exponent of r' with X' at x', which proves conclusion (2). Conclusion (3) is an immediate consequence of (2). □

Theorem 3.9.2 (Stability Theorem 2) *Consider a smooth complex-analytic space Z, a closed complex subspace X of Z, a smooth closed complex subspace E of X, and a smooth closed complex subspace W of Z containing E. Let π, p, q be, respectively, the blow-ups of Z, X, W with center E; we have a commutative diagram*

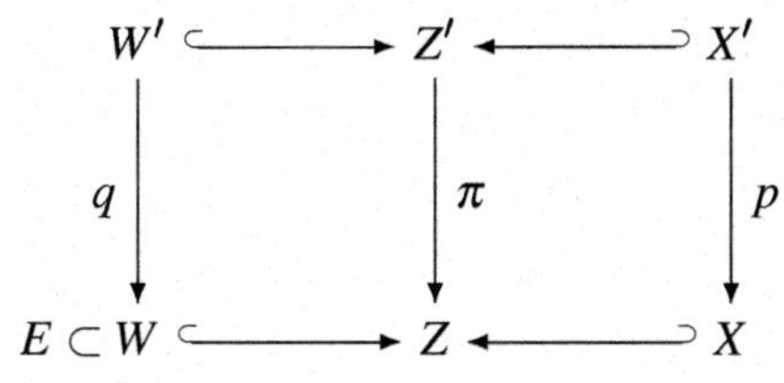

Let $r : Z \to W$ be a retraction, and take $x \in X \cap W$ and $x' \in p^{-1}(x)$. Assume that the following conditions are satisfied:

(a) *The retraction r has maximal contact with X at x.*
(b) $H_{X',x'} = H_{X,x}$.
(c) *The space X is normally flat along E at x.*

Then $x' \in W'$ and the transform r' of r by π has maximal contact with X' at x'.

For the proof we need some preliminary lemmas.

Lemma 3.9.3 *With the notation and hypothesis of Theorem* 3.9.2, *without assuming* (b) *and* (c), *we have $T_{X,x} \subset T_{W,x}$.*

Proof Since r is perpendicular to X at x, we have

$$T_{X,x} \cap T_{r^{-1}(x),x} \subset T_{X \cap r^{-1}(x),x}. \tag{3.37}$$

Assume that $T_{X,x} \not\subset T_{W,x}$; then there is a vector $\mathbf{v} \in T_{X,x} \setminus T_{W,x}$. Put $\mathbf{v}' = (dr)_x(\mathbf{v})$; then $\mathbf{v}' \in T_{X,x}$. In fact, if $\mathbf{v}' \in T_{X,x}$ then $\mathbf{v} - \mathbf{v}' \in T_{X,x} \cap T_{r^{-1}(x),x}$. But, by (3.37), $T_{X,x} \cap T_{r^{-1}(x),x} = (0)$, so $\mathbf{v} = \mathbf{v}' \in T_{W,x}$, which is absurd.

Let Γ be a smooth complex curve, defined around x in Z, and such that $\mathbf{v} \in T_{\Gamma,x}$. Since $\mathbf{v} \notin T_{r^{-1}(x),x}$, by (3.37), $r|_\Gamma$ is, locally around x, an embedding of Γ in W. By localizing around x, we may assume that Γ is a disk $\mathbb{D}$, that $x = 0$ and that $r|_\Gamma$ is a locally closed embedding. Write $\Gamma' = r(\Gamma)$, and denote by $h : \mathbb{D} \to W$ the composition of the inclusion $\Gamma \to Z$ with r.

By basis extension by h, we obtain the diagram

$$\begin{array}{ccccc} X_h & \hookrightarrow & Z_h & \longleftarrow & \Gamma_h \\ & & r_h \downarrow \uparrow & & \\ & & W_h = \mathbb{D} \ni x_h = 0 & & \end{array}$$

The projection $Z_h \to Z$ induces an isomorphism

$$X_h \cap r_h^{-1}(0) \to X \cap r^{-1}(x),$$

so

$$T_{X_h \cap r_h^{-1}(x),x_h} = T_{X \cap r^{-1}(x),x} = (0),$$

hence

$$\dim T_{X_h,x_h} \leq 1.$$

Now, since $T_{\Gamma,x} \subset T_{X,x}$, one has $T_{\Gamma_h,x_h} \subset T_{X_h,x_h}$, and, by the above result, $T_{X_h,x_h} = T_{\Gamma_h,x_h}$, hence $\delta_{x_h}(X_h, \Gamma_h) > 1$.

On the other hand, $T_{\Gamma_h,x_h} \neq T_{W_h,x_h}$, since h is an embedding; therefore

$$T_{X_h,x_h} \neq T_{W_h,x_h},$$

and so

$$\delta_{x_h}(X_h, W_h) = 1,$$

which contradicts the assumption that r has maximal contact with X at x. □

Corollary 3.9.1 *With the notation and hypothesis of Theorem* 3.9.2, *we have* $x' \in W'$.

Proof By Lemma 3.9.1, $x' \in \mathbb{P}(T_{X,x}/T_{E,x})$. But Lemma 3.9.3 implies

$$\mathbb{P}(T_{X,x}/T_{E,x}) \subset \mathbb{P}(T_{W,x}/T_{E,x}) = q^{-1}(x),$$

so $x' \in q^{-1}(x) \subset W'$. □

Lemma 3.9.4 *With the notation and hypothesis of Lemma* 3.9.2, r' *is perpendicular to* X' *at* x'.

Proof By Theorem 3.9.1, r' is a transversal to X' at x'. Moreover, the isomorphism of conclusion (2) of Lemma 3.9.1 induces an isomorphism

$$C_{X \cap r^{-1}(x),x} \to C_{X' \cap r'^{-1}(x'),x'},$$

whose existence easily implies the lemma. □

Lemma 3.9.5 *With the notation and hypothesis of Lemma* 3.9.2, *without assuming* (a), (b), *and* (c), *suppose that* r *is transversal to* X *at* x, *and let* $h : \mathbb{D} \to W$ *be a test-map at* x. *Set* $\delta = \delta_{x_h}(X_h, W_h)$. *There exists a positive integer* m *satisfying these conditions:*

(1) *If* $\delta < \infty$ *then* $\delta_{x_h}(X_h, W_h) = \delta_{x_g}(X_g, W_g)$ *for any test-map* $g : \mathbb{D} \to W$ *at* x *such that* $h \equiv g \pmod{\lambda^{\delta m+1}\mathscr{O}_{\mathbb{D}}}$, *where* λ *is the parameter on* $\mathbb{D}$.
(2) *If* $\delta = \infty$ *then* $\delta_{x_g}(X_g, W_g) \geq \alpha/m$ *for every positive integer* α *and every test-map* $g : \mathbb{D} \to W$ *at* x *such that* $h \equiv g \pmod{\lambda^{m+1}\mathscr{O}_{\mathbb{D}}}$.

Proof By Theorem 3.3.2, there is a neighborhood of x in W and an idealistic exponent $\mathscr{E} = ((\mathbf{J}, b))$ such that

$$\delta_{x_h}(X_h, W_h) = \nu_0(\mathbf{J}_g)/b,$$

where $\mathbf{J}_g$ is identified with $\mathbf{J} \cdot \mathscr{O}_{\mathbb{D}}$ via $g : \mathbb{D} \to W$. If $h \equiv g \pmod{\lambda^t \mathscr{O}_{\mathbb{D}}}$ for some t, we have

$$\mathbf{J}_g + (\lambda^t) \cdot \mathscr{O}_{\mathbb{D}} = \mathbf{J}_h + (\lambda^t) \cdot \mathscr{O}_{\mathbb{D}}.$$

Both parts of the lemma follow immediately if we take for m the b-part of the representative of $\mathscr{E}$. □

Lemma 3.9.6 *With the notation and hypothesis of Lemma* 3.9.5, *assume that* r *is perpendicular to* X *at* x, *and let* Y *be a closed complex subspace of* W *that is nowhere dense in some neighborhood of* x *in* W. *Then* r *has maximal contact with* X *at* x *if and only if*

$$\delta_{x_h}(X_h, W_h) = \delta_{x_h}(X_h)$$

for every test-map $h : \mathbb{D} \to W$ *at* x *such that* $h(\mathbb{D} \setminus \{0\}) \subset W \setminus Y$.

Proof It is only necessary to prove the "if" part, the converse being trivial.

Let $g : \mathbb{D} \to W$ be a test-map at x, and assume, to obtain a contradiction, that $\delta = \delta_{x_h}(X_g, W_g) < \delta_{x_h}(X_g)$. There must exist a curve in Z_g passing through x_g and such that $\delta_{x_h}(X_g, \Gamma) > \delta$ (hence $\delta_{x_h}(X_g, \Gamma) > 1$). Since $\dim T_{X_g, x_g} \leq 1$, this implies that

$$T_{X_g, x_g} = T_{\Gamma, x_g}$$

and

$$T_{r_g^{-1}(x_g), x_g} \cap T_{\Gamma, x_g} = (0).$$

Therefore $r_g|_\Gamma$ is, locally at x_g, an isomorphism, and one can assume, without loss of generality, that $r_g|_\Gamma$ is an isomorphism $\Gamma \simeq W_g = \mathbb{D}$.

Now write $\tilde{Z} = Z \times \mathbb{D}$, $\tilde{X} = X \times \mathbb{D}$, $\tilde{W} = W \times \mathbb{D}$, $\tilde{r} = r \times \mathrm{id}_{\mathbb{D}}$, $\tilde{g} = (g, \mathrm{id}_{\mathbb{D}})$, and $\tilde{h} = ((r_g|_\Gamma)^{-1}, id_{\mathbb{D}})$; then $\tilde{r}$ is a retraction $\tilde{Z} \to \tilde{W}$, $\tilde{g}$ is a test-map of W at $x \times \mathbf{0}$, and h is an embedding. Also, let $\tilde{\Gamma}$ denote the smooth curve in $\tilde{Z}$ that is the image of h.

Then, $\tilde{g} = \tilde{r} \cdot \tilde{h}$ and, since $\tilde{r}|_{\tilde{\Gamma}}$ is an embedding, $\tilde{g}$ is also an embedding and one can find locally a complex subspace V of Z such that

(a) $\tilde{V}$ contains a neighborhood of $x \times 0$ in $\tilde{\Gamma}$, and
(b) $\tilde{r}$ induces an isomorphism between $\tilde{V}$ and an open neighborhood N of $x \times 0$ in W.

Therefore, one can choose a number $a \in \mathbb{R}_+$ small enough that, if $\varphi : \mathbb{D} \to \mathbb{D}$ is defined by $\varphi(\lambda) = a\lambda$, then $\tilde{h} \cdot \varphi$ has image in $\tilde{V}$, and hence, $\tilde{g} \cdot \varphi$ has an image in N. Then, if $\gamma : \tilde{Z} \to Z$ is the projection, γ induces isomorphisms

$$\tilde{Z}_{\tilde{g}} \to Z_g \quad \text{and} \quad \tilde{Z}_{\tilde{g}\varphi} \to Z_{g\varphi}$$

which, in turn, induce isomorphisms

$$\tilde{X}_{\tilde{g}\varphi} \to X_{g\varphi} \quad \text{and} \quad \tilde{W}_{\tilde{g}\varphi} \to W_{g\varphi}.$$

From this, by writing $\tilde{x} = x \times 0$, we deduce that

$$\delta_{\tilde{x}_{\tilde{g}\varphi}}(\tilde{X}_{\tilde{g}\varphi}, \tilde{W}_{\tilde{g}\varphi}) = \delta_{x_{g\varphi}}(X_{g\varphi}, W_{g\varphi}) = \delta_{x_g}(X_g, W_g) = \delta. \tag{3.38}$$

In a similar way, if we consider the retraction $\tilde{s} : \tilde{r}^{-1}(W) \to \tilde{V}$, which exists by condition (b) above, and if we consider $\tilde{h}\varphi$ as a test-map of $\tilde{V}$ at $\tilde{x}$, we have

$$\delta_{\tilde{x}_{\tilde{h}\varphi}}(\tilde{X}_{\tilde{h}\varphi}, \tilde{V}_{\tilde{h}\varphi}) = \delta_{x_g}(X_g, \Gamma) > \delta. \tag{3.39}$$

We now apply Lemma 3.9.5 to the retraction $\tilde{r}|_{\tilde{r}^{-1}(N)} = \tilde{r}\tilde{s}$. Then, taking into account (3.38) and (3.39), one can find a positive integer α such that, for every test-map $f : \mathbb{D} \to V$ satisfying

$$f \equiv \tilde{h}\varphi \mod ((\lambda^{\alpha}) \cdot \mathcal{O}_{\mathbb{D}}),$$

one has

$$\delta_{\tilde{x}_{\tilde{r}f}}(\tilde{X}_{\tilde{r}f}, \tilde{W}_{\tilde{r}f}) = \delta \tag{3.40}$$

and

$$\delta_{\tilde{x}_{\tilde{f}}}(\tilde{X}_{\tilde{f}}, \tilde{V}_{\tilde{f}}) > \delta. \tag{3.41}$$

Now, by the assumption made on Y, the set $\tilde{r}^{-1}(Y \times \mathbb{D})$ is nowhere dense in some neighborhood of $\tilde{x}$ in $\tilde{Z}$. Hence one can choose f in such a way that

$$f(\mathbb{D} \setminus \{0\}) \subset \tilde{Z} \setminus \tilde{r}^{-1}(Y \times \mathbb{D}).$$

Setting $g' = \gamma\tilde{r}f$, we have

$$g'(\mathbb{D} \setminus \{0\}) \subset W \setminus Y, \tag{3.42}$$

and, from (3.40), we obtain

$$\delta_{x_{g'}}(X_{g'}, W_{g'}) = \delta.$$

On the other hand, the basis extensions by $\tilde{r}$ and $\tilde{s}$ carry with them an isomorphism $\tilde{Z}_{\tilde{r}f} \to \tilde{Z}_{g'}$ and an identification $\tilde{Z}_{\tilde{r}} = \tilde{Z}_{\tilde{f}}$, respectively. Then $\Gamma' = \gamma f(\mathbb{D})$ is a smooth curve in $Z_{g'}$, and from (3.41) we deduce that

$$\delta_{x_{g'}}(X_{g'}, \Gamma') = \delta_{x_f}(\tilde{X}_f, \tilde{V}_f) > \delta.$$

But, by (3.42), the assumption of the lemma establishes that

$$\delta = \delta_{x_{g'}}(X_{g'}, W_{g'}) = \delta_{x_g}(X_{g'}),$$

which contradicts the preceding inequality. This concludes our proof. □

Proof of Theorem 3.9.2 By Corollary 3.9.1, x' is in W'. By Lemma 3.9.4, r' perpendicular to X' at x'. It remains only to show that r' has maximal contact with X' at x'.

Let $Y = q^{-1}(E)$; then Y is nowhere dense in W'. By Lemma 3.9.6, our theorem will be proved if we show that

$$\delta_{x'_{h'}}(X'_{h'}, W'_{h'}) = \delta_{x'_{h'}}(X'_{h'})$$

for every test-map $h' : \mathbb{D} \to W'$ at x' such that $h'(\mathbb{D} \setminus \{0\}) \subset W' - Y$. For this, it is always possible to assume that

$$\delta_{x'_{h'}}(X'_{h'}, W'_{h'}) < \infty.$$

Let $\mathscr{E}$ be the contact exponent of r with X at x. By Theorem 3.9.1, $\mathscr{E}' = q^*(\mathscr{E})$ is the contact exponent of r' with X' at x'. By construction, $\mathscr{E}'|_{W' \setminus Y}$ is the pull-back of $\mathscr{E}|_{W \setminus E}$ under the isomorphism induced by q. Then, if $h = qh'$ is considered as a test-map of W at x, the fact that $\delta_{x_h}(X'_{h'}, W'_{h'}) < \infty$ implies that $\delta_{x_h}(X_h, W_h) < \infty$.

We now compare the characteristic cones, in order to analyze the contact of r with X at x. Take local coordinate systems $(\mathbf{z}, \mathbf{w})$, $(\mathbf{z}', \mathbf{w}')$, and normalized standard bases $(f_1, \ldots, f_m)$, $(f'_1, \ldots, f'_m)$, as in the proof of Theorem 3.9.1. Also take an integer α such that $\delta' = \alpha \cdot \delta_{x'_{h'}}(X'_{h'}, W'_{h'})$ and $\delta = \alpha \cdot \delta_{x_h}(X_h, W_h)$ are integers. If we denote by $\varphi : \mathbb{D} \to \mathbb{D}$ the morphism defined by $\varphi(\lambda) = \lambda^\alpha$, we have

$$\delta' = \delta_{x'_{h'\varphi}}(X'_{h'\varphi}, W'_{h'\varphi}),$$

$$\delta = \delta_{x_{h\varphi}}(X_{h\varphi}, W_{h\varphi}).$$

Then

$$(h\varphi)^*(\mathscr{E}) = ((\lambda^\delta \mathscr{O}_{\mathbb{D}}, 1)),$$

$$(h'\varphi)^*(\mathscr{E}') = ((\lambda^{\delta'} \mathscr{O}_{\mathbb{D}}, 1)).$$

Denote by K and K', respectively, the complex subspace of $W_{h\varphi} = \mathbb{D}$ and $W'_{h'\varphi} = \mathbb{D}$ defined by $\lambda^\delta \cdot \mathscr{O}_{\mathbb{D}}$ and $\lambda^{\delta'} \cdot \mathscr{O}_{\mathbb{D}}$. The characteristic cones of (X, r) with respect to $(\mathscr{E}, h\varphi)$ and of (X', r') with respect to $(\mathscr{E}', h'\varphi)$ are defined in $\mathrm{gr}_K(Z_{h\varphi})$ and $\mathrm{gr}_{K'}(Z'_{h'\varphi})$, respectively, by the initial forms of $(f_1\gamma, \ldots, f_m\gamma)$ and $(f'_1\gamma', \ldots, f'_m\gamma')$, where $\gamma : Z_{h\varphi} \to Z$ and $\gamma' : Z'_{h'\varphi} \to Z'$ are the projections.

Now, we have $f'_1 = u_1^{-\nu_i} f_i$ for every $i = 1, \ldots, m$, and $\delta' = \delta - \nu_0(u_1 h)$, so the isomorphism

$$\mathrm{gr}_K(Z_{h\varphi})(x_{h\varphi}) \to \mathrm{gr}_{K'}(Z'_{h'\varphi})(x'_{h'\varphi}),$$

defined by assigning the classes of $\mathbf{z}'$ and $\lambda^{\delta'}$ to the classes of $\mathbf{z}$ and λ^δ, respectively, maps $\mathrm{in}_K(f_i, \gamma)$ onto $\mathrm{in}_{K'}(f'_i, \gamma')$ for $i = 1, \ldots, m$; therefore it induces an isomorphism

$$C_{h'\varphi, x'_{h'\varphi}} \to C_{h\varphi, x_{h\varphi}}$$

between the fibers of the characteristic cones. Then, since r has maximal contact with X at x, no vector of $C_{Z_{h\varphi}, K, x_{h\varphi}}$ leaves $C_{h\varphi, x_{h\varphi}}$ invariant, and this property is transmitted to $C_{h'\varphi, x'_{h'\varphi}}$ by isomorphism. There $r'_{h'\varphi}$ has maximal contact with $X'_{h'\varphi}$ at $x'_{h'\varphi}$, which implies that $X'_{h'}$ has maximal contact with $X'_{h'}$ at $x'_{h'}$, and thus

$$\delta_{x'_{h'}}(X'_{h'}, W'_{h'}) = \delta_{x'_{h'}}(X'_{h'}),$$

as was to be shown. □

Chapter 4
Groves and Polygroves

4.1 Definitions

This section is the natural continuation of Sect. 1.4 of Chap. 1. Recall Definition 1.4.2 for $\mathbb{C}$-trees, Definition 1.4.4 for ambient expansion and contraction of $\mathbb{C}$-trees, and Remark 1.4.4 for the definitions of **s**-trees, **s**-morphisms and the category ((**s**-trees)).

Definition 4.1.1 Let **s** and **s**$'$ be $\mathbb{C}$-elements, and let $\mathbf{h} : \mathbf{s}' \to \mathbf{s}$ be a horizontal morphism.

(1) An **s-grove** (or simply a **grove**, if **s** is understood by the context) is a map $B : |B| \to \text{ob}((\mathbf{s}\text{-trees}))$, where $|B|$ is a set, called the **index set** of the **s**-grove; **s** is called the **soil** of B. (Here, as usual, ob( ) denotes the class of objects of the category.)
(2) Let B' be an **s**$'$-grove and B an **s**-grove. We shall say that B is an **ambient expansion** of B' by **h**, or that B' is a **ambient contraction** of B by **h**, if the following conditions hold:
   (a) For every $b \in |B|$, there exists $b' \in |B'|$ such that $B(b)$ is an ambient expansion of $B'(b')$ by **h**.
   (b) For every $b' \in |B'|$, there exists $b \in |B|$ such that $B'(b')$ is an ambient contraction of $B(b)$ by **h**.

Remark 4.1.1 Let **s** be a model, V a smooth $\mathbb{C}$-space and $\tau : V \to (s)^0$ a smooth morphism. Given an **s**-grove, we can construct a $\tau^*(\mathbf{s})$-grove $\tau^*(B)$ in the following way (see Example 1.4.1 in Chap. 1 for the notation):

(1) $|\tau^*(B)| = |B|$.
(2) For every $b \in |\tau^*(B)|$, $\big(\tau^*(B)\big)(b) = \tau^*(B(b))$.

J. M. Aroca et al., *Complex Analytic Desingularization*,
https://doi.org/10.1007/978-4-431-49822-3_4

In the same way as in Example 1.4.1 of Chap. 1, we shall consider two special cases:

(a) The set V is an open set in $(s)^0$ and $\tau : V \to (s)^0$ is the inclusion. Then $\tau^*(B)$ is an $\mathbf{s}|_V$-grove, which will be denoted by $B|_V$, and called the **restriction** of B to V.
(b) Let V be a smooth complex-analytic space and

$$\tau : (s)^0 \times V \to (s)^0$$

the projection. In this case, $\tau^*(B)$ is an $\mathbf{s} \times V$-grove, which will be denoted by $B \times V$.

Definition 4.1.2 Given a horizontal morphism $\mathbf{h} : \mathbf{s}' \to \mathbf{s}$, an $\mathbf{s}'$-grove B', and an $\mathbf{s}$-grove B, we say that B is **h-equivalent** to B' if there exists an open set U in $(s)^0$ such that $\mathbf{h}$ induces a morphism $\mathbf{g} : \mathbf{s}' \to \mathbf{s}|_U$ and B' is an ambient contraction of $B|_U$ by g.

Definition 4.1.3 Let $\mathscr{V}$ be the category of smooth connected complex-analytic spaces countables at infinity and smooth morphisms, and let $\mathbf{s}$ be a $\mathbb{C}$-element. An **s-polygrove** is a class of groves $\mathbf{P} = \{B_V\}_{V \in \mathrm{ob}(\mathscr{V})}$ satisfying the following conditions:

(1) For every $B \in \mathrm{ob}(\mathscr{V})$, B_V is an $\mathbf{s} \times V$-grove.
(2) For every $w \in \mathrm{Hom}_{\mathscr{V}}(V', V)$ and for every $b \in |B_V|$, there exists $b' \in |B_{V'}|$ such that

$$B_{V'}(b') \simeq (\mathrm{id}_{(s)^0} \times w)^*\big(B_V(b)\big)$$

in $((\mathbf{s} \times V'\text{-trees}))$.

If $\mathbf{P} = \{B_V\}_{V \in \mathrm{ob}(\mathscr{V})}$ is an $\mathbf{s}$-polygrove and A is an $\mathbf{s}$-tree, we say that A **belongs to P** if, for every $V \in \mathrm{ob}(\mathscr{V})$, there is $b \in |B_V|$ such that $B_V(b)$ isomorphic to $A \times V$ in $((s \times V\text{-trees}))$.

Definition 4.1.4 Let $\mathbf{h} : \mathbf{s}' \to \mathbf{s}$ be a horizontal morphism, $\mathbf{P} = \{B_V\}_{V \in \mathrm{ob}(\mathscr{V})}$ an $\mathbf{s}$-polygrove, and $\mathbf{P}' = \{B'_V\}_{V \in \mathrm{ob}(\mathscr{V})}$ an $\mathbf{s}'$-polygrove. We say that $\mathbf{P}'$ is an **ambient contraction** of $\mathbf{P}$ by $\mathbf{h}$, or that $\mathbf{P}$ is an **ambient expansion** of $\mathbf{P}'$ by $\mathbf{h}$, if, for every $V \in \mathrm{ob}(\mathscr{V})$, B'_V is an ambient contraction of B_V by

$$\mathbf{h} \times \mathrm{id}_V = ((h)^0 \times \mathrm{id}_V,\ (h)_0 \times \mathrm{id}_V).$$

If $\mathbf{P}$ is an $\mathbf{s}$-polygrove and U is an open set in $(s)^0$, the **restriction** of $\mathbf{P}$ to U will be the $(\mathbf{s}|_U)$-polygrove $\mathbf{P}|_U$ defined by

$$\mathbf{P}|_U = \{B'_V\}_{V \in \mathrm{ob}(\mathscr{V})}, \quad B'_V = B_V|_{U \times V}.$$

Definition 4.1.5 Let $\mathbf{h} : \mathbf{s}' \to \mathbf{s}$ be a horizontal morphism. We say that an $\mathbf{s}'$-polygrove $\mathbf{P}'$ is $\mathbf{h}$-equivalent (or equivalent over $(h)^0$) to an $\mathbf{s}$-polygrove $\mathbf{P}$ if there exists an open set U of $(s)^0$ such that $\mathbf{h}$ induces a morphism $g : \mathbf{s}' \to \mathbf{s}|_U$ and $\mathbf{P}'$ is an ambient contraction of $\mathbf{P}|_U$ by $\mathbf{g}$.

An $\mathbf{s}'$-polygrove $\mathbf{P}'$ is said to be equivalent to an $\mathbf{s}$-polygrove $\mathbf{P}$ if $\mathbf{P}'$ is $\mathbf{h}$-equivalent to $\mathbf{P}$ for some horizontal morphism $\mathbf{h} : \mathbf{s}' \to \mathbf{s}$.

Example 4.1.1 Let X be an analytic curve in $\mathbb{C}^2$ with a singular point at $x = \mathbf{0}$. Then we may construct the $\mathbb{C}$-element

$$\mathbf{s}_0 : (s_0)_0 = \{x\} \to (s_0)^0 = \mathbb{C}^2.$$

We may take $U_0 = \mathbb{C}^2$, $E_0 = \{x\}$ and apply the quadratic transformation with center x. If $p : X' \to X$ is the blow-up of X corresponding to such a transformation and $p^{-1}(x) = \{x_1\}$, we may choose $(s_1)_0 = \{x_1\}$ and call $\mathbf{s}_1$ the **inclusion** of $\{x_1\}$ in $(s_1)^0$. Next we take $U_1 = (s_1)^0$, $E_1 = \{x_1\}$ and repeat the process until we have no more ν-fold points infinitely near to x in X. The singular diagram of this tree is, essentially, a sequence of points whose number of elements is the integral part of $\delta_x(X)$.

This example is fairly general. In our application of the notion of a tree the singular diagram of a tree will be a diagram of infinitely near singularities, in which our interest is primarily focused.

Example 4.1.2 Let X be a complex-analytic space and $\mathbf{t} : X|_U \to Z$ a local $\mathbb{C}$-element of X. If S is a Samuel stratum of X with the reduced structure, then $\mathbf{t}$ induces a local $\mathbb{C}$-element of S

$$\mathbf{t}|_S : S|_U \to Z - \mathbf{t}\big(\bar{S}|_U - S|_U\big).$$

This $\mathbb{C}$-element will be called a **t-local** $\mathbb{C}$-element of S or a $\mathbb{C}$-element of S **induced by t**, or, if H is the Hilbert–Samuel function of X along S, a $\mathbb{C}$-element of (X, H) **induced by t**.

Assume that $S|_U$ is closed in $X|_U$ and consider the natural inclusion $S|_U \hookrightarrow X|_U$. We can construct a diagram

$$\begin{array}{ccccccc}
z = (s_0)^0 & \xleftarrow{(a_1)^0} & (s_1)^0 & \longleftarrow \cdots & \xleftarrow{(a_m)^0} & (s_m)^0 \\
\uparrow & & \uparrow & & & \uparrow \\
X|_U = X_0 & \xleftarrow{p_1} & X_1 & \longleftarrow \cdots & \xleftarrow{p_m} & (X_m) \\
\uparrow & & \uparrow & & & \uparrow \\
S|_U = (s_0)_0 & \xleftarrow{(a_1)_0} & (s_1)_0 & \longleftarrow \cdots & \xleftarrow{(a_m)_0} & (s_m)_0
\end{array}$$

in the following way:

(1) Take an open subset U_{i-1} of $(s_{i-1})^0$, for $i \geq 1$, and a smooth closed complex subspace E_{i-1} of $U_{i-1} \cap (s_{i-1})^0$. Then $(a_i)^0$ is defined as the composition of the blow-up of U_{i-1} with center E_{i-1} and the natural inclusion of U_{i-1} in $(s_{i-1})^0$.
(2) X_i is the strict transform of $X_{i-1}|_{U_{i-1}}$ by the above blow-up.
(3) The set $(s_i)^0$ is the Samuel stratum of X_i corresponding to the fixed Hilbert–Samuel function H.

This diagram gives rise to a tree when the middle sequence is forgotten. Such a tree is called a **local tree** associated with the space X, the Hilbert–Samuel function H, and the $\mathbb{C}$-element $\mathbf{t}$, or associated with $(\mathbf{t}, X, H)$, for brevity.

More generally, a tree $A = \{a_i, U_{i-1} E_{i-1}\}_{1 \leq i \leq m}$ is said to be **associated with** $(\mathbf{t}, X, H)$ if A is a $(\mathbf{t}|_S)$-tree satisfying the following condition: If $X_0 = X|_U$ and X_i is the strict transform of X_{i-1} by the blow-up of $(s_{i-1})^0$ with center E_{i-1}, then $\mathbf{s}_i$ induces an isomorphism of $(s_i)_0$ onto

$$S_i = \{y \in X_i \mid H_{X_i, y} = H\}$$

with the reduced structure.

Let $\mathbf{t}'$ be another local $\mathbb{C}$-element of X such that there exists a horizontal morphism $\mathbf{g} : \mathbf{t}' \to \mathbf{t}$. If $A = \{\mathbf{a}_i, U_{i-1}, E_{i-1}\}_{1 \leq i \leq m}$ is a tree associated with $(\mathbf{t}, X, H)$, then $A' = ((g)^0)^*(A)$ is a tree associated with $(\mathbf{t}', X, H)$ and is uniquely determined (up to isomorphisms) by A and $\mathbf{g}$. Moreover, A' and A are $\mathbf{g}$-equivalent and the morphism that defines the $\mathbf{g}$-equivalence is also uniquely determined by A and $\mathbf{g}$.

A grove **associated with** $(\mathbf{t}, X, H)$ is a grove B satisfying these conditions:

(1) For every $b \in |B|$, $B(b)$ is associated with $(\mathbf{t}, X, H)$.
(2) For every $\mathbf{t}|_S$-tree A associated with $(\mathbf{t}, X, H)$ there exists $b \in |B|$ such that $B(b)$ is isomorphic to A in the category $((\mathbf{t}|_S\text{-tree}))$.

There always exists such a grove; it is enough to take all the $(\mathbf{t}|_S)$-trees associated with $(\mathbf{t}, X, H)$ and the embedding of this set in $((\mathbf{t}|_S\text{-trees}))$.

Let $\mathbf{t}$ and $\mathbf{t}'$ be two local $\mathbb{C}$-elements of X and let $\mathbf{f} : \mathbf{t} \to \mathbf{t}'$ be a horizontal morphism. If B is a grove associated with $(\mathbf{t}, X, H)$ and B' is a grove associated with $(\mathbf{t}', X, H)$, then $\mathbf{f}$ induces a horizontal morphism $h : \mathbf{t}|_S \to \mathbf{t}'|_S$ and B is $\mathbf{h}$-equivalent to B'.

Now consider $V \in \mathrm{ob}(\mathscr{V})$. Then $S \times V$ is a Samuel stratum of $X \times V$ and $t \times V$ is a local model of $X \times V$. We shall denote by $H \times V$ the Hilbert–Samuel function of $S \times V$ in $X \times V$. We can construct a $(\mathbf{t}|_S \times V)$-grove associated with $(\mathbf{t} \times V, X \times V, H \times V)$, and it is easily checked that $P = \{B_V\}_{V \in \mathrm{ob}(\mathscr{V})}$ is a $(\mathbf{t}|_S)$-polygrove. We will call this a polygrove **associated with** $(\mathbf{t}, X, H)$.

4.2 ν-Quasi-ordinary Singularities

The idea of this section is to justify the following assertion. It is important to consider the polygroves associated with $(\mathbf{t}, X, H)$ rather than just groves associated with $(\mathbf{t}, X, H)$. To illustrate this we take the case of plane curves. Let's say that two curves are **equivalent** if one can connect one with the other by means of a finite chain of ambient expansions and contractions. Now consider various complex-analytic curves X in neighborhoods of the origin x in $\mathbb{C}^2$ such that x is the only singular point of X. With every curve X we can associate a grove, and equivalence of groves means equality of the integral parts of the first characteristic exponent. But equivalence of polygroves will mean equality of these exponents.

Let $Y \subset \mathbb{C}^3$ a hypersurface, locally defined around $x \in Y$ by an equation $f = 0$. Let $\{t, y, z\}$ be a local coordinate system on $\mathbb{C}^3$ centered at x, and let

$$f = \sum_{i,j,k \in \mathbb{Z}_0} a_{i,j,k} t^i y^j z^k, \quad \text{with } a_{i,j,k} \in \mathbb{C}, \tag{T}$$

be the Taylor expansion of f in this coordinate system. Set $\nu = \nu_x(f)$ and

$$\Delta(f) = \{(i, j, k) \in \mathbb{C}_0^3 \mid a_{i,j,k} \in \mathbb{C} \setminus \{0\}\};$$

without loss of generality, we may assume that $(0, 0, \nu) \in \Delta(f)$. Let Δ_0 be the projection of $\Delta(f)$ into the (i, j)-plane from the point $(0, 0, \nu)$; and denote by $\nabla(f)$ the convex hull, in the plane, of the set $\Delta_0 + \mathbb{R}_0^2$.

Definition 4.2.1 Assume that x is a singular point of Y. We say that x is a ν-**quasi-ordinary singular point** of Y if there exists a local coordinate system $\{t, y, z\}$ on $\mathbb{C}^3$, centered at x, satisfying these properties:

(1) The projection $(t, y, z) \to (t, y)$ is transversal to Y at x, that is, $(0, 0, \nu) \in \Delta(f)$.
(2) There exists a point $(\alpha, \beta) \in \mathbb{R}_0^2$ such that $\nabla(f) = (\alpha, \beta) + \mathbb{R}_0^2$.
(3) The characteristic polynomial $F(t, y, z)$ of f is *not* a power of a polynomial of the form

$$a(z - bt^{\alpha/n} y^{\beta/n}),$$

where $a, b \in \mathbb{C} \setminus \{0\}$. (By definition,

$$F(t, y, z) = \sum_{(i,j,k) \in L} a_{i,j,k} t^i y^j z^k,$$

where L is the line connecting $(0, 0, \nu)$ and $(\alpha, \beta, 0)$.)

The pair (α, β) will be called the **characteristic pair** of the singularity.

Remark 4.2.1 The concept of a ν-quasi-ordinary surface singularity is different from that of a quasi-ordinary singularity in the sense of Zariski [128], as we now show. Let $x \in Y \subset \mathbb{C}^3$ be a hypersurface, locally defined around x by an equation $f = 0$, and analytically irreducible at x. Assume that x is a singular point of Y. In Zariski's nomenclature, x is a quasi-ordinary singular point of Y if there exists a projection $q : \mathbb{C}^3 \to \mathbb{C}^2$ such that:

(1) The retraction q is transversal to Y at x.
(2) The critical locus of this projection (that is, the reduced discriminant locus) is either smooth or a normal crossing. By the latter we mean that the curve is a union of two irreducible components, both smooth, and having different tangent lines.

Let $\nu = \nu_x(f)$; the existence of such a projection q is equivalent to the following. There exists a local coordinate system $\{t, y, z\}$ in $\mathbb{C}^3$ centered at x and satisfying these conditions:

(1) If $f(t, y, z)$ is the Taylor expansion of f in this coordinate system, then $\nu = \nu_{\mathbf{0}}(f(0, 0, z))$. In this case, we may assume that $f(t, y, z)$ is a Weierstrass polynomial

$$f(t, y, z) = z^\nu + a_1(t, y)z^{\nu-1} + \cdots + a_{\nu-1}(t, y)z + a_\nu(t, y).$$

(2) If $D(y, z)$ is the discriminant of f with respect to z, then the reduced locus of $D = 0$ is either smooth or a normal crossing.

A well-known condition equivalent to quasi-ordinary singularity is the existence of a quasi-ordinary Puiseux parametric representation. Let t, y be variables and consider the ring $R_n = \mathbb{C}\{t^{1/n}, y^{1/n}\}$ and its field of quotients k_n. Then k_n is a finite normal extension of k, with Galois group isomorphic to $(\mathbb{Z}/\mathbb{Z} \cdot n) \times (\mathbb{Z}/\mathbb{Z} \cdot n)$. A non-unit in R_n is called a **Puiseux power series**. Let $\mathrm{Gal}(k_n/k)$ be the Galois group of k_n over k, and let $h \in R_n$ be a non-unit. We say that h is a **quasi-ordinary Puiseux power series** if for every $\sigma \in \mathrm{Gal}(k_n/k)$ we have $\sigma(h) - h = M_\sigma u$, where M_σ is a monomial in R_n and $u \in R_n$ is a unit.

With this notation, the fact that x is a quasi-ordinary singularity of Y is equivalent to the following. There exists a local coordinate system $\{t, y, z\}$ in $\mathbb{C}^3$ centered at x and satisfying these conditions:

(1) If $f(t, y, z)$ is the Taylor expansion of f in this coordinate system, then $\nu = \nu_{\mathbf{0}}(f(0, 0, z))$. In this case, we may assume that $f(t, y, z)$ is a Weierstrass polynomial

$$f(t, y, z) = z^\nu + a_1(t, y)z^{\nu-1} + \cdots + a_{\nu-1}(t, y)z + a_\nu(t, y).$$

(2) Some (and hence every) root of the equation $f(t, y, z) = 0$ lies in some ring R_n, and is a quasi-ordinary Puiseux power series.

Take a quasi-ordinary singularity as above, and consider the change of coordinates given by $z' = z - 1/\nu a_1$. Then the equation $f(t, y, z) = 0$ acquires the form

$$0 = z^\nu + b_2(t, y)z^{-1} + \cdots + b_\nu(t, y)$$

and its roots (in some R_n) do not have an "integral part" in the obvious sense. Thus we see that this singularity is ν-quasi-ordinary. Therefore quasi-ordinary singularities are ν-quasi-ordinary.

The converse is not true, as seen by the following example. The hypersurface in $\mathbb{C}^3$ given by the equation $f(t, y, z) = 0$, where $f(t, y, z)$ is the minimal polynomial over k of the Puiseux power series

$$t^{7/8}y^{9/8} + t^{13/8}y^{15/8} + t^{15/8}y^{13/8}$$

has at the origin a singularity that is ν-quasi-ordinary but not quasi-ordinary.

Proposition 4.2.1 *Let $x \in Y \in \mathbb{C}^3$ be a hypersurface locally defined at x by an equation $f = 0$, and let $\nu = \nu_x(f)$. Assume that x is a ν-quasi-ordinary singular point of Y, and that $\{t, y, z\}$ a local coordinate system in $\mathbb{C}^3$, centered at x, and satisfying the conditions of Definition* 4.2.1. *Finally, let (α, β) be the corresponding characteristic pair. Then the Samuel stratum S containing x in a neighborhood of this point is as follows:*

(1) *If $\alpha \geq \nu$ and $\beta \geq \nu$, then $S = \{z = t = 0\} \cup \{z = y = 0\}$.*
(2) *If $\alpha \geq \nu$ and $\beta < \nu$ or $\alpha < \nu$ and $\beta \geq \nu$, then $S = \{z = t = 0\}$.*
(3) *If $\alpha < \nu$ and $\beta < \nu$, then $S = \{x\}$. (Note that we always have $\alpha + \beta > \nu$.)*

Proof Let $S_0 = \{z = t = 0\} \cup \{z = y = 0\}$. We will prove that $S \subset S_0$ in all cases. For this, it is important to notice that, since we are in the hypersurface case, the Samuel stratum of x is uniquely determined by the multiplicity. Assume that f has the form as in the equation (T) , and consider a covering $h : \mathbb{C}^3 \to \mathbb{C}^3$ of the type

$$h(t_1, y_1, z_1) = (t_1^m, y_1^m, z),$$

where m is a positive integer. Let $Y_1 = h^{-1}(Y)$, and let x_1 be the point in Y_1 such that $h(x_1) = x$. Then it is easy to see that x_1 is a ν-quasi-ordinary singular point of Y_1 with characteristic pair $(m\alpha, m\beta)$ with respect to (t_1, y_1, z_1). If S_1 is the Samuel stratum of Y_1 containing x_1, then $h(S_1)$ contains S in a neighborhood of x. Therefore, to prove that $S \subset S_0$, it is enough to show that

$$S_1 \subset \{z_1 = t_1 = 0\} \cup \{z_1 = y_1 = 0\}.$$

If we take $m = n!$, with $n > 0$, both $m\alpha/\beta$ and $m\beta/\alpha$ are integers; we choose h satisfying this condition.

We can always choose an equation for Y_1 of the form

$$0 = f_1 = z_1^\nu + A_{\nu-1}(t_1, y_1) z_1^{\nu-1} + \cdots + A_0(t_1, y_1),$$

which is a Weierstrass polynomial. The condition that x_1 is a ν-quasi-ordinary singular point of Y_1 implies these facts:

(a) $A_{\nu-i}(t_1, y_1) = t_1^{i\alpha'} y_1^{i\beta'} A'_{\nu-i}(t_1, y_1)$ for $1 \le i \le \nu$, where $\alpha' = m\alpha/\nu$ and $\beta' = m\beta/\nu$.
(b) $A'_{\nu-i}(t_1, y_1) \in \mathbb{C}\{t_1, y_1\}$ for all $i = 1, \dots, \nu$, and $A'_{\nu-i}(0,0) \neq 0$ for some i.
(c) $z_1^\nu + A'_{\nu-1}(0,0) z_1^{\nu-1} + \cdots + A'_0(0,0)$ is not of the form $(z_1 - p)^\nu$ for $p \in \mathbb{C}$.

Condition (c) implies that there exists a neighborhood U of $(0,0)$ in $\mathbb{C}^2$ such that, for every $(\xi, \eta) \in U$,

$$z_1^\nu + A'_{\nu-1}(\xi, \eta) z_1^{\nu-1} + \cdots + A'_0(\xi, \eta)$$

is not of the form $(z_1 - c)^\nu$ for any c. It follows that, for every pair (ξ, η) not in $(t_1 = 0) \cup (y_1 = 0)$,

$$z_1^\nu + A_{\nu-1}(\xi, \eta) z_1^{\nu-1} + \cdots + A_0(\xi, \eta) = f_1(\xi, \eta, z_1)$$

is not of the form $(z_1 - c\xi^{\alpha'}\eta^{\beta'})^\nu$ for any c. Hence, for any $(\xi, \eta, z_1) \in Y_1$ with $(\xi, \eta) \in U \setminus \big((t_1 = 0) \cup (y_1 = 0)\big)$, we have $\nu_{(\xi,\eta,z_1)}(\delta_1) = \nu$. In other words,

$$S_1 \subset (Y_1 \cap \{t_1 = 0\}) \cup (Y_1 \cap \{y_1 = 0\}) = \{z_1 = t_1 = 0\} \cup \{z_1 = y_1 = 0\}.$$

We have thus proved that $S \subset S_0$, locally around x, which is the only real difficulty in the proposition. The conclusion can be easily reached now. □

We now apply the quadratic transformation with center x, and consider the diagram

$$\begin{array}{ccc} Y' & \hookrightarrow & Z' \\ \downarrow p & & \downarrow \pi \\ x \in Y & \hookrightarrow & Z = \mathbb{C}^3 \end{array}$$

where Z' is the quadratic transform of Z and Y' is the strict transform of Y. We are interested in points $x' \in Y'$ such that $\nu_{x'}(Y') = \nu_x(Y) = \nu$.

Lemma 4.2.1 *Let $x' \in p^{-1}(x)$ be a point such that $\nu_{x'}(Y') = \nu_x(Y) = \nu$. Then x' is contained in the strict transform of the plane $z = 0$.*

Proof We may assume that Y is defined, locally around x, by a Weierstrass equation

$$0 = f = z^\nu + a_{\nu-1}(t, y)z^{\nu-1} + \cdots + a_0(t, y),$$

where

$$\nu_{(0,0)}(a_{\nu-i}) \geq \frac{(\alpha+\beta)i}{\nu} \geq i,$$

for all $i = 1, \ldots, \nu$. It is clear that, to look for the points of $p^{-1}(x)$, we must work in the charts $(t, y/t, z/t)$ and $(t/y, y, z/y)$ in Z' but not in the chart $(t/z, y/z, z)$. Take the chart $(t, y/t, z/t)$; the analysis for the other chart is similar. In this chart, Y' is defined by the equation

$$0 = f' = (z/t)^\nu + a'_{\nu-1}(t, y/t)(z/t)^{\nu-1} + \cdots + a'_0(t, y/t),$$

where $a'_{\nu-i}(t, y/t) = a_{\nu-i}(t, y)/t^i$. If $\nu_{x'}(a'_{\nu-i}) > 0$ for $1 \leq i \leq \nu$, we must have $\nu_{x'}(z/t) > 0$ and then x' is contained in the strict transform of $z = 0$. Suppose that $\nu_{x'}(a'_{\nu-i}) = 0$ for some i in the range $0 \leq i \leq \nu$; that is, $(\alpha+\beta)i/\nu = i$.

Let $z/t = u$ and $y/t = v$ at x'. Because x is a ν-quasi-ordinary singularity, we have

$$a'_{\nu-i}(tv) = \frac{a_{\nu-i}(t, y)}{t^i} = \frac{a_{\nu-i}(ty)}{(t^\alpha y^\beta)^i} v^{(\beta i/\nu)},$$

whence

$$a'_{\nu-i}(0v) = \left(\frac{a_{\nu-i}(t, y)}{(t^\alpha y^\beta)^i}\right)_{t=0} v^{(\beta i/\nu)},$$

for $1 \leq i \leq \nu$ (here this value is zero unless both $\alpha i/\nu$ and $\beta i/\nu$ are integers and $i = (\alpha+\beta)i/\nu$). It follows that

$$u^\nu + a'_{\nu-1}(0, v)u^{\nu-1} + \cdots + a'_0(0, v)$$

is not of the form $(u - e)^\nu$ for any $e \in \mathbb{C}$. This implies that $\nu_{x'}(Y') < \nu$, leading to a contradiction with the lemma's assumption. □

Remark 4.2.2 Since the Samuel stratum containing x lies in $E_1 \cup E_2$, where $E_1 = \{z = t = 0\}$ and $E_2 = \{z = y = 0\}$, the Samuel stratum of Y' corresponding to the same multiplicity is contained in the strict transform of $E_1 \cup E_2$. Set $F = p^{-1}(x)$ and let E'_1, E'_2 be the strict transforms of E_1, E_2 under π; then $E'_i \cap F = \{x'_i\}$ a simple point. The aim of this remark is to study the singularity of Y' at the points of $E'_1 \cup E'_2 \cup F$, in order to compute later the Samuel stratum of Y' corresponding to the multiplicity ν.

We consider the following cases for a point $x' \in E_1' \cup E_2' \cup F$:

Case 1 $x' \in (E_1' \setminus \{x_1'\}) \cup (E_2' \setminus \{x_2'\})$. Here, since p is an isomorphism outside F, the character of Y' at x' is the same as the character of Y at $p(x')$. This means that, if $x' \in E_1' \setminus \{x_1'\}$, then x' is a ν-quasi-ordinary singular point of Y' with characteristic pair $(\alpha, 0)$, always with respect to the corresponding coordinate system. If $x' \in E_2' - \{x_2'\}$, then x' is a ν-quasi-ordinary singular point of Y' with characteristic pair $(0, \beta)$.

Case 2 $x' \in F \setminus \{x_1', x_2'\}$. In this case, x' is the origin of the local coordinate system (t, y_1, z_1), where $y_1 = (y/t) - m$ with $m \in \mathbb{C} \setminus \{0\}$ and $z_1 = z/t$. A local equation of Y' is given by

$$0 = f'(t, y_1, z_1) = \sum_{i,j,k \in \mathbb{Z}_0} a_{i,j,k} t^{i+j+k-\nu} (y_1 + m)^j z_1^k$$

and x' is a ν-quasi-ordinary singular point of Y' with characteristic pair $(\alpha, \alpha + \beta - \nu)$ with respect to the coordinate system.

Case 3 $x' = x_1'$. In this case, x_1' is the origin of the local coordinate system (t_1, y, z_1) with $t_1 = t/y$, $z_1 = z/y$. A local equation of Y' is given by

$$0 = f'(t_1, y, z_1) = \sum_{i,j,k \in \mathbb{Z}_0} a_{i,j,k} y_{i+j+k-\nu} t_1^i z_1^k,$$

and x_1' is a ν-quasi-ordinary singular point of Y' with characteristic pair $(\alpha, \alpha + \beta - \nu)$ with respect to this coordinate system.

Case 4 $x' = x_2'$. This case is symmetric with case 3; x' is a ν-quasi-ordinary singular point of Y' with characteristic pair $(\alpha + \beta - \nu, \alpha)$ with respect to the corresponding coordinate system.

In view of this study, the Samuel stratum S' corresponding to the multiplicity ν, in Y', is obtained by simply comparing the values of ν, α, and β. If $x' \in E_1' \setminus \{x_1'\}$, then $x' \in S'$ if and only if $\nu \le \alpha$. If $x' \in E_2' \setminus \{x_2'\}$, we have $x' \in S'$ if and only if $\nu \le \beta$. If $x' \in F \setminus \{x_1', x_2'\}$, then $x' \in S'$ if and only if $\alpha + \beta - \nu \ge \nu$. Finally, $x_1' \in S'$ if and only if $2\alpha + \beta - \nu \ge \nu$, and $x_2' \in S'$ if and only if $\alpha + 2\beta - \nu \ge \nu$.

Remark 4.2.3 We now look at an application of the preceding result to singularities of plane curves. Let X be a complex-analytic curve defined around $x = \mathbf{0}$ in $\mathbb{C}^2$, and let U be a neighborhood of x in $\mathbb{C}^2$ that does not contain any singular point of X other than x. Choose a coordinate system (y, z) in $\mathbb{C}^2$, centered at x, such that the curve $z = 0$ has maximal contact with X at x. Set $Y_0 = \mathbb{C} \times X \subset \mathbb{C}^3$, let t be a coordinate function in $\mathbb{C}$, and set $x_0 = (0, x)$. Then x_0 is a ν-quasi-ordinary singular point of Y with characteristic pair $(0, \nu\delta)$, where $\nu = \nu_x(X)$ and $\delta = \delta_x(X)$.

Now consider the sequence of quadratic transformations

$$\begin{array}{ccccccc} \mathbb{C}\times U = Z_0 & \xleftarrow{\ \pi_1\ } & Z_1 & \longleftarrow & \cdots & \xleftarrow{\ \pi_n\ } & Z_n \\ \uparrow & & \uparrow & & & & \uparrow \\ \cup & & \cup & & & & \cup \\ Y_0 & \xleftarrow{\ p_1\ } & Y_1 & \longleftarrow & \cdots & \xleftarrow{\ p_n\ } & Y_n \end{array}$$

and the sequence of points

$$x_0 \longleftarrow x_1 \longleftarrow \cdots \longleftarrow x_n$$

where Z_i is the quadratic transform of Z_{i-1} and Y_i is the strict transform of Y_{i-1} by π_i, each x_i is a point of multiplicity ν in Y_i, and the center of π_i is x_{i-1}. Let t be the $\mathbb{C}$-element $\mathbf{t} : X|_U \hookrightarrow \mathbb{C}^2$ and H the Hilbert–Samuel function of a curve corresponding to multiplicity ν. If we delete the central row of the preceding diagram, the resulting diagram is a tree associated with $(\mathbf{t} \times \mathbb{C},\ X \times \mathbb{C},\ H \times \mathbb{C})$.

Let S_i be the Samuel stratum of multiplicity ν of Y_i; in particular, $S_0 = \mathbb{C} \times \{x\}$. Denote by S_{01} the strict transform of S_0 in Y_1 and, inductively, denote by $S_{0,i}$ the strict transform of $S_{0,i-1}$ in Y_i. In the same way, let $F_i = p_{i+1}^{-1}(x_i)$ and observe that $S_{0i} \subset S_i$ for all i.

For every integer $m \geq 0$, we construct a tree in the following way:

(a) For all $i \leq m$, we choose $x_i \in Y_i$ as $\{x_i\} = S_{0i} \cap F_i$. The construction can be easily described by a picture, where the pairs of numbers mean characteristic pairs of ν-quasi-ordinary singularities.

Now, by Remark 4.2.2, $F_i \subset S_i$ if and only if $i\nu(\delta - 1) \geq \nu$, that is, if and only if $\delta \geq 1+1/i$. Let b be the smallest integer such that $\delta \geq 1+1/b$. If $b \geq m$, then (3) in Fig. 4.1 represents the connected component of S_m containing x_m. If $b < m$, this connected component is S_0.

(b) If $F_m \not\subset S_m$, then our process is ended. If $F_m \subset S_m$, we choose x_{m+1} as a generic point in F_{m+1}, in the sense that it is contained neither in $S_{0,m+1}$ nor in the strict transform of F_m. Now, for every index j, write $F_j = F_{0,j}$, define $F_{1,j}$ as the strict transform of $F_{0,j}$ by π_{j+1}, and inductively define $F_{i,j}$ as the strict transform of $F_{i-1,j}$ by π_{j+i}.

Then we continue our process, by choosing x_i, for all $i > m+1$, as a generic point in F_i, in the sense that x_i is not contained in any $F_{j,i}$ for $j < i$, nor in $S_{0,i}$. The generic point of F_i is a ν-quasi-ordinary singularity of Y_i, with second component in the characteristic pair equal to zero. Each time we apply a transformation, we make the first component drop by ν. This implies that the process cannot be continued indefinitely.

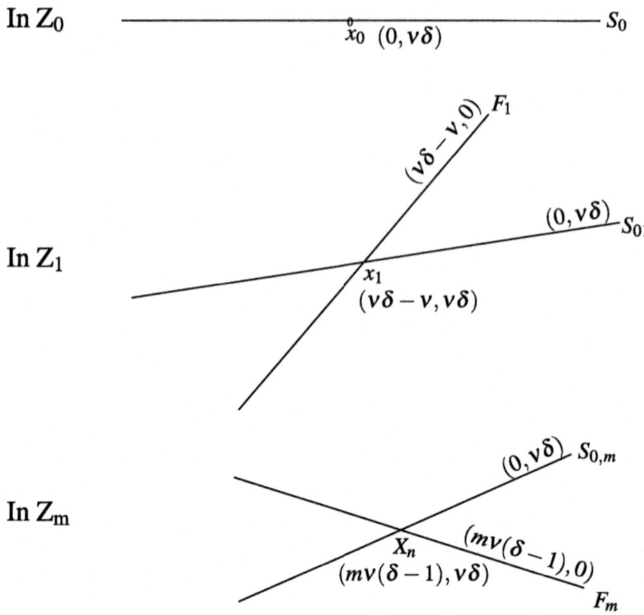

Fig. 4.1 Succession of the connected component of the Samuel strata containing the infinitely near points to x_0

Let $q(m)$ be the smallest integer such that $x_{m+q(m)+1}$ no longer has multiplicity ν. If $q \geq 1$, we know that x_{m+1} is a ν-quasi-ordinary singular point of Y_{m+1} with characteristic pair

$$((m+1)\nu(\delta-1) - (q-1)\nu,\ 0).$$

Then $q = q(m)$ is equivalent to

$$(m+1)\nu(\delta-1) - (q-1)\nu < 2\nu,$$

which in turn is equivalent to

$$q \leq (m+1)(\delta-1) < q+1,$$

or again $q(m) = (m+1)(\delta-1)$. If we know the function defined by q for all m, we know the number δ, since

$$\frac{q(m)}{m+1} + 1 \leq \delta \leq \frac{q(m)}{m+1} + 1 + \frac{1}{m+1},$$

and, therefore,

$$\delta = \lim_{m\to\infty} \frac{q(m)+m+1}{m+1}.$$

Now, if we know the polygrove associated with $(\mathbf{t}, X, H)$, we know the function $q(m)$, and therefore we are able to compute $\delta_x(X) = \delta$. Both the function $q(m)$ and the number δ are invariant under equivalence of polygroves.

With these arguments we have proved the following result.

Theorem 4.2.1 *Let X_1, X_2 be complex-analytic curves in $\mathbb{C}^2$, and let x be a point in $X_1 \cap X_2$. If, for some neighborhood of x in $\mathbb{C}^2$, there exists a polygrove of (X_1, x) equivalent to a polygrove of (X_2, x), then $\delta_x(X_1) = \delta_x(X_2)$.*

4.3 Polygroves and Ambient Changes of Idealistic Exponents

In this section we associate with every idealistic exponent a polygrove that describes the behavior of the idealistic exponent under a sequence of blow-ups, and which is independent of the space on which the idealistic exponent is defined.

Let W be a smooth complex-analytic space, and let $\mathscr{E} = ((\mathbf{J}, b))$ be an idealistic exponent on W. As in Definition 3.3.6, Chap. 3, we shall denote by $\mathrm{Sing}(\mathscr{E})$ the set $\{x \in W \mid \nu_x(\mathscr{E}) \geq 1\}$. Then, with the reduced structure induced by W, $\mathrm{Sing}(\mathscr{E})$ is a locally closed complex analytic subspace of W.

Definition 4.3.1 The **soil** of $\mathscr{E}$ is defined as the $\mathbb{C}$-element $\mathbf{s} : (s)_0 \to (s)^0$, where $(s)_0 = \mathrm{Sing}(\mathscr{E})$, $(s)^0 = W$, and $\mathbf{s}$ is the natural inclusion.

Remark 4.3.1 Let $\mathbf{s}$ be the soil of $\mathscr{E}$, let U be an open subset in $(s)^0$, E a smooth closed complex subspace of $(s)_0 \cap U$, and $\pi_1 : (s_1)^0 \to U$ the blow-up of U with center E. The ideal $\mathbf{P}$ of $\pi_1^{-1}(\mathscr{E})$ in $(s_1)^0$ is invertible of rank 1 and, since $E \subset \mathrm{Sing}(\mathscr{E})$, we have $\nu_E(\mathbf{J}) \geq b$ and hence $\mathbf{J} \cdot \mathscr{O}_{(s_1)^0} \subset \mathbf{P}^b$. Since $(s_1)^0$ is smooth, $\mathscr{E}_1 = ((\mathbf{P}^{-b} \cdot \mathbf{J} \cdot \mathscr{O}_{(s_1)^0}, b))$ is independent of the representative of $\mathscr{E}$ we have chosen for the construction. This $\mathscr{E}_1$ will be called the **transform** of $\mathscr{E}$ by the blow-up of U with center E.

Definition 4.3.2 Let W be a complex-analytic space, $\mathscr{E}$ an idealistic exponent on W, and $\mathbf{s}$ the soil of $\mathscr{E}$. We say a tree $A = \{\mathbf{a}_i, U_{i-1}, E_{i-1}\}_{1\leq i\leq m}$ is **associated with** $\mathscr{E}$ if it satisfies these conditions:

(1) The tree A is an $\mathbf{s}$-tree.
(2) If we write $\mathscr{E}_0 = \mathscr{E}$ and define inductively $\mathscr{E}_i$ as the transform of $\mathscr{E}_{i-1}$ by the blow-up of U_{i-1} with center E_{i-1} , then $\mathbf{s}_i : (s_i)_0 \to (s_i)^0$ induces an isomorphism $\mathbf{s}_i : (s_i)_0 \to \mathrm{Sing}(\mathscr{E}_i)$.

Remark 4.3.2 Let X be a complex-analytic space and x a point in X. Consider a local $\mathbb{C}$-element $\mathbf{t} : (t)_0 \to (t)^0$ of X such that $x \in (t)_0$, and let S be the Samuel stratum of x in X and $\mathbf{s}' : (s')_0 \to (s')^0$ the local $\mathbb{C}$-element of S induced by $\mathbf{t}$. Then, as a consequence of the theorems of existence, continuity and stability of the maximal contact, we shall see later, in the idealistic description theorem, that there exists an idealistic exponent $\mathscr{E}$ that describes this circumstances.

Definition 4.3.3 A grove **associated with** $\mathscr{E}$ is an **s**-grove, where **s** is the soil of $\mathscr{E}$, satisfying the following conditions:

(1) For every $b \in |B|$, $B(b)$ is associated with $\mathscr{E}$.
(2) For every **s**-tree A associated with $\mathscr{E}$, there exists an element $b \in |B|$ such that $B(b)$ is isomorphic to A in ((**s**-trees)).

Remark 4.3.3

4.3.3.1 There is always a grove associated with $\mathscr{E}$. Just take all the trees associated with $\mathscr{E}$. Two groves associated with $\mathscr{E}$ are always $\mathrm{id}_{\mathbf{s}}$-equivalent.
4.3.3.2 Let $V \in \mathrm{ob}(\mathscr{V})$; if $\mathscr{E} = ((\mathbf{J}, b))$ is an idealistic exponent on V, we can construct another idealistic exponent $\mathscr{E} \times V$ on $W \times V$ by taking $\mathscr{E} \times V = ((\mathbf{J} \cdot \mathscr{O}_{W\times V}, b))$. This $\mathscr{E} \times V$ is clearly independent of the representative $((\mathbf{J}, b))$ of $\mathscr{E}$. If $\mathbf{P} = \{B_V\}_{V\in \mathrm{ob}(\mathscr{V})}$ is a family of groves such that, for every $V \in \mathrm{ob}(\mathscr{E})$, B_V is an $\mathbf{s}\times V$-grove associated with $\mathscr{E}\times V$, then $\mathbf{P}$ is a polygrove. Such a polygrove always exists and any two-polygroves constructed in this way are always $\mathrm{id}_{\mathbf{s}}$-equivalent.

Definition 4.3.4 Any polygrove $\mathbf{P}$ obtained through the construction just described will be called a polygrove **associated with** $\mathscr{E}$, and will be denoted by $\mathbf{P}(\mathscr{E})$.

Theorem 4.3.1 *Let W and W' be complex-analytic spaces, $e : W \to W'$ a closed embedding, and $\mathscr{E} = ((\mathbf{J}, b))$ an idealistic exponent on W. There exists an idealistic exponent $\mathscr{E}'$ on W' with these properties:*

(1) *The embedding e induces a horizontal morphism $\mathscr{E}$ from the soil of $\mathscr{E}$ to the soil of $\mathscr{E}'$.*
(2) *The polygroves $\mathbf{P}(\mathscr{E})$ and $\mathbf{P}(\mathscr{E}')$ are e-equivalent.*

For the proof of this theorem, we need a technical result:

Lemma 4.3.1 *Let $\mathbf{L}_1$ and $\mathbf{L}_2$ be ideal sheafs on W, and let m and b be positive integers. Then*

$$((\mathbf{L}_1^m + \mathbf{L}_2^m, b)) = \big(\big((\mathbf{L}_1 + \mathbf{L}_2)^m, b\big)\big).$$

Proof By Theorem 3.3.1 in Chap. 3, in order to prove the lemma, it is enough to take a disk $\mathbb{D}$ in $\mathbb{C}$, around the origin, a morphism $h : \mathbb{D} \setminus W$, and to check that

$$\nu_0((\mathbf{L}_1^m + \mathbf{L}_2^m)^b \cdot \mathscr{O}_{\mathbb{D}}) = \nu_0((\mathbf{L}_1 + \mathbf{L}_2)^{mb} \cdot \mathscr{O}_{\mathbb{D}}).$$

For this, it is enough to check that

$$\nu_0((\mathbf{L}_1^m + \mathbf{L}_2^m) \cdot \mathscr{O}_{\mathbb{D}}) = \nu_0((\mathbf{L}_1 + \mathbf{L}_2)^m \cdot \mathscr{O}_{\mathbb{D}}).$$

Now, we have

$$\nu_0((\mathbf{L}_1^m + \mathbf{L}_2^m) \cdot \mathcal{O}_{\mathbb{D}}) = m \min\{\nu_0(\mathbf{L}_1 \cdot \mathcal{O}_{\mathbb{D}}), (\mathbf{L}_2 \cdot \mathcal{O}_{\mathbb{D}})\}$$
$$= m\, \nu_0((\mathbf{L}_1 + \mathbf{L}_2) \cdot \mathcal{O}_{\mathbb{D}}) = \nu_0((\mathbf{L}_1 + \mathbf{L}_2)^m \cdot \mathcal{O}_{\mathbb{D}}).$$

This proves the lemma. □

Proof of Theorem 4.3.1 For any $y \in W$, take a local retraction $r_y : W' \to W$ with domain U_y such that $e(y) \in U_y$. Let Diff_W^ν be the $\mathcal{O}_W$-module of the holomorphic differential operators of order not greater than ν (see Remark 1.6.1 in Chap. 1). If $\mathbf{L}$ is an ideal on W, then for every open set T in W, each element D of the set $\Gamma(T, \mathrm{Diff}_W^\nu)$ of sections of Diff_W^ν on T induces a map $D : (T, \mathbf{L}) \to (T, \mathcal{O}_W)$. Then, for every $\nu \geq 0$, we may construct the ideal $\mathrm{Diff}_W^\nu(\mathbf{L})$, taking as $\Gamma(T, \mathrm{Diff}_W(\mathbf{L}))$ the ideal generated by the union of the images of all the maps $\Gamma(T, \mathbf{L}) \to \Gamma(T, \mathcal{O}_W)$ as before. If $\mathbf{I}$ is the ideal of $e(W)$ in W' ($\mathbf{I}$ is coherent since e is a closed embedding) and if $((\mathbf{J}, b))$ is a representative of $\mathcal{E}$, we write

$$(r_{y,b}^{\#}\mathbf{J})|_{U_y} = (\mathbf{I}^b|_{U_y}) + \sum_{\nu \geq 0} (\mathbf{I}^\nu|_{U_y}) \cdot r_y^{-1}(\mathrm{Diff}_W^\nu(\mathbf{J})),$$

where $r_y^{-1}(\mathrm{Diff}_W^\nu(\mathbf{J}))$ is identified with $\mathrm{Diff}_W^\nu(\mathbf{J}) \cdot (\mathcal{O}_{W'}|_{U_y})$ via r_y.

With this notation, we shall prove these facts:

(1) One can construct an ideal $r_b^{\#}(\mathbf{J})$ on W' by gluing together the $(r_{y,b}^{\#}\mathbf{J})|_{U_y}$ and then we have the idealistic exponent

$$\mathcal{E}' = ((r_b^{\#}(\mathbf{J}), b)) \text{ on } W'.$$

(2) The idealistic exponent $\mathcal{E}'$ is independent of the retractions r_y we have used to construct it.
(3) The idealistic exponent $\mathcal{E}'$ is also independent of the representative $((\mathbf{J}, b))$ of $\mathcal{E}$.
(4) The idealistic exponent $\mathcal{E}'$ satisfies conditions (1) and (2) in the theorem.

Proof of (1) *and* (2) Let $r_y' : W' \to W$ be another local retraction with domain U_y' such that $x = e(y) \in U_y'$. Then there exist an open neighborhood U of x in $U_y \cap U_y'$ and two local coordinate systems on W', say $(t_1 \dots t_c, u_1 \dots u_d)$ and $(t_1 \dots t_c, u_1' \dots u_d')$, centered at x, consisting of functions that are holomorphic at U and such that, when everything is restricted to U, the following conditions are satisfied:

(a) The family $t_1, \dots, t_c$ generates the ideal $\mathbf{I}$, and likewise $t_1', \dots, t_c'$.
(b) The pair $(\mathbf{t}, \mathbf{u})$ is associated with r_y and $(\mathbf{t}', \mathbf{u}')$ is associated with r_y'.
(c) For every $i = 1, \dots, d$, $w_i = u_i'|_W = u_i|_W$ (that is, $u_i' - u_i \in \mathbf{I}$) and $(w_1, \dots, w_d)$ is a local coordinate system on W around y.

With this notation we will prove that $(r^{\#}_{y,b}\mathbf{J})|_U = (r'^{\#}_{y,b}\mathbf{J})|_U$. By operating locally on U, it is enough to prove that

$$\mathbf{I}^{\nu} \cdot (r'_y)^{-1}(\mathrm{Diff}^{\nu}_W(\mathbf{J})) \subset r^{\#}_{y,b}\mathbf{J}$$

for every $\nu \geq 0$.

Let U_1 be an open subset of U. Let f be a section of $\mathbf{J}$ on U_1, and D a section of Diff^{ν}_W on U_1. Write $\phi = Df$ and $\psi(\mathbf{u}') = \phi \cdot r'_y(\mathbf{u}')$; then, if $\boldsymbol{\delta} = \mathbf{u}' - \mathbf{u}$, we have

$$\psi(\mathbf{u}') = \psi(\mathbf{u} + \boldsymbol{\delta}) = \sum_{A \in \mathbb{Z}_0^d} ((D^A\phi) \cdot r_y)\boldsymbol{\delta}^A,$$

where $D^A = \frac{1}{A!}\frac{\partial^{|A|}}{\partial \mathbf{w}^A}$, $A! = a_1! \cdots a_d!$, and $A = (a_1, \dots, a_d)$. Therefore

$$\mathbf{I}^{\nu} \cdot \psi(\mathbf{u}') \subset r^{\#}_{y,b}(\mathbf{J}),$$

proving (1) and (2).

Proof of (3) Let $((\mathbf{J}, b)) = ((\mathbf{J}', b'))$; then $\mathbf{J}^{b'}$ and $\mathbf{J}'^{b}$ are integrally equivalent. By the construction of $r^{\#}_b$, so are $r^{\#}_{bb'}(\mathbf{J}^{b'})$ and $r^{\#}_{bb'}(\mathbf{J}'^{b})$. Then, if we prove that $(r^{\#}_b, (\mathbf{J}))^m$ is integrally equivalent to $(r^{\#}_{mb}(\mathbf{J}))^m$, for all m and b, we will have shown that $(r^{\#}_b(\mathbf{J}))^{b'}$ and $(r^{\#}_{b'}(\mathbf{J}'))^b$ are integrally equivalent; therefore $((r^{\#}_b(\mathbf{J}), b)) = ((r^{\#}_{b'}(\mathbf{J}'), b'))$ and our proof will be completed.

Let $r : W' \to W$ be a local retraction of domain U. By restricting everything to U, we may write

$$(r^{\#}_b\mathbf{J})^m = \left(\mathbf{I}^b + \sum_{\nu \geq 0} \mathbf{I}^{\nu} r^{-1}(\mathrm{Diff}^{\nu}_W(\mathbf{J}))\right)^m,$$

and the right-hand side is integrally equivalent to

$$\mathbf{I}^{mb} + \left(\sum_{\nu \geq 0} \mathbf{I}^{\nu} r^{-1}(\mathrm{Diff}^{\nu}_W(\mathbf{J}))\right)^m$$

by Lemma 4.3.1. But this latter ideal is also equal to

$$\begin{aligned}
\mathbf{I}^{mb} + \sum_{\nu_1,\dots,\nu_n \geq 0} \mathbf{I}^{\nu_1+\cdots+\nu_n} r^{-1}\left(\prod_{i=1}^{n} \mathrm{Diff}^{\nu_i}_W(\mathbf{J})\right) \\
&= \mathbf{I}^{mb} + \sum_{\mu \geq 0} \mathbf{I}^{\mu} r^{-1}\left(\sum_{\nu_1+\cdots+\nu_n=\mu} \prod_{i=1} \mathrm{Diff}^{\nu_i}_W(\mathbf{J})\right) \\
&= \mathbf{I}^{mb} + \sum_{\mu \geq 0} \mathbf{I}^{\mu} r^{-1}(\mathrm{Diff}^{\mu}_W(\mathbf{J}^m)) = r^{\#}_{mb}\mathbf{J}^m,
\end{aligned}$$

which proves (3).

Proof of (4) If we prove that $e(\mathrm{Sing}(\mathscr{E})) = \mathrm{Sing}(\mathscr{E}')$, then the horizontal morphism $\mathbf{e}$ is automatically constructed by writing $(e)_0 = e|_{\mathrm{Sing}(\mathscr{E})}$ and $(e)^0 = e$. That $e(\mathrm{Sing}(\mathscr{E})) = \mathrm{Sing}(\mathscr{E}')$ is a trivial consequence of two facts:

(a) $\mathrm{Sing}(\mathscr{E}') \subset e(W)$. In fact,

$$\begin{aligned} x \in \mathrm{Sing}(\mathscr{E}') &\Longrightarrow \nu_x(\mathscr{E}') \geq b \Longrightarrow \nu_x(\mathbf{I}^b) \geq b \\ &\Longrightarrow \nu_x(\mathbf{I}) \geq 1 \Longrightarrow x \in e(W). \end{aligned}$$

(b) If $y \in W$, then $\nu_y(\mathbf{J}) \geq b$ is equivalent to $\nu_{e(y)}(r_b^{\#}\mathbf{J}) \geq b$. In fact, locally around $e(y)$, we have $\mathbf{J} \cdot \mathscr{O}_{W'} = r^{-1}(\mathrm{Diff}_W^0(\mathbf{J})) \subset r_b^{\#}\mathbf{J}$, hence $\nu_{e(y)}(r_b^{\#}\mathbf{J}) \geq b$ implies $\nu_y(\mathbf{J}) \geq b$.

To prove the converse just observe that, if $y \in W$, then $\nu_{e(y)}(\mathbf{I}) \geq 1$ and

$$\nu_y(\mathrm{Diff}_W^{\nu}(\mathbf{J})) \geq \nu_y(\mathbf{J}) - \nu, \ \forall \nu \geq 0.$$

We now prove that $P(\mathscr{E})$ and $P(\mathscr{E}')$ are e-equivalent. It is enough to prove that, if B is a grove associated with $\mathscr{E}$ and B' is a grove associated with $\mathscr{E}'$, then B and B' are e-equivalent. In fact, let $V \in \mathrm{ob}(\mathscr{V})$; the embedding $e : W \to W'$ induces a closed embedding $e \times V : W \times V \to W' \times V$. Every local retraction $r : W' \to W$ induces a local retraction $r \times V : W' \times V \to W \times V$ in such a way that the diagram

$$\begin{array}{ccc} W \times V & \underset{r\times V}{\overset{e\times V}{\rightleftarrows}} & W' \times V \\ \Big\downarrow q & & \Big\downarrow q' \\ W & \underset{r}{\overset{e}{\rightleftarrows}} & W' \end{array}$$

is locally commutative and

$$(r \times V)_b^{\#}(\mathbf{J} \cdot \mathscr{O}_{W\times V}) = (r_b^{\#}\mathbf{J}) \cdot \mathscr{O}_{W'\times V},$$

that is,

$$(\mathscr{E} \times V)' = \mathscr{E}' \times V.$$

Therefore, if we prove that any grove associated with $\mathscr{E}$ is e-equivalent to any grove associated with $\mathscr{E}'$, we have also proved that any grove associated with $\mathscr{E} \times V$ is $(e \times V)$-equivalent to any grove associated with $(\mathscr{E} \times V)' = \mathscr{E}' \times V$, so $P(\mathscr{E})$ and $P(\mathscr{E}')$ are e-equivalent.

In a further reduction of the problem, if we prove that the equality $\mathrm{Sing}(\mathscr{E}) = \mathrm{Sing}(\mathscr{E}')$ is preserved by blowing-ups then it is clear that any grove associated with

$\mathscr{E}$ will be e-equivalent to any grove associated with $\mathscr{E}'$. To prove that the equality $\mathrm{Sing}(\mathscr{E}) = \mathrm{Sing}(\mathscr{E}')$ is preserved by a blow-up on an open set U in W', we can restrict the ambient to U and assume that $U = W'$.

Consider the diagram

$$\begin{array}{ccc} W_1 & \xrightarrow{\;e_1\;} & W_1' \\ {\scriptstyle p}\downarrow & & \downarrow{\scriptstyle \pi} \\ E \subset W & \xrightarrow[\;e\;]{} & W' \end{array}$$

where π is the blow-up of W' with center E and W_1 is the strict transform of W. Take an idealistic exponent $\mathscr{E}$ on W, the corresponding idealistic exponent $\mathscr{E}'$ on W', the transform $\mathscr{E}_1$ of $\mathscr{E}$ by p, and the transform $\mathscr{E}_1'$ of $\mathscr{E}'$ by π. Let r be a local retraction for e, set $F = r^{-1}(E)$, and let F_1 be the strict transform of F by π; then $e_1(W_1) \cap F_1 = \varnothing$ and there exists a local retraction $r_1 : W_1' \setminus F_1 \to W_1$ such that the diagram

$$\begin{array}{ccc} W_1 & \xrightarrow{\;e_1\;} & W_1' \setminus F_1 \\ {\scriptstyle p}\downarrow & & \downarrow{\scriptstyle \pi} \\ E \subset W & \xleftarrow[\;r\;]{} & W' \end{array}$$

is locally commutative. Then, by changing the retraction r, and by a process similar to the construction of $\mathscr{E}'$ from $\mathscr{E}$, we construct an idealistic exponent $\mathscr{E}_1''$ from $\mathscr{E}_1$. For these conditions, we shall prove that

$$\mathrm{Sing}(\mathscr{E}_1) \supset \mathrm{Sing}(\mathscr{E}_1') \supset \mathrm{Sing}(\mathscr{E}_1'') \tag{4.1}$$

and, in view of (b) on the preceding page,

$$\mathrm{Sing}(\mathscr{E}_1) = \mathrm{Sing}(\mathscr{E}_1') = \mathrm{Sing}(\mathscr{E}_1''). \tag{4.2}$$

Now, we have $\mathrm{Sing}(\mathscr{E}_1) = \mathrm{Sing}(\mathscr{E}_1')$, and we have, with $\mathscr{E}_1$ and $\mathscr{E}_1''$, the same conditions as with $\mathscr{E}$ and $\mathscr{E}'$ in the beginning, in order to apply new blow-ups. Therefore, if we prove (4.1), the proof of our theorem will be complete.

Recall that $\mathscr{E} = ((\mathbf{J}, b))$ on W. Let $\mathbf{P}$ be the ideal of $\pi^{-1}(E)$ in $\mathscr{O}_{W_1'}$ and $\mathbf{L} = \mathbf{P} \cdot \mathscr{O}_{W_1'}$ the ideal of $p^{-1}(E)$ in $\mathscr{O}_{W_1'}$. If we fix a local retraction $r : W' \to W$ and restrict everything to the domain of r, then $\mathscr{E}_1$, $\mathscr{E}_1'$ and $\mathscr{E}_1''$ are defined by

$$\mathcal{E}_1 = ((\mathbf{L}^{-b} \cdot \mathbf{J} \cdot \mathcal{O}_{W_1'}, b)),$$
$$\mathcal{E}_1' = ((\mathbf{P}^{-b} \cdot (r_b^{\#}\mathbf{J}) \cdot \mathcal{O}_{W_1'}, b)),$$
$$\mathcal{E}_1'' = (((r_1)_b^{\#}(\mathbf{L}^{-b} \cdot \mathbf{J} \cdot \mathcal{O}_{W_1'}), b)) \quad \text{on } W_1' \setminus F_1.$$

Now we check that

$$\mathbf{P}^{-b} \cdot (r_b^{\#}\mathbf{J}) \cdot \mathcal{O}_{W_1' \setminus F_1} \subset (r_1)_b^{\#}(\mathbf{L}^{-b} \cdot \mathbf{J} \cdot \mathcal{O}_{W_1'}). \tag{4.3}$$

Since $\mathbf{L} = \mathbf{P} \cdot \mathcal{O}_{W_1}$ and $\mathbf{I}$ is the ideal of W in $\mathcal{O}_{W'}$,

$$\mathbf{I}_1 = \mathbf{P}^{-1} \cdot \mathbf{I} \cdot \mathcal{O}_{W_1' \setminus F_1}$$

is the ideal of W_1 in $\mathcal{O}_{W_1' \setminus F_1}$. Taking into account this and the definition of $r_b^{\#}$, we see that to prove (4.3) it is enough to prove that

$$\mathbf{L}^{-b+\nu} \cdot (\mathrm{Diff}_W^{\nu}(\mathbf{J})) \cdot \mathcal{O}_{W_1} \subset \mathrm{Diff}_{W_1}^{\nu}(\mathbf{L}^{-b} \cdot \mathbf{J} \cdot \mathcal{O}_{W_1}) \tag{4.4}$$

for every $\nu \geq 0$. This relation is trivial for $\nu = 0$, and if we prove that

$$\mathbf{L}^{-b+1} \cdot (\mathrm{Diff}_W^{1}(\mathbf{J})) \cdot \mathcal{O}_{W_1} \subset \mathrm{Diff}_{W_1}^{1}(\mathbf{L}^{-b} \cdot \mathbf{J} \cdot \mathcal{O}_{W_1}), \tag{4.5}$$

(4.4) will be obtained from (4.5) by replacing $\mathbf{J}$ by $\mathrm{Diff}_W^1(\mathbf{J})$ and b by $b-1$, and repeating the process ν times. Therefore we prove (4.5).

Let y be any point in W such that $e(y)$ belongs to the domain of r and $y_1 \in p^{-1}(y)$. We choose a local coordinate system

$$(\mathbf{t}, \mathbf{u}) = (t_1, \ldots, t_\alpha, u_1, \ldots, u_\beta)$$

on W centered at y in such a way that $t_1 = \cdots = t_\alpha = 0$ defines E and:

$$(t_1, t_2/t_1, \ldots t_\alpha/t_1, u_1, \ldots, u_\beta) = (t_1', \ldots, t_\alpha', u_1', \ldots, u_\beta')$$

is a local coordinate system on W_1 centered at y_1. Now, $\mathbf{L} = t_1' \cdot \mathcal{O}_{W_1}$. For every section f of $\mathbf{J}$, we write

$$f'(\mathbf{t}', \mathbf{u}') = \frac{f(t_1, t_2' t_1', \ldots, t_\alpha' t_1, \mathbf{u}')}{t_1^b}.$$

Then, to prove (5), it is enough to prove that, for $1 \leq i \leq \alpha$ and $1 \leq j \leq \beta$, the set on the right-hand side of (5) contains $t_1'^{-b+1}(\partial f/\partial t_i)$ and $t_1'^{-b+1}(\partial f/\partial u_j)$.

We must consider three cases:

Case 1: $i = 1$. Then

$$t'^{-b+1}_1(\partial f/\partial t_i) = t'^{-b+1}_1 \frac{\partial(f' t'^b_1)}{\partial t_1} = t'^{-b+1}_1 \sum_{i=1}^{\alpha} \frac{\partial(f' t'^b_1)}{\partial t'_i} \cdot \frac{\partial t'_i}{\partial t_1}$$

$$= bf' + t'_1(\partial f'/\partial t'_i) + \sum_{i=2}^{\alpha} t'_i(\partial f'/\partial t'_i) \in \mathrm{Diff}^1_{W_1}(\mathbf{L}^{-b}\mathbf{J}\mathscr{O}_{W_1}).$$

Case 2: $\alpha \geq i \geq 2$. Then

$$t'^{-b+1}_1(\partial f/\partial t_i) = t'_1(\partial f'/\partial t'_i)\,\mathrm{Diff}^1_{W_1}(\mathbf{L}^{-b}\mathbf{J}\mathscr{O}_{W_1}).$$

Case 3: $\beta \geq j \geq 1$. Then

$$t'^{-b+1}_1(\partial f/\partial u_j) = t'_1(\partial f'/\partial u'_j) \in \mathrm{Diff}^1_{W_1}(\mathbf{L}^{-b}\mathbf{J}\mathscr{O}_{W_1}).$$

This proves (4.3).

Now, let $\mathbf{H}$, $\mathbf{H}'$, and $\mathbf{H}''$ be the ideals of $\mathscr{O}_{W'_1\setminus F_1}$ given by

$$\begin{aligned}
\mathbf{H} &= \mathbf{I}^b_1 + \mathbf{P}^{-b}\cdot r_1^{-1}(\mathbf{J}\cdot\mathscr{O}_{W_1}) = \mathbf{I}^b_1 + r_1^{-1}(\mathbf{L}^{-b}\cdot\mathbf{J}\cdot\mathscr{O}_{W_1}),\\
\mathbf{H}' &= \mathbf{P}^{-b}\cdot(r^{\#}_b\mathbf{J})\cdot\mathscr{O}_{W'_1\setminus F_1},\\
\mathbf{H}'' &= (r_1)^{\#}_b(\mathbf{L}^{-b}\cdot\mathbf{J}\cdot\mathscr{O}_{W_1}).
\end{aligned}$$

Clearly $\mathbf{H} \subset \mathbf{H}'$ and, by (3), $\mathbf{H}' \subset \mathbf{H}''$. Therefore in $W'_1 \setminus F_1$ one has

$$\begin{aligned}
\mathrm{Sing}(\mathscr{E}_1) &= \{x \in W_1 \mid \nu_x(\mathbf{H}) \geq b\} \supset \{x \in W_1 \mid \nu_x(\mathbf{H}') \geq b\},\\
&= \mathrm{Sing}(\mathscr{E}'_1) \supset \{x \in W_1 \mid \nu_x(\mathbf{H}'') \geq b\} = \mathrm{Sing}(\mathscr{E}''_1).
\end{aligned}$$

This ends the proof of the theorem. □

Remark 4.3.4 We have studied ambient expansion for idealistic exponents. Now we study ambient contraction for the same objects.

Let W be a smooth complex-analytic space and $\mathscr{E} = ((\mathbf{J}, b))$ an idealistic exponent on W. Let $x \in \mathrm{Sing}(\mathscr{E})$; if $\mathfrak{m}_{W,x}$ represents the maximal ideal of $\mathscr{O}_{W,x}$, then $\mathbf{J}_x \subset \mathfrak{m}^b_{W,x}$ and

$$(\mathbf{J}_x + \mathfrak{m}^{b+1}_{W,x})/\mathfrak{m}^{b+1}_{W,x} \subset gr^b_x(W).$$

The homogeneous ideal in $gr_x(W)$ generated by this subgroup of $gr^b_x(W)$ determines a cone in $T_{W,x}$, which will be denoted by $C_{\mathscr{E},b,x}$. Also, we denote by $T_{\mathscr{E},x}$ the set

$$T_{\mathscr{E},x} = \{\mathbf{v} \in T_{W,x} \mid C_{\mathscr{E},b,x} + \mathbf{v} = C_{\mathscr{E},b,x}\},$$

and we call $T_{\mathscr{E},x}$ the **tangent space** to $\mathscr{E}$ at x.

By the construction of $C_{\mathscr{E},b,x}$ and the properties of integral dependence of ideals we can easily check that $C_{\mathscr{E},b,x}$ does not depend of the ideal $\mathbf{J}$ such that $((\mathbf{J}, b)) = \mathscr{E}$; more precisely, if $((\mathbf{J}, b)) = ((\mathbf{J}', b'))$, the cone constructed from $\mathbf{J}'$, coincides with the one constructed from $\mathbf{J}$.

On the other hand, $T_{\mathscr{E},x}$ depends only on $\mathscr{E}$ and on x. Indeed, since

$$(\mathbf{J}_x + \mathfrak{m}_{W,x}^{b+1})^m = (\mathbf{J}_x + \mathfrak{m}_{W,x}^{mb+1}),$$

we can compare $C_{\mathscr{E},b,x}$ and $C_{\mathscr{E},mb,x'}$ and we find that $T_{\mathscr{E},x}$ is the same for both cones.

Finally, we observe that $T_{\mathscr{E},x} \subset C_{\mathscr{E},b,x} \subset T_{W,x}$ and that $T_{\mathscr{E},x} = T_{W,x}$ if and only if $\nu_x(\mathscr{E}) > 1$.

Remark 4.3.5 In the circumstances just described, we have

$$T_{\mathscr{E},x} \neq T_{W,x} \iff \nu_x(\mathscr{E}) = 1 \iff \nu_x(\mathbf{J}) = b \iff \nu_x(f) = b \text{ for some } f \in \mathbf{J}_x.$$

By the existence of maximal contact (Theorem 3.2.3 of Chap. 3), there exists a local coordinate system $(z, y_1 \dots y_m)$ on W centered at x such that

$$f = z^b + f_2(y)z^{b-2} + \cdots + f_b(y).$$

Let X be the hypersurface $f = 0$ and S the Samuel stratum of X of multiplicity b. Then, $x_1 = (\xi, \alpha_1, \dots, \alpha_m) \in S$ if and only if ξ is a multiple root of order b of the equation

$$0 = z^b + f_2(\boldsymbol{\alpha})z^{b-2} + \cdots + f_b(\boldsymbol{\alpha}).$$

Therefore $b\xi = 0$ and $\xi = 0$.

Thus S is contained, around x, in the hypersurface Y of W with equation $z = 0$. Then, for $x_1 \in W$,

$$\begin{aligned} x_1 \in S &\iff \nu_{x_1}(f) = b \iff \nu_{x_1}(z) \geq 1 \text{ and } \nu_{x_1}(f_i z^{b-i}) \geq b \\ &\iff x_1 \in Y \text{ and } \nu_{x_1}(f_i) \geq i. \end{aligned}$$

Theorem 4.3.2 *Let W be a smooth complex-analytic space, $\mathscr{E} = ((\mathbf{J}, b))$ an idealistic exponent on W, and $x \in \mathrm{Sing}(\mathscr{E})$. Then there exist an open neighborhood U of x in W, a smooth closed complex subspace $Y \subset U$, and an idealistic exponent $\mathscr{K}$ on Y satisfying the following conditions:*

(1) $T_{Y,x} = T_{\mathscr{E},x}$.

(2) *The inclusion* $Y \to W|_U$ *extends to an embedding* $\mathbf{h}$ *of the soil of* $\mathscr{K}$ *into the soil of* $\mathscr{E}$, *and the polygroves* $\mathbf{P}(\mathscr{K})$ *and* $\mathbf{P}(\mathscr{E})$ *are* $\mathbf{h}$*-equivalent.*

Proof If $T_{\mathscr{E},x} = T_{W,x}$, the theorem is trivial, for it is enough to take $Y = U = W$ and $\mathscr{K} = \mathscr{E}$. We assume, then, that $T_{\mathscr{E},x} \neq T_{W,x}$, and we prove the theorem by induction on $\dim_x(W) = \dim T_{W,x}$.

Since $T_{\mathscr{E},x} \neq T_{W,x}$, by Remark 4.3.5, there exists $f \in \mathbf{J}_x$ with $\nu_x(f) = \nu_x(\mathbf{J}) = b$. If X is the hypersurface of equation $f = 0$, locally around x, then $T_{\mathscr{E},x} \subset T_{X,x}$. On the other hand, $T_{X,x}$ is the strict tangent space to the stratum S of maximum multiplicity of X, and S is contained in the hypersurface Y of equation $z = 0$, locally around x. Therefore

$$T_{\mathscr{E},x} \subset T_{Y,x} \subset T_{W,x}$$

and $\dim T_{Y,x} = \dim T_{W,x} - 1$. Hence, if we prove that there exists an idealistic exponent $\mathscr{K}$ on Y such that the inclusion $Y \to W$, locally around x, extends to a restrictive morphism $\mathbf{h}$ of the soil of $\mathscr{K}$ into the soil of $\mathscr{E}$, such that $\mathbf{P}(\mathscr{K})$ and $\mathbf{P}(\mathscr{E})$ are $\mathbf{h}$-equivalent, our theorem will be proved. So, let's prove this result.

In our circumstances, we have a finite set of data, denoted by

$$(\mathscr{E}, b, f, z, \mathbf{y})_{U,x}.$$

Here $\mathscr{E} = ((\mathbf{J}, b))$ is an idealistic exponent on an open neighborhood of x in W; $(z, \mathbf{y}) = (z, y_1, \ldots, y_m)$ is a local coordinate system on W centered at x, consisting of functions which are holomorphic on U, and such that Y is defined in U by $z = 0$; and f is a section of $\mathbf{J}$ on U such that $\nu_x(f) = b$ and

$$f = z^b + f_2(\mathbf{y})z^{b-2} + \cdots + f_b(\mathbf{y}).$$

These properties are stable under restriction to every open neighborhood of x contained in U. Therefore, they can be considered as properties of relative to germs, and hence we may suppress any reference to the open set U. These properties are also stable under a biholomorphic coordinate change among the $\mathbf{y}$.

Now, we can construct explicitly the idealistic exponent $\mathscr{K}$ beginning from the data $(\mathscr{E}, b, f, z, \mathbf{y})_x$ in the following way. Let $\{f^{(1)}, \ldots, f^{(s)}\}$ be a system of generators of $\mathbf{J}_x$; then, on a suitable neighborhood U of x in W, we may write each $f^{(i)}$ as

$$f^{(i)} = \sum_{\alpha \leq b} f_\alpha^{(i)} z^{b-\alpha},$$

where the $f^{(i)}$ depend only on the $\mathbf{y}$. Write $b' = b!$ and let $\mathbf{J}'$ be the ideal sheaf on $Y \cap U$ generated by $\{(f_\alpha^{(i)})^{b'/\alpha}\}_{1 \leq i \leq s,\, 1 \leq \alpha \leq b}$. Let $\mathscr{K} = ((\mathbf{J}', b'))$ on $Y \cap U$. Then $\mathscr{K}$ is independent of the system of generators used in the construction, provided we take a sufficiently small neighborhood of x (so we have, in a certain sense, a "germ of idealistic exponent"). Therefore, $\mathscr{K}$ depends only on the data $(\mathscr{E}, b, f, z, \mathbf{y})_x$.

To assert that $\mathscr{K}$ has the required properties we need to prove three facts:

(1) If V is any smooth connected complex-analytic space, $v \in V$, and if $(\mathbf{t}) = (t_1, \ldots, t_e)$ is a local coordinate system on V centered at v, then we consider the data

$$(\mathscr{E} \times V, b, f, z, (\mathbf{y}, \mathbf{t}))_{(x,v)}$$

on $W \times V$, where f is considered as a function on $W \times V$ by composition with the first projection. Then the properties of $(\mathscr{E}, b, f, z, \mathbf{y})_x$ are stable under this base change. Moreover, $\mathscr{K} \times V$ is the idealistic exponent associated with these new data.

(2) $\mathrm{Sing}(\mathscr{K}) = \mathrm{Sing}(\mathscr{E})$.

(3) The properties of $(\mathscr{E}, b, f, z, \mathbf{y})_x$ are stable under application of any blow-up with a smooth center $E \subset \mathrm{Sing}(\mathscr{E})$.

If we suppose these properties to be proved, the theorem follows by the following argument. Property (2) means that the inclusion of Y in W around x can be extended to an embedding $\mathbf{h}$ of the soil of $\mathscr{K}$ into the soil of $\mathscr{E}$. Then, by property (1), to prove that there exists an $\mathbf{h}$-equivalence between $\mathbf{P}(\mathscr{K})$ and $\mathbf{P}(\mathscr{E})$, it is enough to prove that there exists an $\mathbf{h}$-equivalence between a grove associated with $\mathscr{K}$ and a grove associated with $\mathscr{E}$, and this result is an immediate consequence of properties (2) and (3).

Property (1) is straightforward. We turn to (2). Let $\{f^{(1)}, \ldots, f^{(s)}\}$ be a system of generators of $\mathbf{J}_x$; by restriction to a suitable neighborhood of x in W, we may assume that

$$\mathbf{J} = (f^{(1)}, \ldots, f^{(s)}) \cdot \mathscr{O}_W.$$

Then, if $x_1 \in W$, we have

$$\begin{aligned} x_1 \in \mathrm{Sing}(\mathscr{E}) \\ &\iff \nu_{x_1}(\mathbf{J}) \geq b \\ &\iff \nu_{x_1}(f^{(i)}) \geq b \text{ for all } i = 1, \ldots, s \\ &\iff \nu_{x_1}(z) \geq 1 \text{ and } \nu_{x_1}(f_\alpha^{(i)}) \geq \alpha \text{ for } i = 1, \ldots, s \text{ and } \alpha = 1, \ldots, b \\ &\iff x_1 \in Y \text{ and } \nu_{x_1}((f_\alpha^{(i)}|_Y)^{b'/\alpha}) \geq b' \text{ for } i = 1, \ldots, s \text{ and } \alpha = 1, \ldots, b \\ &\iff \nu_{x_1}(\mathbf{J}') \geq b' \iff x_1 \in \mathrm{Sing}(\mathscr{K}). \end{aligned}$$

Finally, we prove property (3). Take a smooth complex subspace $E \subset \mathrm{Sing}(\mathscr{E})$; by a coordinate change among the $\mathbf{y}$, we may assume that E is defined by the equations

$$z = y_1 = \cdots = y_e = 0, \quad \text{for } 1 \leq e \leq m.$$

Then we take the blow-up $\pi : W_1 \to W$ of W with center E, and we have the diagram:

$$\begin{array}{ccc} Y_1 & \longrightarrow & W_1 \\ P \Big\downarrow & & \Big\downarrow \pi \\ E \subset Y & \longrightarrow & W \end{array}$$

Let $x_1 \in \pi^{-1}(x)$ be an arbitrary point and $\mathbf{P}$ the ideal of $\pi^{-1}(\mathscr{E})$ on $\mathscr{O}_{W_1}$. Then we have the transform

$$\mathscr{E}_1 = ((\mathbf{P}^{-b} \cdot \mathbf{J} \cdot \mathscr{O}_{W_1}, b))$$

of $\mathscr{E}$. If X is the hypersurface of equation $f = 0$, the strict transform X_1 of X by π has multiplicity at x_1 equal to b, because

$$P^{-b}.f.\mathscr{O}_{W_1} \subset \mathbf{P}^{-b} \cdot \mathbf{J} \cdot \mathscr{O}_{W_1}$$

and $x \in \mathrm{Sing}(\mathscr{E})$. Then $x_1 \in Y_1$, and we may assume that

$$(z/y_1, y_1, y_2/y_1, \ldots, y_e/y_1, y_{e+1}, \ldots, y_m)$$

is a local coordinate system on W_1 centered at x_1, where Y_1 is defined by $z/y_1 = 0$.

Write $z^{(1)} = z/y_1$, $y_1^{(1)} = y_1$, $y_i^{(1)} = y_i/y_1$ for $i = 2, \ldots, e$, and $y_j^{(j)} = y_j$ for $j = e+1, \ldots, m$; then the equation of X_1 is

$$f^{(1)} = f/y_1^b = (z^{(1)})^b + f_2^{(1)}(y^{(1)})(z^{(1)})^{b-2} + \cdots + f_b^{(1)}(y^{(1)}),$$

where $f_2^{(1)}(y^{(1)}) = f_i/y_1^i$. We conclude that $(\mathscr{E}_1, b, f^{(1)}, z^{(1)}, y^{(1)})_{x_1}$ has the same properties as $(\mathscr{E}, b, f, z, \mathbf{y})_x$, proving property (3) and the theorem. □

4.4 Idealistic Description

As far as the desingularization theorem and related results are concerned, the goal of the local theory of singularities is to get a manageable description of the totality of infinitely near singularities with a fixed Samuel function, on a complex space locally around each point. The description should be thought of as a generalization of the fact that, for a plane curve, the number of infinitely near m-fold points at an m-fold point is equal to the integral part of the first characteristic exponent of the singular point of the curve.

The aim of this section is to compare the gardening of a Samuel stratum of a complex-analytic space with the gardening of the idealistic exponent which, by the theorems of existence, continuity, and stability of the maximal contact, can be associated with this stratum.

Definition 4.4.1

4.4.1.1 A **point** of a $\mathbb{C}$-situation G (see Definition 1.2.3 in Chap. 1) is a pair (g, x), where $g \in \mathrm{ob}([G])$ and $x \in (G(g))_0$.

4.4.1.2 Two points (g, x) and (g', x') of a $\mathbb{C}$-situation G will be said to be **equivalent**, and we will write $(g, x) \sim (g', x')$, if there exists a finite sequence $(g_0, \ldots, g_n)$ of objects in $[G]$ and a sequence $(\alpha_1, \ldots, \alpha_n)$ of morphisms in $[G]$ satisfying these conditions:

(1) $g_0 = g$ and $g_n = g'$.
(2) $\alpha_i \in \mathrm{Hom}_{[G]}(g_i, g_{i-1}) \cup \mathrm{Hom}_{[G]}(g_{i-1}, g_i)$ for $i = 1, \ldots, n$.
(3) For $i = 1, \ldots, n$, set

$$\varepsilon_i = \begin{cases} 1 & \text{if } \alpha_i \in \mathrm{Hom}_{[G]}(g_{i-1}, g_i), \\ -1 & \text{if } \alpha_i \in \mathrm{Hom}_{[G]}(g_i, g_{i-1}). \end{cases}$$

Then

$$x' \in (G(\alpha_n))_0^{\varepsilon_n} \cdots (G(\alpha_1))_0^{\varepsilon_1}(x).$$

4.4.1.3 A **forest** is a pair $(G, \mathbf{J})$, where G is a $\mathbb{C}$-situation and $\mathbf{J} = \{\mathbf{J}(g)\}_{g \in \mathrm{ob}([G])}$ is a family such that:

(1) For every $g \in \mathrm{ob}([G])$, $\mathbf{J}(g)$ is a $G(g)$-polygrove.
(2) For every $\beta \in \mathrm{Hom}_{[G]}(g', g)$, $\mathbf{J}(g')$ is $G(\beta)$-equivalent to $\mathbf{J}(g)$.

4.4.1.4 Let X be a complex-analytic space and G a $\mathbb{C}$-situation. We say that G is a $\mathbb{C}$-**situation** of X if the following conditions hold:

(1) For every $g \in \mathrm{ob}([G])$, $G(g)$ is a local $\mathbb{C}$-element of X.
(2) The family $\{(G(g))_0\}_{g \in \mathrm{ob}([G])}$ is a covering of X.
(3) If (g, x) and (g', x) are two points of G, there exists a point (g'', x) of G and two morphisms

$$\beta \in \mathrm{Hom}_{[G]}(g'', g), \quad \beta' \in \mathrm{Hom}_{[G]}(g'', g')$$

such that:

(a) $(G(g''))_0 \subset (G(g))_0 \cap (G(g'))_0$.

(b) The morphisms $(G(\beta))_0$ and $(G(\beta'))_0$ are respectively the inclusions $(G(g''))_0 \to (G(g))_0$ and $(G(g''))_0 \to (G(g'))_0$.

4.4.1.5 A $\mathbb{C}$-situation G of X is **full** if it has these properties:

(1) For every local $\mathbb{C}$-element $\mathbf{t}$ of X, there exists $g \in \text{ob}([G])$ such that there is a isomorphism $\alpha : G(g) \to \mathbf{t}$ in $((\text{el} \mid \text{hor}))$, where $(\alpha)_0$ is the identity map.
(2) The functor G is a full functor, that is,

$$G : \text{Hom}_{[G]}(g_1, g_2) \to \text{Hom}_{((\text{el} \mid \text{hor}))}(G(g_1), G(g_2))$$

is surjective, for any $g_1, g_2 \in \text{ob}([G])$.

When a $\mathbb{C}$-situation G is a $\mathbb{C}$-situation of X, we often denote this by writing G_X instead of G.

Remark 4.4.1

4.4.1.1 There always exists a full $\mathbb{C}$-situation G_X of X. Just take these data:

(1) The category $[G]$ is the category whose objects are the pairs $(U, \mathbf{h})$, where U is an open subset of X and $\mathbf{h}$ is a closed embedding $\mathbf{h} : X|_U \to Z$, where Z is a smooth complex-analytic space.
(2) $\text{Hom}_{[G]}((U, \mathbf{h}), (U', \mathbf{h}')) = \text{Hom}_{((\text{el} \mid \text{hor}))}(\mathbf{h}, \mathbf{h}')$.
(3) $G : [G] \to ((\text{el} \mid \text{hor}))$ as the functor defined by $G(U, \mathbf{h}) = \mathbf{h}$ if $(U, \mathbf{h}) \in \text{ob}([G])$, and $G(\boldsymbol{\alpha}) = \boldsymbol{\alpha}$ if $\boldsymbol{\alpha} \in \text{Hom}_{[G]}((U, \mathbf{h}), (U', \mathbf{h}'))$.

4.4.1.2 Let S be a Samuel stratum of X and let H be the Hilbert–Samuel function of X along S. Then every $\mathbb{C}$-situation G_X of X induces a $\mathbb{C}$-situation $G_{X,S}$ of S defined by

(1) $[G_{X,S}] = [G_X]$.
(2) The local element $G_{X,S}(g)$ is the local $\mathbb{C}$-element of S induced by $G_X(g)$.

$G_{X,S}$ is called an X-**situation** of S, and if G_X is a full $\mathbb{C}$-situation of X, then $G_{X,S}$ is called an X-**full situation** of S. If S corresponds to the Hilbert function H, also we shall denote by $G_{X,H}$ the situation $G_{X,S}$.

4.4.1.3 We shall say that a forest $(G, \mathbf{J})$ is a **forest of** (X, S), or a **forest of** (X, H), or an X-**forest of** S, if there exists a $\mathbb{C}$-situation G_X of X such that $G = G_{X,S}$ and, for every $g \in \text{ob}([G])$, $\mathbf{J}(g)$ is a $G_{X,S}(g)$-polygrove associated with $(G_X(g), X, S)$.

The existence of a forest of (X, S) is shown as follows. Pick any $\mathbb{C}$-situation G_X of X. For each $g \in \text{ob}([G_X])$, pick any polygrove $\mathbf{J}(g)$ associated with $(G_X(g), X, S)$. That $(G_{X,S}, \mathbf{J})$ is a forest of (X, S) follows from these facts:

(1) Every morphism $\beta : g \to g'$ in $[G_X]$ induces a horizontal morphism

$$G_{X,S}(\beta) : G_{X,S}(g) \to G_{X,S}(g').$$

(2) The stability of the Samuel stratum under localization and under products by smooth connected complex-analytic spaces implies that, to prove that $\mathbf{J}(g)$ is $G_{X,S}(\beta)$-equivalent to $\mathbf{J}(g')$, it is enough so show that any $G_{X,S}(g)$-grove is $G_{X,S}(\beta)$-equivalent to a $G_{X,S}(g')$-grove, and the latter statement is immediate from the definitions.

Every X-forest of S will be denoted by $(G_X, G_{X,S}, \mathbf{J})$ just to put in evidence the $\mathbb{C}$-situation G_X where $G_{X,S}$ comes from. An X-forest $(G_X, G_{X,S}, J)$ of S will be called an X-**full forest** of S if G_X is a full $\mathbb{C}$-situation.

Definition 4.4.2 We say that a forest $(G, \mathbf{J})$ is **idealistic** if, for every point (g, x) of G, there exists a neighborhood U of x in $(G(g))^0$ and an idealistic exponent $\mathscr{E}$ on U such that the polygrove $\mathbf{J}(g)|_U$ is associated with $\mathscr{E}$.

Theorem 4.4.1 *Let X be a complex-analytic space and S a Samuel stratum of X. Then every X-full forest $\mathscr{F}_{X,S} = (G_X, G_{X,S}, \mathbf{J})$ of S is idealistic.*

Proof The theorem amounts to the following assertion: Let $\mathbf{s}'_0 : X|_U \to Z$ be a local $\mathbb{C}$-element of X; then, for every point $x \in U \cap S$, there exists an open neighborhood V of x in Z and an idealistic exponent $\mathscr{K}$ on V such that the soil of $\mathscr{K}$ is $\mathbf{s}'_0|_V$ and such that, if for $g \in \mathrm{ob}([G_X])$ there exists an isomorphism $\boldsymbol{\alpha} : G_X(g) \to \mathbf{s}'_0|_V$ where $(\alpha)_0$ is the identity map, then the following condition is satisfied:

(1) The $G_{X,S}(g)$-polygrove $\mathbf{J}(g)$ is $\boldsymbol{\alpha}$ equivalent to the polygrove $\mathbf{P}(\mathscr{K})$.

To prove (1) it is enough to show that there is a smooth closed complex subspace W of V such that $S \cap V \subset W$ (let us denote by $h : W \to V$ the natural inclusion) and, if we denote by $\mathbf{s} : S \cap V \to W$ the restriction to W of $(\mathbf{s}'_0|_V)$ via h, then there exists an idealistic exponent $\mathscr{E}$ on W whose soil is $\mathbf{s}$ and satisfying the following condition:

(2) The restriction to $\mathbf{s}_0$ of a $(\mathbf{s}'_0|_S)$-polygrove associated with $(\mathbf{s}'_0, X, H)$ is $\mathrm{id}_{\mathbf{s}_0}$-equivalent to $P(\mathscr{E})$.

In fact, Theorem 4.3.1 shows that if there exists $\mathscr{E}$ on W satisfying condition (2) for the restriction to $\mathbf{s}_0$ of the $G_{X,S}(g)$-polygrove $J(g)$, then there exists an idealistic exponent $\mathscr{K}$ on V satisfying (1). Note that we can always replace the $G_{X,S}(g)$-polygrove $J(g)$ by any $(\mathbf{s}'_0|_S)$-polygrove associated with $(\mathbf{s}'_0, X, H)$ because any two of them are equivalent.

This can be proved as follows. Given $\mathbf{s}_0$ as above, we know from Theorem 3.4.1 in Chap. 3 that there exist an open neighborhood V of x in Z, a smooth complex subspace W of V and a retraction $r : Z|_V \to W$ that has maximal contact with $X|_V$ at x. Replacing V by a suitable neighborhood of x in V, we may assume that there exists an idealistic exponent $\mathscr{E}$ on W that is the contact idealistic exponent of r with X at x (see Chap. 3, Theorem 3.3.2 and Definition 3.3.7). Moreover, by the continuity theorem (Chap. 3, Sect. 3.8), we may further assume that $S \cap V = \mathrm{Sing}\,\mathscr{E}$, that r has maximal contact with X at every point of $S \cap V$, and that $\mathscr{E}$ is the contact exponent of r with X at every point of $S \cap V$.

Now, with V, W, and $\mathscr{E}$ as just described, we claim that assertion (2) holds. Note that, in fact, it is enough to prove (2) for groves instead of polygroves. The reason is the following.

Assume that a retraction $r : Z \to W$ has maximal contact with X at every point of a Samuel stratum S of $X|_V$ and that $S \subset W$. Assume further that we have an idealistic exponent $\mathscr{E}$ on W such that $\mathrm{Sing}(\mathscr{E}) = S$ and $\mathscr{E}$ is the contact exponent of r with X at every point of S. Then we claim that, for every smooth connected complex space P, these properties remain valid when each of the data is replaced by the product with P. In fact, it is easy to see that $S \times P$ is a Samuel stratum of $X \times P$ and that the retraction $r \times P : Z \times P \to W \times P$ is perpendicular to $X \times P$ at every point of $S \times P$; and if $\mathscr{E} \times P$ denotes the pull-back of $\mathscr{E}$ by the projection $W \times P \to W$, we clearly have $S \times P = \mathrm{Sing}(\mathscr{E} \times P)$. The maximal contact and contact exponent conditions can be checked using the following remark.

Let (y, p) be any point of $S \times P$, and let $h' : \mathbb{D} \to W \times P$ be any test-map at (y, p). Let h be the composition of h' and the projection $W \times P \to W$. Then h is a test-map into W at y. Moreover, we then have canonical isomorphisms $(Z \times P)_{h'} \to Z_h$, $(X \times P)_{h'} \to X_h$, and $(W \times P)_{h'} \to W_h$, and, by this last isomorphism, the pull-back of $\mathscr{E} \times P$ by h' coincides with that of $\mathscr{E}$ by h.

We now resume the proof of assertion (2) with the already chosen V, W, and $\mathscr{E}E$, in the case of groves. Consider any finite sequence

$$\delta = \{U_i,\ E_i,\ \pi_i : Z_{i+1} \to Z_i\}_{0 \leq i \leq m}$$

of local blow-ups over $Z_0 = Z|_V$, where each U_i is an open subset of Z_i, each E_i is a closed smooth complex subspace of U_i and π_i is the composition of the blow-up $Z_{i+1} \to U_i$ with center E_i and the inclusion $U_i \subset Z_i$. Then assertion (2) is equivalent to the following statement:

(3) Conditions (A) and (B) being defined below, δ satisfies (A) if and only if it satisfies (B).
 Condition (A): Let X_0 be the restriction to V of the image of $X|_U$ by $\mathbf{s}_0'$. Define $X_i \subset Z_i$, for $1 \leq i \leq m$, by letting X_i be the strict transform of X_{i-1} by the blow-up π_{i-1} (with respect to the center E_{i-1}). Let S_i be the Samuel stratum of X_i corresponding to the Samuel function of X at a point of S. Then $E_i \subset S$ for $0 \leq i \leq m$.
 Condition (B): Let $W_0 = W$ and define $W_i \subset Z_i$, for $1 \leq i \leq m$, by letting W_i be the strict transform of W_{i-1} by π_{i-1} (with respect to the center E_{i-1}). Then:

 (a) For $0 \leq i \leq m$, $E_i \subset W_i$.
 (b) For $0 \leq i \leq m$ is $E_i \subset \mathrm{Sing}(\mathscr{E}_i)$, where $\mathscr{E}_0 = \mathscr{E}$ and $\mathscr{E}_i$ is the transform of $\mathscr{E}_{i-1}$ by the blow-up $W_i \to W_{i-1}$ with center E_{i-1}, for $i \geq 1$.

The equivalence statement (3) for all $\mathscr{E}$ as above, implies that, for every δ as above, we have:

(4) The set S_i of (A) coincides with $\mathrm{Sing}(\mathscr{E}_i)$ of (B).

Statements (3) and (4), and hence Theorem 4.4.1, follow immediately from the stability theorems of Chap. 3 (Theorems 3.9.1 and 3.9.2). □

Remark 4.4.2 We have proved that every X-full forest is idealistic by proving that, given a point x in a Samuel stratum of a complex-analytic space X, the X-forest associated with (X, S) is described around x by an idealistic exponent $\mathscr{E}$. Such an exponent is not unique, and is determined only by its polygrove. In this section we are going to study those properties of an idealistic exponent which depend on its polygrove, that is, properties shared by all idealistic exponents describing the infinitely near singularities of x in X.

Let W be a smooth complex-analytic space and let $\mathscr{E} = ((\mathbf{J}, b))$ be an idealistic exponent on W.

Definition 4.4.3 Take $\mu \in \mathbb{R}_+$. Denote by $\mathrm{Sing}^{\mu}(\mathscr{E})$ the set $\{x \in W \mid \nu_x(\mathscr{E}) \geq \mu\}$, together with the reduced structure of the complex structure induced by W. In this way $\mathrm{Sing}^{\mu}(\mathscr{E})$ is a closed complex subspace of W.

Lemma 4.4.1 *Let W be a smooth complex-analytic space, and let $\mathscr{E}$ and $\mathscr{E}'$ be two idealistic exponents of W with the same soil $\mathbf{s}$ and such that their groves are* $\mathrm{id}_{\mathbf{s}}$*-equivalent. If $x \in \mathrm{Sing}(\mathscr{E})$ and $\nu_x(\mathscr{E}) > 1$, we have*

$$\left|\nu_x(\mathscr{E}) - \nu_x(\mathscr{E}')\right| < 1,$$

where bars mean absolute value.

Proof Let $\mathscr{K} = ((\mathbf{J}, b))$ be any idealistic exponent on W. Take a sequence of quadratic transformations

$$\begin{array}{ccccccccc}
W = W_0 & \xleftarrow{\pi_1} & W_1 & \xleftarrow{\pi_2} & \cdots & \xleftarrow{\pi_m} & W_m & \xleftarrow{\pi_{m+1}} & W_{m+1} \\
\uparrow & & \uparrow & & & & \uparrow & & \\
\cup & & \cup & & & & \cup & & \\
\{x\} = \{x_0\} & \longleftarrow & \{x_1\} & \longleftarrow & \cdots & \longleftarrow & \{x_m\} & &
\end{array}$$

where x_i is the center of π_{i+1} and such that x_i is a generic point of $\pi_i^{-1}(x_{i-1})$. This sequence appears as the ambient part of a tree associate with $\mathscr{K}$ if and only if, writing $\mathscr{K}_0 = \mathscr{K}$ and denoting by $\mathscr{K}_i$ the transform of $\mathscr{K}_{i-1}$, one has $x_i \in \mathrm{Sing}(\mathscr{K}_i)$ for all $i = 0, 1, \ldots, m$. Set $\nu = \nu_x(\mathscr{K})$ and let $\mathbf{P}_1$ be the ideal of $\pi_1^{-1}(x_0)$ in $\mathscr{O}_{W_1}$. Then the transform $\mathscr{K}_1$ of $\mathscr{K}$ is

$$\mathscr{K}_1 = ((\mathbf{P}_1^{-b}\mathbf{J}\mathscr{O}_{W_1}, b)).$$

Since $x_1 \in \pi_1^{-1}(x)$ is generic, we have $\mathbf{J}\mathscr{O}_{W_1} = \mathbf{P}_1^{b\nu}$. Therefore, locally around x_1, one has

$$\mathscr{K}_1 = ((\mathbf{P}_1^{b\nu - b}, b)).$$

After applying this process m times, we arrive at an expression

$$\mathscr{K}_m = ((\mathbf{P}_m^{b\nu - mb}, b)),$$

where $\mathbf{P}_m$ is the ideal of $\pi_m^{-1}(x_{m-1})$ in $\mathscr{O}_{W_m}$. Hence, saying that $x_m \in \operatorname{Sing}(\mathscr{K}_m)$ is equivalent to saying that $b\nu - mb \geq b$, or, equivalently, that $\nu \geq m + 1$.

Coming back to the assumptions in the lemma, if the groves of $\mathscr{E}$ and $\mathscr{E}'$ are equivalent, a sequence such as the one in the beginning of this proof is the ambient part of a tree associated with $\mathscr{E}$ if and only if there exists a sequence of the same length that is the ambient part of a tree associated with $\mathscr{E}'$. Therefore, for any integer m, $\nu_x(\mathscr{E}) \geq m + 1$ implies $\nu_x(\mathscr{E}') \geq m + 1$. Hence the integral parts of $\nu_x(\mathscr{E})$ and $\nu_x(\mathscr{E}')$ coincide, and the lemma follows. □

Lemma 4.4.2 *With the assumptions of Lemma* 4.4.1, *assume further that*

$$\dim_x(\operatorname{Sing}^{\nu}(\mathscr{E})) > 0,$$

where $\nu = \nu_x(\mathscr{E})$. *Then* $\nu_x(\mathscr{E}) \leq \nu_x(\mathscr{E}')$.

Proof Consider a sequence of quadratic transformations

$$\begin{array}{ccccccccc}
W = W_0 & \xleftarrow{\pi_1} & W_1 & \xleftarrow{\pi_2} & \cdots & \xleftarrow{\pi_m} & W_m & \xleftarrow{\pi_{m+1}} & W_{m+1} \\
\uparrow & & \uparrow & & & & \uparrow & & \\
\cup & & \cup & & & & \cup & & \\
\{x\} = \{x_0\} & \longleftarrow & \{x_1\} & \longleftarrow & \cdots & \longleftarrow & \{x_m\} & &
\end{array}$$

satisfying the following requirement. Let $F_0 = \operatorname{Sing}^{\nu}(\mathscr{E})$ and denote by F_i the strict transform of F_{i-1} by π_i, for $i = 1, \ldots, m$; then $x_0 \in F_0$. Choose x_i as a generic point in F_i, for $i = 1, \ldots, m$.

Let $\mathbf{P}_i$ be the ideal of $\pi_i^{-1}(x_{i-1})$ in $\mathscr{O}_{W_i}$, set $\mathscr{E} = \mathscr{E}_0 = ((\mathbf{J}, b))$, and, for $i = 1, \ldots, m + 1$, let $\mathscr{E}_i = ((\mathbf{J}_i, b))$ be the transform of $\mathscr{E}_{i-1}$ by π_i, with

$$\mathbf{J}_i = \mathbf{P}_i^{-b} \cdot \mathbf{J}_{i-1} \cdot \mathscr{O}_{W_i}.$$

Then we can construct a sequence $\tilde{\mathbf{J}}_0, \tilde{\mathbf{J}}_1, \ldots, \tilde{\mathbf{J}}_{m+1}$, where $\tilde{\mathbf{J}}_0 = \mathbf{J}$ and $\tilde{\mathbf{J}}_i = \mathbf{P}_i^{-b\nu} \cdot \tilde{\mathbf{J}}_{i-1} \cdot \mathscr{O}_{W_i}$.

We now prove by induction that $\nu_{x_i}(\tilde{\mathbf{J}}_i) = \nu b$. Since $\nu = \nu_x(\mathscr{E})$, we have $\nu_{x_0}(\tilde{\mathbf{J}}_0) = \nu_x(\mathbf{J}) = \nu b$, so the equality holds for $i = 0$. Assume that $\nu_{x_{i-1}}(\tilde{\mathbf{J}}_{i-1}) = \nu b$; we have $\nu_{x_i}(\tilde{\mathbf{J}}_i) \leq \nu_{x_{i-1}}(\tilde{\mathbf{J}}_{i-1})$. On the other hand, since π_i is an isomorphism outside $\pi_i^{-1}(x_{i-1})$, we have

$$\nu_y(\tilde{\mathbf{J}}_i) = \nu_{\pi_1(y)}(\tilde{\mathbf{J}}_{i-1})$$

around x_i, for all $y \in F_i \setminus \pi_i^{-1}(x_{i-1})$. By assumption, x_0 is not an isolated point in F_0; then x_i is not isolated in F_i. Therefore, around x_i the closure

$$\overline{F_i \setminus \pi_i^{-1}(x_{i-1})}$$

contains F_i. Now, $\nu_y(\tilde{\mathbf{J}}_i) \leq \nu_{x_i}(\tilde{\mathbf{J}}_i)$, since order is upper semicontinuous. Since x_{i-1} is a generic point in F_{i-1}, there exists, for every neighborhood of x_{i-1} in F_{i-1}, at least one point $y \in F_i \setminus \pi_i^{-1}(x_{i-1})$ such that $\nu_{x_{i-1}}(\tilde{\mathbf{J}}_{i-1}) = \nu_{\pi_i(y)}(\tilde{\mathbf{J}}_{i-1})$. Therefore $\nu_{x_i}(\tilde{\mathbf{J}}_i) = \nu b$.

From the definition of $\mathbf{J}_i$ and $\tilde{\mathbf{J}}_i$ we have

$$\nu_{x_m}(\mathbf{J}_m) = \nu_{x_m}(\mathbf{J}_m) + \nu_{x_m}\left(\prod_{i=1}^{m} \mathbf{P}_i^{b(\nu-1)} \mathcal{O}_{W_m}\right) = \nu b + \sum_{i=1}^{m} b(\nu-1)\nu_{x_m}(\mathbf{P}_i \mathcal{O}_{W_m}).$$

Now, if we apply the same process to $\mathscr{E}' = ((\mathbf{J}', b))$ with $\nu' = \nu_x(\mathscr{E}')$, we get two sequences

$$\mathbf{J}' = \mathbf{J}'_0,\ \mathbf{J}'_1, \ldots,\ \mathbf{J}'_m,\ \mathbf{J}'_{m+1},\ \mathbf{J}'_i = \mathbf{P}_i^{-b} \cdot \mathbf{J}'_{i-1} \cdot \mathcal{O}_{W_i}$$

and

$$\mathbf{J}' = \tilde{\mathbf{J}}'_0,\ \tilde{\mathbf{J}}'_1, \ldots,\ \tilde{\mathbf{J}}'_m,\ \tilde{\mathbf{J}}'_{m+1},$$

where $\mu_{i-1} = \nu_{x_{i-1}}(\tilde{\mathbf{J}}'_{i-1})$ and $\tilde{\mathbf{J}}'_i = \mathbf{P}_i^{-\mu_{i-1}} \tilde{\mathbf{J}}'_{i-1} \cdot \mathcal{O}_{W_i}$. In this case, we can only assert that

$$\nu'_b = \mu_0 \geq \mu_1 \geq \cdots \geq \mu_m.$$

As before, we get

$$\nu_{x_m}(\mathbf{J}'_m) = \nu_{x_m}(\tilde{\mathbf{J}}'_m) + \sum_{i=1}^{m} (\mu_{i-1} - b)\nu_{x_m}(\mathbf{P}_i \mathcal{O}_{W_m}).$$

Assume that $\nu' < \nu$ and look for a contradiction. Then

$$\nu b > \nu' b = \mu_0 \geq \mu_1 \cdots \geq \mu_m$$

and

$$\begin{aligned} \nu_{x_m}(\mathbf{J}_m) - \nu_{x_m}(\mathbf{J}'_m) &= (\nu b - \mu_m) + \sum_{i=1}^{m} (\nu b - \mu_{i-1})\nu_{x_m}(\mathbf{P}_i \cdot \mathcal{O}_{W_m}) \\ &\geq (\nu b - \nu' b) + m(\nu b - \nu' b) = b(m+1)(\nu - \nu'). \end{aligned}$$

Since m can be arbitrarily large because F_m is always non-empty, we can assume that

$$\nu_{x_m}(\mathbf{J}_m) - \nu_{x_m}(\mathbf{J}'_m) > b,$$

by taking $m \geq 1/(\nu - \nu')$. Since the groves associated with $\mathscr{E}$ and $\mathscr{E}'$ are equivalent, the same is true for $\mathscr{E}_m$ and $\mathscr{E}'_m$, and, since $\nu_{x_m}(\mathscr{E}_m) > 1$, we can apply Lemma 4.4.1 to get

$$|\nu_{x_m}(\mathscr{E}_m) - \nu_{x_m}(\mathscr{E}'_m)| < 1.$$

This is the desired contradiction. □

Lemma 4.4.3 *With the assumptions of Lemma* 4.4.2, $\nu_x(\mathscr{E}) = \nu_x(\mathscr{E}')$.

Proof To prove this, we are going to apply Lemma 4.4.2 to each W_i, $\mathscr{E}_i$, $\mathscr{E}'_i$, x_i. This can be done because $x_i \in F_i \subset \mathrm{Sing}^{\nu}(\mathscr{E}_i)$ and x_i is not an isolated point in $\mathrm{Sing}^{\nu}(\mathscr{E}_i)$, and $\mathscr{E}_i$, $\mathscr{E}'_i$ have equivalent associated groves. Therefore

$$\frac{\mu_i}{b} = \nu_{x_i}(\mathscr{E}'_i) \geq \nu_{x_i}(\mathscr{E}_i) = \nu,$$

so $\mu_i \geq \nu b$ for every i.

On the other hand,

$$\nu_{x_m}(\mathbf{J}_m) \geq \nu b + \sum_{i=1}^{m} b(\nu - 1),$$

and $m > (2 - \nu)/(\nu - 1)$ implies $\nu_{x_m}(\mathbf{J}_m) > 2b$. Moreover,

$$\nu_{x_m}(\mathbf{J}'_m) - \nu_{x_m}(\mathbf{J}_m) = (\mu_m - \nu b) + \sum_{i=1}^{m}(\mu_{i-1} - \nu b)\nu_{x_m}(\mathbf{P}_i \cdot \mathscr{O}_{W_m}).$$

Now, assuming that $\nu' > \nu$, we are going to look for a contradiction. Since $\mu_0 = \nu' b$ and $\mu_i \geq \nu b$ for every $i > 0$, the preceding equation gives

$$\nu_{x_m}(\mathbf{J}'_m) - \nu_{x_m}(\mathbf{J}_m) > 0.$$

Further, let x_{m+1} be a generic point in $\pi_{m+1}^{-1}(x_m)$. Then, if $\beta = \nu_{x_m}(\mathbf{J}_m)$ and $\beta' = \nu_{x_m}(\mathbf{J}_m)$, around x_m, we have

$$\mathbf{J}_{m+1} = \mathbf{P}_{m+1}^{\beta - b} \quad \text{and} \quad \mathbf{J}'_{m+1} = \mathbf{P}_{m+1}^{\beta' - b}.$$

Therefore $\beta' - b = \nu_{x_{m+1}}(\mathbf{J}'_{m+1}) > \nu_{x_{m+1}}(\mathbf{J}_{m+1}) = \beta - b > b$. An application of Lemma 4.4.2 to $W_{m+1}, \mathscr{E}_{m+1}, \mathscr{E}'_{m+1}$, and x_{m+1} (allowed because $\nu_{x_{m+1}}(\mathbf{J}_{m+1}) > b$), gives $\beta - b \geq \beta' - b$, which is the desired contradiction. □

Theorem 4.4.2 *Let W be a smooth complex-analytic space, and let $\mathscr{E}$, $\mathscr{E}'$ be idealistic exponents on W with the same soil* $\mathbf{s}$ *and such that* $\mathbf{P}(\mathscr{E})$ *and* $\mathbf{P}(\mathscr{E}')$ *are* $\mathrm{id}_{\mathbf{s}}$*-equivalent. Then:*

(1) *For every $x \in \mathrm{Sing}(\mathscr{E})$, we have $\nu_x(\mathscr{E}) = \nu_x(\mathscr{E}')$.*
(2) *For every $\nu \geq 0$, we have $\mathrm{Sing}^{\nu}(\mathscr{E}) = \mathrm{Sing}^{\nu}(\mathscr{E}')$.*

Proof This follows easily from the lemmas, since, for every smooth connected complex-analytic space V and every $\nu \geq 1$, we have $\mathrm{Sing}^{\nu}(\mathscr{E} \times V) = \mathrm{Sing}^{\nu}(\mathscr{E}) \times V$. Taking V with dimension greater that 1, the additional assumption of Lemma 4.4.2 always holds for $\mathscr{E} \times V$. □

Example 4.4.1 The additional condition of Lemma 4.4.2 is necessary, in view of the following example. Let $W = \mathbb{C}$, let z be a coordinate function on $\mathbb{C}$, and set

$$\mathscr{E} = \big(((z^5), 4)\big), \quad \mathscr{E}' = \big(((z^6), 4)\big).$$

The groves associated with $\mathscr{E}$ and $\mathscr{E}'$ are equivalent, but $\dim_0 \mathrm{Sing}(\mathscr{E}) = 0$, and $\nu_0(\mathscr{E}) = \frac{5}{4}$, whereas $\nu_0(\mathscr{E}') = \frac{6}{4}$. This example also shows that it may happen that the groves associated with $\mathscr{E}$ and $\mathscr{E}'$ are equivalent, but not the polygroves.

Remark 4.4.3 Our aim now is to establish a globality theorem guaranteeing the existence of enough closed smooth subspaces of a smooth complex-analytic space that can be considered as the centers of blowing-up in the desingularization process.

To properly state the theorem, we need to construct a more precise measure of the complexity of an idealistic exponent than the usual order.

Proposition 4.4.1 *Consider a smooth complex-analytic space W, smooth closed complex subspaces Z_1, Z_2 of W, a point $x \in Z_1 \cap Z_2$, and idealistic exponents $\mathscr{E}_1$ and $\mathscr{E}_2$ on Z_1 and Z_2, respectively. Assume that:*

(1) $x \in \mathrm{Sing}(\mathscr{E}_1)$,
(2) $\dim_x Z_1 = \dim_x Z_2$, *and*
(3) *there exists a polygrove* $\mathbf{P}$ *on W that is an ambient expansion of $\mathscr{E}_i$ for the inclusions $Z_i \to W$, for $i = 1, 2$.*

Then $\nu_x(\mathscr{E}_1) = \nu_x(\mathscr{E}_2)$.

Proof The scene being local, by restricting ourselves to a small enough neighborhood of x we may assume that there exist retractions $r_1 : W \to Z_1$ and $r_2 : W \to Z_2$ and an isomorphism $\beta : Z_1 \to Z_2$ such that $\beta r_1 = r_2$ and $\beta(x) = x$. Since β is an isomorphism, we can construct the pull-back $\beta^*(\mathscr{E}_2) = \mathscr{E}'$; then $\nu_x(\mathscr{E}') = \nu_x(\mathscr{E}_2)$. On the other hand, $\mathscr{E}_1$ and $\mathscr{E}'$ have equivalent polygroves over β; therefore $\mathscr{E}_1$ and $\mathscr{E}'$ are two idealistic exponents on Z_1 whose polygroves are ambient contractions of $\mathbf{P}$ by the inclusion $Z_1 \to W$; hence these polygroves are equivalent. By Theorem 4.4.2, $\nu_x(\mathscr{E}_1) = \nu_x(\mathscr{E}')$, which proves the proposition. □

Definition 4.4.4 Let W be a smooth complex-analytic space, $\mathscr{E}$ an idealistic exponent on W, and $x \in \mathrm{Sing}(\mathscr{E})$. If $d \in \mathbb{Z}_0$, we define $\nu_x^d(\mathscr{E})$ as follows:

(1) If $d > \dim_x(W)$ then $\nu_x^d(\mathscr{E}) = 1$.
(2) If $d \leq \dim_x(W)$ and if there exist $U, Z, \mathscr{K}$ such that:

   (a) the set U is an open neighborhood of x in W,
   (b) the complex analytic space Z is a smooth closed subspace of W with $\dim_x(Z) = d$, and
   (c) the idealistic exponent on Z, $\mathscr{K}$, verifies that $\mathbf{P}(\mathscr{K})$ and $\mathbf{P}(\mathscr{E}|_U)$ are equivalent over the inclusion of Z in U,

   then $\nu_x^d(\mathscr{E}) = \nu_x(\mathscr{K})$.
(3) If $d < \dim_x(W)$ and there are no $(U, Z, \mathscr{K})$ as in (2), then $\nu_x^d(\mathscr{E}) = \infty$.

Remark 4.4.4

4.4.4.1 Proposition 4.4.1 implies that, in case (2) of Definition 4.4.4, then number $\nu_x^d(\mathscr{E})$ is independent of the choice of $(U, Z, \mathscr{K})$.
4.4.4.2 In case (1) of Definition 4.4.4, we could have defined $\nu_x^d(\mathscr{E})$ by using ambient expansion, just as we used ambient contraction in (2). That is, if $d > \dim_x(W)$ and if there exist an open neighborhood U of x in W, a smooth complex analytic space Z such that there is a closed embedding $j : U \to Z$, and an idealistic exponent $\mathscr{K}$ on Z such that $\mathbf{P}(\mathscr{K})$ is an ambient expansion of $\mathbf{P}(\mathscr{E})$ by j, we could have set $\nu_x^d(\mathscr{E}) = \nu_x(\mathscr{K})$.

By Theorem 4.3.1, such a triple $(U, Z, \mathscr{K})$ always exists; moreover, we claim that $\nu_x(\mathscr{K}) = 1$. To prove this, we may assume, since the scene is local, that there is local retraction $r : Z \to U$. Let $\mathscr{E} = ((\mathbf{J}, b))$ and $\mathscr{E}' = ((r^\#(\mathbf{J}), b))$, as in Theorem 4.3.1. Then $P(\mathscr{E}')$ is equivalent to $P(\mathscr{E})$ and hence to $P(\mathscr{K})$. Therefore $\nu_x(\mathscr{K}) = \nu_x(\mathscr{E}') = \nu_x(r_b^\#(\mathbf{J}))/b$. Now, if $\mathbf{I}$ is the ideal of $j(U)$ in Z, then $\nu_x(\mathbf{I}) = 1$, which proves our assertion.

Thus, in case (1) of Definition 4.4.4, it doesn't matter whether we define $\nu_x^d(\mathscr{K})$ in the original way or using ambient expansions as just described.

4.4.4.3 ν_x^d does not change under ambient expansions and contractions; this follows from the definition and from Proposition 4.4.1. Moreover, ν_x^d does not depend on the idealistic exponent; only on the polygroves associated with it. Therefore, if P is an $\mathbf{s}$-polygrove, if $x \in (s)_0$ and if there is an idealistic exponent $\mathscr{E}$ on an open neighborhood U of x in $(s)^0$ such that $P(\mathscr{E})$ is $\mathrm{id}_{\mathbf{s}}$-equivalent to P, it makes sense to write $\nu_x^d(P)$ as $\nu_x^d(\mathscr{E})$.

Definition 4.4.5 Let $\mathscr{F} = (G, J)$ be an idealistic forest. For all g in $\mathrm{ob}([G])$, $d \in \mathbb{Z}_0$ and $\mu \in \mathbb{R}_0 \cup \{\infty\}$, set

$$\mathrm{Sing}_\mu^d(\mathscr{F})(g) = \{x \in (G(g))_0 \mid \nu_x^d(J(g)) \geq \mu\}.$$

Remark 4.4.5 One can define on the category ((el)) – and, by restriction, on the category ((el | hor)) – two covariant functors

$$((\mathrm{el})) \xrightarrow{(\ )_0} ((\mathbb{C}\text{-spaces}))$$

and

$$((\mathrm{el})) \xrightarrow{(\ )^0} ((\mathbb{C}\text{-spaces}))$$

with values in the category (($\mathbb{C}$-spaces)) of complex spaces, as follows:

$$(\ )_0(\mathbf{s}) = (s)_0, \quad (\ )_0(\mathbf{a}) = (a)_0,$$

$$(\ )^0(\mathbf{s}) = (s)^0, \quad (\ )^0(\mathbf{a}) = (a)^0.$$

Then, if G is a $\mathbb{C}$-situation, one can assign to G two covariant functors $G_0 = (\ )_0 G$ and $G^0 = (\ )^0 G$ on $[G]$, with values in (($\mathbb{C}$-spaces)).

Definition 4.4.6

4.4.6.1 Let H be a covariant functor

$$H : [H] \longrightarrow ((\mathbb{C}\text{-spaces})).$$

A **subset** of H is a family $C = \{C(h)\}_{h \in \mathrm{ob}([H])}$ such that $C(h)$ is a subset of $H(h)$. We say that a subset C of H is **compact** if $C(h) = \varnothing$ except for a finite number of $h \in \mathrm{ob}([H])$, and $C(h)$ is compact for every $h \in \mathrm{ob}([H])$.

4.4.6.2 A **complex subspace** of H is a functor

$$H' : [H] \longrightarrow ((\mathbb{C}\text{-spaces}))$$

satisfying these conditions:

(1) For every $h \in \mathrm{ob}([H])$, $H'(h)$ is a complex subspace of $H(h)$.
(2) For every $\beta \in \mathrm{Hom}_{[H]}(h, h')$, $H'(h) = H(\beta)^{-1}(H'(h'))$ and $H'(\beta)$ is induced by $H(\beta)$.

4.4.6.3 If $\mathfrak{p}$ is a property applicable to complex subspaces of a complex-analytic space, we say that a complex subspace H' of H **satisfies** $\mathfrak{p}$ if, for every $h \in [H]$, $H'(h)$ satisfies $\mathfrak{p}$ in $H(h)$. For instance, we say that H' is **closed** if $H'(h)$ is closed in $H(h)$, for all $h \in \mathrm{ob}([H])$.

Definition 4.4.7 Given a forest $\mathscr{F} = (G, \mathbf{J})$ and a point (g, x) of G, the **ambient dimension** of $\mathscr{F}$ (or of G) at (g, x) is the number

$$\dim_{(g,x)}(\mathscr{F}) = \dim_{(g,x)}(G) = \min_{(g_1,x_1)\sim(g,x)} \{\dim_{x_1}(G(g_i))^0\}.$$

The **ambient dimension** of $\mathscr{F}$ (or of G) is the number

$$\dim(\mathscr{F}) = \dim(G) = \sup_{(g,x)} \{\dim_{(g,x)}(G)\}.$$

Theorem 4.4.3 *Let $\mathscr{F} = (G, J)$ be an idealistic forest and d an integer such that $d \geq \dim(\mathscr{F})$. Then, for every real number $\mu \geq 1$, the functor*

$$\mathrm{Sing}^d_\mu(\mathscr{F}) : [G] \longrightarrow ((\mathbb{C}\text{-}spaces))$$

is defined by

(1) *for every $g \in \mathrm{ob}([G])$,*

$$(\mathrm{Sing}^d_\mu(\mathscr{F}))(g) = \mathrm{Sing}^d_\mu(\mathscr{F})(g)$$

and

(2) *for every $\alpha \in \mathrm{Hom}_{[G]}(g', g)$, $\mathrm{Sing}^d_\mu(\mathscr{F})(\alpha)$ is the morphism induced by $(G(\alpha))_0$*

is a closed complex subspace of $(G)_0$.

Proof The theorem follows from three facts:

(i) For every $g \in \mathrm{ob}([G])$, $\mathrm{Sing}^d_\mu(\mathbf{J}(g))$ is a closed complex subspace of $(G(g))^0$ with the induced and reduced structure.
(ii) $\mathrm{Sing}^d_\mu(\mathscr{F})$ is a functor.
(iii) For every $\alpha \in \mathrm{Hom}_{[G]}(g, g')$,

$$\left((G(\alpha))_0\right)^{-1}(\mathrm{Sing}^d_\mu(\mathbf{J}(g'))) = \mathrm{Sing}^d_\mu(\mathbf{J}(g)).$$

The last two facts are trivial; we prove only (i).

Since $d \geq \dim(\mathscr{F})$, we have $d \geq \dim_{(g,x)}(G)$ for every point (g, x) of G. Fix (g, x); then there exists $(g_1, x_1) \sim (g, x)$ with $d \geq \dim_{x_1}(G(g_1))^0$. Since $\mathscr{F}$ is idealistic, and by Theorems 4.3.1 and 4.3.2, there exist an open neighborhood U of x in $(G(g))^0$, a smooth closed complex subspace W of U (which may coincide with U if $d \geq \dim_x(G(g))^0$), and an idealistic exponent $\mathscr{E} = ((\mathbf{J}, b))$ on W such that $\mathbf{P}(\mathscr{E})$ is $\mathrm{id}_{G(g)}$-equivalent to $\mathbf{J}(g)$. Then

$$U \cap \mathrm{Sing}^d_\mu(J(g)) = \{y \in W \mid \nu_x(\mathscr{E}) \geq \mu\} = \mathrm{Sing}^\mu(\mathscr{E}),$$

which is a smooth closed complex subspace of W with the induced and reduced structure. □

Remark 4.4.6 Let X be a complex-analytic space and Γ_X the pseudogroup of local automorphisms of X: the elements of Γ_X are isomorphisms of $X|_U$ onto $X|_V$, where U and V are open sets in X. We say that a complex subspace $Y \subset X$ is Γ_X-**invariant** if, for every two open sets U, V in X, every isomorphism $X|_U \simeq X|_V$ induces an isomorphism $Y|_U \simeq Y|_V$.

Given a Samuel stratum S in X, there is a one-to one correspondence between the closed complex-analytic subspaces of (G_X, S) and the closed complex subspaces of S that are Γ_X-invariant. Then the globality theorem implies that, for every integer $d \geq 0$ and every real $\mu \geq 1$, there exists a closed complex subspace Y^d of S that is Γ_X-invariant.

4.5 Gardens

To globalize the results of the preceding chapters, we need a new concept: gardens. This section is devoted to the definition of gardens and to the study of their behavior under three elementary operations: completion, intersection, and modification.

Remark 4.5.1 Let Z be a smooth complex-analytic space and let $\mathbf{E} = \{E_\lambda\}_{\lambda \in \Lambda}$ be a family of closed hypersurfaces in Z. We say that $\mathbf{E}$ is **locally finite** if, for every $x \in Z$, there exists an open neighborhood U of x in Z such that only finitely many of E_λ intersect U. In particular, $\Lambda_x = \{\lambda \in \Lambda \mid x \in E_\lambda\}$ is finite.

Let $\mathbf{E}$ be a locally finite family of closed hypersurfaces as above, Z' a complex space, $i : Z' \hookrightarrow Z$ a locally closed embedding, $x' \in Z'$ and $x = i(x')$. We say that i has **normal crossings** with $\mathbf{E}$ at x' if there exists a local coordinate system $z = (z_1, \ldots, z_d)$ on Z centered at x, a subset $I = \{i_1, \ldots, i_r\} \subset \{1, \ldots, d\}$, and an injective map $j : \Lambda_x \to \{1, \ldots, d\} \setminus I$, in such a way that:

(1) The complex analytic set $i(Z')$ is defined, around x, by $z_{i_1} = \cdots = z_{i_r} = 0$, and
(2) For every $\lambda \in \Lambda_{x'}$, E_λ is defined around x by $z_{j(\lambda)} = 0$.

If Z' is a complex subspace of Z, we say that Z' has **normal crossings with** $\mathbf{E}$ at $x' \in Z'$ if the inclusion $i : Z' \hookrightarrow Z$ has normal crossings with $\mathbf{E}$ at x'. We say that Z' has **normal crossings with E** if it has normal crossings with $\mathbf{E}$ at every point $x' \in Z'$. We say that $\mathbf{E}$ has **normal crossings at** $x \in Z$ in Z if $\{x\}$ has normal crossings with $\mathbf{E}$.

Let Z' be a smooth subspace of Z and let $\mathbf{E} = \{E_\lambda\}_{\lambda \in \Lambda}$ be as above. Assume that Z' is not contained, locally, in any of these hypersurfaces. Then $\mathbf{E}$ induces a locally finite family of hypersurfaces in Z', say $\mathbf{E}' = \{E'_\lambda\}_{\lambda \in \Lambda}$.

Note that Z' has normal crossings with $\mathbf{E}$ at a point $x \in Z'$ if and only if both $\mathbf{E}$ in Z and $\mathbf{E}'$ in Z' have normal crossings at x. In particular, when $\mathbf{E}$ has normal crossings (everywhere) in Z, Z' has normal crossings with $\mathbf{E}$ if and only if $\mathbf{E}'$ has normal crossings in Z'.

Definition 4.5.1 Let G be a $\mathbb{C}$-situation. By a **hypersurface** of G we mean a complex hypersurface of $(G)^0$.

An **allée** in G is a family $\boldsymbol{\Delta} = \{\Delta_\lambda\}_{\lambda\in\Lambda}$ satisfying these conditions:

(1) For every $\lambda \in \Lambda$, Δ_λ is a smooth closed hypersurface in G.
(2) For every $g \in \mathrm{ob}([G])$, $\{\Delta_\lambda(g)\}_{\lambda\in\Lambda}$ is a locally finite family in $(G(g))^0$.
(3) For every $g \in \mathrm{ob}([G])$, $\{\Delta_\lambda(g)\}_{\lambda\in\Lambda}$ has normal crossings in $(G(g))^0$.

An allée in a forest (G, J) is an allée in G.

A **garden** is a triple $(G, \mathbf{J}, \boldsymbol{\Delta})$, where $(G, \mathbf{J})$ is a forest and $\boldsymbol{\Delta}$ is an allée in $(G, \mathbf{J})$. We shall say that a garden $(G, \mathbf{J}, \boldsymbol{\Delta})$ is idealistic if its forest $(G, \mathbf{J})$ is idealistic. We shall say that a garden $(G, \mathbf{J}, \boldsymbol{\Delta})$ is associated with (X, S) if $(G, \mathbf{J})$ is an X-forest of S, where S is a Samuel stratum of X.

The **dimension** of a garden $(G, \mathbf{J}, \boldsymbol{\Delta})$ is the dimension of G.

Remark 4.5.2 We have seen that, given a complex-analytic space X and a Samuel stratum S of X, one can construct an X-full forest of S and an idealistic forest $(G, \mathbf{J})$ associated with (X, S). Then, by taking an allée in $(G, \mathbf{J})$ to be the empty set, we always obtain an idealistic garden associated with (X, S). This will be denoted by $\mathscr{G}_{X,S}$.

Definition 4.5.2 Let $\mathscr{G} = (G, \mathbf{J}, \boldsymbol{\Delta})$ be a garden, take $g \in \mathrm{ob}([G])$, and let $\mathbf{h} : \mathbf{s} \to G(g)$ be a horizontal morphism of $\mathbb{C}$-elements. We say that $\mathscr{G}$ is **contractible** by $\mathbf{h}$ (or by $(h)^0$) if:

(1) There exists an open set U in $(G(g))^0$ such that $\mathbf{h}$ induces a morphism $\mathbf{h} : \mathbf{s} \to G(g)|_U$ and $\boldsymbol{J}(g)|_U$ admits an ambient contraction by $\mathbf{h}$, and
(2) $\left((h)^0\right)^{-1}(\boldsymbol{\Delta}(g)) = \{\left((h)^0\right)^{-1}(\Delta_\lambda(g)) \mid \lambda \in \Lambda\}$ is a system of smooth closed hypersurfaces with normal crossings in $(s)^0$.

We say that $\mathscr{G}$ is **complete** if, for every $g \in \mathrm{ob}([G])$ and every positive integer m, the following conditions hold:

(1) There exists $g_1 \in \mathrm{ob}([G])$, $\alpha \in \mathrm{Hom}_{[G]}(g, g_1)$, and an isomorphism

$$f : (G(g_1))^0 \to (G(g))^0 \times \mathbb{C}^m$$

such that $f \cdot (G(\alpha))^0 : (G(g))^0 \to (G(g))^0 \times \mathbb{C}^m$ induces the canonical isomorphism $(G(g))^0 \to (G(g))^0 \times \{0\}$, and $f(\Delta_\lambda(g_1)) = \Delta_\lambda(g) \times \mathbb{C}^m$ for every $\lambda \in \Lambda$.
(2) If $\mathbf{h} : \mathbf{s} \to G(g)$ is a horizontal morphism such that $\mathscr{G}$ is contractible by $\mathbf{h}$, there exists $g' \in \mathrm{ob}([G])$, $\alpha \in \mathrm{Hom}_{[G]}(g', g)$, and an isomorphism $\mathbf{h}' : G(g') \to \mathbf{s}$ such that $G(\alpha) = \mathbf{h} \cdot \mathbf{h}'$.

Remark 4.5.3 (Completion of a garden) Let $\mathscr{G} = (G, \mathbf{J}, \boldsymbol{\Delta})$ be an arbitrary garden. We construct, starting from $\mathscr{G}$, a complete garden $\hat{\mathscr{G}} = (\hat{G}, \hat{\mathbf{J}}, \hat{\boldsymbol{\Delta}})$, called the **completion** of $\mathscr{G}$.

We start by defining the category $[\hat{G}]$ in the following way:

(a) The objects of $[\hat{G}]$ are the quadruples (g, m, U, W) consisting of an element $g \in \mathrm{ob}([G])$, a positive integer m, an open set U in $(G(g))^0 \times \mathbb{C}^m$, and a smooth closed complex subspace W of U, and satisfying the following property: if $k : (G(g))^0 \to (G(g))^0 \times \mathbb{C}^m$ is the morphism defined by $k(x) = (x, \mathbf{0})$ and if $k' : W \to U$ is the inclusion, there exists an ambient expansion of $\mathbf{J}(g)$ by k which, restricted to U, admits an ambient contraction $\hat{\mathbf{J}}(g, m, U, W)$ by k', and $\{k'^{-1}(\Delta_\lambda(g) \times \mathbb{C}^m)\}_{\lambda \in \Lambda}$ is a system of smooth hypersurfaces with normal crossings in W.

(b) If $(g, m, U, W), (g', m', U', W') \in \mathrm{ob}[\hat{G}]$, the elements of

$$\mathrm{Hom}_{[G]}\big((g', m', U', W'), (g, m, U, W)\big)$$

are the pairs (β, j), where $\beta \in \mathrm{Hom}_{[G]}\ (g', g)$ and $j : \mathbb{C}^{m'} \to \mathbb{C}^m$ is an injective linear map such that

$$U' \subset (G(\beta))^0 \times j^{-1}(U)$$

and

$$W' \subset (G(\beta))^0 \times j^{-1}(W).$$

It can easily be checked that, with these data, $[\hat{G}]$ is a category.

We then define a $\mathbb{C}$-situation $\hat{G} : [\hat{G}] \to ((\mathrm{el} \mid \mathrm{hor}))$ in the following way:

(i) For every $(g, m, U, W) \in \mathrm{ob}[\hat{G}]$,

$$\hat{G}(g, m, U, W) : W \cap ((G(g))_0 \times \{\mathbf{0}\}) \to W$$

is a natural inclusion.

(ii) For every

$$(\beta, j) \in \mathrm{Hom}_{[\hat{G}]}((g', m', U', W'), (g, m, U, W)),$$

$(\hat{G}(\beta, j))^0$ and $(\hat{G}(\beta, j))_0$ are the morphisms induced by $(\hat{G}(\beta))^0 \times j$.

It is easily seen that $\hat{G}$ is, in fact, a $\mathbb{C}$-situation.

Then, for every $(g, m, U, V) \in \mathrm{ob}[\hat{G}]$ we denote by $\hat{\mathbf{J}}(g, m, U, V)$ the polygrove defined in (a) from $\mathbf{J}(g)$. With these data, $(\hat{G}, \hat{\mathbf{J}})$ is a forest.

Finally, we denote by $\hat{\boldsymbol{\Delta}} = \{\hat{\Delta}_\lambda\}_{\lambda \in \Lambda}$ the allée in $(\hat{G}, \hat{\mathbf{J}})$ defined by

$$\boldsymbol{\Delta}_\lambda(g, m, U, W) = k'^{-1}(\Delta_\lambda(g) \times \mathbb{C}^m).$$

In this way, $\hat{\mathscr{G}} = (\hat{G}, \hat{\mathbf{J}}, \hat{\boldsymbol{\Delta}})$ is a complete garden, called the **completion** of $\mathscr{G}$. Observe that the category $[G]$ can be embedded, in a natural way, in $[\hat{G}]$, so that, every point of G is equivalent to a point whose object belongs to $[\hat{G}]$.

Proposition 4.5.1 *If $\mathscr{G}$ is an idealistic garden, so is $\hat{\mathscr{G}}$.*

Proof Let $((g, m, U, W), x')$ be a point of $\hat{G}$; then

$$x' \in (\hat{G}(g, m, U, W))_0 = W \cap \big((G(g)) \times \{\mathbf{0}\}\big).$$

Hence there exists $x \in (G(g))_0$ such that $x' = (x, \mathbf{0})$. Therefore (g, x) is a point of G and, since $\mathcal{G}$ is an idealistic garden, there exist an open neighborhood U of x in $(G(g))^0$ and an idealistic exponent $\mathcal{E}$ on U such that $\mathbf{J}(g)|_U$ is associated with $\mathcal{E}$.

Now, by considering the morphism $k : (G(g))^0 \to (G(g))^0 \times \mathbb{C}^m$ of Remark 4.5.3(a), and by applying Theorem 4.3.1, we can assert that the ambient expansion of $J(g)|_U$ by k is associated with the idealistic exponent $\mathcal{E} \times \mathbb{C}^m$. Then, since $J(g, m, U, W)$ is an ambient contraction by k' of such an expansion, by Theorem 4.3.2, there exist an open neighborhood U' of x' in $(G(g, m, U, W))^0$ and an idealistic exponent $\mathcal{E}'$ on U' such that $J(g, m, U, W)|_{U'}$ is associated with $\mathcal{E}'$. This proves the proposition. □

Remark 4.5.4 (Intersection) Let $\{\mathbf{s}_i\}_{1 \le i \le n}$ be a finite family of $\mathbb{C}$-elements such that $(s_i)^0 = W$ for $i = 1, \ldots, n$. We form a new $\mathbb{C}$-element

$$\mathbf{s} : \left(\bigcap_{i=1}^{n} (s_i)^0\right)_{\text{red}} \hookrightarrow W,$$

called the **reduced intersection** of the family $\{\mathbf{s}_i\}_{1 \le i \le n}$ and denoted by $\mathbf{s} = \bigcap_{i=1}^{n} \mathbf{s}_i$.

Let $A_i = \{\mathbf{a}_{i,j}, U_{i,j-1}, E_{i,j-1}\}_{1 \le j \le m_i}$, for $1 \le i \le \alpha$, be a finite family of $\mathbf{s}_{i,0}$-trees such that $(s_{i,0})^0 = W$ for $1 \le i \le \alpha$. Set $\mathbf{s}_0 = \bigcap_{i=1}^{\alpha} \mathbf{s}_{i,0}$; by definition, the reduced intersection of the family $\{A_i\}_{1 \le i \le \alpha}$, denoted by $A = \bigcap_{i=1}^{\alpha} A_i$, is an $\mathbf{s}_0$-tree $A = \{\mathbf{a}_l, U_{l-1}, E_{l-1}\}_{1 \le l \le m}$ constructed in the following way:

(1) $m = \min_{1 \le i \le \alpha}\{m_i\}$, and U_0 is the largest open set in $\bigcap_{i=1}^{\alpha} U_{i,0}$ such that $U_0 \cap E_{i,0} = E_0$ is independent of i.
(2) For $1 \le i \le \alpha$ and $1 \le j \le m$ we have a morphism $(s_{i,j})^0 \to (s_{i,0})^0 = W$, which is the composition of all the $(a_{i,j})^0$. Then we can construct the fiber product

$$M_j = \prod_{1 \le i \le \alpha} {}_W (s_{ij})^0 = (s_{1j})^0 \times_W (s_{2j})^0 \times_W \cdots \times_W (s_{\alpha j})^0$$

and a canonical morphism $M_j \to M_i$ if $i < j$. We identify M_0 with W in a natural way.
(3) We construct the blow-up $\pi_1 : (s_1)^0 \to U_0$ with center E_0. Then $(s_1)^0$ can be embedded in M_1 in such a way that the canonical map $M_1 \to M_0$ induces the composition of π_1 with the inclusion $U_0 \to W$. We denote this morphism by $(a_1)^0$. We denote by $(s_1)_0$ the subspace

$$\left((s_1)^0 \cap \prod_{1 \le i \le \alpha} {}_W (s_{i,0})_0\right)_{\text{red}},$$

where $(s_1)^0$ is defined with its image in M_1. Finally, denote by $(a_1)_0$ the restriction

$$(a_1)_0 = (a_1)^0|_{(s_1)_0}.$$

Then $\mathbf{a}_1 : \mathbf{s}_1 \to \mathbf{s}_0$ is a vertical morphism.

(4) Let us assume that we have constructed $\{\mathbf{a}_i, U_{i-1}, E_{i-1}\}_{1\leq i\leq k}$ in such a way that each $(s_i)^0$ can be embedded in M_i and the canonical morphism $M_i \to M_{i-1}$ induces $(a_i)^0$, for $1 \leq i \leq k$. We denote by $E'_{j,k}$ the inverse image of E_{j,k_0} by the projection $M_k \to (s_{j,k})^0$, for $1 \leq j \leq \alpha$. We let U_k be the largest open set in

$$(s_k)^0 \cap \prod_{\substack{W\\1\leq i\leq\alpha}} U_{i,k}$$

such that $E_k = U_k \cap E'_{j,k}$ does not depend on i.

(5) The blow-up $\pi_{k+1} : (s_{k+1})^0 \to U_k$ is such that $(s_{k+1})^0$ can be embedded in M_{k+1} and π_{k+1} is the restriction to $(s_{k+1})^0$ of the canonical morphism $M_{k+1} \to M_k$. Then, we put

$$(s_{k+1})_0 = \left((s_{k+1})^0 \cap \prod_{\substack{W\\1\leq i\leq\alpha}} (s_{i,k})_0 \right)_{\mathrm{red}},$$

we take $(a_{k+1})^0$ as the composition of π_{k+1} with the inclusion $U_k \to (s_k)^0$, and we set $(a_{k+1})_0 = (a_{k+1})^0|_{(s_{k+1})_0}$. In this way, we have constructed $\mathbf{a}_{k+1} : \mathbf{s}_{k+1} \to \mathbf{s}_k$, which is a vertical morphism.

This concludes the inductive construction of the tree A.

Let $\mathscr{F}_i = (G_i, \mathbf{J}_i)$, for $1 \leq i \leq \alpha$, be forests such that $[G_i] = [G_j]$ and $(G_i)^0 = (G_j)^0$ for every $i, j = 1, \ldots, \alpha$. The **(reduced) intersection** of the family $\{\mathscr{F}_i\}_{1\leq i\leq\alpha}$ is the forest $\mathscr{F} = (G, \mathbf{J})$, also denoted by $\bigcap_{i=1}^{\alpha} \mathscr{F}_i$, defined by the following properties:

(i) For $1 \leq i \leq \alpha$, $[G] = [G_i]$ and $(G)^0 = (G_i)^0$.

(ii) $G(g) = \bigcap_{i=1}^{\alpha} G_i(g)$ for every $g \in \mathrm{ob}([G])$, and $(G(h))_0$ is the morphism induced by $(G(h))^0$, for every $h \in \mathrm{Hom}_{[G]}(g, g')$.

(iii) For every $g \in \mathrm{ob}([G])$, the $G(g)$-polygrove $\mathbf{J}(g)$ is defined as

$$\{\mathbf{J}(g)_V\}_{V\in\mathrm{ob}(\mathscr{V}},$$

where $\mathbf{J}(g)_V$ is a $G(g)$-grove defined by

$$|\mathbf{J}(g)_V| = \prod_{i=1}^{\alpha} |\mathbf{J}_i(g)_V|$$

and, for every $\mathbf{b} = (b_i, \ldots, b_\alpha) \in |\mathbf{J}(g)_V|$,

$$\mathbf{J}(g)_V(\mathbf{b}) = \bigcap_{i=1}^{\alpha} \mathbf{J}_i(g)_V(b_i).$$

If the forests $\{\mathscr{F}_i\}_{1\le i\le \alpha}$ are idealistic and if their intersection $\mathscr{F}$ exists, then $\mathscr{F}$ is idealistic. To see this, write $\mathscr{F} = (G, \mathbf{J})$ and let (g, x) be a point in G. Then

$$x \in (G(g))_0 = \left(\bigcap_{i=1}^{\alpha}(G_i(g))_0\right)_{red},$$

so $x \in (G_i(g))_0$ for $1 \le i \le \alpha$ and (g, x) is a point of G_i for $1 \le i \le \alpha$. Since the $\mathscr{F}_i$ are idealistic, there exist neighborhoods $\{U_i\}_{1\le i\le\alpha}$ of x in $(G_i(g))^0$ and idealistic exponents $\mathscr{E}_i = ((\mathbf{J}_i', b_i)) = ((\mathbf{J}_i, b))$ (it is always possible to choose representatives with the same b), such that $\mathbf{J}_i(g)|_{U_i}$ is associated with $\mathscr{F}_i$. Now write $U = \bigcap_{i=1}^{\alpha} U_i$, which is an open neighborhood of x in $(G(g))^0$, and set

$$\mathscr{E}_i'' = ((\mathbf{J}_i'', b)) = ((\mathbf{J}_i|_U, b)).$$

By the definition of an intersection, $\mathbf{J}(g)|_U$ is associated with

$$\mathscr{E} = ((\mathbf{J}_1'' + \cdots + \mathbf{J}_\alpha'', b)),$$

so $(G, \mathbf{J})$ is idealistic.

Remark 4.5.5 (Modifications) Let $\mathbf{a} : \mathbf{s}' \to \mathbf{s}$ be a vertical morphism defined by U and E. Let $A = \{\mathbf{a}_i, U_{i-1}, E_{i-1}\}_{1\le i\le m}$ be an $\mathbf{s}$-tree such that the following condition holds:

(C) There is a horizontal morphism $\mathbf{f} : \mathbf{s}_1 \to \mathbf{s}'$ such that $(\mathrm{id}_{\mathbf{s}}, \mathbf{f})$ is a morphism of trees $\{\mathbf{a}_1, U_0, E_0\} \to \{\mathbf{a}, U, E\}$.

We will construct a tree $A' = \{\mathbf{a}_i', U_{i-1}', E_{i-1}'\}_{1\le i\le m-1}$, called the **transform** of A by $\mathbf{a}$.

Using this condition, we write:

(1) $\mathbf{a}_1' = \mathbf{f}\cdot\mathbf{a}_1 : \mathbf{s}_2 \to \mathbf{s}'$, $U_0' = (f)^0(U_1)$, and $E_0' = (f)^0(E_1)$. These relations make sense because f^0 is open.
(2) $\mathbf{a}_1' = \mathbf{a}_{i+1}$, $U_{i-1}' = U_i$, and $E_{i-1}' = E_i$, for $2 \le i \le m-1$.

If the morphism $\mathbf{f}$ in condition (C) exists, it is unique, so A' depends only on A and $\mathbf{a}$.

Under these assumptions, let B be an $\mathbf{s}$-grove. We define the **transform** of B by $\mathbf{a}$ as the $\mathbf{s}'$-grove constructed in the following way. The set of indices $|B'|$ is the set of the elements $b \in |B|$ such that $B(b)$ satisfies (C). If $b \in B'$, we denote by $B'(b)$ the $\mathbf{s}'$-tree that is the transform of B by $\mathbf{a}$.

With the same assumptions, let **P** be a **s**-polygrove. We define the **transform** of **P** by A as the $\mathbf{s}'$-polygrove $\mathbf{P}' = \{B'_V\}_{V \in \mathrm{ob}(\mathscr{V})}$, where B'_V is the $\mathbf{s}' \times V$-grove transform of the $\mathbf{s} \times V$ grove B_V by the vertical morphism

$$\{\mathbf{a} \times V,\ U \times V,\ E \times V\}.$$

We can check that P' is actually a polygrove. In fact, condition (1) of Definition 4.1.3 is trivial, and condition (2) is stable by modification. To see this, consider a vertical morphism $\mathbf{a}_1 : \mathbf{s}_1 \to \mathbf{s}$ such that there is a morphism $\mathbf{f} : \mathbf{s}_1 \to \mathbf{s}'$ satisfying condition (C). Then $\mathbf{f} \times V$ satisfies condition (C) with respect to $\mathbf{a}_1 \times V$ and $\mathbf{a} \times V$.

Let G be a $\mathbb{C}$-situation; since ((el | hor)) is a subcategory of ((el)), we can compose G with the inclusion functor, and consider G as a functor

$$G : [G] \to ((\mathrm{el})).$$

Let Y be a smooth closed complex subspace of $(G)_0$. A pair (G', π) is a **blow-up**, or **modification**, of G, with center Y, if

(1) G' is a $\mathbb{C}$-situation, with $[G'] = [G]$,
(2) π is a morphism of functors $\pi : G' \to G$, and
(3) for every $g \in \mathrm{ob}([G])$, $(\pi(g))^0 : (G'(g))^0 \to (G(g))^0$ is the blow-up with center $Y(g)$.

Given G and Y as above, the following (G', π) is an example of a modification of G with center Y. Define G' by taking $(G'(g))^0$ and $(\pi(g))^0$ to be those given in (3) above and $(G'(g))_0$ to be the strict transform of $(G(g))_0$ by $(\pi(g))^0$. Another example is obtained by taking $(G'(g))_0 = \varnothing$ for all $g \in \mathrm{ob}([G])$. These examples show that (G', π) is not uniquely determined (even up to an isomorphism) by G and Y. The essential non-uniqueness is in the choice of $(G')_0$.

Usually, if (G', π) is a modification of G with center Y, we shall denote by $(\pi(g), Y(g))$ the $G(g)$-tree $\{\pi(g), (G(g))^0, Y(g)\}$. That is it really a $G(g)$-tree follows from (2) and (3) above.

Definition 4.5.3 Let $\mathscr{G} = (G, \mathbf{J}, \boldsymbol{\Delta})$ be a garden and Y a complex subspace of $(G)_0$. We say that Y is a **permissible center** for $\mathscr{G}$ if the following conditions hold:

(1) The complex space Y is smooth and closed.
(2) There exists a modification (G', π) of G with center Y such that, for every point (g, x) of G, there is an open neighborhood U of x in $(G(g))^0$ such that the tree $\big(\pi(g), Y(g)\big)|_U = \big(\pi(g)'_U, U, Y(g)|_U\big)$ belongs to $\mathbf{J}(g)|_U$.
(3) For every $g \in \mathrm{ob}([G])$, $Y(g)$ has normal crossings with $\boldsymbol{\Delta}(g)$ everywhere.

For these conditions, we shall also say that the modification or blow-up of (2) is **permissible** for $\mathscr{G}$.

Proposition 4.5.2 *If* $\mathscr{G} = (G, \mathbf{J}, \boldsymbol{\Delta})$ *is an idealistic garden, and* Y *is a complex subspace of* $(G)_0$ *which satisfies the conditions* (1) *and* (3) *above,* Y *is permissible for* $\mathscr{G}$. *Moreover, if* (G'_1, π_1) *and* (G'_2, π_2) *are two permissible modifications of* G

with center Y for π, there exists an isomorphism of functors $\boldsymbol{\omega} : G'_1 \to G'_2$ such that $\pi_2\boldsymbol{\omega} = \pi_1$.

Proof Since Y satisfies (1) for every $g \in \mathrm{ob}([G])$, $(\pi(g))^0 : (G'(g))^0 \to (G(g))^0$ is the blow-up of $(G(g))^0$ with center $Y(g)$. Therefore, the only problem is to construct $(G'(g))_0$ so that condition (2) holds.

Let $(U_\alpha, \mathscr{E}_\alpha)_{\alpha \in \Lambda}$ be the set of all the pairs that consist of an open set U_α of $(G(g))^0$ and an idealistic exponent $\mathscr{E}_\alpha$ on U_α such that $\mathbf{J}(g)|_{U_\alpha}$ is associated with $\mathscr{E}_\alpha$. Clearly $(G(g))_0 \subset \bigcup_{i=1} U_\alpha$ since, by assumption, $\mathscr{G}$ is an idealistic garden. If $\mathscr{E}'_\alpha$ is the transform of $\mathscr{E}_\alpha$ by $(\pi(g))^0|_U$, there exists a unique complex subspace of $(G'(g))^0$, denoted by $(G'(g))_0$, such that

$$(G'(g))_0 \cap ((\pi(g))^0)^{-1}(U_\alpha) = \mathrm{Sing}\,\mathscr{E}'_\alpha,$$

for all $\alpha \in \Lambda$.

We construct a modification (G', π) of G with center Y with the following data:

(1) The morphism $(\pi(g))_0 : (G'(g))_0 \to (G(g))_0$ is induced by $(\pi(g))^0$.
(2) The morphism $G'(g)$ is the inclusion $(G'(g))_0 \to (G'(g))^0$.
(3) For every $\alpha \in \mathrm{Hom}_{[G]}(g_1, g_2)$, $(G'(\alpha))^0$ is the unique morphism

$$(G'(\alpha))^0 : (G'(g_1))^0 \to (G'(g_2))^0$$

such that the diagram

$$\begin{array}{ccc} (G'(g_1))^0 & \longrightarrow & (G'(g_2))^0 \\ {\scriptstyle (\pi(g_1))^0}\downarrow & & \downarrow{\scriptstyle (\pi(g_2))^0} \\ (G(g_1))^0 & \longrightarrow & (G(g_2))^0 \end{array}$$

is commutative (by the definition of Y as a complex subspace of $(G)_0$).
(4) $(G'(\alpha))_0$ is the induced morphism.

This modification satisfies condition (2) of the proposition. In fact, for every pair $(U_\alpha, \mathscr{E}_\alpha)$, the tree $(\pi(g), Y(g))|_{U_\alpha} = (\pi(g)|_{U_\alpha}, U_\alpha, Y(g)|_{U_\alpha})$ is associated with $\mathscr{E}_\alpha$ for every $\alpha \in \Lambda$, and belongs to $J(g)|_{U_\alpha}$.

The uniqueness of the modification (G', π), up to isomorphisms, is an immediate consequence of the above construction, and can also be deduced from condition (2). □

Remark 4.5.6 Let $(G, \mathbf{J})$ be a forest, Y a smooth closed complex subspace of G, and (G', π) a modification of G with center Y. By the definition of a modification,

$$\pi(g) : G'(g) \to G(g)$$

is a vertical morphism for every $g \in \mathrm{ob}([G])$; therefore it makes sense to speak, for every $g \in \mathrm{ob}([G])$, of the $G'(g)$-polygrove transform of the $G(g)$-polygrove $J(g)$ by $\pi(g)$. Thus we define the transform of the forest $(G, \mathbf{J})$ by the modification or blow-up (G', π) with center Y as the forest $(G', \mathbf{J}')$, where $\mathbf{J}'(g)$ is the transform of $\mathbf{J}(g)$ by the vertical morphism $\pi(g)$. That (G', J') is a forest can be seen from the fact that equivalence of polygroves is stable by modifications.

In this case, if $\boldsymbol{\Delta} = \{\Delta_\lambda \mid \lambda \in \Lambda\}$ is an allée in G, we can construct, for every $\lambda \in \Lambda$ and every $g \in \mathrm{ob}([G])$, the strict transform $\Delta'_\lambda(g)$ of $\Delta_\lambda(g)$ by $(\pi(g))^0 : (G'(g))^0 \to (G(g))^0$. This transform is always a smooth closed hypersurface; moreover, for every $g \in \mathrm{ob}([G])$, the hypersurfaces $\{\Delta'_\lambda(g) \mid \lambda \in \Lambda\}$ have normal crossings, and they form a locally finite system.

Denote by $\Delta'_\alpha(g)$ the smooth hypersurface

$$\left((\pi(g))^0\right)^{-1}(Y(g))$$

of $(G'(g))^0$. Clearly, $\Delta'(g) = \{\Delta'_\lambda(g) \mid \lambda \in \Lambda\} \cup \Delta'_\alpha(g)$ is a locally finite system with normal crossings, for every $g \in \mathrm{ob}([G])$. Therefore, the family $\boldsymbol{\Delta}' = \{\Delta'_\lambda\}_{\lambda\in\Lambda} \cup \{\Delta'_\alpha\}$ is an allée in G', which will be called the **transform of** $\boldsymbol{\Delta}$ by (G', π).

Definition 4.5.4 Let $\mathscr{G} = (G, \mathbf{J}, \boldsymbol{\Delta})$ be a garden, and (G', π) a modification of $\mathscr{G}$ with a center Y permissible for $\mathscr{G}$. The **transform of** $\mathscr{G}$ by the modification (G', π) with respect to the center Y is the garden $\mathscr{G}' = (G', \mathbf{J}', \boldsymbol{\Delta}')$, where $(G', \mathbf{J}')$ and $\boldsymbol{\Delta}'$ are the transforms of $(G, \mathbf{J})$ and $\boldsymbol{\Delta}$ by the blow-up (G', π) with center Y.

Remark 4.5.7 If $\mathscr{G} = (G, \mathbf{J}, \boldsymbol{\Delta})$ is an idealistic garden and Y is a permissible center for $\mathscr{G}$, the transform $\mathscr{G}' = (G', \mathbf{J}', \boldsymbol{\Delta}')$ of $\mathscr{G}$ by the blow-up with center Y is also idealistic.

In fact, for every pair (U_α, E_α) as in the proof of Proposition 4.5.2, the construction used there shows that

$$\mathbf{J}'(g)|_{(\pi(g)^0)^{-1}}(U_\alpha)$$

is associated with the transform $\mathscr{E}'_\alpha$ of $\mathscr{E}$ by $((\pi(g))^0|_U$. Moreover, since the family $\{U_\alpha\}$ is a covering of $(G(g))_0$, the forest $(G', \mathbf{J}')$ is idealistic, hence so is the garden $\mathscr{G}'$.

Remark 4.5.8 Let X be a complex-analytic space, S a Samuel stratum of X, and $\mathscr{G} = (G, \mathbf{J}, \boldsymbol{\Delta})$ a garden associated with (X, S). Let Y be a permissible center for $\mathscr{G}$ and let Y_0 be the unique smooth complex subspace of S such that $Y_0 \cap (G(g))_0 = Y(g)$ for every $g \in \mathrm{ob}([G])$. Assume that Y_0 is closed in X. Denote by $p_0 : X' \to X$ the blow-up of X with center Y_0, and let S' be the Samuel stratum of X' corresponding to the Samuel function of X along S.

Since $(G, \mathbf{J})$ is an X-forest of S, there exists a $\mathbb{C}$-situation G_X of X that induces G on S and such that, for every $g \in \mathrm{ob}([G]) = \mathrm{ob}([G_X])$, $\mathbf{J}(g)$ is a $G(g)$-polygrove associated with $(G_X(g), X, S)$. Then we can construct a modification $(G'_X, \mathbf{p})$ of

G_X with center Y such that $(G'_X(g))_0$ is the strict transform of $(G_X(g))_0$ by $(\mathbf{p}(g))^0$ with center $Y(g)$, for every $g \in \mathrm{ob}([G])$.

Denote by G' the $\mathbb{C}$-situation of S' induced by G'_X and by $\pi : G' \to G$ the morphism of functors induced by $\mathbf{p}$. Then:

(1) (G', π) is a modification of G with center Y. This fact is trivial by construction.
(2) (G', π) is a permissible modification for $\mathcal{G}$. It is enough to observe that, by the construction of π, the tree $(\pi(g), Y(g))$ is associated with $(G(g), X, S)$ for every $g \in \mathrm{ob}([G])$.
(3) The transform $\mathcal{G}' = (G', \mathbf{J}', \boldsymbol{\Delta}')$ of $\mathcal{G}$ by (G', π) is a garden associated with (X', S'). In other words, the forest $(G', \mathbf{J}')$ is associated with (X', S'). Since G' is induced by G'_X it is enough to show that G'_X is a $\mathbb{C}$-situation of X' and that, for every $g \in \mathrm{ob}([G]) = \mathrm{ob}([G'_X])$, $\mathbf{J}'(g)$ is a $G'(g)$-polygrove associated with $(G'_X(g), X', S')$. The first point is a consequence of the construction of X' and the fact that G_X is a $\mathbb{C}$-situation of X; the second is a consequence of the definition of $\mathbf{J}'(g)$.

Chapter 5
The Induction Process

5.1 The Statement of the Induction Process

Definition 5.1.1 A $\mathbb{C}$-situation G is said to be **countable at infinity** if there exists a sequence $\{F_k\}_{k\geq 1}$ of compact subsets of G^0 (see Definition 4.4.6 in Chap. 4) such that, for every point (g, x) of G, there is a point (g', x') of G and an integer $k \geq 1$ such that $(g, x) \sim (g', x')$ and $(g', x') \in F_k$, that is, $x' \in F_k(g')$. A garden $\mathscr{G} = (G, J, \Delta)$ is **countable at infinity** if G is also.

Remark 5.1.1 Let X be a complex-analytic space and S a Samuel stratum of X. Then, if S is countable at infinity, every X-situation G of S is countable at infinity, but the converse is not true, in general.

In fact, if S is countable at infinity, there exists a sequence $\{C_k\}_{k\geq 1}$ of compact subsets of S such that $C_k \subset C_{k+1}$, for every $k \geq 1$, and $S = \bigcup_{k=1}^{\infty} C_k$. Then, if G is a $\mathbb{C}$-situation of (X, S), there exists a $\mathbb{C}$-situation G_X of X that induces G on S. By condition (2) in the definition of a $\mathbb{C}$-situation of X (Definition 4.1.3 in Chap. refchap4), $\{(G_X(g))_0\}_{g\in \mathrm{ob}([G])}$ is an open covering of X, and hence $\{(G(g))_0\}_{g\in \mathrm{ob}([G])}$ is an open covering of S. Then, with every compact set C_k, we can associate a finite family $\{g_{k,i}\}_{1\leq i\leq r_k}$ of objects in $[G]$ in such a way that

$$C_k \subset \bigcup_{i=1}^{r_k} (G(g_{k,i}))_0,$$

as subsets of S, and we can choose a compact subset F_k of $(G)^0$, for each $k \geq 1$, satisfying these conditions:

(1) For $1 \leq j \leq r_k$, $F_k(g) = \varnothing$ if $g \neq g_{k,j}$.
(2) For $1 \leq j \leq r_k$, $F_k(g_{k,j})$ is the closure of $C_k \cap (G(g_{k,j}))_0$ in $(G(g_{k,j}))^0$.

J. M. Aroca et al., *Complex Analytic Desingularization*,
https://doi.org/10.1007/978-4-431-49822-3_5

One sees easily that the family $\{F_k\}_{k\geq 1}$ satisfies the conditions in Definition 5.1.1.

It is easy to check that, if a garden $\mathscr{G}$ is countable at infinity, so is its completion $\hat{\mathscr{G}}$ in the sense of Remark 4.5.3 in Chap. 4. In fact, we can construct a family of compact subsets $\hat{F}_k \subset \hat{G}^0$, for $k \geq 1$, satisfying the conditions in Definition 5.1.1, as follows:

$$\hat{F}_k(g,\, 0,\, (G(g))^0 \times \{0\},\, (G(g))^0 \times \{0\}) = F_k(g)$$

and, in all other cases,

$$\hat{F}_k(g, m, U, W) = \varnothing.$$

Proposition 5.1.1 *Let X be a complex-analytic space, S a Samuel stratum of X, $\mathscr{G} = (G, \mathbf{J}, \boldsymbol{\Delta})$ a garden associated with (X, S), and $\hat{\mathscr{G}}$ the completion of $\mathscr{G}$. Then*

$$\dim(\hat{\mathscr{G}}) = \max_{x\in S}\{\dim T_{X,x}\} \leq \dim X,$$

where $T_{X,x}$ denotes the Zariski tangent space to X at x (see Remark 2.5.2 *in Chap.* 2*).*

Proof The garden $\hat{\mathscr{G}}$ is idealistic. In fact, since G is a $\mathbb{C}$-situation of (X, S), there exists a $\mathbb{C}$-situation G_X of X inducing G on S. Therefore G_X induces G on S, and G_X is X-full. By Theorem 4.4.1 of Chap. 4, G_X is idealistic, and therefore so is $\hat{\mathscr{G}}$.

The proposition will be proved if we show that, if $\mathscr{G} = (G, \mathbf{J}, \boldsymbol{\Delta})$ is a garden such that G is a X-full $\mathbb{C}$-situation associated with (X, Y),

$$\dim(\hat{\mathscr{G}}) = \max_{x\in S}\{\dim T_{X,x}\} \leq \dim X.$$

Let (g, x) be a point in G. Since $\hat{\mathscr{G}}$ is idealistic, there exist an open neighborhood U of $x \in (G(g))^0$ and an idealistic exponent $\mathscr{E}$ on U such that $J(g)|_U$ is a polygrove associated with $\mathscr{E}$. This $\mathscr{E}$ is constructed, as in the proof of Theorem 4.4.1 of Chap. 4, in the following way.

Take a $\mathbb{C}$-situation G_X of X inducing G on S and such that $\mathbf{J}(g)$ is a $G(g)$-polygrove $G_X(g)$-associated with (X, S), for every $g \in \mathrm{ob}([G])$. We may assume that, at every point (g, x) of G, the strict tangent space $T_{X,x}$ is a vector subspace of $T_{(G(g))^0,\,x}$. In this case there exists a smooth closed complex subspace W of $(G(g))^0$, around x, such that $\dim_x(W) = \dim T_{X,x}$ and such that there is a projection $\mathbf{r}$ around x from $(G(g))^0$ to W that is perpendicular to X at x. By the theorems of existence, continuity and stability of the maximal contact, there exists an idealistic exponent $\mathscr{K}$ on a neighborhood of x in W such that $T_{\mathscr{K},x} = T_{W,x} = T_{X,x}$. By Theorem 4.3.1 of Chap. 4, the idealistic exponent $\mathscr{E}$ is an ambient expansion of $\mathscr{K}$ and $\dim T_{\mathscr{E},x} = \dim T_{\mathscr{K},x} = \dim T_{X,x}$. Then, by the definition of dimension,

$$\dim_{(g,x)} G = \dim T_{\mathscr{E},x} = \dim T_{X,x}$$

and

$$\dim \hat{\mathscr{G}} = \sup_{x \in S} T_{X,x}.$$

Moreover, since $T_{X,x} \subset C_{X,x}$ and $\dim_x(X) = \dim C_{X,x}$, one has

$$\sup_{x \in S} \dim T_{X,x} \leq \dim X.$$

This proves the proposition. □

Definition 5.1.2 Let $H : [H] \to ((\mathbb{C}\text{-spaces}))$ be a functor. We say that a complex subspace Y of H is **restrictive** if, for every $\beta \in \mathrm{Hom}_{[H]}(h', h)$, $Y(\beta) : Y(h') \to Y(h)$ is a locally isomorphic locally closed embedding. If Y is a restrictive subspace of H, the **dimension** of Y is the (possibly infinite) maximum of the dimensions of $Y(h)$, for $h \in \mathrm{ob}([H])$:

$$\dim Y = \max_{h \in \mathrm{ob}([H])} \{\dim Y(h)\}.$$

Remark 5.1.2 If G is a $\mathbb{C}$-situation, every complex subspace of $(G)_0$ is a restrictive subspace of $(G)^0$, as one can easily check from the definitions.

Definition 5.1.3 We say that a $\mathbb{C}$-situation G is **homodimensional** if, for every integer $n \geq \dim G$ and for every point (g, x) of G, there exists a point (g_1, x) of G equivalent to (g, x) (in the sense of Definition 4.4.1.2 in Chap. 4) and such that

$$\dim_{(g_1,x_1)} G = n.$$

If $\mathscr{G}$ is a garden over the $\mathbb{C}$-situation G, we say that $\mathscr{G}$ is homodimensional if G is.

Remark 5.1.3 By construction, the completion of a homodimensional garden is also homodimensional.

Definition 5.1.4 A **resolution datum** of type $\mathbb{I}^n$ is a family

$$(\mathscr{G}; X_1, \ldots, X_q)^n_{\mathbb{I}},$$

with $q \geq 1$, where:

(1) the garden $\mathscr{G} = (G, \mathbf{J}, \boldsymbol{\Delta})$ is an idealistic garden that is countable at infinity, homodimensional, and such that $\dim \mathscr{G} \leq n$, and
(2) the family $(X_1, \ldots, X_q)$ is a finite system of closed, restrictive complex subspaces of $(G)^0$ such that $\dim X_i < n$ for $i = 1, \ldots, q$.

A resolution datum of type $\mathbb{III}^n$ is a family $(\mathscr{G}; L, M)^n_{\mathbb{III}}$, where:

(1) the garden $\mathscr{G} = (G, \mathbf{J}, \boldsymbol{\Delta})$ is an idealistic garden that is countable at infinity, homodimensional, and such that $\dim \mathscr{G} \leq n$, and

(2) (L, M) is an ordered pair of closed complex subspaces of $(G)_0$, such that $(G)_0 = L \cup M$ (that is, such that $(G)_0(g) = L(g) \cup M(g)$ for every $g \in \mathrm{ob}([G])$).

In each of these definitions, we say that G is the **underlying $\mathbb{C}$-situation** of the resolution datum. A resolution datum **over** a $\mathbb{C}$-situation G means one of which G is its underlying $\mathbb{C}$-situation.

Definition 5.1.5 Let G be a $\mathbb{C}$-situation and Y a restrictive complex subspace of $(G)^0$. A **point of** Y is a point (g, x) of G such that $x \in Y(g)$. If (g, x) is a point of Y, the function $H_{Y(g),x}$ is called the **Hilbert–Samuel function** of Y at (g, x) (see Definition 2.4.3 in Chap. 2).

Given a function $H : \mathbb{Z}_0 \to \mathbb{Z}_0$, we can construct a reduced, restrictive, complex subspace $S(H)$ of $(G)^0$ (we also write simply S when the reference to H is clear), in the following way:

(1) For every $g \in \mathrm{ob}([G])$, $S(g) = \{y \in Y(g) \mid H_{Y(g),y} = H\}$.
(2) For every $\alpha \in \mathrm{Hom}_{[G]}(g', g)$, $S(\alpha) = Y(\alpha)|_{S(g')}$.

We call $S(H)$ the **Samuel stratum of** Y corresponding to the function H.

Definition 5.1.6 Consider a resolution datum R over a $\mathbb{C}$-situation G, a smooth closed complex subspace Y of $(G)_0$, and a modification (G', π) with center Y. We say that (G', π) (or simply π, if the reference to G' is clear) is **permissible** for R if the following conditions hold:

(1) The pair (G', π) is permissible for the garden $\mathcal{G}$ of R (see Definition 4.5.3 in Chap. 4).
(2) If R is a resolution datum of type $\mathbb{I}^n$, then Y is contained in a single Samuel stratum of X_i, for every i. If R is a resolution datum of type $\mathbb{III}^n$, then Y is contained in L.

Remark 5.1.4 If (G', π) is a modification of G with center Y and F is a complex subspace of $(G)^0$, we can define a complex subspace F' of $(G')^0$ in the following way:

(i) For every $g \in \mathrm{ob}([G])$, set $F'(g) = ((\pi(g))^0)^{-1}(F(g))$.
(ii) For every $\alpha \in \mathrm{Hom}_{[G]}(g, g')$, $F'(\alpha) : F'(g) \to F'(g')$ is the morphism induced by $(G'(\alpha))^0$.

We call F' the **total transform** of F by $(\pi)^0$.

In a similar way, we can construct another complex subspace F^* of $(G)_0$, as follows:

(i) For every $g \in \mathrm{ob}([G])$, set $F^*(g) = F'(g) \cap (G'(g))_0$.
(ii) For every $\alpha \in \mathrm{Hom}_{[G]}(g, g')$, set $F^*(\alpha) = F'(\alpha)|_{F^*(g)}$.

We call F^* the **total transform** of F by $(\pi)_0$. If F is restrictive, so is F^*, by the construction of G' (Remark 4.5.5 of Chap. 4). But F' is not necessarily restrictive when F is, for the reason that morphisms in $(G)^0$ needed not be locally isomorphic.

The distinction between both total transform, one by $(\pi)^0$ and other by $(\pi)_0$, is very important in the later propositions.

We say that a complex subspace F'' of $(G')^0$ is the **strict transform** of F by $(\pi)^0$ if these conditions hold:

(i) $F''(g)$ is the strict transform of $F(g)$ by $(\pi(g))^0$, with center $Y(g)$, for every $g \in \mathrm{ob}([G])$.
(ii) If $\alpha \in \mathrm{Hom}_{[G]}(g, g')$, then $F''(\alpha)$ is induced by $(G'(\alpha))^0$. Moreover, $F''(\alpha)$ is the unique morphism $F''(g) \to F''(g')$ such that

$$(\pi(g'))^0 F''(\alpha) = F(\alpha)(\pi(g))^0.$$

Such an F'' does not always exist as a complex subspace of $(G')^0$. In fact the condition (2) of Definition 4.4.6.2 (Chap. 4) may not hold in general. It exists, for instance, when F is restrictive, in which case it follows trivially from the definition that F'' is also restrictive.

Definition 5.1.7 Let R be a resolution datum and (G', π) a modification with center Y permissible for R. The **transform** of R by this modification is the resolution datum R' obtained in the following way:

(i) The garden $\mathscr{G}'$ of R' is the transform of the garden $\mathscr{G}$ of R by (G', π).
(ii) If $R = (\mathscr{G}; X_1, \dots, X_q)^n_{\mathrm{I}}$, then $R' = (\mathscr{G}'; X'_1, \dots, X'_q)^n_{\mathrm{I}}$, where X'_i is the strict transform of X_i by $(\pi)^0$, for $i = 1, \dots, q$.
(iii) If $R = (\mathscr{G}; L, M)^n_{\mathrm{III}}$, then $R' = (\mathscr{G}'; L', M')^n_{\mathrm{III}}$, where L' and M' are the total transforms of L and M by $(\pi)_0$ (not the ones by $(\pi)^0$).

Now consider the following setup. Let Λ be a well-ordered set. As usual, we denote its first element by 0, and the successor of $\beta \in \Lambda$ by $\beta + 1$; moreover, if β has a predecessor α, we write $\alpha = \beta - 1$.

For every $\alpha \in \Lambda$, let there be given a $\mathbb{C}$-situation G_α with $[G_\alpha] = [G]$, where $[G]$ is a fixed category and Y_α is a smooth closed complex subspace of $(G_\alpha)_0$.

For every $\alpha, \beta \in \Lambda$ with $\alpha > \beta$, let there be given a morphism $\pi_{\alpha,\beta} : G_\alpha \to G_\beta$, considered as an ((el))-valued functor in such a way that:

(a) for every $\alpha, \beta, \gamma \in \Lambda$ with $\alpha > \beta > \gamma$, we have $\pi_{\alpha\gamma} = \pi_{\beta\gamma}\pi_{\alpha\beta}$, and
(b) for every $\beta \in \Lambda$, $(G_{\beta+1}, \pi_{\beta+1,\beta})$ is a modification of G_β with center Y_β.

Assume that the $\{\pi_{\alpha\beta}\}_{\alpha,\beta\in\Lambda}$ are a locally finite family, that is, for every compact subset F of $(G_0)_0$, there exists a finite subset $\Lambda(F) \subset \Lambda$ such that, for every $\beta \in \Lambda - \Lambda(F)$ and every $g \in \mathrm{ob}([G])$,

$$Y_\beta(g) \cap ((\pi_{\beta,0}(g))^0)^{-1}(F(g)) = \varnothing.$$

Finally, assume that, for every $\beta \in \Lambda$ that does not have predecessor, the family $(\pi_{\beta\alpha})_{\alpha<\beta}$ defines an isomorphism of ((el))-valued functors

5.1.7.1 $$\pi_\beta : G_\beta \to \lim_{\longleftarrow}\{\pi_{\alpha,\lambda} : G_\alpha \to G_\lambda\}_{\beta>\alpha>\lambda}.$$

Under the above conditions of local finiteness we say that the set of data $\bar{\pi}$ consisting of the families $\{\pi_{\alpha,\beta}\}$, $\{G_\alpha\}$, and $\{Y_\alpha\}$ is a **proper succession of modifications of $\mathbb{C}$-situations**.

Remark 5.1.5 Since projective limits exist in the category ((el)), it makes sense to speak about the projective limit of the family of $\mathbb{C}$-situations $\{G_\alpha\}_{\alpha<\beta}$ with the $\pi_{\alpha\lambda}$ as morphisms.

By definition of a projective limit, the condition of local finiteness in the definition of a proper succession (and hence the existence of the isomorphisms of 5.1.7.1 in Definition 5.1.7) can be replaced by the following condition.

For every $\beta \in \Lambda$ that does not have a predecessor, every $\lambda \in \Lambda$ with $\beta > \lambda$, and every $g \in \mathrm{ob}([G])$, $(\boldsymbol{\pi}_{\beta\lambda}(g))^0$ and $(\boldsymbol{\pi}_{\beta\lambda}(g))_0$ is a proper morphism and, for every relatively compact open subset U of $(G_\beta(g))^0$, there exist $\alpha \in \Lambda$ satisfying $\alpha < \beta$ and an open subset V_α of $(G_\alpha(g))^0$ such that

$$U = ((\pi_{\beta\alpha}(g))^0)^{-1}(V_\alpha)$$

and $\boldsymbol{\pi}_{\beta\alpha}$ induces an isomorphism $G_\beta(g)|_U \to G_\alpha(g)|_{V_\alpha}$.

Definition 5.1.8 Let $\bar{\boldsymbol{\pi}} = \{\boldsymbol{\pi}_{\alpha\beta}, G_\alpha, Y_\alpha\}_{\alpha,\beta\in\Lambda}$ be a proper succession of modifications of $\mathbb{C}$-situations and let R_0 be a resolution datum over G_0. We say that $\bar{\boldsymbol{\pi}}$ is **permissible** for the resolution datum R_0, or that $\bar{\boldsymbol{\pi}}$ is a proper succession of permissible modifications for R_0, if, for every $\beta \in \Lambda$, there exists a resolution datum R_β with garden $\mathcal{G}_\beta = (G_\beta, \mathbf{J}_\beta, \boldsymbol{\Delta}_\beta)$ satisfying these conditions:

(i) For every $\beta \in \Lambda$, the modification $(G_{\beta+1}, \boldsymbol{\pi}_{\beta+1,\beta})$ with center Y_β is permissible for R_β and $R_{\beta+1}$ is the transform of R_β by this modification.

(ii) For every $\beta \in \Lambda$ not having a predecessor, every $g \in \mathrm{ob}([G])$, and every relatively compact open subset U of $(G_\beta(g))^0$, there exists $\alpha_0 < \beta$ with the property that, for all $\alpha \in \Lambda$ satisfying $\alpha_0 \le \alpha < \beta$, there exist an open subset V_α of $(G(g))^0$ and an idealistic exponent $\mathcal{E}_\alpha$ on V_α satisfying the following conditions:

  (a) The morphism $(\boldsymbol{\pi}_{\beta\alpha}(g))^0$ induces an isomorphism $U \to V_\alpha$ and $\mathbf{J}_\beta(g)|_U$ is associated with the pull-back of $\mathcal{E}_\alpha$ by $(\boldsymbol{\pi}_{\beta\alpha}(g))^0$.

  (b) $\mathbf{J}_\alpha(g)|_{V_\alpha}$ is associated with $\mathcal{E}_\alpha$.

(iii) For every $\beta \in \Lambda$ not having a predecessor, the ingredients of R_β other than the polygroves are the projective limits of the corresponding ingredients of R_λ, for all $\lambda < R$, by means of $(\boldsymbol{\pi}_{\alpha,\lambda})^0$ for all $\lambda < \alpha < \beta$. (See next remark.)

Remark 5.1.6

5.1.6.1 Since $\boldsymbol{\pi}$ is a proper succession of modifications, for every $\beta \in \Lambda$ not having a predecessor we have an isomorphism

$$G_\beta \to \varprojlim\{\boldsymbol{\pi}_{\alpha,\lambda} : G_\alpha \to G_\lambda\}_{\beta>\alpha>\lambda}.$$

By the condition in Remark 5.1.5, we may assume that α_0 and V_α in condition (ii) of Definition 5.1.8 have the following property: for every $g \in \mathrm{ob}([G])$ and every $\alpha \in \Lambda$ with $\alpha_0 \leq \alpha < \beta$, we have an isomorphism

$$G_\beta(g)|_U \to G_\alpha(g)|_{V_\alpha}$$

and $U = ((\boldsymbol{\pi}_{\beta,\alpha}(g))^0)^{-1}(V_\alpha)$.

5.1.6.2 We elaborate a bit on condition (iii) of Definition 5.1.8, regarding allées. Let $\boldsymbol{\Delta}_\lambda = \{\Delta_{\lambda,\mu}\}_{\mu \in \Gamma_\lambda}$ be the allée corresponding to the resolution datum R_λ. By condition (i) in Definition 5.1.8, $R_{\lambda+1}$ is the transform of R_λ by a modification with center Y_λ. Therefore, $\Gamma_{\lambda+1} = \Gamma_\lambda \cup \{\sigma_{\lambda+1}\}$ and there is an injective map $\gamma_{\lambda+1} : \Gamma_\lambda \to \Gamma_{\lambda+1}$ (the natural inclusion). In addition, for every $t \in \Gamma_\lambda$ and every $g \in \mathrm{ob}([G])$, $\Delta_{\lambda+1,\gamma_{\lambda+1}(t)}(g)$ is the strict transform of $\Delta_{\lambda,t}(g)$ by $(\boldsymbol{\pi}_{\lambda+1,\lambda}(g))^0$.

Denote by $\delta_{\lambda,t}(g)$ the blow-up of $\Delta_{\lambda,t}(g)$ with center $Y_\lambda(g)$. Then condition (iii) of Definition 5.1.8 means the following:

(a) For every $\alpha, \lambda \in \Lambda$ with $\alpha > \lambda$, there is an injective map $\gamma_{\alpha\lambda} : \Gamma_\lambda \to \Gamma_\alpha$ satisfying these conditions:

(1) For every $\alpha \in \Lambda$, the map $\gamma_{\alpha+1,\alpha}$ is the natural inclusion, as above.
(2) For every $\alpha > \mu > \lambda$, we have $\gamma_{\alpha\mu}\gamma_{\mu\lambda} = \gamma_{\alpha\lambda}$.
(3) For every $\beta \in \Lambda$ not having a predecessor, and setting

$$\Gamma_\beta = \varinjlim\{\gamma_{\alpha\lambda} : \Gamma_\lambda \to \Gamma_\alpha \mid \lambda < \alpha < \beta\},$$

the map $\gamma_{\beta\lambda} : \Gamma_\lambda \to \Gamma_\beta$ is the canonical map of the inductive limit, for every $\lambda < \beta$.

(b) For every $\beta \in \Lambda$ not having a predecessor, every $r \in \Gamma_\beta$, and every $g \in \mathrm{ob}([G])$,

$$\Delta_{\beta,r}(g) = \varprojlim_{\beta>\lambda,\, \gamma_{\beta\lambda}(t)=r} \delta_{\lambda,t}(g)^{-1}(\Delta_{\lambda,t}(g)).$$

5.1.6.3 Condition (iii) in Definition 5.1.8, regarding the subspaces of the resolution data, means that, for every $\beta \in \Lambda$ not having a predecessor, we have:

(a) If the resolution data are of type $\mathbb{I}^n$, then R_β is of the form

$$R_\beta = (\mathscr{G}_\beta; X_{\beta,1}, \ldots, X_{\beta,q})^n_{\mathbb{I}},$$

where q and n are independent of β. For every $\alpha \in \Lambda$ and $g \in \mathrm{ob}([G])$, denote by $h_{\alpha,i} : X_{\alpha+1,i}(g) \to X_{\alpha,i}(g)$ the blow-up with center $Y_\alpha(g)$. Then, for $1 \leq i \leq q$, we have

$$X_{\beta i}(g) = \varprojlim_{\beta>\lambda} h^{-1}_{\alpha,i}(X_{\alpha,i}(g)).$$

(b) If the resolution data are of type III^n, then R_β is of the form

$$R_\beta = (\mathcal{G}_\beta;\, L_\beta, M_\beta)^n_{\mathrm{III}}$$

and L_β, M_β are constructed in a similar way to the $X_{\beta i}$ in the case (a), but using this time total transforms and restrictions to the singular part of the model, instead of the strict transforms.

Remark 5.1.7 In Definition 5.1.8, for every $\beta \in \Lambda$, the components of R_β are explicitly constructed from those of R_α with $\alpha < \beta$, with the exception of $\mathbf{J}_\beta$, for which we have the existence and uniqueness only up to equivalences. But one can give a constructive process to obtain a $\mathbf{J}_\beta$, for every $\beta \in \Lambda$, which satisfies condition (ii) of Definition 5.1.8. This process is as follows.

Let $g \in \mathrm{ob}([G_0])$ and $V \in \mathrm{ob}(\mathcal{V})$, and take $\beta > 0$ having no predecessor. For every $\alpha < \beta$, denote by $N_\alpha(g)$ the subset of $|\mathbf{J}_\alpha(g)_V|$ consisting of all the b such that, if $\mathbf{J}_\alpha(g)_V(b) = \{\mathbf{a}_i, U_{i-1}, E_{i-1}\}_{1\le i\le m}$, one has

$$U_0 \cap \big((\boldsymbol{\pi}_{\lambda\alpha}(g))^0(Y_\lambda(g)) \times V\big) = \varnothing$$

for every λ satisfying $\alpha \le \lambda < \beta$. Hence, if $b \in N_\alpha(g)$, there exists a unique restrictive morphism $\mathbf{f}_{b,\alpha} : G_\alpha(g)|_{U_0} \to G_\beta(g)$ such that $\boldsymbol{\pi}_{\beta'\alpha}\mathbf{f}_{b,\alpha}$ is the inclusion $G_\alpha(g)|_{U_0} \to G_\alpha(g)$.

Then, we construct $\mathbf{J}_\beta$ as follows:

(a) Set $|\mathbf{J}_\beta(g)_V| = \bigcup_{\alpha<\beta} N_\alpha(g)$. This is a disjoint union.
(b) For every $b \in N_\alpha(g)$, set $\mathbf{J}_\beta(g)_V(b) = \{\mathbf{a}'_i, U'_{i-1}, E'_{i-1}\}_{1\le i\le m}$, where $\mathbf{a}'_i = \mathbf{a}_i$, $U'_{i-1} = U_{i-1}$, and $E'_{i-1} = E_{i-1}$, for $2 \le i \le m$, and

$$\mathbf{a}'_1 = (\mathbf{f}_{b,\alpha} \times \mathbf{1}_V)\boldsymbol{\alpha}_1 : \mathbf{s}_1 \to G_\beta(g) \times V,$$

where $\boldsymbol{\alpha}_1$ is the blow-up of U_0 with center E_0 (that is, the same as $\mathbf{a}_1$ except for the inclusion of U_0 in $(s_0)^0$).

With this construction, it can be immediately checked that $\mathbf{J}_\beta$ satisfies condition (ii) in Definition 5.1.8, so we have a constructive process for R_β, not just an existence argument.

Remark 5.1.8 If $\bar{\boldsymbol{\pi}}$ is a proper succession of modifications of $\mathbb{C}$-situations starting from G_0, one can associate with it a transformation $(G', \boldsymbol{\pi})$ of $G = G_0$, which will be called the limit of the succession $\bar{\boldsymbol{\pi}}$. Namely, G' is the projective limit of the system $\bar{\boldsymbol{\pi}}$. By a similar construction to that in Remarks 5.1.6 and 5.1.7, one can speak about the transform of a resolution datum R_0 by a proper succession $\bar{\boldsymbol{\pi}}$ of permissible modifications.

If G_0 is countable at infinity, the Y_β are necessarily empty (and hence the $\boldsymbol{\pi}_{\beta+1,\beta}$ are isomorphisms), except for countable many indices β. Therefore, by suppressing isomorphisms arising from empty Y_β, every proper succession of permissible modifications, with G_0 countable at infinity, can be equivalently reduced to a countable succession (but not necessarily to a sequence indexed by positive integers).

Let R_0 be a resolution datum over a $\mathbb{C}$-situation G. We can construct the completion $\hat{G}$ of G and extend, in a natural way, the other components of R_0 to $\hat{G}$. In this way, we get a new resolution datum, of the same type as R_0, which will be denoted by $\hat{R}_0$ and called the **completion** of R_0. If $\bar{\pi}$ is a proper succession of strictly permissible modifications for R_0, then $\bar{\pi}$ extends, in a natural way, to a proper succession $\bar{\pi}$ of strictly permissible modifications for $\hat{R}_0$.

Theorem 5.1.1 (Fundamental Theorem $\mathbb{I}^n$) *Given a resolution datum $R = (\mathcal{G}; X_1, \ldots, X_q)^n_{\mathbb{I}}$ over a $\mathbb{C}$-situation G, there exists a proper succession $\bar{\pi}$ of permissible modifications for R such that, if we denote by $R' = (\mathcal{G}'; X'_1, \ldots, X'_q)^n_{\mathbb{I}}$ the transform of R by $\bar{\pi}$, then*

$$(G')_0 \cap \left(\bigcap_{i=1}^{q} X'_i \right) = \varnothing,$$

where G' is the $\mathbb{C}$-situation of $\mathcal{G}'$.

Theorem 5.1.2 (Fundamental Theorem $\mathbb{III}^n$) *Given a resolution datum $R = (\mathcal{G}; L, M)^n_{\mathbb{III}}$ over a $\mathbb{C}$-situation G, there exists a proper succession $\bar{\pi}$ of permissible modifications for R such that, if $R' = (\mathcal{G}'; L', M')^n_{\mathbb{III}}$ is the transform of R by $\bar{\pi}$, then $(G')_0 \subset M'$, where G' is the $\mathbb{C}$-situation of $\mathcal{G}'$.*

We prove these theorems in the next two sections, by showing that $\mathbb{III}^{n-1} + \mathbb{I}^n \Rightarrow \mathbb{III}^n$ and that $\mathbb{III}^{n-1} \Rightarrow \mathbb{I}^n$, for every $n > 0$.

5.2 $\mathbb{III}^{n-1} + \mathbb{I}^n \Rightarrow \mathbb{III}^n$

Definition 5.2.1 Let $R = (\mathcal{G}; L, M)^n_{\mathbb{III}}$, where $\mathcal{G} = (G, \mathbf{J}, \boldsymbol{\Delta})$ with $\boldsymbol{\Delta} = \{\Delta_\lambda\}_{\lambda \in \Lambda}$. Denote by $\mathbb{Q}_0$ the set of non-negative rational numbers. A **divisor** of R is a map $e : \Lambda \to \mathbb{Q}_0$ such that, for every point (g, x) of G where $\dim_x(G(g))^0 = n$, there exists an open neighborhood U of x in $(G(g))^0$ and an idealistic exponent $\mathcal{E}$ on U satisfying these properties:

(a) $\mathbf{J}(g)|_U$ is associated with $\mathcal{E}$.
(b) There is a representative $((\mathbf{I}, b))$ of $\mathcal{E}$, and an ideal $\mathbf{I}_0$ in

$$\mathcal{O}_{(G(g))^0}|_U = \mathcal{O}_U$$

satisfying the following conditions, where we set

$$\Lambda(g, U) = \{\lambda \in \Lambda \mid \Delta_\lambda(g) \cap U \neq \varnothing\} :$$

(1) For $\lambda \in \Lambda(g, U)$, $e(\lambda).b$ is an integer.
(2) If $\mathbf{H}_\lambda$ is the ideal in $\mathcal{O}_U$ defining the hypersurface $\Delta_\lambda(g)$, for $\lambda \in \Lambda$, then

$$Is\mathbf{I} = \mathbf{I}_0. \prod_{\lambda \in \Lambda(g,U)} \mathbf{H}_\lambda^{e(\lambda).b}$$

Definition 5.2.2 Let R be a resolution datum as above. The **total exponent** of R at the point (g, x) of G is the number

$$\nu_{(g,x)}(R) = \nu_x^n(J(g)).$$

If (R, e) is a pair consisting of a resolution datum as above and a divisor e of R, the **total exponent** of (R, e) at the point (g, x) of G is the number

$$\nu_{(g,x)}(R, e) = \nu_{(g,x)}(R).$$

The **exponent** of e at (g, x) is the number

$$\nu_{(g,x)}(e) = \sum_{\lambda \in \Lambda(g,x)} e(\lambda),$$

where $\Lambda_{(g,x)} = \{\lambda \in \Lambda \mid x \in \Delta_\lambda(g)\}$. Finally, the **residual exponent** of (R, e) at (g, x) is

$$\omega_{(g,x)}(R, e) = \nu_{(g,x)}(R, e) - \nu_{(g,x)}(e).$$

Definition 5.2.3 Let (R, e) be a pair consisting of a resolution datum R of type $\mathbb{III}^n$ over a $\mathbb{C}$-situation G and a divisor e of R. Let Y be a smooth closed complex subspace of $(G)_0$. We say that a modification $(G', \boldsymbol{\pi})$ of G with center Y is **strictly permissible** for (R, e) if the following conditions hold:

(i) The modification $(G', \boldsymbol{\pi})$ is permissible for R.
(ii) If $\boldsymbol{\Delta} = \{\Delta_\lambda\}_{\lambda \in \Lambda}$ is the allée of the garden of R, then for every $\lambda \in \Lambda$, either we have $Y \subset \Delta_\lambda$ or $Y \cap \Delta_\lambda$ is nowhere dense in Y.
(iii) For almost every point (g, x) of Y, $\omega_{(g,x)}(R, e)$ is constant, that is, there is a closed complex subspace B of Y that is nowhere dense in Y and such that $\omega_{(g,x)}(R, e)$ is constant for every point of $Y \setminus B$.

Remark 5.2.1 We can define the concept of proper succession of modifications strictly permissible for (R, e), and the transform of (R, e) by such a succession, in a way similar to the one used to define the concept of a proper succession of modifications permissible for R and the transform of R by such a succession.

Consider a resolution datum R of type $\mathbb{III}^n$ with garden $\mathcal{G} = (G, J, \Delta)$, a smooth closed complex subspace Y of $(G)_0$, and a modification $(G', \boldsymbol{\pi})$ of G with center Y permissible for R. If $\boldsymbol{\Delta} = \{\Delta_\lambda\}_{\lambda \in \Lambda}$, for every $\lambda \in \Lambda$ we denote by Y_λ the complex subspace of $(G)_0$ such that, for every $g \in \text{ob}([G])$, $Y_\lambda(g)$ is the union of all the connected components of $Y(g)$ contained in $\Delta_\lambda(g)$. We write $Y_0 = Y \setminus \bigcup_{\lambda\in\Lambda} Y_\lambda$, meaning that Y_0 is the subspace such that $Y_0(g)$ is the union of all the connected components of $Y(g)$ that are not contained in any $\Delta_\lambda(g)$. Then we have a family $\{Y_\lambda\}_{\lambda\in\Lambda\cup\{0\}}$ of smooth closed complex subspaces of $(G)_0$, all of which satisfy condition (ii) of Definition 5.2.3.

Let $g \in \mathrm{ob}([G])$ and let F be a connected component of $Y(g)$. Then $\omega_{(g,x)}(R, e)$ is constant for almost all $x \in F$; denote this constant by $\omega(g, F)$. Let Ω be the set of the rational numbers $\omega(g, F)$, for all g and F as above. Set $\Lambda' = (\Lambda \cup \{0\}) \times \Omega$ and, for each $(\lambda, \alpha) \in \Lambda'$, define $Y_{(\lambda,\alpha)}$ by letting $Y_{(\lambda,\alpha)}(g)$ be equal to the union of those connected components F of $Y(g)$ such that $\omega(g, F) = \alpha$.

It is then easy to prove that, for every $\lambda' \in \Lambda'$, $Y_{\lambda'}$ is a smooth closed complex subspace of $(G)_0$ that satisfies conditions (ii) and (iii) of Definition 5.2.3. Moreover, $\{Y_{\lambda'}\}_{\lambda'\in\Lambda'}$ is locally finite in G (in the sense that $\{Y_{\lambda'}(g)\}_{\lambda'\in\Lambda'}$ is locally finite in $(G(g))^0$ for every $g \in \mathrm{ob}([G])$), and Y is a disjoint union of the $Y_{\lambda'}$ with $\lambda' \in \Lambda'$.

Take a well-ordering on Λ', with first element 0. We can then construct a proper succession of modifications $\bar{\boldsymbol{\pi}} = \{\boldsymbol{\pi}_{\alpha,\beta}, G_\alpha, Y'_\alpha\}_{\alpha,\beta\in\Lambda'}$, where $G_0 = G$, G_α is the modification of G with center $\bigcup_{\alpha>\beta} Y_\beta$, and Y'_α is the strict transform of Y_α by such a blow-up (and then $Y'_\alpha \to Y_\alpha$). In this way, the given modification $(G', \boldsymbol{\pi})$ is the limit of the succession $\bar{\boldsymbol{\pi}}$; that is, it factors into a proper succession of strictly permissible modifications for (R, e).

Definition 5.2.4 Let $(G', \boldsymbol{\pi})$ be a strictly permissible modification for (R, e) with center Y. By the **transform** of (R, e) by $(G', \boldsymbol{\pi})$ we mean a pair (R', e'), where:

(i) R' is the transform of R by π', and
(ii) If $\boldsymbol{\Delta} = \{\Delta_\lambda\}_{\lambda\in\Lambda}$ is the allée of the garden of R and

$$\boldsymbol{\Delta}' = \{\Delta'_\lambda\}_{\lambda\in(\Lambda\cup\{0\})}$$

is the allée of the garden of R', then $e'(\lambda) = e(\lambda)$ for $\lambda \in \Lambda$ and

$$e'(0) = \sum_{\Delta_\lambda \supset Y} e(\lambda) + \omega - 1,$$

where ω is the constant value of $\omega_{(g,x)}(R, e)$ for almost every point (g, x) of Y.

Remark 5.2.2 If ω is as in Definition 5.2.4, and if (g, x) is a point in Y such that $\omega_{(g,x)}(R, e) = \omega$, then, for every $x' \in \big((\boldsymbol{\pi}(g))_0\big)^{-1}(x)$, one has $\omega_{(g,x')}(R', e') \leq \omega_{(g,x)}(R, e)$. But this inequality is not valid in general for other points (g, x) in Y. For a point $(g, y) \in Y$ with $\omega_{(g,y)}(R, e) > \omega$ a corresponding $\omega_{(g,y')}(R', e')$ can be greater than $\omega_{(g,x)}(R, e)$ by as much as $\omega_{(g,y)}(R, e) - \omega$.

Note also that, in Definition 5.2.4, e' is again a divisor of R'. In fact, using the notation of Definition 5.2.1, let $U' = \big(\boldsymbol{\pi}(g)^0\big)^{-1}(U)$. Let $\mathbf{H}'_\gamma$ be the ideal of $\Delta'_\gamma(g)$ in $\mathscr{O}_{U'}$, for each $\gamma \in \Lambda'$. For $\lambda \in \Lambda$, we have

$$\mathbf{H}_\lambda \mathscr{O}_{U'} = \begin{cases} \mathbf{H}'_\lambda & \text{if } Y(g) \not\subset \Delta_\lambda(g), \\ \mathbf{H}'_\lambda.\mathbf{H}'_0 & \text{if } Y(g) \subset \Delta_\lambda(g). \end{cases}$$

On the other hand, by the definition of $\omega_{(g,x)}(R, e)$ we have

$$\omega b = \nu_x(\mathbf{I}_0)$$

for almost all points $x \in Y(g)$; hence there exists an ideal $\mathbf{I}'_0$ in $\mathscr{O}_{U'}$ such that

$$\mathbf{I}_0\mathscr{O}_{U'} = \mathbf{I}'_0(\mathbf{H}'_0)^{\omega b}.$$

Therefore

$$\mathbf{I}\mathscr{O}_{U'} = \mathbf{I}'_0(\mathbf{H}'_0)^{e'(o)b+b} \prod_{\lambda \in \Lambda(g,U)} (\mathbf{H}'_\lambda)^{e'(\lambda)b}.$$

By definition, the transform $\mathscr{E}'$ of $\mathscr{E}$ by $(\boldsymbol{\pi}(g))^0$ is represented by $((\mathbf{I}', b))$, where

$$\mathbf{I}\mathscr{O}_{U'} = \mathbf{I}'(\mathbf{H}'_0)^b.$$

It follows that

$$\mathbf{I}' = \mathbf{I}'_0 \prod_{\gamma \in \Lambda'} (\mathbf{H}'_\gamma)^{e'(\gamma)b}.$$

Remark 5.2.3 Let ω be a non-negative rational number, let $R = (\mathscr{G}; L, M)^n_{\mathrm{III}}$ be a resolution datum over G of type III_n with $\mathscr{G} = (G, \mathbf{J}, \boldsymbol{\Delta})$, where

$$\boldsymbol{\Delta} = \{\Delta_\lambda\}_{\lambda \in \Lambda},$$

and let e be a divisor of R. We denote by $\mathrm{Sing}^\omega(R, e)$ the subset of $(G)_0$ defined, for every $g \in \mathrm{ob}([G])$, by

$$\mathrm{Sing}^\omega(R, e)(g) = \{x \in (G(g))_0 \mid \omega_{(g,x)}(R, e) = \omega\}.$$

By an argument similar to the one used for the globality theorem, we can make $\mathrm{Sing}^\omega(R, e)$ in a natural way into a reduced complex subspace of $(G)_0$, which will be called the ω-**stratum** of (R, e).

If F is a subset of $(G)^0$, the (F, ω)-**stratum** of (R, e) will be the smallest closed complex subspace $\mathrm{Sing}^\omega(R, e)_F$ of $\mathrm{Sing}^\omega(R, e)$ that coincides locally with $\mathrm{Sing}^\omega(R, e)$ around every point in F. Obviously, $\mathrm{Sing}^\omega(R, e)_F \subset L$ if $F \subset (G)^0 \setminus M$.

Then, for every $g \in \mathrm{ob}([G])$ we write:

(a) $(G_\omega(g))^0 = (G(g))^0$,
(b) $(G_\omega(g))_0 = \{x \in (G(g))_0 \mid \omega_{(g,x)}(R, e) \geq \omega\}$, with the reduced complex structure, and
(c) $G_\omega(g) : (G_\omega(g))_0 \to (G_\omega(g))^0$ for the embedding induced by $G(g)$.

Also, if $\alpha \in \mathrm{Hom}_{[G]}(g, g')$, we denote by

$$G_\omega(\alpha) : G_\omega(g) \to G_\omega(g')$$

the horizontal morphism of given by

$$(G_\omega(\alpha))^0 = (G(\alpha))^0, \quad (G_\omega(\alpha))_0 = (G(\alpha))_0|_{(G_\omega(g))_0}.$$

In this way, we obtain a $\mathbb{C}$-situation G_ω with the same index category as G. Clearly $G_\omega = G$ when $\omega = 0$. From now on, we consider the case $\omega > 0$.

Proposition 5.2.1 *Under the assumptions of Remark 5.2.3, let* $\omega = \alpha/\beta$ *with positive integers* α *and* β, *then there exists a forest* $(G_\omega, \mathbf{J}_\omega)$ *having the following property:*

If (g, x) *is any point in* G *such that:*

(i) $\dim_x(G(g))^0 = n$;
(ii) *the set* U *is a pure* n*-dimensional open neighborhood of* x *in* $(G(g))^0$ *(that is, one such that* $\dim_y(G(g))^0 = n$ *for every* $y \in U$*), and* $\mathscr{E}$ *is an idealistic exponent on* U *such that* $\mathbf{J}(g)|_U$ *is associated with* $\mathscr{E}$*;*
(iii) *the idealistic exponent* $((\mathbf{I}, b))$ *is a representative of* $\mathscr{E}$ *and* $\mathbf{I}_0$ *is an ideal in* $\mathscr{O}_U$ *such that:*

   (a) *for every* $\lambda \in \Lambda$ *such that* $\Delta_\lambda(g)$ *intersects* U, $e(\lambda)b$ *is an integer and*
   (b) $\mathbf{I} = \mathbf{I}_0 . \prod_\lambda \mathbf{H}_\lambda^{e(\lambda)b}$, *where the product is taken over the* $\lambda \in \Lambda$ *such that* $\Delta_\lambda(g)$ *intersects* U, *and where* $\mathbf{H}_\lambda$ *is the ideal of* $\Delta_\lambda(g)$ *in* $\mathscr{O}_U$*;*

then $\mathbf{J}_\omega(g)|_U$ *is the* $G_\omega(g)$*-polygrove associated with the idealistic exponent* $\mathscr{E}_\omega = ((\mathbf{I}^\alpha + \mathbf{I}_0^\beta, \alpha b))$ *on* U.

The proof of this proposition is a consequence of the following lemmas.

Lemma 5.2.1 *The assumptions and the notation being as above, we have*

$$(G_\omega(g))_0 \cap U = \mathrm{Sing}((\mathbf{I}_\alpha + \mathbf{I}_0^\beta, \alpha b)).$$

Proof By the definition of G_ω, a point y lies in $(G_\omega(g))_0 \cap U$ if and only if

$$y \in (G(g))_0 \cap U \text{ and } \omega_{(g,y)}(R, e) \geq \omega.$$

By the definition of $\omega_{(g,y)}(R, e)$, we have

$$\omega_{(g,y)}(R, e) = \nu_y^n(\mathscr{E}) - \sum_{\lambda \in \Lambda_{(g,y)}} e(\lambda) = \nu_y(\mathbf{I})/b - \sum_{\lambda \in \Lambda_{(g,y)}} e(\lambda).$$

Therefore, for $y \in (G(g))_0 \cap U$,

$$\nu_y(\mathbf{I}) \geq \frac{\alpha b}{\beta} + \sum_{\lambda \in \Lambda_{(g,y)}} e(\lambda) b \Longleftrightarrow \omega_{(g,y)}(R, e) \geq \omega.$$

Moreover, for $y \in U$, assumption (ii) of the proposition implies that $y \in (G(g))_0 \cap U$ if and only if $y \in \mathrm{Sing}(\mathscr{E})$, which in turn is equivalent to $\nu_y(\mathbf{I}) \geq b$. On the other hand, for $y \in U$, assumption (iii)(b) gives

$$\nu_y(\mathbf{I}) = \nu_y(j_0) + \sum_{\lambda \in \Lambda(g,U)} e(\lambda) b \nu_y(H_\lambda) = \nu_y(\mathbf{I}_0) + \sum_{\lambda \in \Lambda_{(g,y)}} e(\lambda) b.$$

This equality and the equivalence in the preceding display imply that, for $y \in (G(g))_0 \cap U$,

$$\nu_y(\mathbf{I}_0) \geq \frac{\alpha b}{\beta} \Longleftrightarrow \omega_{(g,y)}(R, e) \geq \omega.$$

Putting all of this together we get

$$\begin{aligned} y \in (G_\omega(g))_0 \cap U &\Longleftrightarrow \nu_y(\mathbf{I}) \geq b \text{ and } \nu_y(\mathbf{I}_0) \geq \frac{\alpha b}{\beta} \Longleftrightarrow \nu_y(\mathbf{I}^\alpha + \mathbf{I}_0^\beta) \geq \alpha b \\ &\Longleftrightarrow y \in \mathrm{Sing}((\mathbf{I}^\alpha + \mathbf{I}_0^\beta, \alpha b)). \end{aligned}$$

□

Lemma 5.2.2 *Let* $\mathbf{P} = \{B_V\}_{V \in \mathrm{ob}(\mathscr{V})}$ *and* $\mathbf{P}' = \{B'_V\}_{V \in \mathrm{ob}(\mathscr{V})}$ *be polygroves satisfying the following conditions:*

(a) $|B'_V| \subset |B_V|$ *for all* $V \in \mathrm{ob}(\mathscr{V})$.
(b) *Let* $V \in \mathrm{ob}(\mathscr{V})$ *and* $b \in |B'_V|$, *and set* $B_V(b) = \{\mathbf{a}_i, U_{i-1}, E_{i-1}\}_{1 \leq i \leq m}$ *and* $B'_V(b) = \{\mathbf{a}'_i, U'_{i-1}, E'_{i-1}\}_{1 \leq i \leq m'}$. *Then:*
    (1) *The ambient diagrams of* $B_V(b)$ *and* $B'_V(b)$ *coincide, that is, we have* $m = m'$, $(s'_i)^0 = (s_i)^0$ *for* $0 \leq i \leq m$, *and* $(a_i)^0 = (a'_i)^0$ *for* $1 \leq i \leq m$; *and*
    (2) $(s'_i)_0 \subset (s_i)_0$ *for* $1 \leq i \leq m$.

Let $\mathbf{h} : \mathbf{t} \to \mathbf{s}_0$ *be a restrictive morphism of models and* $\mathbf{h}' : \mathbf{t} \to \mathbf{s}_0$ *the composition of* $\mathbf{h}$ *with the morphism induced by the identification of* $(s'_0)^0$ *and* $(s_0)^0$.

Then, if there is an ambient contraction of $\mathbf{P}$ *by* $\mathbf{h}$, *there exists an ambient contraction of* $\mathbf{P}'$ *by* $\mathbf{h}'$.

Proof The lemma is easy to prove. In fact, if $V \in \mathrm{ob}(\mathscr{V})$ and $\widetilde{B_V}$ is an ambient contraction of B_V by $\mathbf{h} \times V$, one can construct $\widetilde{B'_V}$ in the following way:

(i) The set $|\widetilde{B'_V}|$ is the set of all $b \in |\widetilde{B_V}|$ such that there is a $b' \in |B'_V|$ such that $\widetilde{B_V}(b')$ is an ambient contraction of $B_V(b')$ by $\mathbf{h} \times V$.

(ii) For every $b' \in |\widetilde{B'_V}|$, $\widetilde{B'_V}(b')$ is the tree with the same ambient diagram as $\widetilde{B_V}(b')$ and whose singular diagram is the inverse image of the singular diagram of $B'_V(b')$ by the identity between ambient spaces.

□

Lemma 5.2.3 *For every $g \in \mathrm{ob}([G])$ such that $(G(g))^0$ has dimension n everywhere, one can construct a $G_\omega(g)$-polygrove $\mathbf{J}_\omega(g)$, unique up to equivalences over the open subsets (see Definition 4.1.5 in Chap. 4), that satisfies the following conditions:*

(1) *For all $V \in \mathrm{ob}(\mathscr{V})$, $|\mathbf{J}_\omega(g)_V|$ is the subset of $|\mathbf{J}(g)_V|$ characterized by the following property. For every $b \in |\mathbf{J}(g)_V|$, one has $b \in |\mathbf{J}_\omega(g)_V|$ if and only if, for each $x \in (G(g))_0$ there is a system $\{U, \mathbf{I}, b, \mathbf{I}_0\}$ that satisfies conditions (ii) and (iii) of Proposition 5.2.1, and the ambient diagram of the tree $\mathbf{J}(g)_V(b)|_{U\times V}$ coincides with the ambient diagram of a tree belonging to a polygrove associated with the idealistic exponent $\mathscr{E}_\omega = ((\mathbf{I}^\alpha + \mathbf{I}_0^\beta, \alpha b))$ on U.*
(2) *For all $b \in |\mathbf{J}_\omega(g)_V|$, the ambient diagram of $\mathbf{J}_\omega(g)_V(b)$ is the same as that of $\mathbf{J}(g)_V(b)$.*
(3) *For all $x \in G(g)_0$, if $\{U, \mathbf{I}, b, \mathbf{I}_0\}$ satisfy conditions (ii) and (iii) of Proposition 5.2.1, then $\mathbf{J}_\omega(g)|_U$ is a polygrove associated with the idealistic exponent $\mathscr{E}_\omega$ of (1).*

Proof Observe first that, for every $V \in \mathrm{ob}(\mathscr{V})$, the set $|J_\omega(g)_V|$ is uniquely determined by condition (1). Therefore, it remains only to show that, for all $b \in |J_\omega(g)_V|$, conditions (2) and (3) allow us to define a tree $J_\omega(g)_V(b)$, unique up to isomorphism.

Let $x \in (G(g))_0$ be an arbitrary point, and let $\{U, \mathbf{I}, b, \mathbf{I}_0\}$ be a system that satisfies conditions (ii) and (iii) of Proposition 5.2.1. Set

$$\mathscr{E} = \omega((\mathbf{I}^\alpha + \mathbf{I}_0^\beta, \alpha b)),$$

and let $\mathbf{P} = \{P_V\}_{V\in\mathrm{ob}(\mathscr{V})}$ be a polygrove associated with $\mathscr{E}_\omega$. Since, by assumption, $\mathbf{J}(g)|_U$ is associated with the idealistic exponent $\mathscr{E} = ((\mathbf{I}^\alpha, \alpha b))$ on U, and since $\mathbf{I}^\alpha \subset \mathbf{I}^\alpha + \mathbf{I}_0^\beta$, we obtain:

Fact 1 *For all $V \in \mathrm{ob}(\mathscr{V})$ and all $d \in |P_V|$, there exists $c \in |\mathbf{J}(g)_V|$ such that the tree $P_V(d)$ is isomorphic (over the identity on $(G(g))^0 \times V$) to a tree whose ambient part is the same as the one of $\mathbf{J}(g)_V(c)|_{U\times V}$.*

Now suppose that $\{U', \mathbf{I}', b', \mathbf{I}'_0\}$ also satisfy conditions (ii) and (iii) of Proposition 5.2.1, and define $\mathscr{E}'_\omega$ and $\mathbf{P}'$ similarly to $\mathscr{E}_\omega$ and $\mathbf{P}$ above.

Fact 2 *If $U'' = U \cap U'$, then $\mathbf{P}|_{U''}$ is equivalent to $\mathbf{P}'|_{U''}$ over the identity on U''.*

In fact, let $V \in \mathrm{ob}(\mathscr{V})$ and $d \in |P_V|$. By Fact 1, we may assume that there is $c \in |\mathbf{J}(g)_V|$ such that $P_V(d)$ and $\mathbf{J}_g(V)(c)|_{U\times V}$ have the same ambient diagram.

Let

$$\mathbf{J}_g(V)(c)|_{U''\times V} = \{\mathbf{a}_i, U_{i-1}, E_{i-1}\}_{1\le i\le m}$$

with $\mathbf{a}_i : \mathbf{s}_i \to \mathbf{s}_{i-1}$; $(s_0)^0 = U'' \times V$, $(s_0)^0 = ((G(g))_0 \cap U'') \times V$.

Put $\mathscr{E}(0) = (\mathscr{E}|_{U''}) \times V$ and $\mathscr{E}'(0) = (\mathscr{E}'|_{U''}) \times V$, and, for $1 \le i \le m$, let $\mathscr{E}(i)$ and $\mathscr{E}'(i)$ be the transforms of $\mathscr{E}(i-1)$ and $\mathscr{E}'(i-1)$ by $(a_i)^0$. Since $\mathscr{E}|_{U''}$ and $\mathscr{E}'|_{U''}$ have a common associated polygrove $\mathbf{J}(g)|_{U''}$, so do $\mathscr{E}(i)$ and $\mathscr{E}'(i)$, for each i. This proves Fact 2.

Therefore, by Theorem 4.4.2 of Chap. 4, one has:

Fact 3 $\mathrm{Sing}(\mathscr{E}(i)) = \mathrm{Sing}(\mathscr{E}'(i)) = (s_i)_0$, *for* $0 \le i \le m$, *and*

$$\nu_{\mathbf{y}}(\mathscr{E}(i)) = \nu_{\mathbf{y}}(\mathscr{E}'(i)),$$

for $0 \le i \le m$ *and* $y \in (s_i)_0$.

Now write $\mathscr{E}_\omega(0) = (\mathscr{E}_\omega|_{U''}) \times V$ and $\mathscr{E}'_\omega(0) = (\mathscr{E}'_\omega|_{U''}) \times V$. Let $\mathscr{E}_\omega(i)$ be the transform of $\mathscr{E}_\omega(i-1)$ by $(\mathbf{a}_i)^0$, so long as this is definable. We may assume in Fact 2, without loss of generality, that $b = b'$ and that $b\omega \in \mathbb{Z}_+$. Then, we can write:

$$\mathscr{E}(0) = ((\mathbf{K}_0\mathbf{L}_0, b)), \quad \mathscr{E}_\omega(0) = ((\mathbf{K}_0^\alpha\mathbf{L}_0^\alpha + \mathbf{L}_0^\beta, \alpha b)),$$
$$\mathscr{E}'(0) = ((\mathbf{K}_0\mathbf{L}'_0, b)), \quad \mathscr{E}'_\omega(0) = ((\mathbf{K}_0^\alpha\mathbf{L}'^\alpha_0 + \mathbf{L}'^\beta_0, \alpha b)),$$

where $\mathbf{L}_0 = (\mathbf{I}_0|_{U''})\mathscr{O}_{U''\times V}$, $\mathbf{L}'_0 = (\mathbf{I}'_0|_{U''}).\mathscr{O}_{U''\times V}$, and

$$\mathbf{K}_0 = \prod_{\lambda\in\Lambda(g,\mathbf{y})} \mathbf{H}^{e(\lambda)b}\mathscr{O}_{U''\times V},$$

all constructed via the first projection $U'' \times V \to U'' = U \cap U'$. (Observe that $\mathbf{K}_0$ does not change if we replace U by U'.)

Let $\mathbf{B}_i$ be the ideal of $\big((a_i)^0\big)^{-1}(E_{i-1})$ in $\mathscr{O}_i = \mathscr{O}_{(s_i)^0}$, for $1 \le i \le m$. Then there are two ideals $\mathbf{L}_i$ and $\mathbf{K}_i$ of $\mathscr{O}_i$ $(1 \le i \le m)$ such that

$$\mathbf{L}_{i-1}\mathscr{O}_i = \mathbf{L}_i\mathbf{B}_i^{\omega b} \quad \text{and} \quad \mathbf{K}_{i-1}\mathscr{O}_i = \mathbf{K}_i\mathbf{B}_i^{b-\omega b} \quad \text{for } 1 \le i \le m.$$

In this way

$$\mathscr{E}(i) = ((\mathbf{K}_i\mathbf{L}_i, b)) \quad \text{and} \quad \mathscr{E}_\omega(i) = ((\mathbf{K}_i^\alpha\mathbf{L}_i^\alpha + \mathbf{L}_i^\beta, \alpha b)) \quad \text{for } 0 \le i \le m.$$

We now prove, by induction, yet another claim:

Fact 4 $\mathscr{E}'_\omega(i)$ is defined for $1 \le i \le m$, and

$$\mathrm{Sing}(\mathscr{E}'_\omega(i)) = \mathrm{Sing}(\mathscr{E}_\omega(i)).$$

If $i = 0$, $\mathscr{E}'_\omega(0)$ is defined and, by Lemma 5.2.1,

$$\mathrm{Sing}(\mathscr{E}_\omega(0)) = (G(g))_0|_{U'' \times V} = \mathrm{Sing}(\mathscr{E}'_\omega(0)).$$

Now assume we have proved Fact 4 for i ranging from 0 to $j-1$ (where $j \leq m$); we shall prove it for $i = j$.

By Fact 3, we have $\nu_{\mathbf{y}}(\mathscr{E}(j-1)) = \nu_{\mathbf{y}}(\mathscr{E}'(j-1))$, for $1 \leq j \leq m+1$ and $\mathbf{y} \in (s_{j-1})_0$. Therefore, $\nu_{\mathbf{y}}(\mathbf{K}_{j-1}\mathbf{L}_{j-1}) = \nu_{\mathbf{y}}(\mathbf{K}_{j-1}\mathbf{L}'_{j-1})$ and $\nu_{\mathbf{y}}(\mathbf{L}_{j-1}) = \nu_{\mathbf{y}}(\mathbf{L}'_{j-1})$, for $1 \leq j \leq m+1$. Since $E_{j-1} \subset (s_{j-1})_0$, for all $j < i$ there exists an ideal $\mathbf{L}'_j$ of $\mathscr{O}_j$ such that $\mathbf{L}'_{j-1}\mathscr{O}_j = \mathbf{L}'_j \mathbf{B}_j^{b-\omega b}$ and

$$\nu_{\mathbf{y}}(\mathbf{L}'_j) = \nu_{\mathbf{y}}(\mathbf{L}_j) \quad \text{for } \mathbf{v} \in (s_{i-1})_0.$$

Then

$$\mathscr{E}'(j) = ((\mathbf{K}_j\mathbf{L}'_j, b)) \text{ and } \mathscr{E}'_\omega(j) = ((\mathbf{K}_j^\alpha \mathbf{K}'^\alpha_j + \mathbf{L}'^\beta_j, \alpha b))$$

and, for all $\mathbf{y} \in (s_i)_0$, one has $\nu_{\mathbf{y}}(\mathscr{E}'_\omega(j)) = \nu_{\mathbf{y}}(\mathscr{E}_\omega(j))$, which proves Fact 4 and the lemma. □

Proof of Proposition 5.2.1 By adding the empty allée to $(G, \mathbf{J})$ we have a garden $\mathscr{G}$. Without loss of generality we may assume $\mathscr{G}$ complete. Now Lemma 5.2.3 allows us to construct $\mathbf{J}_\omega(g)$, provided that $(G(g))^0$ has dimension n everywhere. Since G is homodimensional, any object g' can be connected to g by a chain of morphisms; then we can construct $\mathbf{J}_\omega(g')$ by taking into account the following observations:

(i) Given a morphism $h : g_1 \to g_2$, and once $\mathbf{J}_\omega(g_1)$ is known, the ambient expansion by h of $\mathbf{J}'_\omega(g_1)$ always exists.
(ii) Given a morphism $h : g_1 \to g_2$, and once $\mathbf{J}_\omega(g_2)$ is known, Lemma 5.2.2 guarantees that there exists the ambient contraction of $\mathbf{J}_\omega(g_2)$ by h.
(iii) The construction of $\mathbf{J}_\omega(g)$ and the fact that $(\mathscr{G})$ is a garden implies that, if g_1 and g_2 are connected by a chain of morphisms in $[G]$, the transfer of $\mathbf{J}_\omega(g_1)$ to g_2 does not depend on the chain.

This ends our proof. □

Remark 5.2.4 (Notation) In the conditions of Remark 5.2.3, and in view of Proposition 5.2.1, we write $\mathscr{G}_\omega = (G_\omega, \mathbf{J}_\omega, \boldsymbol{\Delta})$ and $R_\omega = (\mathscr{G}_\omega; L_\omega, M_\omega)^n_{\mathbb{III}}$, with $L_\omega = (G_\omega)_0 \cap L$ and $M_\omega = (G_\omega)_0 \cap M$.

Remark 5.2.5 Let $(G, \mathbf{J})$ be a forest and let H be a smooth closed hypersurface of $(G)^0$. We can construct a forest $(G_1, \mathbf{J}_1)$, in the following way:

(i) $[G_1] = [G]$, and G_1 is a $\mathbb{C}$-situation defined by:

(a) $(G_1(g))^0 = H(g)$, $(G_1(g))_0 = H(g) \cap (G(g))_0$, and

$$G_1(g) = G(g)|_{(G_1(g))_0},$$

for $g \in \mathrm{ob}([G])$.

(b) $(G_1(\alpha))^0$ and $(G_1(\alpha))_0$ are the morphisms induced by $(G(\alpha))^0$ and $(G(\alpha))_0$, for every morphism α in $[G]$.

(ii) For every $g \in \mathrm{ob}([G])$ and every $V \in \mathrm{ob}(\mathscr{V})$, $|\mathbf{J}_1(g)_V|$ is the subset of $|\mathbf{J}(g)_V|$ consisting of the elements b such that, if

(5.2.5.1) $$\mathbf{J}(g)_V(b) = \{\mathbf{a}_i, U_{i-1}, E_{i-1}\}_{1\leq i\leq m},$$

with $\mathbf{a}_i : \mathbf{s}_i \to \mathbf{s}_{i-1}$, then E_i is contained in T_i for $i = 0, 1, \dots, m-1$, where $T_0 = H(g)x \times V$ and the T_i, for $i > 0$, are the successive strict transforms of T_0 by the $(a_i)^0$.

(iii) For every $b \in |\mathbf{J}_1(g)_V|$, if $\mathbf{J}(g)_V(b)$ is of the form (5.2.5.1), then

$$\mathbf{J}_1(g)_V(b) = \{\mathbf{c}_i, U'_{i-1}, E'_{i-1}\}_{1\leq i\leq m},$$

where $\mathbf{c}_i : \mathbf{t}_i \to \mathbf{t}_{i-1}$, $(t_0)^0 = H(g) \times V$, and $(t_i)^0$ is the strict transform of $(t_{i-1})^0$ by $(a_i)^0$ (that is, $U'_{i-1} = U_{i-1} \cap (t_{i-1})^0$, $E'_{i-1} = E_{i-1}$, $(t_i)_0 = (t_i)^0 \cap (s_i)_0$, and $\mathbf{c}_i$ is the morphism induced by $\mathbf{a}_i$).

Then $\mathbf{J}_1(g)$ is a $G(g)$-polygrove for every $g \in \mathrm{ob}([G_1])$, and

$$h \in \mathrm{Hom}_{[G_1]}(g, g')$$

implies that $\mathbf{J}_1(g)$ is $G_1(h)$-equivalent to $\mathbf{J}_1(g')$, because $\mathbf{J}(g)$ is $G(h)$-equivalent to $\mathbf{J}(g')$ and $(G(h)^0)^{-1}(H(g')) = H(g)$.

Definition 5.2.5 The forest $(G_1, \mathbf{J}_1)$ of Remark 5.2.5 will be called the **section** of $(G, \mathbf{J}, \boldsymbol{\Delta})$ by H, and will be denoted by $(G, \mathbf{J}) \cap H$. If $\mathscr{G} = (G, \mathbf{J}, \boldsymbol{\Delta})$ is a garden, with $\boldsymbol{\Delta} = \{\Delta_\lambda\}_{\lambda\in\Lambda}$, and if $\tau \in \Lambda$, the τ-**section** of $\mathscr{G}$ we shall understand the garden $\mathscr{G}\cap\Delta_\tau = (G_1, \mathbf{J}_1, \boldsymbol{\Delta}^*)$, where $(G_1, \mathbf{J}_1) = (G, \mathbf{J})\cap\Delta_\tau$ and $\boldsymbol{\Delta}^* = \{\Delta_\tau\cap\Delta_\lambda\}_{\lambda\in\Lambda\setminus\{\tau\}}$.

Remark 5.2.6 If $(G, \mathbf{J})$ is an idealistic forest, then $(G_1, \mathbf{J}_1) = (G, \mathbf{J}) \cap H$ is also idealistic. In fact, if (g, x) is a point in G_1, then $x \in H(g)\cap(G(g))_0$ and there exist an open neighborhood U of x in $(G(g))^0$ and an idealistic exponent $\mathscr{E} = ((\mathbf{I}, b))$ on U in such a way that $\mathbf{J}(g)|_U$ is associated with $\mathscr{E}$. Let $\mathbf{K}$ be a ideal of $H(g)$ in U and define an idealistic exponent $\mathscr{E}'_1 = \big(((\mathbf{I} + \mathbf{K}^b), b)\big)$ on U. Then

$$T_{\mathscr{E}'_1,x} \subset T_{H(g),x},$$

and hence there exist an open neighborhood U' of x in U and an idealistic exponent $\mathscr{E}_1$ on $U_1 = U'\cap H(g)$ such that $\mathscr{E}'_1|_{U'}$ is an ambient expansion of $\mathscr{E}_1$ by the inclusion

$U_1 \to U'$. It is easy to see that $\mathbf{J}_1(g)|_{U_1}$ is a polygrove associated with $\mathscr{E}_1$. This proves that $(G_1, \mathbf{J}_1)$ is also idealistic.

Remark 5.2.7 Now we prove that $\mathbb{III}^{n-1}$ (that is, Theorem 5.1.2 for $n-1$) and $\mathbb{I}^n$ (Theorem 5.1.1 for n) together imply $\mathbb{III}^n$. Therefore we shall assume that $\mathbb{III}^{n-1}$ and $\mathbb{I}^n$ hold.

Let $R = (\mathscr{G}; L, M)^n_{\mathbb{III}}$ be a resolution datum of type $\mathbb{III}_n$, with $\mathscr{G} = (G, J, \Delta)$, and let Y be the subset of $(G)_0$ defined, for every $g \in \mathrm{ob}([G])$, by

$$Y(g) = \{x \in (L(g))_0 \mid \dim_x(G(g))_0 \geq n\}.$$

Since $\dim \mathscr{G} = n$,

$$Y(g) = \{x \in (L(g))_0 \mid \dim_x(G(g))_0 = n\}$$

for every $g \in \mathrm{ob}([G])$, and so Y can be naturally made into a smooth closed complex subspace of $(G)_0$. Moreover, since $\mathscr{G}$ is an idealistic garden, one can easily check that Y is a permissible center for R.

Let e be a divisor of R. By Remark 5.2.1, there exists a proper succession $\bar{\pi}$ of strictly permissible modifications of (R, e) that produces on R the same effect as the permissible modification with center Y. Then, by the construction of Y and the definition of e, if (R', e') is the transform of (R, e) by $\bar{\pi}$, one has $\dim \mathrm{Sing}^{\omega}(R', e') < n$ for $\omega \in \mathbb{Q}_+$, and $\mathrm{Sing}^{\infty}(R', e') = \varnothing$.

Lemma 5.2.4 *With the assumptions of Remark* 5.2.7, *let S be a closed complex subspace of L such that:*

(a) $\dim(S) < n$, *(that is,* $\dim(S(g)) < n$ *for all* $g \in \mathrm{ob}([G])$*).*
(b) *There exists a positive rational number ω such that $S \setminus \mathrm{Sing}^{\omega}(R, e)$ is nowhere dense in S.*

Then there exists a proper succession $\bar{\pi}$ of strictly permissible modifications for (R, e) with the following properties:

(i) *The centers of $\bar{\pi}$ are contained in the successive total transforms of S (constructed as in Definition* 5.1.8*).*
(ii) *The strict transform of S by the modification limit of $\bar{\pi}$, which we denote by (G', π), is contained in the ω-stratum of the transform (R', e') of (R, e) by $\bar{\pi}$.*

Proof Let $\{\omega(i)\}_{i \in \mathbb{Z}_+}$ be a strictly decreasing sequence of rational numbers with limit ω. We construct the garden $\mathscr{G}_{\omega(1)}$ (see Proposition 5.2.1) and consider the resolution datum

$$R_1 = (\mathscr{G}_{\omega(1)}; S)^n_{\mathbb{I}}.$$

By applying Theorem $\mathbb{I}^n$ to R_1 and by Remark 5.2.1, we deduce that there exists a proper succession $\bar{\pi}(1)$ of strictly permissible modifications for (R, e) with centers contained in S such that, if $(R(1), e(1))$ is the strict transform of (R, e) by $\bar{\pi}(1)$, and if $S(1)$ is the strict transform of S by $\bar{\pi}(1)$ (i.e., by the modification limit $(G(1), \pi(1))$ of $\bar{\pi}(1)$), one has

$$(G(1))_0 \cap S(1) = \varnothing.$$

Together with Remark 5.2.7, this inequality implies that

$$\omega(1) > \omega_{(g,x)}(R(1), e(1)) \geq \omega$$

for every point (g, x) of $S(1)$.

Using induction, assume that we have constructed $(R(i), e(i))$, with

$$R(i) = (\mathscr{G}(i); L(i), M(i))^n_{\mathbb{III}}.$$

Applying Theorem $\mathbb{I}^n$ to

$$R_{i+1} = (\mathscr{G}(i)_{\omega(i+1)}, S(i))^n_{\mathbb{I}},$$

we obtain, by a process similar to that used in the construction of $R(1)$, a proper succession $\bar{\pi}(i+1)$ of strictly permissible modifications for $(R(i), e(i))$ such that, if $(R(i+1), e(i+1))$ is the transform of $(R(i), e(i))$ by $\bar{\pi}(i+1)$ and if $S(i+1)$ is the strict transform of $S(i)$, one has

$$\omega(i+1) > \omega_{(g,x)}(R(i+1), e(i+1)) \geq \omega, \tag{5.2.4.1}$$

for every point (g, x) of $S(i+1)$.

Let Λ_i be the set of indices of $\bar{\pi}(i)$. Set $\Lambda = \bigcup_{i \in \mathbb{Z}_+} \Lambda_i$ (disjoint union) and define on Λ an order relation as follows. If $\alpha \in \Lambda_i$ and $\beta \in \Lambda_j$, say that $\alpha > \beta$ if and only if either $i > j$ or $i = j$ and $\alpha > \beta$ in Λ_i. One can then construct the union $\bar{\pi}$ of the $\bar{\pi}(i)$:

$$\bar{\pi} = \bigcup_{i \in \mathbb{Z}_+} \bar{\pi}(i) = \{\pi_{\alpha,\beta}\}_{\beta \in \Lambda_\alpha},$$

the $\pi_{\alpha,\beta}$ being modifications of G_α with center Y_α and being defined in the obvious way.

The succession $\bar{\pi}$ constructed in this way satisfies the conditions of the lemma: (i) holds by the construction of the $\bar{\pi}(i)$, and (ii) holds by inequality (5.2.4.1). □

Lemma 5.2.5 *With the notation and hypothesis of Remark 5.2.7, let F be an arbitrary subset of $(G)^0 \setminus M$, ω a positive rational number, $S_0 = \mathrm{Sing}^{\omega}(R, e)_F$, $V_0 = \mathrm{Sing}^{\omega}(R, e) \setminus S_0$, S the closure of S_0, and V the closure of V_0. If*

$$F \cap (S \setminus S_0) = F \cap V = \varnothing,$$

there exists a proper succession $\bar{\pi}$ of strictly permissible modifications for (R, e) with the following properties:

(i) *The centers of $\bar{\pi}$ are contained in the successive total transforms of $(V \cap S) \cup (S - S_0)$.*
(ii) *If (R', e), with $R' = (\mathcal{G}'; L', M')$, is the transform of (R, e) by $\bar{\pi}$, and if $F' = \big((\pi)^0\big)^{-1}(F)$, where (G', π) is the modification limit of $\bar{\pi}$, then*

$$S' = \mathrm{Sing}^{\omega}(R', e')_{F'}, \ \mathrm{Sing}^{\omega}(R', e')_{S'}$$

are closed subspaces of L'.

Proof Being $\dim \mathcal{G} = n$, we have:

$$\dim \mathrm{Sing}^{\omega}(R, e) < n, \ \forall \omega \in \mathbb{Q}_{+}, \text{ and } \dim \mathrm{Sing}^{\infty}(R, e) = n.$$

Since $F \cap M = \varnothing$, we have $S \subset L$ (see Remark 5.2.3), so conditions (a) and (b) of Lemma 5.2.4 hold for (R, e), S and ω. The lemma then yields a proper succession $\bar{\pi}$ of strictly permissible modifications for (R, e), all of which have centers contained in the successive total transforms of $S \setminus S_0$, and which have the property that the strict transform of S by $\bar{\pi}$ is contained in the ω-stratum of (R', e').

Now, from the definition of $\mathrm{Sing}^{\omega}(R', e')_{F'}$, if S' is the strict transform of S by $\bar{\pi}$, the intersection of S' with the ω-stratum of (R', e') is $\mathrm{Sing}^{\omega}(R', e')$. Therefore, in this case, $\mathrm{Sing}^{\omega'}(R', e')_{F'} = S'$ is the strict transform of S by $\bar{\pi}$, and so it is closed in $(G')_0$.

Thus, to prove our lemma, we can assume that S_0 is closed (that is, that $S = S_0$); otherwise, by applying a succession $\bar{\pi}$ of modifications that are strictly permissible for (R, e) and satisfy conditions (i) and (ii) in the lemma, we return to the initial conditions, but this time with S_0 closed.

Assuming $S_0 = S$, then, and applying $\mathbb{I}^n$ and Remark 5.2.1 to the resolution datum $R_1 = (\mathcal{G}; V, S)^n_{\mathbb{I}}$, we can obtain a succession $\bar{\pi}$ of strictly permissible modifications for (R, e) that satisfies (i) and such that, if S^* and V^* are the strict transforms of S and V by $\bar{\pi}$, we have

$$S^* \cap V^* \cap (G)_0 = \varnothing.$$

This is equivalent to

$$S^* \cap V^* = \varnothing,$$

since, by construction, $S^* \subset (G)_0$. Since $F \cap V = \varnothing$, the disjointness of S^* and V^* implies that $S^* = S'$, so $T' = \mathrm{Sing}^{\omega}(R', e')_{S'}$ is contained in the total transform of S by $\bar{\pi}$, which is closed in $(G')^0$. Applying Remark 5.2.2, we conclude that T' is closed in $(G')_0$.

Therefore, this succession $\bar{\pi}$ is the one we wanted. □

Lemma 5.2.6 *With the notation and hypothesis of Remark 5.2.7, let F be an arbitrary subset of $(G)^0$, ω a positive rational number, $S = \mathrm{Sing}^{\omega}(R,e)_F$, $T = \mathrm{Sing}^{\omega}(R,e)_S$, and let $\boldsymbol{\Delta} = \{\Delta_\lambda\}_{\lambda\in\Lambda}$ be the allée of R. If T is contained in L and is closed in L, there exists a proper succession $\bar{\boldsymbol{\pi}}^{(1)}$ of strictly permissible modifications for (R,e) satisfying these conditions:*

(i) *The centers of $\bar{\boldsymbol{\pi}}^{(1)}$ are contained in the ω-stratum of the successive transforms of (R,e) by the modifications of $\bar{\boldsymbol{\pi}}^{(1)}$.*
(ii) *If $(G^{(1)}, \bar{\pi}^{(1)})$ is the modification limit of $\boldsymbol{\pi}^{(1)}$ and if we set $F^{(1)} = \big((\bar{\pi}^{(1)})^0\big)^{-1}(F)$, then, for all $\lambda \in \Lambda$, the strict transform of Δ_λ does not meet the $(F^{(1)}, \omega)$-stratum of the transform of (R,e) by $\boldsymbol{\pi}^{(1)}$.*

Proof Since, by assumption, G is contable at infinity, and since the family $\boldsymbol{\Delta}$ is locally finite, we may assume that Λ is countable, and so, without loss of generality, that $\Lambda = \mathbb{Z}_0$.

Write $\mathscr{G}_{(0)} = \mathscr{G}_\omega \cap \Delta_0$ (see Remark 5.2.4 and Definition 5.2.5 for the notation). Set $L_{(0)} = S \cup T = T$, and denote by $M_{(0)}$ the closure of $(G_{(0)})_0 \setminus L_{(0)}$ in $(G_{(0)})_0$. Since, by construction, $\dim \mathscr{G}_{(0)} = n-1$, we can construct the resolution datum

$$R_{(0)} = (\mathscr{G}_{(0)}, L_{(0)}, M_{(0)})_{\mathbb{I}}^{n-1}$$

over $G_{(0)}$, of type $\mathbb{III}^{n-1}$. Now, by applying Theorem $\mathbb{III}^{n-1}$, we conclude that there exists a proper succession $\bar{\boldsymbol{\pi}}'$ of strictly permissible modifications for $R_{(0)}$ such that, if $R'_{(0)} = (\mathscr{G}'_{(0)}; L'_{(0)}, M'_{(0)})_{\mathbb{III}}^{n-1}$ is the transform of $R_{(0)}$ by $\bar{\boldsymbol{\pi}}'$, with $\mathscr{G}'_0 = (G'_0, \mathbf{J}'_0, \boldsymbol{\Delta}'_0)$, then $(G'_{(0)})_0 \subset M'_{(0)}$.

By applying an ambient expansion by the inclusion

$$(G_{(0)})_0 \hookrightarrow (G)^0,$$

we can extend all the elements above. By applying Remark 5.2.1 we obtain a proper succession $\bar{\boldsymbol{\pi}}(1)$ of strictly permissible modifications for (R,e), with centers contained in the ω-strata of the successive transforms of (R,e), and such that, if $(R(1), e(1))$ is the transform of (R,e) by $\bar{\boldsymbol{\pi}}(1)$, if $(G(1), \pi(1))$ is the modification limit of $\bar{\boldsymbol{\pi}}(1)$, and if $F(1) = \big((\pi(1))^0\big)^{-1}(F)$, then the strict transform of $\boldsymbol{\Delta}_0$ by $\pi(1)$ does not meet the $(F(1),\omega)$-stratum of $(R(1), e(1))$. Thus the assumptions of the lemma stay true for $F(1)$ and $(R(1), e(1))$ instead of F and (R,e).

We now proceed inductively in the following way. Let $i \in \mathbb{Z}_0$ be a number such that, for every $j \in \mathbb{Z}_0$ with $j \le i$, we have constructed $\bar{\boldsymbol{\pi}}(j)$, $(R(j), e(j))$ and $F(j)$ in such a way that:

(a) The sequence $\bar{\boldsymbol{\pi}}(j)$ is a proper succession of strictly permissible modifications for $(R(j-1),\ e(j-1))$ with centers contained in the ω-strata of the successive transforms of $(R(j-1),\ e(j-1))$ by the modifications of $\bar{\boldsymbol{\pi}}(j)$.
(b) If $(G(j), \pi(j))$ is the modification limit of $\bar{\boldsymbol{\pi}}(j)$, then the strict transform of $\boldsymbol{\Delta}_{j-1}$ by $\pi(j)\pi(j-1)\ldots\pi$ does not meet the $(F(j),\omega)$-stratum of the transform $(R(j), e(j))$ of $(R(j-1),\ e(j-1))$ by $\bar{\boldsymbol{\pi}}(j)$.

(c) The assumptions of the lemma stay true for $(R(j), e(j))$, $F(j)$, and

$$S(j) = \mathrm{Sing}^{\omega}(R(j), e(j))_{F(j)}, T_j = \mathrm{Sing}^{\omega}(R(j), e(j))_{S(j)}.$$

In these conditions we define

$$(G_{(i)}, \mathbf{J}_{(i)}, \boldsymbol{\Delta}_{(i)}) = \mathcal{G}_{(i)} = (\mathcal{G}(i))_{\omega} \cap \boldsymbol{\Delta}, \quad L_{(i)} = T(i),$$

and $M_{(i)}$ as the closure of $(G_{(i)})_0 - T(i)$ in $(G_{(i)})_0$. Since $\dim \mathcal{G}_{(i)} = n-1$ we can construct the resolution datum

$$R_{(i)} = (\mathcal{G}_{(i)}, L_{(i)}, M_{(i)})^{n-1}_{\mathrm{III}}$$

of type III^{n-1}. Now, by application of Theorem III^{n-1}, by ambient expansion by $(G_{(i)})^0 \hookrightarrow (G(i))^0$ and by application of Remark 5.2.1, we obtain a proper succession $\bar{\pi}(i+1)$ of strictly permissible modifications for $(R(i), e(i))$ such that, if $(R(i+1), e(i+1))$ is the transform of $(R(i), e(i))$ by $\bar{\pi}(i+1)$, if $(G(i+1), \pi(i+1))$ is the modification limit of $\bar{\pi}(i+1)$, and if $F(i+1) = \big((\pi(i+1))^0\big)^{-1}(F(i))$, then conditions (a), (b) and (c) above hold for $i+1$ also.

Then, by composing the $\bar{\pi}(i)$ in a way similar to that used in Lemma 5.2.5, we obtain a proper succession $\bar{\pi}^{(1)}$ of strictly permissible modifications for (R, e), satisfying the conditions of the lemma. □

Lemma 5.2.7 *Under the assumptions of Lemma 5.2.6, let $\bar{\pi}^{(1)}$ be the proper succession of strictly permissible modifications for (R, e) which satisfies (i) and (ii) in that lemma, and whose existence is asserted there. Let $(R(1), e(1))$ be the transform of (R, e) by $\bar{\pi}^{(1)}$. Then there exists a proper succession $\bar{\pi}^{(2)}$ of strictly permissible modifications for $(R^{(1)}, e^{(1)})$ satisfying these conditions:*

(i) *The centers of $\bar{\pi}^{(2)}$ are contained in the ω-strata of the successive transforms of $(R^{(1)}, e^{(1)})$ by the modification of $\bar{\pi}^{(2)}$.*
(ii) *If $(R^{(2)}, e^{(2)})$ is the transform of $(R^{(1)}, e^{(1)})$ by $\bar{\pi}^{(2)}$, if $(G^{(2)}, \pi^{(2)})$ is the modification limit of $\bar{\pi}^{(2)}$, and $F^{(2)} = \big((\pi^{(2)})^0\big)^{-1}(F^{(2)})$, the $(F^{(2)}, \omega)$-stratum of $(R^{(2)}, e^{(2)})$ is empty.*

Proof With the notation and hypothesis of Lemma 5.2.6, let $\mathcal{G}^{\omega} = (G_{\omega}, \mathbf{J}_{\omega}, \boldsymbol{\Delta}_{\omega})$ be the garden introduced in Notation 5.2.4. Denote by B the closure of $(G_{\omega})_0 \setminus T$ in $(G_{\omega})_0$, and define a $\mathbb{C}$-situation G_0, with the same index category as G, by

$$(G_0)^0 = (G_{\omega})^0 - B \quad \text{and} \quad (G_0)_0 = (G_{\omega})_0 - B.$$

(With this, $S \subset (G_0)^0$.) Since $(G_0)^0$ is an open subspace of $(G_{\omega})^0 = (G)^0$, we can consider the restriction $\mathbf{J}_0$ of the polygrove $\mathbf{J}_{\omega}$ to $(G_0)^0$, and so $(G_0, \mathbf{J}_0)$ is a forest. Also, we shall consider the garden $\mathcal{G}_0 = (G_0, \mathbf{J}_0, \varnothing)$ and its completion $\hat{\mathcal{G}}_0 = (\hat{G}_0, \hat{\mathbf{J}}_0, \varnothing)$.

If (g, x) is a point of G_0, we have $\nu_x^n(\mathbf{J}(g)) = 1$; hence, if V is an open neighborhood of x in $(G(g))^0$ such that there exists an idealistic exponent $\mathscr{E}$ on V such that $\mathbf{J}(g)|_v$ is associated with $\mathscr{E}$, we get

$$\nu_x^n(\mathscr{E}) = 1,$$

and so

$$T_{\mathscr{E},x} \subset\neq T_{(G(g))^0,x}.$$

Therefore, by Theorem 4.3.2 of Chap. 4, and since $\hat{\mathscr{G}}_0$ is the completion of $\mathscr{G}_0$, there exists a point $((g, \mathbf{0}, U, Y), x)$ in $\hat{G}_0$ equivalent to

$$((g,\ 0,\ (G(g))^0 \times \{\mathbf{0}\},\ (G(g))^0 \times \{\mathbf{0}\}, x\}$$

and such that

$$\dim_{((g,0,U,Y),x)} \hat{G}_0 = \dim T_{\mathscr{E},x} < n,$$

and thus

$$\dim \hat{\mathscr{G}}_0 < n.$$

Now consider the resolution data

$$R_0 = (\mathscr{G}_0;\ (G_0)_0,\ \varnothing)_{\mathrm{III}}^n$$

and

$$\hat{R} = (\hat{\mathscr{G}}_0;\ (\hat{G})_0,\ \varnothing)_{\mathrm{III}}^{n-1}.$$

$\bar{\boldsymbol{\pi}}^{(1)}$ induces a proper succession $\bar{\boldsymbol{\pi}}_0^{(1)}$ of permissible modifications for R_0, which, by Remark 5.1.8, can be naturally extended to a proper succession $\hat{\boldsymbol{\pi}}^{(1)}$ of permissible modifications for $\hat{R}$.

Let $R_0^{(1)} = (\mathscr{G}_0^{(1)}; (G_0^{(1)})_0, \varnothing)$ and $\hat{R}^{(1)} = (\hat{\mathscr{G}}^{(1)}; (\hat{G}^{(1)})_0, \varnothing)_{\mathrm{III}}^{n-1}$ be the transforms of R_0 and $\hat{R}$ by $\bar{\boldsymbol{\pi}}_0^{(1)}$ and $\hat{\boldsymbol{\pi}}^{(1)}$, respectively. Let $S^{(1)}$ be the $(F^{(1)}, \omega)$-stratum of $(R^{(1)}, e^{(1)})$: by the construction of $\bar{\boldsymbol{\pi}}_0^{(1)}$, $S^{(1)}$ induces a complex subspace $S_0^{(1)}$ of $(G_0^{(1)})_0$ and, by the construction of $\hat{\boldsymbol{\pi}}^{(1)}$, $S_0^{(1)}$ induces a complex subspace $\hat{S}^{(1)}$ of $(\hat{G}^{(1)})_0$. Denote by N the closure of $(\hat{G}^{(1)})_0 \setminus \hat{S}^{(1)}$ in $(\hat{G}^{(1)})_0$, and write

$$\tilde{R}^{(1)} = (\hat{\mathscr{G}}^{(1)},\ \hat{S}^{(1)},\ \hat{N})_{\mathrm{III}}^{n-1}.$$

Now apply Theorem III^{n-1} to $\tilde{R}^{(1)}$, and let $\hat{\pi}$ be the proper succession of permissible modifications for $\tilde{R}^{(1)}$ given in that theorem. By restricting ourselves to $R_0^{(1)}$, we conclude that $\hat{\pi}$ induces a proper succession $\bar{\pi}_0^{(2)}$ of permissible modifications for $R_0^{(1)}$. We see also that all the centers of $\bar{\pi}_0^{(2)}$ are contained in the successive total transforms of $S_0^{(1)}$.

On the other hand, since $S \subset (G_0)_0$ and S is closed in $(G)_0$, we conclude that $S_0^{(1)}$ is, by construction, canonically isomorphic to $S^{(1)}$ and $S^{(1)}$ is closed in $(G^{(1)})_0$. Therefore $\bar{\pi}_0^{(2)}$ gives rise to a proper succession $\bar{\pi}^{(2)}$ of permissible modifications for $R^{(1)}$, and, by Remark 5.2.1, we may assume that $\bar{\pi}^{(2)}$ is a proper succession of strictly permissible modifications for $(R^{(1)}, e^{(1)})$.

Let $N^{(1)}$ be the closure of $(G_\omega^{(1)}) - S^{(1)}$ in $(G^{(1)})_0$. Then $N^{(1)}$ and $\hat{N}$ induce the same complex subspace in $(G_0^{(1)})_0$ and, by the construction of $\hat{\pi}$, if $(R^{(2)}, e^{(2)})$ is the transform of $(R^{(1)}, e^{(1)})$ by $\bar{\pi}^{(2)}$ and if $F^{(2)} = \big((\pi^{(2)})^0\big)^{-1}(F^{(1)})$, then the $(F^{(2)}, \omega)$-stratum $S^{(2)}$ of $(R^{(2)}, e^{(2)})$ is contained in the total transforms of $N^{(1)}$ and $S^{(1)}$ by $\bar{\pi}^{(2)}$. But, by construction, $N^{(1)} \cap (S^{(1)} \cap F^{(1)}) = \varnothing$, so $S^{(2)} = \varnothing$, which proves our lemma. □

Lemma 5.2.8 *With the notation and hypothesis of Remark 5.2.7, let D be a compact subset of $(G)^0$ and set $F = D \setminus M$. If $\omega_{(g,x)}(R, e) = \mathbf{0}$ for every point (g, x) of $F \cap (G)_0$, there exists a proper succession $\bar{\pi}'$ of strictly permissible modifications for (R, e) such that, if (G', π') is the modification limit of $\bar{\pi}'$ and $F' = \big((\pi')^0\big)^{-1}(F)$, the sets $(G')_0$ and F' are disjoint.*

Proof Let

$$P = \left\{ A \subset \Lambda \;\middle|\; F \cap \Big(\bigcap_{\lambda \in A} \Delta_\lambda \Big) = \varnothing \right\}.$$

Since Δ_λ meets F only for a finite number of $\lambda \in \Lambda$, all the elements $A \in P$ are finite sets. For every $A \in P$ we write

$$p(A) = \mathrm{card}(A), \quad q(A) = \sum_{\lambda \in A} e(\lambda) \geq 1.$$

Also set $p(R, e, D) = \min_{A \in P}\{p(A)\}$, $P_1 = \{A \in P \mid p(A) = p(R, e, D)\}$, $q(R, e, D) = \max_{A \in P_1}\{q(A)\}$, and

$$Q(R, e, D) = \{A \in P_1 \mid q(A) = q(R, e, D)\}.$$

Let A be an arbitrary element in $Q(R, e, D)$, and, for all $g \in \mathrm{ob}([G])$, denote by $Y(g)$ the closure in $(G)_0$ of the set

$$\left\{ x \in \bigcup_{\lambda \in A} \Delta_\lambda(g) \setminus M(g) \;\middle|\; \omega_{(g,x)}(R, e) = 0 \right\} \cap (G(g))_0.$$

The family $\{Y(g)\}_{g\in \mathrm{ob}([G])}$, together with the morphisms induced by the ones in $(G)_0$, defines a reduced subspace Y of $(G)_0$ that is open and closed in $\bigcap_{\lambda\in\Lambda} \Delta_\lambda$. Hence Y is a connected component of this intersection, and so smooth. By construction, $Y \subset L$, and there exists a modification (G^*, f) with center Y that is strictly permissible for (R, e). Let (R^*, e^*) be the transform of (R, e) by this modification and let $D^* = ((f)^0)^{-1}(D)$. Then

$$p(R^*, e^*, D^*) \geq p(R, e, D);$$

if equality holds, we have

$$q(R^*, e^*, D^*) \leq q(R, e, D);$$

and if this, too, is an equality, we have

$$Q(R^*, e^*, D^*) = Q(R, e, D) \setminus \{A\}.$$

This is just a reformulation of the following simple case analysis: If $P_1 = \{A\}$, we have $p(R^*, e^*, D^*) > p(R, e, D)$. If, on the contrary, $P_1 \neq \{A\}$, then $p(R^*, e^*, D^*) = p(R, e, D)$, and there are two cases: when $Q_1 = \{A\}$ is $q(R^*, e^*, D^*) < q(R, e, D)$, and when $Q_1 \neq \{A\}$ we have $q(R^*, e^*, D^*) = q(R, e, D)$.

By repeating this process of modification a finite number of times, our lemma is proved. □

Proof of Theorem III^n *(Theorem* 5.1.2*)* Let $R = (\mathcal{G}; L, M)^n_{\mathrm{III}}$ be a resolution datum of type III^n, with $\mathcal{G} = (G, \mathbf{J}, \boldsymbol{\Delta})$. Since G is countable at infinity, in order to prove Theorem III^n it is enough to show that, for every compact subset D of $(G)^0$, there exists a proper succession $\bar{\boldsymbol{\pi}}'$ of permissible modifications for R, with modification limit $(G', \boldsymbol{\pi}')$, such that, if $R' = (\mathcal{G}', L', M')^n_{\mathrm{III}}$ is the transform of R by $\boldsymbol{\pi}'$ with $\mathcal{G}' = (G', \mathbf{J}', \boldsymbol{\Delta}')$, and if $D' = \left((\pi')^0\right)^{-1}(D)$, then $D' \cap (G')_0 \subset M'$.

To prove this, let e be a divisor of R: for instance, $e = 0$. By applying a strictly permissible modification for (R, e), such as the one in Remark 5.2.7, we may assume that $\dim \mathrm{Sing}^\omega(R, e) < n$, for $\omega \in \mathbb{Q}_+$, and that $\mathrm{Sing}^\infty(R, e)$ is empty.

Set

$$\omega = \max\{\omega_{(g,x)}(R, e) \mid (g, x) \text{ point in } F = D - M\}.$$

Since D is compact, ω always exists. If $\omega = 0$, Lemma 5.2.8 implies our theorem. Assume that $\omega > 0$; by applying Lemma 5.2.5 to (R, e), with these F and ω, we may suppose that, after applying a proper succession of strictly permissible modifications for (R, e), we are in a position similar to the initial one, but this time with $S = \mathrm{Sing}^\omega(R, e)_F$ and $T = \mathrm{Sing}^\omega(R, e)_S$ contained and closed in L.

By applying Lemmas 5.2.6 and 5.2.7, and always after a proper succession of strictly permissible modifications, we conclude that $\mathrm{Sing}^\omega(R, e)_S = \varnothing$. Then, starting again from the new value of ω, smaller than the preceding one, and repeating the process as many times as we need, we arrive at $\omega = 0$. By Lemma 5.2.8, the theorem is proved. □

5.3 $\mathbb{III}^{n-1} \Rightarrow \mathbb{I}^n$

Throughout this section we consider a fixed resolution datum of type $\mathbb{I}^n R = (\mathscr{G}; X_1, \dots, X_q)^n_{\mathbb{I}}$, with $\mathscr{G} = (G, \mathbf{J}, \boldsymbol{\Delta})$, and we shall assume that Theorem $\mathbb{III}^{n-1}$ holds.

Definition 5.3.1 We denote by $\mathrm{Sing}(R)$ the closed complex subspace of $(G)_0$ given by

$$\mathrm{Sing}(R) = \bigcap_{i=1}^{q} X_i \cap (G)_0,$$

with the reduced structure.

Given a point (g, x) of $\mathrm{Sing}(R)$, we define the **Samuel function** of R at (g, x) as the system

$$\mathbf{H}_{(R,g,x)} = (H_1, \dots, H_q),$$

where

$$H_i = H_{X_i(g),x} \quad \text{for } 1 \le i \le q.$$

Also, given a system of functions $\mathbf{H} = (H_1, \dots, H_q)$, with $H_i : \mathbb{Z}_0 \to \mathbb{Z}_0$ for $1 \le i \le q$, we can consider the subset Y of $(G)_0$ given, for all $g \in \mathrm{ob}([G])$, by

$$Y(g) = \{x \in \mathrm{Sing}(R)(g) \mid H_{X_i(g),x} = H_i \text{ for } 1 \le i \le q\};$$

if we endow Y with the reduced structure induced by $(G)_0$ we have a reduced complex subspace of $(G)_0$, which we call the **H-stratum** of R, and denote by $\mathrm{Sing}^{\mathbf{H}}(R)$.

If F is a subset of $(G)^0$ and $\mathbf{H}$ is as above, we define the **$(F, \mathbf{H})$-stratum** of R as the minimum closed complex subspace $\mathrm{Sing}^{\mathbf{H}}(R)_F$ of $\mathrm{Sing}^{\mathbf{H}}(R)$ that coincides with $\mathrm{Sing}^{\mathbf{H}}(R)$ around every point of F. This $\mathrm{Sing}^{\mathbf{H}}(R)_F$ always exists, and its closure in $(G)_0$ is also a closed complex subspace of $(G)_0$.

Let $\mathbf{H}_1 = (H_1^1, \dots, H_q^1)$ and $\mathbf{H}_2 = (H_1^2, \dots, H_q^2)$ be q-tuples of functions $H_1^i, H_2^i : \mathbb{Z}_0 \to \mathbb{Z}_0$. We write $\mathbf{H}_1 > \mathbf{H}_2$ if and only if, for some index $k = 1, \dots, q$ and some $m \ge 0$, the following conditions are satisfied:

(a) $H_j^1(p) = H_j^2(p)$, for every $j < k$ and $p \ge 0$.

(b) $H_k^1(p) = H_k^2(p)$, for every $p < m$.
(c) $H_k^1(m) > H_k^2(m)$.

If F is a subset of $(G)^0$, we defined the F-**largest Samuel function** of R to be a system $\mathbf{H}$ as above satisfying these additional assumptions:

(i) There exists a point (g, x) of $F \cap \mathrm{Sing}(R)$ such that $\mathbf{H} = \mathbf{H}_{(R,g,x)}$.
(ii) For every point (g', x') of $F \cap \mathrm{Sing}(R)$, one has $\mathbf{H} \geq H_{(R,g',x')}$.

Clearly, if F is compact and $F \cap \mathrm{Sing}(R) \neq \varnothing$, the F-largest Samuel function of R always exists.

Lemma 5.3.1 *Using the conditions of Definition 5.3.1, let F be a subset of $(G)^0$ such that:*

(a) *The F-largest Samuel function $\mathbf{H}$ of R exists.*
(b) *$S = \mathrm{Sing}^{\mathbf{H}}(R)_F$ and $T = \mathrm{Sing}^{\mathbf{H}}(R)_S$ are closed in $(G)_0$.*

Then there exists a proper succession bar$\boldsymbol{\pi}^{(1)}$ of permissible modifications for R, with modification limit $(G^{(1)}, \boldsymbol{\pi}^{(1)})$, having these properties:

(i) *The centers of the modifications of $\bar{\boldsymbol{\pi}}^{(1)}$ are contained in the $\mathbf{H}$-strata of the successive transforms of R by the modifications of $\bar{\boldsymbol{\pi}}^{(1)}$.*
(ii) *If $R^{(1)}$ is the transform of R by $\bar{\boldsymbol{\pi}}^{(1)}$, if $F^{(1)} = ((\boldsymbol{\pi}^{(1)})^0)^{-1}(F)$ and if $\boldsymbol{\Delta} = \{\Delta_\lambda\}_{\lambda \in \Lambda}$ is the allée of $\mathscr{G}$, then, for all $\lambda \in \Lambda$ the strict transforms $\Delta_\lambda^{(1)}$ of the Δ_λ do not meet the $(F^{(1)}, \mathbf{H})$-stratum of $R^{(1)}$.*

Proof By a similar argument to that of Lemma 5.2.6, we may assume $\Lambda = \mathbb{Z}_0$. We start with Δ_0 and suppose that $\Delta_0 \cap S \neq \varnothing$.

Set $\mathbf{H} = (H_1, \ldots, H_q)$. For $i = 1, \ldots, q$, we have a restrictive closed complex subspace B_i of $(G)^0$ given, for all $g \in \mathrm{ob}([G])$, by

$$B_i(g) = \{x \in (G(g))^0 \mid H_{X_i(g),x} > H_i\}.$$

Now define, for $i = 1, \ldots, q$, a $\mathbb{C}$-situation $G^{(i)}$ in the following way:

(a) $[G^{(i)}] = [G]$.
(b) $(G^{(i)})^0 = (G)^0 \setminus \bigcup_{j=1}^q B_j$.
(c) $(G^{(i)})_0$ is the restrictive complex subspace of $(G^{(i)})^0$ such that, for every $g \in \mathrm{ob}([G])$, $(G^{(i)})_0(g)$ is the restriction to $(G)_0 \setminus \bigcup_{j=1}^q B_j$ of the Samuel stratum of $X_i(g)$ corresponding to H_i.

For $i = 1, \ldots, q$, we can construct from $G^{(i)}$ a forest $(G^{(i)}, \mathbf{J}^{(i)})$ in such a way that, for every $g \in \mathrm{ob}([G])$, $\mathbf{J}^{(i)}(g)$ is a $G^{(i)}(g)$-polygrove associated with $(X_i(g), H_i)$. We also define a forest $(\bar{G}, \bar{\mathbf{J}})$ in the following way:

(a) $[\bar{G}] = [G]$.
(b) For every $g \in \mathrm{ob}([G])$, we have $\bar{G}(g) = G(g)|_{(G(g))^0 \setminus \bigcup_{j=1}^q B_j}$ (hence $(\bar{G})^0 = (G)^0 \setminus \bigcup_{j=1}^q B_j$).

(c) For every $g \in \text{ob}([\bar{G}])$, $\bar{\mathbf{J}}(g) = J(g)|_{(\bar{G})^0(g)}$.

Now, since $[G^{(i)}] = [\bar{G}]$ and $(G^i)^0 = (\bar{G})^0$ for $i = 1, \ldots, q$, we can construct the intersection

$$(G^0, \mathbf{J}^0) = \bigcap_{i=1}^{g} (G^{(i)}, \mathbf{J}^{(i)}) \cap (\bar{G}, \bar{\mathbf{J}}).$$

Denote by $\mathbf{E}$ the restriction of $\boldsymbol{\Delta}$ to $(G^0)^0$. We define a garden $\mathscr{G}^0$ as

$$\mathscr{G}^0 = (G^0, \mathbf{J}^0, \mathbf{E}).$$

Also, if we put $L^0 = T$ (which is, by construction and our assumptions, a closed complex subspace of $(G^0)_0$), and if we denote by M^0 the closure of $(G^0)_0 - L^0$ in $(G^0)_0$, we obtain a resolution datum of type $\mathbb{III}^n$:

$$R^0 = (\mathscr{G}^0, L^0, M^0)^n_{\mathbb{III}}.$$

Let E_0 be the hypersurface of $(G^0)^0$ induced by Δ_0, and put $\mathscr{G}^1 = (G^1, J^1, E^1) = \mathscr{G}^0 \cap E_0$ (see Definition 5.2.5). Hence $\dim \mathscr{G}^1 \leq n - 1$. Also set $L^1 = L^0 \cap (G^1)_0$, $M^1 = M^0 \cap (G^1)_0$, and

$$R^1 = (\mathscr{G}^1, L^1, M^1)^{n-1}_{\mathbb{III}}.$$

Now apply Theorem $\mathbb{III}^{n-1}$ to R^1 and let $\bar{\boldsymbol{\pi}}^1$ be the proper succession of permissible modifications for R^1 given by that theorem. By an ambient extension by $(G^1)^0 \subset (G^0)^0$, $\bar{\boldsymbol{\pi}}^1$ can be naturally extended to a proper succession $\bar{\boldsymbol{\pi}}^0$ of permissible modifications for R^0. Also, since $L^0 = T$ is closed in $(G)_0$, and by the definition of R^0, $\bar{\boldsymbol{\pi}}^0$ induces a proper succession $\bar{\boldsymbol{\pi}}$ of permissible modifications for R.

Since $\bar{\boldsymbol{\pi}}^1$ satisfies the conditions of theorem $\mathbb{III}^{n-1}$, we conclude that $\bar{\boldsymbol{\pi}}$ satisfies property (i) in the conclusion of our lemma. Moreover, since $M^0 \cap S = \varnothing$, the strict transform of Δ_0 by $\bar{\boldsymbol{\pi}}$ does not meet the (F', H)-stratum of the transform R' of R by $\bar{\boldsymbol{\pi}}$, where $(G', \boldsymbol{\pi})$ is the modification limit of $\bar{\boldsymbol{\pi}}$ and $F' = \big((\boldsymbol{\pi})^0\big)^{-1}(F)$.

Now, taking the hypersurface Δ_1 of $\boldsymbol{\Delta}$, denoting by Δ_1' the strict transform of Δ_1 by $\bar{\boldsymbol{\pi}}$, and by applying the above process to R' and Δ_1', and so on, we obtain the desired $\bar{\boldsymbol{\pi}}^{(1)}$, in a way similar to that of Lemma 5.2.6. □

Lemma 5.3.2 *Under the assumptions of Lemma* 5.3.1, *let* $\bar{\boldsymbol{\pi}}^{(1)}$ *be a proper succession of permissible modifications for* R *as provided by the lemma. There exists a proper succession* $\bar{\boldsymbol{\pi}}^{(2)}$ *of permissible modifications for* $R^{(1)}$, *with modification limit* $(G^{(2)}, \boldsymbol{\pi}^{(2)})$, *satisfying the following conditions:*

(i) *The centers of* $\bar{\boldsymbol{\pi}}^{(2)}$ *are contained in the* **H**-*strata of the successive transforms of* $R^{(1)}$ *by the modifications of* $\bar{\boldsymbol{\pi}}^{(2)}$.

(ii) *If* $R^{(2)}$ *is the transform of* $R^{(1)}$ *by* $\bar{\pi}^{(2)}$ *and if* $F^{(2)}=((\pi^{(2)})^0)^{-1}(F^{(1)})$, *the* $(F^{(2)}, \mathbf{H})$*-stratum of* $R^{(2)}$ *is empty.*

Proof Let (G^0, J^0) be the forest constructed in the proof of Lemma 5.3.1, and consider the garden

$$\mathscr{G}_0 = (G^0, J^0, \varnothing)$$

and its completion

$$\hat{\mathscr{G}}_0 = (\hat{G}^0, \hat{J}^0, \varnothing).$$

By applying Proposition 5.1.1, and taking into account that $\dim X_i < n$, we get

$$\dim \mathscr{G}_0 < n.$$

Therefore, we can define a resolution datum of type $\mathbb{III}^{n-1}$ as follows:

$$\hat{R}_0 = (\hat{\mathscr{G}}_0; (\hat{G}^0)_0, \varnothing)^{n-1}_{\mathbb{III}}.$$

By Remark 5.1.8, the proper succession $\bar{\pi}^{(1)}$ induces, in a natural way, a proper succession $\hat{\pi}^{(1)}$ of permissible modifications for $\hat{R}_0$. Let

$$\hat{R}^{(1)} = (\hat{\mathscr{G}}^{(1)}; (\hat{G}^{(1)})^0, \varnothing)^{n-1}_{\mathbb{III}}$$

be the transform of $\hat{R}_0$ by $\hat{\pi}^{(1)}$ and let $\hat{S}^{(1)}$ be the closed complex subspace of $(\hat{G}^{(1)})_0$ induced by the $(F^{(1)}, \mathbf{H})$-stratum $S^{(1)}$ of $R^{(1)}$. Denote by $\hat{N}$ the closure of $(\hat{G}^{(1)})_0 \setminus \hat{S}^{(1)}$ in $(\hat{G}^{(1)})_0$, and write

$$\hat{R} = (\hat{\mathscr{G}}^{(1)}; \hat{S}^{(1)}, \hat{N})^{n-1}_{\mathbb{III}}.$$

By applying theorem $\mathbb{III}^{n-1}$ to $\hat{R}$, we obtain a proper succession $\hat{\pi}$ of permissible modifications for $\hat{R}$ that satisfies the conditions in the conclusion of the theorem. Then $\hat{\pi}$ induces the desired proper succession of permissible modifications for $R^{(1)}$. □

Lemma 5.3.3 *With the assumptions of Lemma* 5.3.1, *let* S *be a closed complex subspace of* $(G)_0$ *contained in all the* X_i, *for* $1 \leq i \leq q$, *and let*

$$\mathbf{H} = (H_1, \dots, H_q), \quad \textit{with } H_i : \mathbb{Z}_0 \to \mathbb{Z}_0,$$

be a system of functions. If $S \cap \mathrm{Sing}^{\mathbf{H}}(R)$ *is everywhere dense in* S, *there exists a proper succession* $\bar{\pi}$ *of permissible modifications for* R, *with modification limit* (G', π), *satisfying these conditions:*

(i) *All the centers of* $\bar{\pi}$ *are contained in the successive strict transforms of* S *by the modifications of* $\bar{\pi}$.
(ii) *For each center* Y_α *of* $\bar{\pi}$ *there exists a family of functions* $\mathbf{H}_\alpha = (H_1^\alpha, \ldots, H_q^\alpha)$, *with* $H_i^\alpha : \mathbb{Z}_0 \to \mathbb{Z}_0$, *such that* $\mathbf{H}_\alpha > \mathbf{H}$ *and that* Y_α *is contained in the* $\mathbf{H}_\alpha$-*stratum of the transform of* R *at the level* α.
(iii) *If* R' *is the transform of* R *by* $\bar{\pi}$ *and* S' *is the strict transform of* S *by* π, *then* S' *is contained in the* $\mathbf{H}$-*stratum of* R'.

Proof We work by induction on $\dim S$. If $\dim S = \mathbf{0}$, the lemma is trivial, so let's assume our result true for $\dim S < s$, where $s < n$, and prove it for $\dim S = s$.

Since G is countable at infinity, we will be done if we show that, for every compact subset D or $(G)^0$ such that

$$D \cap (S \setminus \operatorname{Sing}^{\mathbf{H}}(R)) \neq \varnothing,$$

there exists a proper succession $\bar{\pi}$ as in the lemma, satisfying (i), (ii), and the following condition:

(3′) If $D' = [(\pi)^0]^{-1}(D)$, then $D' \cap (S' = \setminus \operatorname{Sing}^{\mathbf{H}}(R')) = \varnothing$.

To construct such a $\bar{\pi}$, we consider on S the reduced complex structure and, since $\dim S < n$, we also consider the resolution datum of type $\mathbb{I}^n$ given by

$$R^* = (\mathscr{G}; X_1, \ldots, X_q, S)^n_{\mathbb{I}}.$$

Let H^* be the D-largest Samuel function of R^*, and let V^* be the closure of the $(D, \mathbf{H}^*)$-stratum of R^* in $(G)_0$; then V^* is a closed complex subspace of $(G)_0$ such that $V^* \subset S \setminus \operatorname{Sing}^{\mathbf{H}}(R)$ and $\dim V^* < s$.

Since $V^* \cap \operatorname{Sing}^{\mathbf{H}^*}(R^*)$ is everywhere dense in V^* and since $\dim V^* < s$, the induction hypothesis implies that there exists a proper succession $\bar{\pi}_1$ of permissible modifications for R^* that satisfies conditions (i), (ii), and (iii) for R^* and V^*.

Clearly, $\bar{\pi}_1$ is also a proper succession of permissible modifications for R and it satisfies conditions (i) and (ii) for R and S. Set $D_1 = \left((\pi_1)^0\right)^{-1}(D)$ and let R_1^* be the transform of R^* by π_1. Then, since D does not intersect $V^* \setminus \operatorname{Sing}^{\mathbf{H}^*}(R^*)$, the $(D, \mathbf{H}^*)$-stratum of R_1^* is the strict transform V_1^* of V^* by π_1. Therefore, if W_1^* is the closure in $(G_1)_0$ of the $(V^*, \mathbf{H})$-stratum of R_1^*, one has $\dim W_1^* < s$.

On the other hand, since $W_1^* \cap \operatorname{Sing}^{\mathbf{H}^*}(R_1^*)$ is everywhere dense in W_1^*, there exists, by the induction hypothesis, a proper succession $\bar{\pi}_2$ of permissible modifications for R_1^* that satisfies conditions (i), (ii), and (iii) for R_1^* and W_1^*. Therefore, the composition of $\bar{\pi}_1$ and $\bar{\pi}_2$ is a proper succession of permissible modifications for R that satisfies (i) and (ii) for R and S. Let R_2^* be the transform of R_1^* by $\bar{\pi}_2$, let (G_2, π_2) be the modification limit of $\bar{\pi}_2$, and set $D_2 = \left((\pi_2)^0\right)^{-1}(D_1)$. Then the $(D_2, \mathbf{H})$-stratum V_2^* of R_2^* and the $(V_2^*, \mathbf{H})$-stratum W_2^* of R_2^* are closed in $(G_2)_0$.

Now we apply Lemmas 5.3.1 and 5.3.2 to D_2 and R_2^*. As a result, we can assert that there exists a proper succession $\bar{\pi}_3$ of permissible modifications for R_2^* satisfying these conditions:

(a) The composition $\bar{\pi}^{(1)}$ of $\bar{\pi}_1, \bar{\pi}_2$ and $\bar{\pi}_3$ satisfies conditions (i) and (ii) for S and R.

(b) If (G_3, π_3) is the modification limit of $\bar{\pi}_3$ and if

$$D_3 = \left((\pi_3)^0\right)^{-1}(D_2),$$

the $(D_3, \mathbf{H}^*)$-stratum of R_3^* (the transform of R_2^* by $\bar{\pi}_3$) is empty.

Let $R^{(1)}$ be the transform of R by $\bar{\pi}^{(1)}$, and $(G^{(1)}, \pi^{(1)})$ the modification limit of $\bar{\pi}^{(1)}$. Also let $S^{(1)}$ be the strict transform of $S^{(1)}$ by $\pi^{(1)}$, and set $D^{(1)} = D_3$. If $D^{(1)} \cap \left(S^{(1)} - \mathrm{Sing}^{\mathbf{H}}(R^{(1)})\right) = \varnothing$, the lemma is proved. On the other hand, if $D^{(1)} \cap \left(S^{(1)} - \mathrm{Sing}^{\mathbf{H}} \neq \varnothing\right.$, we can repeat the initial process and construct a new proper succession $\bar{\pi}^2$ of permissible modifications for $R^{(1)}$ that satisfies conditions (a) and (b) for $R^{(1)}$ and $S^{(1)}$. In this way, we obtain new data $R^{(2)}$, $S^{(2)}$, and $D^{(2)}$ and we are in similar position as with $R^{(1)}$, $S^{(1)}$, and $D^{(1)}$. By repeating this process we obtain a sequence of modifications with the same properties. This sequence cannot be infinite, by finiteness theorem. The composition of this finite sequence gives us the desired $\bar{\pi}$ that satisfies (i), (ii), and (iii′). □

Lemma 5.3.4 *Let D be a compact subset of $(G)^0$ and let $\mathbf{H}$ be the D-largest Samuel function of R. Then there exists a proper succession $\bar{\pi}^*$ of permissible modifications for R, with modification limit (G^*, π^*), such that if R^* is the transform of R by $\bar{\pi}^*$ and $D^* = \left((\pi^*)^0\right)^{-1}(D)$, then the $(D*, \mathbf{H})$-stratum of R^* is empty.*

Proof **Step 1.** Let S be the closure of $\mathrm{Sing}^{\mathbf{H}}(R)_D$ in $(G)_0$. Then S and $\mathbf{H}$ satisfy the assumptions of Lemma 5.3.3, and we can construct a proper sequence $\bar{\pi}$ of permissible modifications for R that satisfy conditions (i), (ii), and (iii) of Lemma 5.3.3.

Step 2. Let (G', π) be the modification limit of $\bar{\pi}$, let R' be the transform of R by $\bar{\pi}$, and set $D' = \left((\pi^0)\right)^{-1}(D)$. Then the $(D', \mathbf{H})$-stratum S' of R' coincides with the strict transform of S by π and, hence, it is closed in $(G')_0$. Denote by T' the closure of the $(S', \mathbf{H})$-stratum of R'. Then T' and $\mathbf{H}$ again satisfy the assumptions of Lemma 5.3.3, and we can construct a proper succession $\bar{\pi}'$ of permissible modifications for R' that satisfies conditions (i), (ii), and (iii) of Lemma 5.3.3 for R'.

Step 3. Let R'' be the transform of R' by $\bar{\pi}'$, let (G'', π') be the modification limit of $\bar{\pi}'$, and set $D'' = \left((\pi')^0\right)^{-1}(D')$. It follows from Steps 1 and 2 that the $(D'', \mathbf{H})$-stratum S'' of R'' and the $(S'', \mathbf{H})$-stratum T'' of R'' are closed in $(G'')_0$. In this way, we can apply successively Lemmas 5.3.1 and 5.3.2 to R'', D'', and $\mathbf{H}$ and so obtain a proper succession $\bar{\pi}''$ of permissible modifications for R'' that satisfies the properties in these lemmas.

We easily see that the composition $\bar{\pi}^*$ of $\bar{\pi}, \bar{\pi}'$, and $\bar{\pi}''$ is the desired succession.

□

Proof of Theorem $\mathbb{I}^n$ *(Theorem* 5.1.1*)* Let $R = (\mathcal{G}; X_1, \dots, X_q)^n_{\mathbb{I}}$ be as in the statement of the theorem. Since G is countable at infinity, in order to prove the theorem it is enough to show that, for every compact D in $(G)^0$, there exists a proper succession $\bar{\boldsymbol{\pi}}$ of permissible modifications for R such that $D' \cap (G')_0 \cap \bigcap_{i=1}^q X'_i = \varnothing$, where $R' = (\mathcal{G}'; X'_1, \dots, X'_q)^n_{\mathbb{I}}$ is the transform of R by $\bar{\boldsymbol{\pi}}$, $(G', \boldsymbol{\pi})$ is the modification limit of $\bar{\boldsymbol{\pi}}$, and $D' = \big((\boldsymbol{\pi})^0\big)^{-1}(D)$. This can be done by applying Lemma 5.3.4 a finite number of times, thanks to the finiteness theorem (Theorem 2.5.2 in Chap. 2). □

5.4 The Desingularization Theorem

Definition 5.4.1 A resolution datum of type $\mathbb{III}$ is a family

$$R = (G_1, \dots, G_q)_{\mathbb{III}},$$

where $q > 0$, of $\mathbb{C}$-situations that are countable at infinity and that all have the same index category (say $[G_i] = [G]$ for $1 \le i \le q$) and the same functor of ambient spaces (say $(G_i)^0 = (G)^0$ for $1 \le i \le q$). We denote by $\mathrm{Sing}(R)$ the restrictive closed complex subspace of $(G)^0$ given by

$$\mathrm{Sing}(R) = \bigcap_{i=1}^{q} (G_i)_0.$$

Definition 5.4.2 If (g, x) is a point of $\mathrm{Sing}(R)$, the **Samuel function** of R at (g, x) is the family

$$\mathbf{H}_{(R,g,x)} = (H_1, \dots, H_q),$$

with $H_i = H_{(G_i(g))_0,x}$.

Given a family $\mathbf{H} = (H_1, \dots, H_q)$ of functions $H_i : \mathbb{Z}_0 \to \mathbb{Z}_0$, the **Samuel H-stratum** of R is the unique reduced complex subspace $\mathrm{Sing}^{\mathbf{H}}(R)$ of $\mathrm{Sing}(R)$ such that, for every $g \in \mathrm{ob}([G])$,

$$\mathrm{Sing}^{\mathbf{H}}(R)(g) = \{x \in \mathrm{Sing}(R)(g) \mid \mathbf{H} = \mathbf{H}_{(R,g,x)}\}.$$

If D is a subset of $(G)^0$, the **$(D, \mathbf{H})$-stratum** of R is the minimum closed complex subspace $\mathrm{Sing}^{\mathbf{H}}(R)_D$ of $\mathrm{Sing}^{\mathbf{H}}(R)$ that coincides with $\mathrm{Sing}^{\mathbf{H}}(R)$ around every point in D.

Among q-tuples $\mathbf{H}$ as above, we define the same order relation as in Definition 5.3.1. If F is a subset of $(G)^0$, we define the F-largest Samuel function of R as Definition 5.3.1.

Definition 5.4.3 If R is a resolution datum of type $\mathbb{III}$, a **permissible center for** R is any smooth and closed complex subspace of $(G)^0$ contained in a **H**-stratum of R.

If Y is a permissible center for R, a **permissible modification** for R with center Y, or simply a permissible modification for R (if the reference to Y is clear in the context), is a pair $(R', \boldsymbol{\pi})$ such that:

(i) $R' = (G'_1, \ldots, G'_q)_{\mathbb{III}}$ is a resolution datum of type $\mathbb{III}$, and
(ii) $\boldsymbol{\pi} = (\boldsymbol{\pi}_1, \ldots, \boldsymbol{\pi}_q)$ is a family such that, for $1 \leq i \leq q$, $(G'_i, \boldsymbol{\pi}_i)$ is a modification of G_i with center Y, satisfying the condition that $(G'_i)_0$ is the strict transform of $(G_i)_0$ by $(\boldsymbol{\pi}_i)^0$.

Definition 5.4.4 Let $R = (G_1, \ldots, G_q)_{\mathbb{III}}$ be a resolution datum of type $\mathbb{III}$. A **proper succession** $\bar{\Pi}$ **of permissible modifications for** R is a set of data

$$\{\boldsymbol{\pi}_{\alpha,\beta}, R_\alpha, Y_\alpha\}_{\alpha,\beta\in\Lambda}$$

satisfying these conditions:

(i) Λ is a well-ordered set with first element 0.
(ii) For every $\alpha \in \Lambda$, $R_\alpha = (G_{\alpha,1}, \ldots, G_{\alpha,q})_{\mathbb{III}}$ is a resolution datum of type $\mathbb{III}$ with $R_0 = R$, $[G_{\alpha,i}] = [G]$, and $(G_{\alpha,i})^0 = (G)^0$, for every $i = 1, \ldots, q$.
(iii) For every $\alpha \in \Lambda$, Y_α is a permissible center for R_α and $\boldsymbol{\pi}_{\alpha+1,\alpha}$ is a modification of R_α with center Y_α.
(iv) For all $\alpha, \beta \in \Lambda$, we have $\boldsymbol{\pi}_{\alpha,\beta} = \{\boldsymbol{\pi}_{\alpha,\beta,1}, \ldots, \boldsymbol{\pi}_{\alpha,\beta,q}\}$ with $\boldsymbol{\pi}_{\alpha\beta i} : G_{\alpha i} \to G_{\beta i}$; moreover, each $\bar{\boldsymbol{\pi}}_i = \{\boldsymbol{\pi}_{\alpha,\beta,i}, G_{\alpha i}, Y_\alpha\}_{\alpha,\beta\in\Lambda}$, for $i = 1, \ldots, q$, is a proper succession of modifications of the $\mathbb{C}$-situation G_i.

Definition 5.4.5 If $\bar{\Pi}$ is a proper succession of permissible modifications for R and G'_i is the transform of G_i by $\boldsymbol{\pi}_i$, for $1 \leq i \leq q$, the resolution datum of type $\mathbb{III}$

$$R' = (G'_1, \ldots, G'_q)_{\mathbb{III}}$$

is called the **transform** of R by $\bar{\Pi}$.

Remark 5.4.1 If $\bar{\Pi} = \{\boldsymbol{\pi}_{\alpha\beta}, R_\alpha, Y_\alpha\}$ is a proper succession of permissible modifications for R, the family $\{\bar{\boldsymbol{\pi}}_i\}_{1\leq i\leq q}$ of proper modifications of G_i (see Definition 5.4.4(iv)) has the following property: If, for every $i = 1, \ldots, q$, we denote by $(G'_i, \boldsymbol{\pi}_i)$ the modification limit of $\bar{\boldsymbol{\pi}}_i$, then $(\boldsymbol{\pi}_i)^0$ is independent of i. This common morphism will be denoted by $(\boldsymbol{\pi})^0$.

Theorem 5.4.1 (Fundamental Theorem $\mathbb{III}$) *Given a resolution datum* $R = (G_1, \ldots, G_q)_{\mathbb{III}}$ *of type* $\mathbb{III}$, *there exists a proper succession* Π *of permissible modifications for* R *such that, if* R' *is the transform of* R *by* Π, *one has* $\mathrm{Sing}(R') = \varnothing$.

The proof consists of an induction process given in the following lemmas.

Lemma 5.4.1 *Let D be a compact subset of $(G)^0$ and let H be the D-largest Samuel function of R. If $\mathrm{Sing}^{\mathbf{H}}(R)_D$ is closed in $(G)^0$, there exists a proper succession $\Pi = \{\pi_{\alpha\beta}, R_\alpha, Y_\alpha\}_{\alpha,\beta\in\Lambda}$ of permissible modifications for R satisfying these conditions:*

(i) *Y_α is contained in the $\mathbf{H}$-stratum of R_α, for every $\alpha \in \Lambda$.*
(ii) *If $D' = ((\pi)^0))^{-1}(D)$ (see Remark 5.4.1), then $\mathrm{Sing}^{\mathbf{H}}(R')_{D'} = \varnothing$.*

Proof Let $\mathbf{H} = (H_1, \dots, H)$ and, for all $i = 1, \dots, q$, let S_i be the H_i-stratum of $(G_i)_0$ and $\bar{S}_i$ the closure of S_i in $(G)^0$. We construct a family of forests $(G_{(i)}, \mathbf{J}_{(i)}) = \mathscr{F}_i$, for $1 \le i \le q$, in the following way:

(a) $[G_{(i)}] = [G]$, for $i = 1, \dots, q$.
(b) $(G_{(i)})^0 = (G)^0 \setminus \bigcup_{j=1}^q (\bar{S}_i \setminus S_i)$, for $i = 1, \dots, q$.
(c) $(G_{(i)})_0 = S_i \cap (G_{(i)})^0$, for $i = 1, \dots, q$.
(d) If $g \in \mathrm{ob}([G])$, we take a $G_i(g)$-polygrove $\mathbf{J}_i(g)$ associated with $((G_i(g))_0, S_i(g))$, and then we take

$$\mathbf{J}_{(i)}(g) = \mathbf{J}_i(g)|_{(G_{(i)}(g))^0}$$

as the restriction of $\mathbf{J}_i(g)$ to the open subset $(G_{(i)}(g))^0$ of $(G_i(g))_0$.

By construction, the intersection

$$\mathscr{F}^0 = \bigcap_{i=1}^{q} \mathscr{F}_{(i)} = (G^0, J^0)$$

is defined, and $(G^0)_0 = \mathrm{Sing}^{\mathbf{H}}(R)$, by the definition of $\mathrm{Sing}^{\mathbf{H}}(R)$.

Now, consider the garden $\mathscr{G}_0 = (G^0, \mathbf{J}^0, \varnothing)$, put $L = \mathrm{Sing}^{\mathbf{H}}(R)_D$, and denote by M the closure of $(G^0)_0 \setminus L$ in $(G^0)_0$. Let $\hat{\mathscr{G}}^0 = (\hat{G}^0, \hat{\mathbf{J}}^0, \varnothing)$ be the completion of $\mathscr{G}$ and let $\hat{L}$ and $\hat{M}$ be the natural extensions of L and M to $(G^0)_0$. In this way, we get a resolution datum

$$\hat{R} = (\hat{\mathscr{G}}_0, \hat{L}, \hat{M})^n_{\mathrm{III}}$$

of type III^n, with

$$n = \max_{x\in D(g)} \{\dim_x(G_i(g))_0\} \ge \max_{x\in \mathrm{Sing}^{\mathbf{H}}(R)(g)} \{\dim_x(G_i(g))_0\}.$$

By applying Fundamental Theorem III_n to $\hat{R}$, we obtain a proper succession $\hat{\pi}$ of permissible modifications for $\hat{R}$ satisfying the conclusion of the theorem. Moreover, $\hat{\pi}$, by restriction to G^0 and by ambient expansion by $(G^0)^0 \to (G_{(i)})^0$, gives a family $\{\bar{\pi}\}_{1\le i\le q}$ of proper modifications with the property that $(\pi_i)^0$ is independent of i, where (G_i, π_i) is the modification limit of $\bar{\pi}_i$. In this case $\Pi = (\pi_i, \dots, \pi_q)$ is a proper succession of permissible modifications for R, which proves the lemma. □

Lemma 5.4.2 *Let S be a closed complex subspace of* $\mathrm{Sing}(R)$ *such that there exists a system* $\mathbf{H}$ *of functions with* $S \setminus \mathrm{Sing}^{\mathbf{H}}(R)$ *nowhere dense in S. Then there exists a proper succession*

$$\Pi = \{\boldsymbol{\pi}_{\alpha,\beta}, R_\alpha, Y_\alpha\}_{\alpha,\beta \in \Lambda}$$

of permissible modifications for R satisfying these conditions:

(i) *For every $\alpha \in \Lambda$, Y_α is contained in the strict transform of S in R_α and in the* $\mathbf{H}_\alpha$*-stratum of R_α, with* $\mathbf{H}_\alpha > \mathbf{H}$.
(ii) *The strict transform S' of S by $(\boldsymbol{\pi})^0$ is contained in* $\mathrm{Sing}(R')$, *where R' is the transform of R by $\bar{\boldsymbol{\pi}}$.*

Proof The assumptions imply that $\dim S$ is finite, so we can prove the lemma by induction on $\dim S$. For $\dim S = 0$ the lemma is trivial, so we assume the lemma true for $\dim S < s$, where $s > 0$, and we prove it for $\dim S = s$. Since G_i is countable at infinity, for $1 \le i \le q$, the lemma will be proved if we show that, for every compact subset D of $(G)^0$, there exists a proper succession $\bar{\boldsymbol{\pi}}$ of permissible modifications for R that satisfies condition (i) and the following condition:

(ii′) If $D' = \big((\boldsymbol{\pi})^0\big)^{-1}(D)$, then $S' \cap D'$ is contained in $\mathrm{Sing}^{\mathbf{H}}(R')$.

Now we look at $S \cap D$: if $S \cap D \subset \mathrm{Sing}^{\mathbf{H}}(R)$, there is no problem. Assume that $S \cap D \not\subset \mathrm{Sing}^{\mathbf{H}}(R)$, and define a $\mathbb{C}$-situation G_{q+1} by setting $[G_{q+1}] = [G]$, $(G_{q+1})^0 = (G)^0$, and $(G_{q+1})_0 = S$. Denote by R^* the resolution datum of type $\mathbb{III}$

$$R^* = (G_1, \dots, G_q, G_{q+1})_{\mathbb{III}}.$$

Let H^* be the D-largest Samuel function of R^* and let S^* be the closure of $\mathrm{Sing}^{\mathbf{H}}(R^*)_D$ in $(G)^0$. Then $\dim S^* < s$ and, by the induction hypothesis, there exists a proper succession $\bar{\boldsymbol{\pi}}_1^*$ of permissible modifications for R^* that satisfies conditions (i) and (ii) for R^* and S^*. Therefore, if we set $D_1^* = \big((\boldsymbol{\pi}_1^*)\big)^{-1}(D)$ (see Remark 5.4.1), we conclude that $\mathrm{Sing}^{\mathbf{H}}(R_1^*)_{D_1}$ is closed in the $\mathbb{C}$-situation of R_1^*, where R_1^* is the transform of R^* by $\bar{\boldsymbol{\pi}}_1^*$.

Applying Lemma 5.4.1, we can construct a proper succession $\bar{\boldsymbol{\pi}}_2^*$ of permissible modifications for R_1^* in such a way that, if $D_2^* = \big((\boldsymbol{\pi}_2^*)\big)^{-1}(D_1^*)$ and if R_2^* is the transform of R_1^* by $\bar{\boldsymbol{\pi}}_2^*$, then $\mathrm{Sing}^{\mathbf{H}}(R_2^*) = \varnothing$.

The composition $\bar{\boldsymbol{\pi}}^*$ of $\bar{\boldsymbol{\pi}}_1^*$ and $\bar{\boldsymbol{\pi}}_2^*$ induces a proper succession $\bar{\boldsymbol{\pi}}_1$ of permissible modifications for R that satisfies condition (i) of the present Lemma 5.4.2. Let S_1 be the strict transform of S by $(\boldsymbol{\pi}_1)^0$ and R_1 the transform of R by $\bar{\boldsymbol{\pi}}_1$. If $S_1 \cap D_1 \subset \mathrm{Sing}(R_1)$, where $D_1 = D_2^*$, there is no problem. If not, we obtain the desired succession by iterating the process a finite number of times, thanks to the finiteness theorem (Theorem 2.5.2 of Chap. 2). □

Proof of Theorem $\mathbb{III}$ (Theorem 5.4.1) Since all the G_i in R are countable at infinity, it is enough to show that, for every compact subset D of $(G)^0$, there exists a proper succession $\bar{\pi}$ of permissible modifications for R such that, if R_1 is the transform of R by $\bar{\pi}$ and $D_1 = ((\pi)^0)^{-1}(D)$, then $D_1 \cap \mathrm{Sing}(R_1) = \varnothing$.

If $\mathbf{H}$ is the D-largest Samuel function of R and S is the closure of $\mathrm{Sing}^{\mathbf{H}}(R)_D$ in $(G)^0$, we can apply Lemmas 5.4.2 and 5.4.1, in that order, to obtain a proper succession of permissible modifications for R, which eliminates $\mathrm{Sing}^{\mathbf{H}}(R)_D$. Then, by taking the new D-largest Samuel function of the result, and by repeating the process, we arrive at the proof of our theorem after a finite number of steps, again thanks to the finiteness theorem (Theorem 2.5.2 of Chap. 2). □

Finally, we arrive at the objective of this book:

Theorem 5.4.2 (Desingularization of a complex-analytic space) *Let X be a complex-analytic space that is countable at infinity, and let S be a closed complex subspace of X containing all the singular points of X. There exists a proper morphism $\pi : X' \to X$ having the following properties:*

(i) *The complex analytic space X' is smooth and π induces an isomorphism*

$$X' \setminus \pi^{-1}(S) \to X - S.$$

(ii) *$\pi^{-1}(S)$ is a hypersurface that is set-theoretically a locally finite union of smooth hypersurfaces having normal crossings everywhere.*
(iii) *If U and V are two open subsets of X and φ is an isomorphism $X|_U \to X|_V$ such that $\varphi(S|_U) = S|_V$, there exists an isomorphism $\varphi' : X'|_{\pi^{-1}(U)} \to X'|_{\pi^{-1}(V)}$ such that $\varphi(\pi|_U) = (\pi|_V)\varphi'$.*

Proof Let G_1 be a full $\mathbb{C}$-situation of X, and consider the resolution datum of type $\mathbb{III}$:

$$R = (G_1)_{\mathbb{III}}.$$

An application of Theorem $\mathbb{III}$ yields a proper succession $\bar{\pi}_1$ of permissible modifications for R such that $\mathrm{Sing}(R') = \varnothing$, where $R' = (G_1')_{\mathbb{III}}$ is the transform of R by $\bar{\pi}_1$. Then $\bar{\pi}_1$ induces a locally finite succession $\bar{\pi}(1)$ of blow-ups of X, with limit $\pi(1) : X(1) \to X$ such that, since $\bar{\pi}_1$ is permissible for R and $\mathrm{Sing}(R') = \varnothing$, one has:

(a) The centers of $\bar{\pi}(1)$ are smooth closed complex subspaces of X, contained in the Samuel strata of the transforms of X.
(b) All the successive transforms of X by the blow-ups of $\bar{\pi}(1)$ satisfy condition (iii) in the statement of the present theorem (because G_1 is full).
(c) $X(1)$ is smooth.

On the other hand, given two closed complex subspaces Y_1 and Y_2 of X such that $Y_1 \cap Y_2 = \varnothing$, we obtain the same result no matter which of the two we choose as the center of the first blow-up and which as the second. Therefore, given a closed subspace S of X containing the singular locus of X, we can suppress in $\bar{\pi}(1)$ all the blow-ups whose centers are disjoint from S. Since the new succession $\bar{\boldsymbol{\pi}}'(1)$ obtained in this way also satisfies properties (a), (b), and (c) above, its modification limit $\boldsymbol{\pi}'(1) : X'(1) \to X$ satisfies conditions (i) and (iii) in the statement of the theorem. Hence, it remains only to modify $X'(1)$ so that condition (ii) also holds.

Since $X'(1)$ is smooth, it is a disjoint union of smooth, connected, complex spaces. So we can easily reduce the problem to the case in which $X'(1)$ is connected. $\dim X'(1) = n < \infty$. We can define a $\mathbb{C}$-situation G in the following way:

(d) ob([G]) is the collection of open subsets of X. If $g, g' \in$ ob([G]), an element of $\mathrm{Hom}_{[G]}(g, g')$ is by definition a morphism $\varphi : X|_g \to X|_{g'}$ such that there exists an open subset g'' of g' such that φ maps $X|_g$ to $X|_{g''}$ and $S|_g$ to $S|_{g''}$.
(e) For every $g \in$ ob([G]), $G(g)$ is the inclusion

$$\boldsymbol{\pi}'(1)^{-1}(S|_g) \to X'(1)|_{\boldsymbol{\pi}'(1)^{-1}(g)},$$

and, for every $\varphi \in \mathrm{Hom}_{[G]}(g, g')$, $G(\varphi)$ is the restrictive morphism of models obtained by lifting φ by $\boldsymbol{\pi}'(1)$. The lifting is possible by (i).

If $\mathbf{I}$ is the ideal of S in X, we define a forest (G, J) as follows:

(f) For every $g \in$ ob([G]), $J(g)$ is the polygrove associated with the idealistic exponent $((\mathbf{I}.\mathscr{O}_{(G(g))^0}, 1)$ on $(G(g))^0 = X'(1)|_{\bar{\boldsymbol{\pi}}'(1)^{-1}(g)}$.

In this way, we construct an idealistic garden

$$\mathscr{G} = (G, \mathbf{J}, \varnothing),$$

its completion

$$\hat{\mathscr{G}} = (\hat{G}, \hat{\mathbf{J}}, \varnothing),$$

and the resolution datum of type III^n

$$\hat{R} = (\hat{\mathscr{G}}, (\hat{G})_0, \varnothing)^n_{\mathrm{III}}.$$

By applying Theorem III^n, we obtain a proper succession $\bar{\boldsymbol{\pi}}$ of permissible modifications for R, with modification limit $(G', \boldsymbol{\pi})$ such that $(G')_0 = \varnothing$. Then $\bar{\boldsymbol{\pi}}$ induces a proper morphism $\pi' : X' \to X'(1)$, and the composition $\boldsymbol{\pi} = \pi'.\boldsymbol{\pi}'(1)$ satisfies the conditions of our theorem. □

Chapter 6
Epilogue: Singularities of Differential Equations

Josè Manuel, I propose to you a problem:
Solve all the differential equations
(H. Hironaka. 1976)

The problem of resolution of singularities of an algebraic or analytic variety is, at least in its local formulation, close related with another problem, the parametrization of a neighborhood of a point on the variety, i.e. the problem of finding a solution, in some sense, of the system of equations defining the variety.

Given a system (S) of algebraic or analytic equations, let V be the variety germ of zeroes of (S) at some point $\mathbf{p}$ (assume we are dealing with well-chosen systems of equations, that is with a basis of the ideal of all functions which vanish on V). Then we have two possibilities: either the implicit function theorem is applicable to S or not. In the first case, i.e. $\mathbf{p}$ is a non-singular point, the implicit function theorem gives us a solution of the system on a neighborhood of $\mathbf{p}$, which means that there are local parametric equations of a representative of V around $\mathbf{p}$.

In the second case $\mathbf{p}$ is a singular point of V. Now the local resolution of the singularities of V at $\mathbf{p}$ gives us parametrizations of some special subsets of V (wedges following Walker's notation [118]).

This idea of solving equations is the original purpose of some classical proofs of the resolution of surface singularities (Jung [77] or Walker [118] for instance) inspired by the construction of local parametrizations by Hensel [60].

The problem of resolution of systems of ordinary differential equations may be reduced to the problem of finding integral varieties of systems of differential forms. If we have, for instance, a system of ordinary differential equations in the complex plane

$$f_i(x, y, y', \ldots, y^{(r)}) = 0,\ 1 \leq\ i\ \leq n,$$

solving the system is equivalent to solving the problem of finding integral curves for the family of differential forms

$$\omega_j = z_j dx - dz_{j-1},\ 1 \leq\ j\ \leq r\ , z_0 = y,$$

J. M. Aroca et al., *Complex Analytic Desingularization*,
https://doi.org/10.1007/978-4-431-49822-3_6

over the variety X with equations

$$f_i(x, z_0, \dots, z_r) = 0, \ 1 \leq i \leq n,$$

and projecting these curves onto the (x, z_0)-plane. X will be, in general a singular variety, but one may solve the singularities of X, and by pull-back, transform the original problem into the following:

Given a complex-analytic manifold Y, a normal crossings divisor D on Y, and a system of differential forms $\mathscr{S} = \{\omega_1, \dots, \omega_r\}$ find integral curves for $\mathscr{S}$ not contained in D.

Then, in order to solve differential equations we need to study differential forms on complex analytic manifolds. There exists a classical theorem (Frobenius theorem), similar to the implicit function theorem, that gives the integral varieties in the non-singular case. Then it is very natural to consider the possibility that the formal apparatus of resolution of singularities will enable us to *solve the singularities* of differential forms, and transform any system in a non-singular system.

The fact that there are more than formal analogies between the problems of solving differential equations and the reduction of singularities, can be easily seen in the case of ordinary differential equations of first order and first degree.

Consider the equation

$$a(x, y) + b(x, y)y' = 0, \qquad (*)$$

with a, b convergent power series in some disk D centered at $(0, 0)$. From a geometric point of view, obtaining this equation is equivalent to obtaining the differential form

$$\omega = a(x, y)d(x) + b(x, y)d(y)$$

or an analytic field of directions over the disk D, by taking on each point (x_0, y_0) the straight line through (x_0, y_0) with the direction of $(-b(x_0, y_0), a(x_0, y_0))$, provided that we are on a regular point, i.e. $(a(x_0, y_0), b(x_0, y_0)) \neq (0, 0)$.

Then solving the equation $(*)$ with the initial conditions (x_0, y_0) is done finding an analytic curve $t \mapsto (x(t), y(t))$ such that $(x(0), y(0)) = (x_0, y_0)$ and

$$a(x(t), y(t)) \frac{d(x(t))}{dt} + b(x(t), y(t)) \frac{d(y(t))}{dt} = 0, \ \text{for all } t.$$

This means that the curve is tangent, at each regular point for the equation, to the line determined on this point by ω. Picard's theorem proves that for each regular point there is one and only one solution.

When $a(x_0, y_0) = b(x_0, y_0) = 0$, ω does not define a line by (x_0, y_0), and we will say that the point is singular. In general the idea "**p** *is a singular point*" is associated with the existence of several different limit directions for sequences of directions associated with sequences of regular points converging to **p**. Blowing-up **p** points we separate the limit directions, simplifying in this way the local structure of the differential equation.

This idea was introduced by Seidenberg [105] in order to prove the existence of a sort of reduction of singularities for differential forms in dimension two. Seidenberg's theorem has been used by C. Camacho and P. Sad [19] to prove the existence of at least one analytic integral curve of a differential form in $\mathbb{C}^2$ at a singular point.

For people working on singularities, there is another reason of interest in the problem of reduction of singularities for differential forms.

The characteristic, for $p > 0$, of the hypersurface singularity defined by the equation

$$z^{p^e} = f(x_1, \dots, x_n)$$

is especially interesting.

The singularities of this hypersurface are the same as the singularities of the differential form

$$\omega = df = \sum_{i=1}^{n} \frac{\partial f}{\partial x_i} dx_i .$$

The behavior under blowing-up of both objects is very similar. This is the basis of a program of J. Giraud [52, 53]:

> A form ω on K^2 has strong similarity with a curve, and helps us to solve the singularities of a surface, closely connected to a form in K^3, etc..

This chapter is devoted to a survey of some results about reduction of singularities of foliations, and existence of integral varieties inspired by the ideas and methods of Hironaka and Giraud, combined with ideas of other people coming from the area of dynamical systems: Camacho, Cerveau, Mattei, Moussu, Sad and others.

6.1 Foliations and Vector Fields

Let $(M, \mathcal{O}_M)$ be a complex-analytic manifold of pure dimension n; we will use the following common notation:

- $T(M)$, and $T^*(M)$: Respectively the *tangent bundle* and *cotangent bundle* of M.
- Θ_M: The sheaf of sections of $T(M)$, i.e. the sheaf of germs of *holomorphic vector fields* on M.
- Ω^r_M: The sheaf of sections of $\bigwedge^r T^*(M)$, i.e. the sheaf of *germs of holomorphic r-forms* on M,

- $d : \Omega^r_M \to \Omega^{r+1}_M$: the *exterior derivative*.
- $<,>: \Omega^1_{M\mathbf{p}} \times \Theta_{M\mathbf{p}} \to \mathcal{O}_{M\mathbf{p}}$: the *canonical pairing* consequence of the equality $\Omega^1_M = \Theta^*_M$.

Let U be an open set of M and $\mathbf{p} \in U$; any section $X \in \Theta_M(U)$ (resp. $\omega \in \Omega^r_M(U)$), has a germ: $X_{\mathbf{p}} \in \Theta_{M\mathbf{p}}$ (resp. $\omega_{\mathbf{p}} \in \Omega^r_{M\mathbf{p}}$) at $\mathbf{p}$ and a value at $\mathbf{p}$, $X(\mathbf{p}) \in T_{\mathbf{p}}(M)$ (resp. $\omega(\mathbf{p}) \in \bigwedge^r T_{\mathbf{p}}(M)^*$).

The canonical pairing induces a relation of *orthogonality* as follows:

- For any coherent $\mathcal{O}_M$-submodule $\mathcal{F} \subset \Omega^1_M$, $\mathcal{F}^\perp$ is the coherent submodule of Θ_M given by:

$$\forall \mathbf{p} \in M,\ \mathcal{F}^\perp_{\mathbf{p}} = \{X \in \Theta_{M\mathbf{p}} \mid <\omega, X>= 0,\ \forall \omega \in \mathcal{F}_{\mathbf{p}}\}.$$

- For any coherent $\mathcal{O}_M$-submodule $\mathcal{T} \subset \Theta_M$, $\mathcal{T}^\perp$ is the coherent submodule of Ω^1_M given by:

$$\forall \mathbf{p} \in M,\ \mathcal{T}^\perp_{\mathbf{p}} = \{\omega \in \Omega^1_{M\mathbf{p}} \mid <\omega, X>= 0,\ \forall X \in \mathcal{T}_{\mathbf{p}}\}.$$

In general:

$$\forall\, \mathcal{F} \subset \Omega^1_M,\ \mathcal{F} \subset \mathcal{F}^{\perp\perp},\ \forall\, \mathcal{T} \subset \Theta_M,\ \mathcal{T} \subset \mathcal{T}^{\perp\perp}.$$

If the equality happens we will say that $\mathcal{F}$ or $\mathcal{T}$ are *saturated*. The orthogonality relation restricted to saturated coherent submodules is bijective.

Let us remember that if $\mathcal{N}$ is a coherent $\mathcal{O}_M$- submodule of a locally free $\mathcal{O}_M$-module $\mathcal{P}$, a point $\mathbf{p} \in M$ is called a *singular point* of $\mathcal{N}$ if and only if $\mathcal{P}_{\mathbf{p}}/\mathcal{N}_{\mathbf{p}}$ is not a free $\mathcal{O}_{M\mathbf{p}}$-module. In the sequel we will denote by $\mathrm{Sing}(\mathcal{F})$ the set of singular points of $\mathcal{F}$.

Definition 6.1.1 A *q-codimensional singular foliation* $\mathcal{F}$ over a complex-analytic manifold M is an $\mathcal{O}_M$-submodule of Ω^1_M such that:

(1) $\mathcal{F}$ is locally free of rank q.
(2) $\mathcal{F}$ is *integrable*, i.e. at any point $\mathbf{p} \in M \setminus \mathrm{Sing}(\mathcal{F})$,

$$d\mathcal{F}_{\mathbf{p}} \subset \Omega^1_{M\mathbf{p}} \wedge \mathcal{F}_{\mathbf{p}}.$$

(3) $\mathcal{F} = \mathcal{F}^{\perp\perp}$.

A *regular foliation* is a foliation with empty singular locus.

The dual of a singular foliation is a *singular distribution*:

Definition 6.1.2 A *q-dimensional singular distribution* $\mathcal{T}$ over a complex-analytic manifold M is a $\mathcal{O}_M$-submodule of Θ_M such that:

(1) $\mathcal{T}$ is locally free of rank q.

(2) $\mathscr{T}$ is involutive, i.e. at any point $\mathbf{p} \in M \setminus \mathrm{Sing}(\mathscr{T})$, the Lie product $[\mathscr{T}_{\mathbf{p}}, \mathscr{T}_{\mathbf{p}}] \subset \mathscr{T}_{\mathbf{p}}$.
(3) $\mathscr{T} = \mathscr{T}^{\perp\perp}$.

A *regular distribution* is a distribution with empty singular locus.

Then orthogonality is a bijection between singular foliations of codimension q and singular distributions of dimension q.

Let $(z_1, \dots, z_n)$ be the coordinate functions in a local chart of M centered at $\mathbf{p} \in M$:

- The stalk $\mathscr{O}_{M\mathbf{p}}$ is isomorphic to the local algebra of convergent power series $\mathbb{C}\{\mathbf{z}\} = \mathbb{C}\{z_1, \dots, z_n\}$.
- The stalk $\Theta_{M\mathbf{p}}$ is a free $\mathscr{O}_{M\mathbf{p}}$-module with basis:

$$\left\{\frac{\partial}{\partial z_1}, \dots, \frac{\partial}{\partial z_n}\right\}.$$

- The stalk $\Omega^1_{M\mathbf{p}}$ is a free $\mathscr{O}_{M\mathbf{p}}$-module with basis:

$$\{dz_1, \dots, dz_n\}.$$

Remark 6.1.1 Let $\mathscr{F}$ be a q-codimensional singular foliation on M; $\mathrm{Sing}(\mathscr{F})$ is a closed complex-analytic subspace of M. We may describe it explicitly by taking a sufficiently small local chart U in M with coordinate functions $(z_1, \dots, z_n)$, and a basis $\{\omega_1, \dots, \omega_q\}$ of $\mathscr{F}(U)$ as $\mathscr{O}_M(U)$-module. Then:

$$\omega_i = \sum_{j=1}^{n} g_{ij} dz_j, \ 1 \le i \le q, \ g_{ij} \in \mathscr{O}_M(U).$$

and:

$$\begin{aligned}\mathrm{Sing}(\mathscr{F}) \cap U &= \{\mathbf{p} \in U \mid rank(g_{ij}(\mathbf{p})) < q\} \\ &= \{\mathbf{p} \in U \mid \omega_1(\mathbf{p}) \wedge \cdots \wedge \omega_q(\mathbf{p}) = 0\}.\end{aligned}$$

The integrability condition means, over this chart U, that for any non-singular point $\mathbf{p} \in U \setminus \mathrm{Sing}(\mathscr{F})$, there exist $\{\eta_{ij}\}_{1 \le i, j \le q} \subset \Omega^1_{M\mathbf{p}}$ such that:

$$d\omega_i = \sum_{j=1}^{q} \eta_{ij} \wedge \omega_j.$$

A classical Frobenius theorem is now applicable.

Theorem 6.1.1 (Frobenius Theorem) *A system of germs of differential forms $L = \{\omega_1, \dots, \omega_q\} \subset \Omega^1_{M\mathbf{p}}$ with the property:*

$$\omega_1(\mathbf{p}) \wedge \cdots \wedge \omega_q(\mathbf{p}) \neq 0$$

is integrable locally at **p** *if and only if there exist germs:*

$$\{f_i\}_{1 \le i \le q} \subset \mathcal{O}_{M\mathbf{p}}, \ \{g_{ij}\}_{1 \le i,j \le q} \subset \mathcal{O}_{M\mathbf{p}}$$

such that:

$$det(g_{ij})(\mathbf{p}) \neq 0, \ and, \ \omega_i = \sum_{j=1}^{q} g_{ij} df_j, \ 1 \le i \le q.$$

The theorem will be applied to a foliation $\mathcal{F}$, a nonsingular point **p**, and a basis $\{\omega_1, \ldots, \omega_q\}$ of $\mathcal{F}_{\mathbf{p}}$. There exists a certain neighborhood V of **p**, such that $\{df_1, \ldots, df_q\}$ is a basis of $\mathcal{F}(V)$, and $\{f_1, \ldots, f_q\}$ are part of a coordinate system on V, because the matrix (g_{ij}) is invertible. Then there is an atlas over $M \setminus \text{Sing}(\mathcal{F})$ compatible with the complex-analytic structure of M, $\{(U_i, \varphi_i)\}_{i \in I}$, each chart (U_i, φ_i) with coordinate functions $(z_{i1}, \ldots, z_{in})$, such that $\{dz_{i1}, \ldots, dz_{iq}\}$ is a basis of $\mathcal{F}(U_i)$. This means that in the change of charts, from (U_i, φ_i) to (U_j, φ_j), $(z_{j1}, \ldots, z_{jq})$ depends only on $(z_{i1}, \ldots, z_{iq})$.

The classical definition of complex-analytic or holomorphic foliation on M is as follows:

Definition 6.1.3 A *q-codimensional foliated holomorphic atlas* on M is an analytic atlas compatible with the structure of M, $\mathcal{F} = \{(U_i, \phi_i)\}_{i \in I}$, such that:

(1) $\phi_i(U_i) = P_i \times D_i$ with P_i a polydisk in $\mathbb{C}^{n-q}$ and D_i a polydisk in $\mathbb{C}^q$, both centered at the origin.
(2) If $\mathbf{x}_i = (x_{i1}, \ldots, x_{in-q})$ and $\mathbf{y}_i = (y_{i1}, \ldots, y_{iq})$ are respectively the coordinates on P_i and D_i, then for all i, j there are holomorphic functions h_{ij}, g_{ij} such that

$$\phi_j \phi_i^{-1}(\mathbf{x}_i, \mathbf{y}_i) = (h_{ij}(\mathbf{x}_i, \mathbf{y}_i), g_{ij}(\mathbf{y}_i)), \forall (\mathbf{x}_i, \mathbf{y}_i) \in \phi_i(U_i \cap U_j).$$

Two foliated atlases are *equivalent* if their union is a foliated atlas. A *foliation* on M is an equivalence class of foliated atlases or a maximal foliated atlas.

Remark 6.1.2 The sets $T_{i,\mathbf{p}} = \phi_i^{-1}(P_i \times \{\mathbf{p}\})$, $\mathbf{p} \in D_i$ are called *plaques* of the foliation. One can define an equivalence relation on M by

$$\mathbf{p} \sim \mathbf{s} \Leftrightarrow \text{There are plaques } T_1, \ldots, T_r \text{ such that}$$
$$T_i \cap T_{i+1} \neq \varnothing, \text{if } 0 < i < r, \ \mathbf{p} \in T_1, \ \mathbf{s} \in T_r.$$

Each equivalence class in this relation is called a *leaf* of the foliation. We will denote by $L_{\mathbf{p}}$ the leaf that contains the point **p**. The leaves are connected analytic manifolds of dimension $n - q$.

Then as we have seen, by the Frobenius theorem, a singular foliation $\mathscr{F}$ on M is exactly a foliation over $M \setminus \mathrm{Sing}(\mathscr{F})$.

The general theory of singular foliations over complex manifolds may be seen in the book of T. Suwa [113] and the extension of the theory to singular foliations over complex-analytic spaces in Saeki [103]. From now on we will restrict ourselves to codimension-one foliations or dimension-one foliations, i.e. rank-1 free submodules of $\Omega^1_{M\mathbf{p}}$ and $\Theta_{M\mathbf{p}}$, because from the point of view of reduction of singularities there exists some work only in these cases.

By following Cano, Cerveau, Deserti [28]:

Definition 6.1.4 A codimension-one singular holomorphic *prefoliation* is a locally free rank-1 submodule $\mathscr{F}$ of the cotangent sheaf Ω^1_M on a complex analytic manifold M, such that $\mathscr{F} \wedge d\mathscr{F} = 0$.

Remark 6.1.3 The definition means that $\mathscr{F}$ is locally generated by the pull-back, via local charts, of a differential 1-form

$$\omega = a_1(\mathbf{z})dz_1 + \cdots + a_n(\mathbf{z})dz_n, \quad a_i(\mathbf{z}) \in \mathbb{C}\{\mathbf{z}\},\ 1 \leq i \leq n,$$

verifying the integrability condition: $\omega \wedge d(\omega) = 0$. Usually we will call these forms *local generators* of $\mathscr{F}$. By the division lemma of K. Saito [104] two differential forms ω and ω' generate the same foliation if and only if $\omega \wedge \omega' = 0$. In the sequel we will identify a germ of foliation on M at $\mathbf{p}$ with its image in a local chart of M centered at $\mathbf{p}$, i.e. with a germ of foliation at the origin on $\mathbb{C}^n$.

The condition of integrability in Definition 6.1.1 is, in codimension one, equivalent to Definition 6.1.4 and, also in codimension one, $\mathscr{F}_\mathbf{p} = \mathscr{F}_\mathbf{p}^{\perp\perp}$ if and only if the $\mathscr{O}_{M\mathbf{p}}$-module $\Omega^1_{M\mathbf{p}}/\mathscr{F}_\mathbf{p}$ is torsion free. This means that a prefoliation is a foliation if and only if for any local generator ω of $\mathscr{F}_\mathbf{p}$, there are $a_1, \ldots, a_n \in \mathbb{C}\{\mathbf{z}\}$ such that

$$\omega = a_1(\mathbf{z})dz_1 + \cdots + a_n(\mathbf{z})dz_n, \text{ and } g.c.d.\{a_1(\mathbf{z}), \ldots, a_n(\mathbf{z})\} = 1.$$

The singular locus of a prefoliation $\mathscr{F}$ is the closed analytic subset of M, locally defined by the ideal of $\mathbb{C}\{\mathbf{z}\}$ generated by $\{a_1, \ldots, a_n\}$. Then if $\mathscr{F}$ is a prefoliation generated at $\mathbf{p}$ by $\omega = \sum_1^n a_i(\mathbf{z})dz_i$, with $g.c.d.\{a_1(\mathbf{z}), \ldots, a_n(\mathbf{z})\} = a(\mathbf{z}),\ \omega/a(\mathbf{z})$ is a local generator of its saturation $\mathscr{F}^{\perp\perp}$, and both foliations coincide over $U \setminus \mathrm{Sing}(\mathscr{F}^{\perp\perp})$.

Globally the singular locus of a prefoliation is defined by a coherent ideal $\mathscr{J}$ of $\mathscr{O}_M$, with stalks constructed as above, and the saturated foliation verifies: $\mathscr{J}\mathscr{F}^{\perp\perp} = \mathscr{F}$.

Sometimes it is convenient to consider the sheaf of meromorphic functions over M, $\mathscr{M}_M$ instead of $\mathscr{O}_M$, then a *singular integrable hyperplane field* on M is a one-dimensional integrable vector subspace of the meromorphic cotangent sheaf. A singular integrable hyperplane field $\mathscr{H}$ can be locally generated by any of its non-zero sections, and there is a coherent ideal sheaf $\mathscr{J}$ such that $\mathscr{F} = \mathscr{J}\mathscr{H}$ is a foliation over M.

Remark 6.1.4 In dimension two, the integrability condition is automatically verified and a singular foliation is locally a germ of a 1-form:

$$\omega = a(x, y)dx + b(x, y)dy, \quad a(x, y), b(x, y) \in \mathbb{C}\{x, y\}, \ g.c.d.(a, b) = 1$$

which is equivalent to the two following ordinary differential equations:

$$f(x, y, y') = a(x, y) + b(x, y)y' = 0, \quad g(x, y, x') = a(x, y)x' + b(x, y) = 0,$$

where

$$y' = \frac{dy}{dx}, \ x' = \frac{dx}{dy}.$$

By duality this foliation is equivalent also to a complex vector field, or more precisely, to a complex-analytic one-dimensional distribution over M, generated by the vector field:

$$X = -b(x, y)\frac{\partial}{\partial y} + a(x, y)\frac{\partial}{\partial x}.$$

Let $\mathscr{F}$ be a regular codimension-one foliation on M; a non-singular analytic hypersurface H of M is called *invariant* by $\mathscr{F}$ if their connected components are leaves of the foliation. If $f \in \mathscr{O}_M(U)$ verifies that $H = \{f = 0\}$ is a regular hypersurface, and if $\mathscr{F}|_U$ is a regular foliation globally defined by $\omega \in \Omega^1_M(U)$, H is invariant by $\mathscr{F}|_U$ if and only if

$$\exists \mu \in \Omega^2_M(U), \ \omega \wedge df = f\mu.$$

Definition 6.1.5 Let $\mathscr{F}$ be a codimension-one foliation; a function $f \in \mathscr{O}_M(M)$ is a *first integral* of $\mathscr{F}$ if $\omega \wedge df = 0$ for all local sections ω of $\mathscr{F}$ (i.e. the leaves of the foliation are the level curves of f, $f =$ constant).

An analytic hypersurface H of M is an *integral hypersurface* of $\mathscr{F}$ if for any point $\mathbf{p} \in M$, h being a local (reduced) equation of H, and ω a generator of $\mathscr{F}$ at $\mathbf{p}$, one has

$$\omega \wedge dh = h\mu, \quad \mu \in \Omega^2_{M,\mathbf{p}}.$$

Remark 6.1.5 In dimension two and locally at a point we may suppose that we are in an open neighborhood of the origin in $\mathbb{C}^2$. An irreducible analytic curve $\mathscr{C}$ given by its reduced equation $h(x, y) = 0$ is an integral curve of $\omega = a(x, y)dx + b(x, y)dy$, if:

$$dh \wedge \omega = fhdx \wedge dy \Rightarrow b\frac{\partial h}{\partial x} - a\frac{\partial h}{\partial y} = hf. \qquad (*)$$

Then there are several interpretations:

- From the geometric point of view, ω defines at any non-singular point (α, β) a straight line passing through (α, β) and orthogonal to $(a(\alpha, \beta), b(\alpha, \beta))$. Then $\mathscr{C}$ is an integral curve of ω if, at any point of $\mathscr{C}$ non-singular neither in $\mathscr{C}$ nor in ω, $\mathscr{C}$ is tangent to the line defined by ω at this point.
- From the point of view of analysis, at any non-singular point $\mathbf{p} \in \mathscr{C}$ is:

$$\left(\frac{\partial h}{\partial x}(\mathbf{p}), \frac{\partial h}{\partial y}(\mathbf{p})\right) \neq (0, 0)$$

  and $\mathscr{C}$ admits an equation in $\mathbf{p}$, either $y - g(x) = 0$ or $x - g(y) = 0$. Then $\mathscr{C}$ is an integral curve if, in the first case $a + bg' = 0$, and in the second $ag' + b = 0$.

One may extend these ideas to formal power series and consider *formal integral hypersurfaces*. A foliation may have formal, non-convergent, integral hypersurfaces; for instance the differential form $x^2dy + (-y + x)dx$ (Euler equation) has $x = 0$ as an integral curve and has also the formal, non-convergent, integral curve

$$y = \sum_{i=0}^{\infty} n!x^{n+1}.$$

We want to study the existence of integral hypersurfaces. Our method will be, first the reduction of the singularities to some which are as simple as possible, and then to show that in these singularities, possibly with some additional conditions, there are solutions. In dimension two there exist always integral curves, but in dimension three we will need additional conditions, because there is a form constructed by Darboux:

$$\omega = (x^n z - y^{n+1})dx + (y^n x - z^{n+1})dy + (z^n y - x^{n+1})dz, n \geq 3,$$

that has no (even formal) integral hypersurfaces (Jouanolou [76]).

6.2 Reduction of Singularities of Foliations in Dimension Two

This section is dedicated to proving that it is possible to reduce, by blowing-up points, the singularities of any foliation in dimension two, to simpler ones which are stable by blowing-up.

The singularities of foliations in dimension two are isolated, but there are many differences between resolution of singularities of plane curves and reduction of singularities of codimension-one foliations in $\mathbb{C}^2$. The greatest difference is that some singularities of foliations do not disappears; we will then talk of reduction of singularities instead of resolution. The theorem of reduction of singularities of foliations in dimension two was first proved by Seidenberg [105]. We will give here the sketch of one proof of F. Cano [28]. Throughout this section, foliation means singular foliation, and we will work only in dimension two.

Let $\mathscr{F}$ be a foliation on a complex-analytic manifold of dimension two, M, and let $\pi : \tilde{M} \longrightarrow M$ be the blow-up of M with center $\mathbf{p}$. $\mathscr{F}$ is generated by a saturated form ω in a local chart centered at $\mathbf{p}$:

$$\omega = a_1(x_1, x_2)dx_1 + a_2(x_1, x_2)dx_2.$$

Then $\pi^*(\mathscr{F})$ is locally generated in the charts of the blowing-up by:

$$\pi_1^*(\omega) = [a_1(u_1, u_1u_2) + u_2a_2(u_1, u_1u_2)]du_1 + u_1a_2(u_1, u_1u_2)du_2,$$

$$\pi_2^*(\omega) = u_2a_1(u_1u_2, u_2)du_1 + u_1[a_1(u_1u_2, u_2) + a_2(u_1u_2, u_2)]du_2.$$

As in the case of plane curves, the first invariant to be considered is the order or multiplicity.

Definition 6.2.1 We will denote the *order* of $\mathscr{F}$ at $\mathbf{p} \in M$ by the number:

$$\nu_{\mathbf{p}}(\mathscr{F}) = max\{r \mid \mathscr{F}_{\mathbf{p}} \subset m_{\mathbf{p}}^r \Omega^1_{M\mathbf{p}}\}.$$

Then if the multiplicity of each a_i is ν_i and the initial form of a_i is $In(a_i)$ one can write:

$$a_i(u_1, u_1u_2) = u_1^{\nu_i} In(a_i)(1, u_2) + u_1^{\nu_i+1} a_i'(u_1, u_2),\ i = 1, 2.$$

Also:

$$\nu_{\mathbf{p}}(\mathscr{F}) = \min\{\nu_1, \nu_2\}.$$

We will let $\nu = \nu_{\mathbf{p}}(\mathscr{F})$ and let $In_\nu(a_i)$ be the homogeneous component of a_i of degree ν, i.e. $In_\nu(a_i) = 0$ if $\nu_i \neq \nu$ and $In_{\nu_i}(a_i) = In(a_i)$ if $\nu = \nu_i$.

With this notation:

$$\pi_1^\star(\omega) = u_1^\nu[P(1, u_2)du_1 + u_1 In_\nu(a_2)(1, u_2)du_2] + u_1^{\nu+1}\omega',$$

$$\pi_2^\star(\omega) = u_2^\nu[P(u_1, 1)du_2 + u_2 In_\nu(a_1)(u_1, 1)du_1] + u_2^{\nu+1}\omega'',$$

with: $P(x_1, x_2) = x_1 In_\nu(a_1) + x_2 In_\nu(a_2)$.

Then there are two possibilities:

(1) Non-dicritical blow-up, $P(x_1, x_2) \not\equiv 0$

The strict transform of $\mathscr{F}$ by π is defined as the foliation $\tilde{\mathscr{F}} = \pi^*(\mathscr{F})^{\perp\perp}$ generated in the first chart by the saturated 1-form:

$$\widetilde{\omega} = \frac{\pi_1^\star(\omega)}{{u_1}^\nu} = P(1, u_2)du_1 + u_1[In_\nu(a_2)(1, u_2)du_2 + \omega'].$$

In this case, the exceptional divisor, given in this chart by $u_1 = 0$, is an integral curve of the new foliation. The new singularities, placed all over the exceptional divisor, are given by the equation $P(1, u_2) = 0$.

Also in this case it can happens that:

$$\exists\, \mathbf{q} \in \pi^{-1}(\mathbf{p}),\ \nu_{\mathbf{q}}(\widetilde{\mathscr{F}}) > \nu_{\mathbf{p}}(\mathscr{F}).$$

For instance, if $\omega = dx_2$ in $\mathbb{C}^2$, $\pi_1^\star(\omega) = d(u_1u_2) = u_1du_2 + u_2du_1$ and $\tilde{\omega} = \pi_1^\star(\omega)$ has a singularity at (0, 0) with $\nu_{(0,0)}(\widetilde{\omega}) = 1 > 0 = \nu_{(0,0)}(\omega)$.

(2) Dicritical blow-up, $P(x_1, x_2) \equiv 0$

The strict transform of $\mathscr{F}$ is now the foliation generated in the first chart by:

$$\widetilde{\omega} = \frac{\pi_1^\star(\omega)}{u_1^{\nu+1}} = In_\nu(a_2)(1, u_2)du_2 + \omega'.$$

The exceptional divisor is not an integral curve of $\widetilde{\mathscr{F}}$, the multiplicity does not increase, and there appears an infinite set of integral curves, transverse to the exceptional divisor.

For instance, $\omega = x_1dx_2 - x_2dx_1$ is dicritical and $\pi^\star(\omega) = u_1^2du_2$. Then

$$\widetilde{\omega} = \frac{\pi_1^\star(\omega)}{u_1^2} = du_2$$

with integral curves $u_2 =$ constant at each point of the exceptional divisor.

Remark 6.2.1 At this point we have already seen the first difference between foliations and plane curves, the multiplicity may grow after blowing-up. Now we will describe the stable singularities.

Given a point $\mathbf{p} \in M$, the foliation $\mathscr{F}$, generated around a point $\mathbf{p}$ by ω gives rise at $\mathbf{p}$ to a germ of a tangent vector field that is a derivation X of $\mathscr{O}_{M\mathbf{p}}$. If $\mathbf{p}$ is a singular point of $\mathscr{F}$, and $m_{\mathbf{p}}$ is the maximal ideal of $\mathscr{O}_{M,\mathbf{p}}$, being $\mathbf{p}$ singular, $X(m_{\mathbf{p}}) \subset m_{\mathbf{p}}$, and X induces a $\mathbb{C}$-linear map

$$L_{X,\mathbf{p}} : m_{\mathbf{p}}/m_{\mathbf{p}}^2 \longrightarrow m_{\mathbf{p}}/m_{\mathbf{p}}^2.$$

In a local chart centered at $\mathbf{p}$,

$$\omega = a_1(x_1, x_2)dx_1 + a_2(x_1, x_2)dx_2, \text{ with } a_1(0, 0) = a_2(00) = 0.$$

Hence

$$X = -a_2(x_1, x_2)\frac{\partial}{\partial x_1} + a_1(x_1, x_2)\frac{\partial}{\partial x_2}$$

and $L_{X,\mathbf{p}}$, in the basis $\{x_1 + m_{\mathbf{p}}^2, x_2 + m_{\mathbf{p}}^2\}$, has the matrix

$$J_{(\mathscr{F},(x_1,x_2))} = \begin{pmatrix} -\frac{\partial a_2}{\partial x_1}(0,0) & \frac{\partial a_1}{\partial x_1}(0,0) \\ -\frac{\partial a_2}{\partial x_2}(0,0) & \frac{\partial a_1}{\partial x_2}(0,0) \end{pmatrix}.$$

A change of coordinates gives rise to a similar matrix, but a change of the generator ω of $\mathscr{F}$ may transform the matrix by multiplying it for some element of $\mathbb{C}^*$. Then the eigenvalues of $J_{(\mathscr{F},(x_1,x_2))}$ are not invariants of $\mathscr{F}$, but their quotients $\{\lambda/\mu, \mu/\lambda\}$, when defined, are invariants.

Definition 6.2.2 A singular point $\mathbf{p}$ of $\mathscr{F}$ is called *presimple*, if $L_{X,\mathbf{p}}$ is not nilpotent, i.e. has a non-zero eigenvalue. A singular point $\mathbf{p}$ of $\mathscr{F}$ is called *simple*, if $L_{X,\mathbf{p}}$ has eigenvalues λ, μ such that one of them, for instance μ, is non-zero, and $\lambda/\mu \notin \mathbb{Q}_+$.

Remark 6.2.2 Let $\mathbf{p}$ be a simple singularity, it is possible to perform a linear change of coordinates in such a way that $J_{(\mathscr{F},(x_1,x_2))}$ becomes diagonal. Then, in the new coordinate system

$$\omega = \lambda x_2 dx_1 - \mu x_1 dx_2 + a(x_1, x_2)dx_1 + b(x_1, x_2)dx_2, \ \ \nu(a) \geq 2, \nu(b) \geq 2.$$

Now, by blowing-up $\mathbf{p}$, the exceptional divisor is an integral curve, because the singularity is non-dicritical, and the strict transform of $\mathscr{F}$ is generated in the first chart of the blowing-up, by

$$\widetilde{\omega}_1 = (\lambda - \mu)u_2 du_1 - \mu u_1 du_2 + \delta u_1 du_1 + h.d.t.$$

The only singular point in this chart is the origin, and the matrix of $L_{X,(00)}$ is

$$\begin{pmatrix} \mu & \delta \\ 0 & \lambda - \mu \end{pmatrix},$$

with quotient of eigenvalues

$$\left\{\frac{\mu}{\lambda - \mu}, \frac{\lambda - \mu}{\mu}\right\} = \left\{\frac{1}{a - 1}, a - 1\right\},$$

where $a = \lambda/\mu$.

In the chart U_2, the situation is the same, with a singularity at the origin and with quotients of eigenvalues $\{1/(b-1), b-1\}$, where $b = \mu/\lambda$, or $\{a/(1-a), (1-a)/a\}$.

Because $\mathbf{p}$ is simple and assuming $\mu \neq 0$ is $\mu \neq \lambda$ ($a \notin \mathbb{Q}_+$), $a - 1 \notin \mathbb{Q}_+$, $a/(a-1) \notin \mathbb{Q}_+$, then by blowing up, each simple singularity gives rise to exactly two simple singularities. Therefore the simple singularities are stable by blowing-up.

Each blow-up adds to the ambient space one exceptional divisor, and each new divisor has a normal crossing with the old ones. Then it is important to consider the foliation in relation with a divisor with normal crossings.

Definition 6.2.3 Let $\mathscr{F}$ be a foliation on M, and E a divisor with normal crossings:

- We will say that an irreducible component H of E is a *dicritical component* of E, if H is not an integral hypersurface of $\mathscr{F}$.
- Assume that $\mathbf{p} \notin \mathrm{Sing}(\mathscr{F})$. Let L be the leaf of $\mathscr{F}$ at $\mathbf{p}$. We will say that $\mathscr{F}$ has a *normal crossing* with E at $\mathbf{p}$ if and only if $L \cup E$ has a normal crossings at $\mathbf{p}$.
- The *adapted singular locus* of $\mathscr{F}$ to E is: the union $\mathrm{Sing}(\mathscr{F}, E)$ of $\mathrm{Sing}(\mathscr{F})$ and the set:

$$\{\mathbf{p} \in M \mid \mathscr{F}, E \text{ dont have normal crossing at } \mathbf{p}\}.$$

  The set $\mathrm{Sing}(\mathscr{F}, E)$ is an analytic subset of M of codimension ≥ 2.
- A point $\mathbf{p} \in \mathrm{Sing}(\mathscr{F}, E)$ will be called a *simple singularity of $\mathscr{F}$ adapted to E* if $\mathbf{p}$ is a simple singularity of $\mathscr{F}$; $\mathbf{p} \in E$ and the irreducible components of E through $\mathbf{p}$ are non-dicritical.

Now we are going to prove the theorem of reduction of singularities of foliations in dimension two.

Theorem 6.2.1 *Let $\mathscr{F}$ be a germ of foliation over $(\mathbb{C}^2, (0,0))$; there is a finite sequence of blow-ups*

$$\pi = \pi_n \cdots \pi_1$$

with π_1 the blow-up of $(0,0) = \mathbf{p}_0$, and π_i the blow-up with center $\mathbf{p}_i \in \pi_i^{-1}(\mathbf{p}_{i-1})$, such that, if $\tilde{\mathscr{F}}$ is the strict transform of $\mathscr{F}$ by π and $\widetilde{E} = \pi^{-1}(0,0)$, then any point $\mathbf{p} \in \mathrm{Sing}(\widetilde{\mathscr{F}}, \widetilde{E})$ is a simple singularity of $\widetilde{\mathscr{F}}$ adapted to $\widetilde{E}$.

The proof will be divided into three steps. Each one is based on the construction of a system of invariants adapted to that step. First we will show how to reduce all the singularities of $\mathscr{F}$ to presimple ones. In the second step we will reduce the singularities to simple ones. As a third step, we will do the adaptation of the singularities by some extra blowing-up, perhaps with nonsingular centers.

6.2.1 Local Invariants

Our first invariant is the order (or multiplicity) of $\mathcal{F}$ at a point $\mathbf{p} \in M$, $\nu_{\mathbf{p}}(\mathcal{F})$ that has been defined in Definition 6.2.1. We can define a better invariant by considering instead the minimum of the order of the coefficients, their contact order.

Definition 6.2.4 If $\omega = a_1(x_1, x_2)dx_1 + a_2(x_1, x_2)dx_2$ is a local generator of $\mathcal{F}$ at $\mathbf{p}$,

$$\mu_{\mathbf{p}}(\mathcal{F}) = \dim_{\mathbb{C}}\Big(\frac{\mathbb{C}\{x_1, x_2\}}{a_1\mathbb{C}\{x_1, x_2\} + a_2\mathbb{C}\{x_1, x_2\}}\Big)$$

is called the *Milnor number* of $\mathcal{F}$ at $\mathbf{p}$.

Remark 6.2.3 It easy to see that:

- $\mu_{\mathbf{p}}(\mathcal{F})$ is the intersection number at $(0, 0)$ of the curves $\{a_1(x_1, x_2) = 0\}$ and $\{a_2(x_1, x_2) = 0\}$; it will be represented by $\sharp(a_1, a_2)_{(0,0)}$, and is independent of the local chart used in its construction. If a_1 and a_2 have no common factors, $\mu_{\mathbf{p}}(\mathcal{F}) < \infty$.
- The conditions, $\nu_{\mathbf{p}}(\mathcal{F}) = 0$, $\mu_{\mathbf{p}}(\mathcal{F}) = 0$ and $\mathcal{F}$ nonsingular at $\mathbf{p}$, are equivalent.
- $\mu_{\mathbf{p}}(\mathcal{F}) = 1 \Rightarrow \mathbf{p}$ is presimple $\Rightarrow \nu_{\mathbf{p}}(\mathcal{F}) = 1$. But, there are singularities with order 1 that are not presimple ($x^2dx + ydy$ at $(0, 0)$) and there are presimple singularities whose Milnor number is greater than $1(xdy + y^2dx$ at $(0, 0)$).
- Noether's formula gives us that after blowing-up:

$$\mu_{\mathbf{p}}(\mathcal{F}) = \sharp(a_1, a_2)_{\mathbf{p}} = \nu_{\mathbf{p}}(a_1)\nu_{\mathbf{p}}(a_2) + \sum_{\mathbf{q}\in\pi^{-1}(\mathbf{p})} \sharp(\widetilde{a}_{1\mathbf{q}}, \widetilde{a_{2\mathbf{q}}})_{\mathbf{q}},$$

  where $\widetilde{a}_{1\mathbf{q}}, \widetilde{a_{2\mathbf{q}}}$ are the strict transforms at $\mathbf{q}$ of a_1 and a_2.

  Then if $\widetilde{\mathcal{F}}$ is the strict transform of $\mathcal{F}$ by the blow-up with center $\mathbf{p}$, $E = \pi^{-1}(0)$, $\nu = \nu_{\mathbf{p}}(\mathcal{F})$, Noether's formula can be rewritten as follows:

(1) If π is not dicritical:

$$\mu_{\mathbf{p}}(\mathcal{F}) = \nu^2 - (\nu + 1) + \sum_{\mathbf{q}\in E} \mu_{\mathbf{q}}(\widetilde{\mathcal{F}}).$$

(2) If π is dicritical:

$$\mu_{\mathbf{p}}(\mathcal{F}) = (\nu + 1)^2 - (\nu + 2) + \sum_{\mathbf{q}\in E} \mu_{\mathbf{q}}(\widetilde{\mathcal{F}}).$$

Then the Milnor number decreases after blowing-up either if $\nu > 1$ or if $\nu = 1$ and the blow-up is dicritical. But it may increase if the blow-up is non-dicritical and

$\nu = 1$. To avoid this last possibility we needed to improve Milnor number by taking account of the exceptional divisor. More precisely:

If we fix an irreducible curve, not necessarily convergent, and consider the sequence of blow-ups that follow a point **p** of the curve, we want to prove that if all the members of the sequence of infinitely near points to **p** are singular points of the foliation, the curve is an integral curve of the foliation, and if also the blow-ups are non-dicritical, we will arrive, in a finite number of steps, to a presimple singularity. To do that we need a new invariant to control the contact of a curve and a foliation.

Given the foliation $\mathscr{F}$, at the point **p** by choosing a local chart, we may suppose as usual that we are working in $\mathscr{C}^2$ locally at $(0, 0)$. Let Γ be a nonsingular (perhaps formal) branch of curve at $(0, 0)$, we may choose a system of coordinates (perhaps formal), such that Γ be defined by $y = 0$.

Definition 6.2.5 If a local generator of $\mathscr{F}$ in this chart is:

$$\omega = a(x, y)dx + b(x, y)dy,$$

we will denote the *restricted order of $\mathscr{F}$ at* **p** *along Γ* by the number

$$t(\mathscr{F}, \Gamma; \mathbf{p}) = \nu(a(x, 0)) \in \mathbb{Z}_{\geq 0} \cup \{\infty\}.$$

Remark 6.2.4 The properties of this number are:

- The number $t(\mathscr{F}, \Gamma; \mathbf{p})$ is independent of the choice of the chart (always with $y = 0$ being the equation of Γ).
- $t(\mathscr{F}, \Gamma; \mathbf{p}) = \infty$ if and only if Γ is an integral curve of $\mathscr{F}$.
- If $\widetilde{\Gamma}$ is the strict transform of Γ by the blow-up π with center **p**, and $\mathbf{p}' = \pi^{-1}(\mathbf{p}) \cap \widetilde{\Gamma}$, an $\nu = \nu_{\mathbf{p}}(\mathscr{F})$, then

$$t(\widetilde{\mathscr{F}}, \widetilde{\Gamma}; \mathbf{p}') = t - \nu, \ \text{if } \pi \text{ is non-dicritical,}$$
$$t(\widetilde{\mathscr{F}}, \widetilde{\Gamma}; \mathbf{p}') = t - \nu - 1, \ \text{if } \pi \text{ is dicritical.}$$

6.2.2 Reduction to Simple Singularities

First we characterize the integral branches by their infinitely near points, by using the last defined invariant.

Let $\mathscr{F} = \mathscr{F}_0$ be a foliation germ at $\{\mathbb{C}^2, (0, 0)\}$, and let $\Gamma = \Gamma_0$ be a regular algebroid branch (i.e. a regular formal curve) at $(0, 0)$; we may consider a sequence of blow-ups that follows Γ, i.e. an infinite sequence of blow-ups

$$U_0 \xleftarrow{\pi_1} U_1 \xleftarrow{\pi_2} U_2 \cdots$$

defined in the following way:

(1) U_0 is a representative of $\{\mathbb{C}^2, 0\}$ contained in the domain of $\mathscr{F}_0$.
(2) π_1 is the blow-up with center $\mathbf{p}_0 = 0$.
(3) π_{i+1} is the blow-up with center $\mathbf{p}_i = \Gamma_i \cap \pi_i^{-1}(\mathbf{p}_{i-1})$ of U_i.
(4) Γ_i is the strict transform of Γ_{i-1} by π_i.

Then, if $\mathscr{F}_i$ is the strict transform of $\mathscr{F}_{i-1}$ by π_i, and $\mathbf{p}_i \in \mathrm{Sing}(\mathscr{F}_i)$, for all i,

$$t(\mathscr{F}_i, \Gamma_i; \mathbf{p}_i) < t(\mathscr{F}_{i-1}, \Gamma_{i-1}; \mathbf{p}_{i-1}), \text{ for all } i \geq 1,$$

and this is incompatible with $t(\mathscr{F}, \Gamma; 0) < \infty$, therefore Γ is an integral curve for $\mathscr{F}$.

Let us suppose now that all the blowingups are non-dicritical, then π_1 being non-dicritical, Γ_1 and $\pi_1^{-1}(0)$ are transversal integral curves of $\mathscr{F}_0$. Then in some local chart (possibly formal), $\mathscr{F}_1$ has a generator

$$\omega_1 = x_1 y_1 \Big(a_1(x_1, y_1)\frac{dx_1}{x_1} + b_1(x_1, y_1)\frac{dy_1}{y_1}\Big),$$

where $\{x_1 = 0\} = \pi_1^{-1}(0, 0)$, $\{y_1 = 0\} = \Gamma_1$, and y_1 does not divide b_1 because π_1 is non-dicritical. If $\nu_1 = \nu(\mathscr{F}_1) = 1$, $\mathbf{p}_1$ is presimple, and, if $\nu_1 > 1$, $\mathscr{F}_2$ is generated by

$$\omega_2 = x_2 y_2 \Big(a_2(x_2, y_2)\frac{dx_2}{x_2} + b_2(x_2, y_2)\frac{dy_2}{y_2}\Big),$$

with

$$b_2 = \frac{b_1(x_2, x_2 y_2)}{x_2^{\nu_1 - 1}}.$$

Now, it being possible to do that only for a finite number of times, because y_1 does not divide b_1, we conclude that in some step, $\nu(\mathscr{F}_r) = 1$ and $\mathbf{p}_r$ is presimple.

Remark 6.2.5 Let us consider a collection (S) of data, constructed as follows:

(1) A sequence of blow-ups

$$U_0 \xleftarrow{\pi_1} U_1 \xleftarrow{\pi_2} U_2 \cdots$$

with centers $\mathbf{p}_i \in U_i$; U_0 is an open neighborhood of $(0, 0)$ on $\mathbb{C}^2$, $\mathbf{p}_0 = (0, 0)$ and $\pi_i(\mathbf{p}_i) = \mathbf{p}_{i-1}$.
(2) $\mathscr{F} = \mathscr{F}_0$ is a foliation germ in U_0 with $\mathrm{Sing}(\mathscr{F}_0) = \{(0, 0)\}$, $\mathscr{F}_i$ is the strict transform of $\mathscr{F}_{i-1}$ by π_i, and $\mathbf{p}_i \in \mathrm{Sing}(\mathscr{F}_i)$, $\forall i$.
(3) $E_i = (\pi_1 \cdots \pi_i)^{-1}(0)$, E_i' is the union of all the non-dicritical components of E_i.

Proposition 6.2.1 *Given (S), there is an index r such that* $\mathbf{p}_r$ *is presimple with respect to* $\mathscr{F}_r$.

Let $e(E'_i, \mathbf{p}_i) \in \{1, 2\}$ denote the number of irreducible components of E'_i that contains $\mathbf{p}_i$. If $e(E'_i, \mathbf{p}_i) = 2$ and $\nu_{\mathbf{p}_i}(\mathscr{F}_i) = 1$, there is a system of local coordinates (x, y) on a neighborhood of $\mathbf{p}_i$ such that $x(\mathbf{p}_i) = y(\mathbf{p}_i) = 0$, E'_i equals $\{xy = 0\}$. Therefore $\mathscr{F}_i$ has a generator

$$\omega = xy\Big(\frac{dx}{x} + b(x, y)\frac{dy}{y}\Big). \qquad (*)$$

Hence $\mathbf{p}_i$ is a presimple singularity of $\mathscr{F}_i$

Let us consider now the number:

$$I_i = \mu_{\mathbf{p}-i}(\mathscr{F}_i) - e(E'_i, \mathbf{p}_i) \in \mathbb{Z}_{\geq -2}$$

and let us suppose that for some index i, $\mathbf{p}_i$ is not presimple. Then:

- $I_{i+1} \leq I_i$.
- If $\nu_{\mathbf{p}_i}(\mathscr{F}_i) \geq 2$,one has $I_{i+1} < I_i$.
- If $\nu_{\mathbf{p}_i}(\mathscr{F}_i) = 1$ and π_{i+1} is dicritical also $I_{i+1} < I_i$.

Then if in the sequence (S) all the points are not presimple there exists an index j such that $\nu_{\mathbf{p}_i}(\mathscr{F}_i) = 1$ and all the π_{i+1} are non-dicritical $\forall i > j$. Perhaps after some more steps, there is r such that also $I_i = I_r$, $\forall i \geq r$. Then $e(E'_i, \mathbf{p}_i) = 1$, for all i, and the π_i's being non-dicritical, the chain of blow-ups gives a formal non-singular branch Γ that must be an integral curve. Hence there is an i such that $\mathbf{p}_i$ is presimple.

Now by using a new invariant we may see that it is possible to transform presimple singularities into simple ones.

Proposition 6.2.2 *If (S) is a collection of data as in Remark* 6.2.5, *and* $(0, 0)$ *is presimple, there is an index i such that* $\operatorname{Sing}(\mathscr{F}_i)$ *consists only of simple singularities.*

Let $\mathbf{p}$ be a presimple singularity of $\mathscr{F}$. Consider the invariant of resonance given by

$$\operatorname{Res}(\mathscr{F}, \mathbf{p}) = \begin{cases} 0 & \text{if } \alpha \notin \mathbb{Q}_+, \\ p+q & \text{if } \alpha = \frac{p}{q} \in \mathbb{Q}_+, \ \text{g.c.d.}(p, q) = 1, \end{cases}$$

where $\alpha = \lambda/\mu$ with λ, μ ($\mu \neq 0$) the eigenvalues of $L_{X,\mathbf{p}}$. By the formula about the behavior of α by blowing-up, we know that $\operatorname{Res}(\mathscr{F}, \mathbf{p})$ decreases after blowing-up. After a finite number of steps, all the singular points of $(\mathscr{F}_i)$ are simple.

Remark 6.2.6 In order to adapt the situation to a convenient divisor, we need only to separate at some simple singularity $\mathbf{p}$ the components of E' through $\mathbf{p}$ and the formal integral curves of $\mathscr{F}$ by $\mathbf{p}$, in order to have normal crossing. But this is a well-known result of the theory of plane curves.

6.3 Existence of Analytic Integral Curves

The result obtained in Sect. 6.2.2 may be read as a proof of the existence at each singular point of a formal integral curve of the foliation. The problem of the existence of analytic solutions is of a different nature, and there is a theorem of Camacho and Sad [19] proving it.

Theorem 6.3.1 (Camacho–Sad) *Any germ of complex-analytic singular foliation over* $(\mathbb{C}^2, 0)$ *has at least a convergent integral curve.*

The first proof of the theorem is not constructive, based on the evolution of a index (Camacho–Sad index). There is a second proof by J. Cano [38] that gives an effective method to find a convergent solution in the spirit of the first proof. There is also a third proof of J. Cano extending Newton polygons to differential equations. We will explain this third proof because it uses methods developed in this book and also because is not limited to first order and first degree equations.

The use of the Newton polygon to build Puiseux solutions of an ordinary differential equation over the complex domain goes back to H. B. Fine [44, 45]; he *proved* that any (sufficiently general in some undetermined sense) ordinary differential equation has a convergent Puiseux solution (see Ince's book [74]). The first complete proof of the existence of a solution for a singular first order and first degree differential equation, by using the Newton polygon, is due to J. Cano [36, 37].

When we try to solve ordinary differential equations by using the Newton polygon, our idea is to construct a Puiseux series solution of the equation at the singular points, the reason for using such a series is the existence of a flow of analytic solutions of the equation in the regular set. But the method is useful too to obtain a more general series solution [57].

As we will see, the results of the application of Newton's method to the construction of solutions of differential equations presents serious differences compared with the case of algebraic or analytic equations. The most interesting are:

- the possibility of non-existence of solutions;
- he existence of divergent formal solutions of analytic equations;
- the existence, sometimes, of an infinite set of solutions, even for first order and first degree equations.

A natural issue is to ask for the possibility of reducing a first order and degree n equation $(n > 1)$, $f(x, y, y') = 0$, to n first order and first degree equations just by solving the algebraic equation in the differential variable. This can be achieved in some cases, for instance in the case of quasi-ordinary equations [5].

Let us recall that an equation $f(x, y, y') = 0$ is called quasi-ordinary with respect to the projection

$$(x, y, y') \longmapsto (x, y),$$

if the projection is transversal to the equation, that is

$$f(x, y, y') = (y')^n + \sum_{r=0}^{n-1} f_r(x, y)(y')^r \text{ with } \mathrm{ord}(f_r) \geq n - r$$

and the discriminant has normal crossings. Under these hypotheses the equation admits a local parametrization:

$$\begin{aligned} x &= u^s, \\ y &= v^s, \\ y' &= g(u, v). \end{aligned}$$

For each local parametrization (because in principle, the solutions depend on the parametrization) our equation gives rise to a first order and first degree equation

$$g(u, v) = y' = \frac{dy}{dx} = \frac{sv^{s-1}dv}{su^{s-1}du} \Rightarrow v' = \frac{dv}{du} = \frac{u^{s-1}g(u, v)}{v^{s-1}},$$

and each solution of the second equation gives a solution of the first one. For more general equations there are several possibilities: either they can be transformed into quasi-ordinary ones by using birrational transformations [5] or can be solved in more general rings [87].

6.3.1 The Newton Polygon of a Differential Equation

Given a polynomial

$$\begin{aligned} f(x, \mathbf{y}) &= f(x, y_0, \ldots, y_n) = \\ &= \sum f_{ij_0\ldots j_n} x^{i/s} y_0^{j_0} \ldots y_n^{j_n} \in \mathbb{C}[[x^{1/s}, y_0]][y_1, \ldots, y_n], \; s \in \mathbb{N}, \end{aligned}$$

we define an operator $\hat{f} : \mathscr{U} \longrightarrow \mathscr{U}'$, where

$$\mathscr{U} = \{\sum_{i=1}^{\infty} a_i x^{\mu_i} \mid a_i \in \mathbb{C}, \; \mu_i \in \mathbb{Q}_{>0}, \quad \mu_i < \mu_{i+1}, \text{ for all } i\},$$

$$\mathscr{U}' = \{\sum_{i=1}^{\infty} a_i x^{\mu_i} \mid a_i \in \mathbb{C}, \; \mu_i \in \mathbb{Q}, \quad \mu_i < \mu_{i+1}, \text{ for all } i\}.$$

By the obvious substitution in f,

$$y_0 = \sum_{i=1}^{\infty} a_i x^{\mu_i}, \quad y_j = \frac{d^j}{dx^j}(\sum_{i=1}^{\infty} a_i x^{\mu_i}) = \sum_{i=j}^{\infty} (\mu_i)_j a_i x^{\mu_i - j},$$

$$(\mu)_j = \mu(\mu-1)\cdots(\mu-j+1),$$

and we will call any $s(x) \in \mathscr{U}$, such that $\hat{f}(s(x)) = 0$, a solution of the differential equation $f(x, \mathbf{y}) = 0$. Clearly we may use convergent equations and convergent solutions in the same way.

Let:

- For all $\mathbf{j} = (j_0, \ldots, j_n)$, $\mid \mathbf{j} \mid = j_0 + \cdots + j_n$, $< \mathbf{j} > = j_1 + 2j_2 + \cdots + nj_n$.
- $\alpha : \frac{1}{s}\mathbb{Z} \times \mathbb{N}^{n+1} \longrightarrow \frac{1}{s}\mathbb{Z} \times \mathbb{N}$, $\alpha(i, \mathbf{j}) = (i - < \mathbf{j} >, \mid \mathbf{j} \mid)$.
- $M_{f,(a,b)}(x, y) = \sum_{\alpha(i,\mathbf{j})=(a,b)} f_{i\mathbf{j}} x^i \mathbf{y}^{\mathbf{j}}$.
- $\Delta(f) = \{(a, b) \mid M_{f,(a,b)}(x, y) \neq 0\} \subset \mathbb{R} \times \mathbb{R}_{\geq 0}$.

We will call $\Delta(f)$ the Newton diagram of $f(x, \mathbf{y})$, and the border of the convex hull, $N(f)$, of

$$\bigcup_{P \in \Delta(f)} (P + \mathbb{R}^2_{\geq 0})$$

the Newton polygon of $f(x, \mathbf{y})$. The sides and vertices of $N(f)$ are defined as the obvious sides and vertices of its border, with the only exception of the vertical side, and the horizontal one (if it is placed on the x-axis). For brevity we will say *straight line of coslope* μ instead of *straight line of slope* $-1/\mu$.

For any $\mu \in \mathbb{Q}_+$ the unique straight line of coslope μ which intersects $N(f)$ either on a side, or on a vertex, will be denoted by $L(f, \mu)$ (Fig. 6.1). Also for any f we will use the polynomials

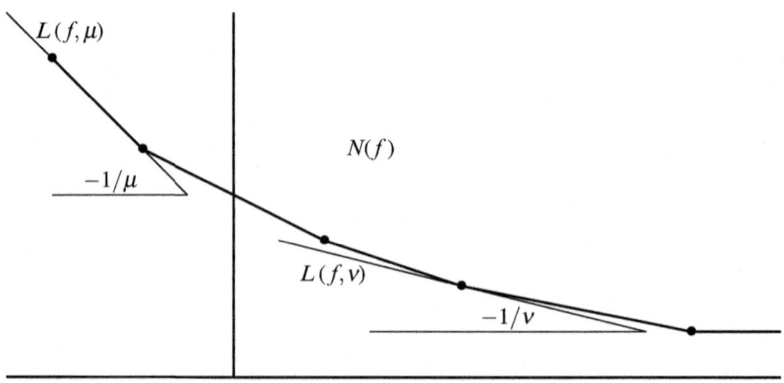

Fig. 6.1 $L(f, \mu)$ intersects $N(f)$ either on a side, or on a vertex

$$\Psi_{f,(a,b)}(u) = \sum_{\alpha(i,\mathbf{j})=(a,b)} f_{i\mathbf{j}}(u)_1^{j_1} \cdots (u)_n^{j_n}, \text{ for all } (a,b) \in \frac{1}{s}\mathbb{Z} \times \mathbb{N},$$

$$\Phi_{f,\mu}(c) = \sum_{(a,b)\in L(f,\mu)} \Psi_{f,(a,b)}(\mu) \cdot c^b.$$

Let $z \in \mathscr{U}$, $z = c_1 x^{\mu_1} + h.d.t.$ (higher degree terms), a straightforward calculation of the terms of minimum degree in $\hat{f}(z)$, proves the following:

Proposition 6.3.1 *Let $z \in \mathscr{U}$, $z = c_1 x^{\mu_1} + h.d.t.$ (higher degree terms), $c_1 \neq 0$, be a solution of f, then:*

(1) *$L(f,\mu_1)$ intersects the border of $N(f)$ either along a side, or in a vertex (a,b) such that there is more than one element $(i,\mathbf{j})$ with $\alpha(i,\mathbf{j}) = (a,b)$ and $f_{i,\mathbf{j}} \neq 0$. Also the sides of $N(f)$ are contained in a single half-plane with respect to $L(f,\mu_1)$.*
(2) *If $L(f,\mu_1)$ intersects $N(f)$ only in a vertex (a,b), then $\Psi_{f,(a,b)}(\mu_1) = 0$.*
(3) *If $L(f,\mu_1)$ intersects $N(f)$ in a side (of coslope μ_1), then $\Phi_{f,\mu_1}(c_1) = 0$.*

Then, in order to have a solution in $\mathscr{U}$ of f, we need either a vertex or a side of $N(f)$ verifying respectively (2) or (3) in Proposition 6.3.1 for some μ_1 and c_1, and then $c_1 x^{\mu_1}$ may be the first term of a solution.

Example 6.3.1 We will give some examples to show the limits of this construction:

(1) The equation

$$f(x,y,y_1,y_2) = xyy_2 - xy_1^2 + yy_1$$

has the Newton polygon of Fig. 6.2. For any line through $(-1,2)$, all of the Newton polygon is contained in the same half-plane, and for the vertex $(-1,2)$

$$\Psi_{f,(-1,2)}(u) \equiv 0.$$

In fact, $y = cx^{\mu}$ is a solution of f for all $\mu \in \mathbb{Q}_+$ and even for all $\mu \in \mathbb{C}$.

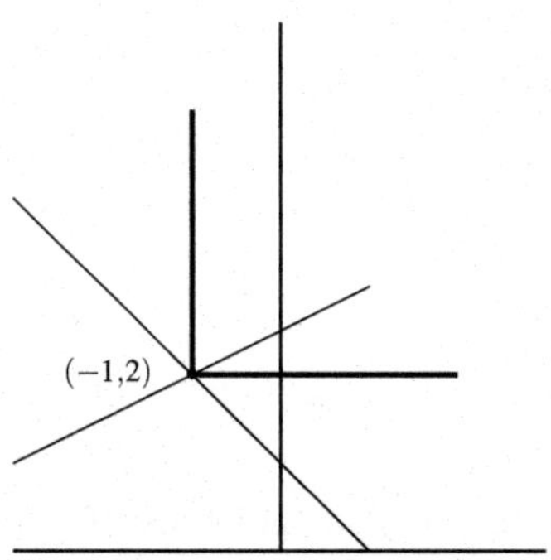

Fig. 6.2 The Newon polygon may have just one vertex with zero characteristic polynomial

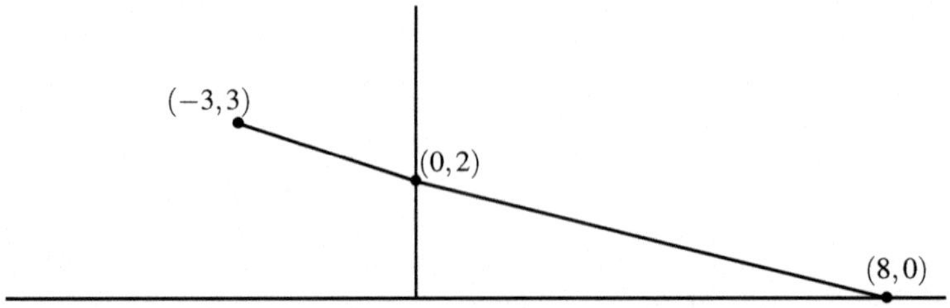

Fig. 6.3 There are equations without series solution

(2) The equation

$$f(x, y, y_1) = y_1^3 + x^2 y_1^2 - 7xyy_1 + 12y^2 + x^8$$

has the Newton polygon of Fig. 6.3. This polygon has two sides with slopes $-1/3$, $-1/4$, three vertices, $(-3, 3)$, $(0, 2)$, $(8, 0)$, and

$$\begin{aligned}\Psi_{f,(-3,3)}(u) &= u^3,\\ \Psi_{f,(0,2)}(u) &= u^2 - 7u + 12 = (u-3)(u-4),\\ \Psi_{f,(8,0)}(u) &= 1,\\ \Phi_{f,3}(c) &= 27c^3,\\ \Phi_{f,4}(c) &= 1.\end{aligned}$$

Then there are neither solutions $y = cx^3 + h.d.t.$ nor $y = cx^4 + h.d.t.$, corresponding to the sides. The only vertex with possibilities according to (1) of Proposition 6.3.1 is $(0, 2)$, and for $(0, 2)$ the only possible values for u are $u = 3$ and $u = 4$, which are not useful. Therefore f has no solution in $\mathscr{U}$.

Remark 6.3.1 Once we have (c_1, μ_1) satisfying either (2) or (3) of Proposition 6.3.1, we may apply a change of variable

$$y = c_1 x^{\mu_1} + z.$$

Then, we get a new equation

$$f[c_1 x^{\mu_1} + z] = f(x, c_1 x^{\mu_1} + z_0, (\mu_1)_1 c_1 x^{\mu_1 - 1} + z_1, \dots, (\mu_1)_n c x^{\mu_1 - n} + z_n).$$

We now want to repeat the process by computing for the new equation, c_2, μ_2, with $\mu_2 > \mu_1$. Observe that:

(1) $\sum_{i=1}^{\infty} c_i x^{\mu_i}$ is a solution of f if and only if $\sum_{i=2}^{\infty} c_i x^{\mu_i}$ is a solution of $f[c_1 x^{\mu_1} + z]$.

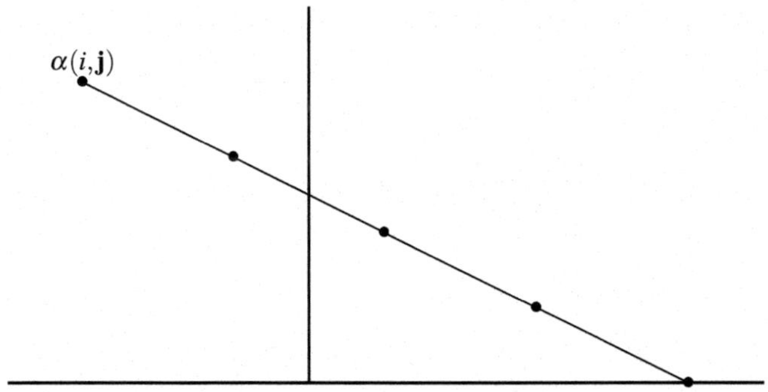

Fig. 6.4 After the change of variable, each monomial gives rise to points placed over a line of coslope μ_1

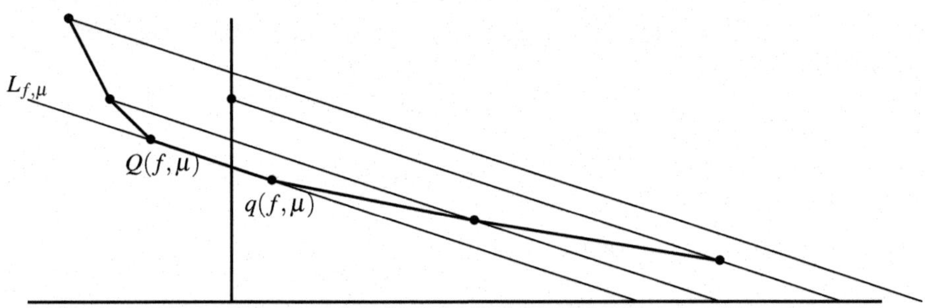

Fig. 6.5 After the change of variable, the part of the Newton polygon placed over the side of coslope μ does not change

(2) If $M_{i\mathbf{j}}(x,\mathbf{y}) = x^i\mathbf{y}^{\mathbf{j}}$, for all $c \in \mathbb{C}$, for all $\mu \in \mathbb{Q}_+$, we have (Fig. 6.4).

$$\alpha(i,\mathbf{j}) \in \Delta(M_{i,\mathbf{j}}[cx^\mu + y]) \subset \{(a,b) \in L(M_{i\mathbf{j},\mu}) | b \leq |\,\mathbf{j}\,|\}.$$

This means that, after the substitution, each point of the Newton polygon of f gives rise to a collection of points, placed on the line of slope $-1/\mu$, passing through the original point, and including it, and points placed below it.

As a consequence, the border of the Newton polygon of $f[cx^\mu + y]$ coincides with the original one on the region placed to the left of the higher point $Q(f,\mu)$ in $L(f,\mu) \cap N(f)$ (Fig. 6.5).

(3) Under the hypothesis that $L(f,\mu)$ corresponds to a side,

$$L(f,\mu) \cap \{y = 0\} = (t,o) \Rightarrow M_{f[cx^\mu + y],(t,0)} = \Phi_{f,\mu}(c)x^t.$$

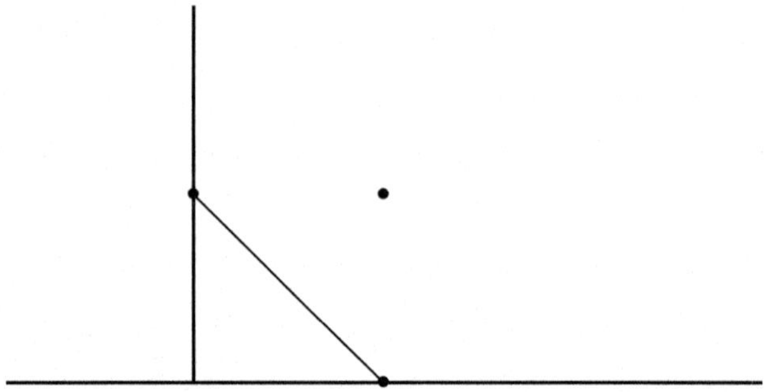

Fig. 6.6 Newton polygon of Euler's equation

Then, in $N(f[cx^\mu + y])$, the side of slope $-1/\mu$ does not reach the x-axis. Consequently, in this new polygon, there is at least a side with coslope higher than μ.

Example 6.3.2 Let

$$f(x, y, y_1) = (y^6 + xy^2 - 3x^2y + 2x^3)y_1 + xy^2 - x^2y + x^5.$$

Then $N(f)$ has a side of coslope 1 and

$$\Phi_{f,1}(c) = c(c-1)^2.$$

For $c = 1$, $\mu = 1$,

$$f[x+y] = (x+y)^6(1+y_1) + xy^2y_1 - x^2yy_1 + 2xy^2 + x^5.$$

Also $N(f[x+y])$ has only a side with coslope $2 > 1$, but

$$\Phi_{f[x+y],2}(c) = 1$$

and f has no solutions $y = x + h.d.t.$.

Thus, there is no guarantee that the process can be continued.

Example 6.3.3 Let us consider the equation (Euler's equation)

$$f(x, y, y_1) = x^2y_1 - y + x.$$

Its Newton polygon is shown in Fig. 6.6.

This equation has the convergent solution $x = 0$ (in the following sense: the equation is really the differential form $x^2dy + (-y+x)dx$ and may also be written as $x^2 + (-y+x)x' = 0$, now it is clear that $x = 0$ is a solution) and also, corresponding to the side of slope -1, the non-convergent solution

$$y = \sum_{i=0}^{\infty} n!x^{n+1}.$$

This example proves that a solution constructed by using the Newton polygon may be a non-convergent series.

Remark 6.3.2

(1) If $z = \sum_{i=1}^{\infty} c_i x^{\mu_i} \in \mathscr{U}$ is a solution of f, we have a sequence of Newton polygons, $N(f)$, $N(f_1), \ldots$, with

$$f_1 = f[c_1 x^{\mu_1} + y], \ f_2 = f_1[c_2 x^{\mu_2} + y], \ \ldots.$$

After each step in the process the y-coordinate of the higher point $Q_i(f_i, \mu_i)$ in $L(f_i, \mu_i) \cap N(f_i)$ either remains the same or decreases. Then, being the y-coordinate of $Q_1(f_1, \mu_1)$ finite, after a finite number of steps, the point $Q_i(f_i, \mu_i)$ remains fixed. This fixed point will be called the pivot point of the solution.

(2) With the same notation as above, if $Q = (a, b)$ is the pivot point, there is an index i_0 such that

$$\text{for all } i \geq i_0, \ (a + \mu_i, b - 1) \in \Delta(f_i),$$

since, otherwise, there would be an infinite set of indices $\mu_i \in \mathbb{Q}_+$ such that

$$M_{f_i[c_{i+1}x^{\mu_{i+1}}+y],(a,b)}(x, \mathbf{y}) = M_{f_i,(a,b)}(x, \mathbf{y}) + T,$$

where all the monomials in T have total exponents on $\mathbf{y}$ not exceeding $b - 2$. It is easy to see, by elementary linear algebra, that this is not possible.

Then, if $i \geq i_0$ and $\Delta(f_i) \subset \frac{1}{s}\mathbb{Z} \times \mathbb{N}$ (Fig. 6.7), after the substitution $A \in \frac{1}{s}\mathbb{Z} \times \mathbb{N}$, $B \in \frac{1}{s}\mathbb{Z} \times \mathbb{N}$ and $\mu_{i+1} \in \frac{1}{s}\mathbb{Z}$. Then if $z = \sum_{i=1}^{\infty} a_i x^{\mu_i} \in \mathscr{U}$ is a solution of f, z is a Puiseux series.

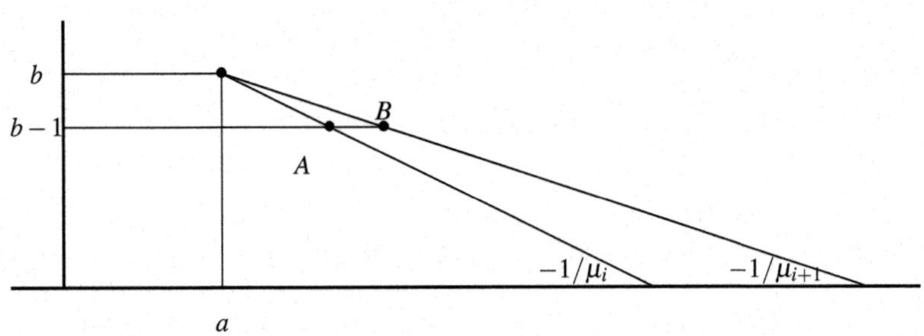

Fig. 6.7 Any series solution with rational growing exponents is a Puiseux series

6.3.2 *Existence of Solutions for First Order and First Degree Equations*

As we have seen, there are examples of equations with no series solutions; our problem now is to find Puiseux solutions of first order and first degree equations, that means integral curves of plane foliations. Let us consider the equation:

$$f(x, y, y') = A(x, y) + B(x, y)y' = \sum_{(i,j)\in 1/s\mathbb{Z}\times\mathbb{N}} (A_{ij}x^i y^j + B_{ij}x^{i+j}y^{j-1}y').$$

As we did before, for any $L(f, \mu)$ corresponding to a side of $N(f)$, we will define $Q(f, \mu)$ and $q(f, \mu)$ respectively as the higher and lower points of $L(f, \mu) \cap N(f)$. We will also denote the side contained in this line by $L(f, \mu)$.

Definition 6.3.1 We will say that $L(f, \mu)$ is a pregood side of f or μ is a pregood coslope of f, if $L(f, \mu)$ contains a side of $N(f)$ and $\Psi_{f;Q(f,\mu)}(u)$ has no roots in $\mathbb{Q}_{\geq\mu}$. $L(f, \mu)$ is a good side (or μ is a good coslope) for f if it is pregood and $\Psi_{f;q(f,\mu)}(u)$ either has no roots or has at least a root in $\mathbb{Q}_{>\mu}$.

Proposition 6.3.2 *Let $L(f, \mu)$ be a pregood side which is not good, and suppose that $y = 0$ is not a solution of f. Then there is $\mu' > \mu$, $\mu' \in \mathbb{Q}_{>0}$ such that $L(f, \mu')$ is a good side.*

In this case $\Psi_{f,(i,j)}(u) = A_{ij} + B_{ij}u$, then $\Psi_{f,(i,j)}(u)$ has no roots if $B_{ij} = 0$, and, if $B_{ij} \neq 0$, its only root is $-A_{ij}/B_{ij}$. The hypothesis means that:

(1) The point $q(f, \mu)$ is not placed on the x-axis.
(2) The next side to the one of coslope μ (which is not parallel to the x-axis because $y = 0$ is not a solution) is pregood.

Then we can follow the polygon until reaching the side with a point on the axis. This side, being pregood, is automatically good.

Remark 6.3.3 It can be proved by direct computation that if $L(f, \mu)$ is a good side of $N(f)$ is possible to choose a root (good root) of $\Phi_{f,\mu}(c) = 0$ such that either $y = 0$ is a solution of $f[c_0x^\mu + y]$ or the side of $N(f[c_0x^\mu + y])$ which follows that of coslope μ is pregood.

Theorem 6.3.2 *Any first order and first degree equation, f, formal or convergent, has a formal Puiseux solution.*

The theorem is true because being the first side $N(f)$ pregood, there is a good side $L(f, \mu_1)$. For this side there is at least a good root c_1. Then, by applying the change of variable $z = c_1x^{\mu_1} + y$, we get a new equation f_1. In $N(f_1)$ the side that follows the one of coslope μ_1 is pregood, and then we may continue the process. (The algorithm has been implemented in Maple by M. Farto [42].)

Remark 6.3.4 The solutions obtained using this algorithm are in general non-convergent. There exists another algorithm which gives us a very precise convergent solution, but there neither exists a characterization of the solutions obtained using this algorithm, nor it is known even if there are convergent solutions which do not come from one Newton polygon.

Theorem 6.3.3 *Any convergent differential equation of first order and first degree has a convergent Puiseux solution.*

Proof We need some more restrictive criteria, for the selection of the good side and the good root in order to get the convergence of the solution.

We will define a c-pregood side as a side $L(f, \mu)$ such that $\Psi_{f,Q(f,\mu)}(u) = 0$ has a root which is not in $\mathbb{Q}_{\geq \mu}$; as above, a c-good side is a c-pregood one with the same additional conditions which make the difference between pregood an good. The only difference in the characterization (Remark 6.3.4) is that $\mathbb{Q}_{>0}$ is replaced by $\mathbb{Q}_{\geq 0}$ in condition 2. In the definition of a good root the only condition to add is

$$\nu_\gamma(\Phi_{f,\mu}) \geq \nu_\gamma(\Phi_{0,1}) + 1,$$

then k is the R.H.S. of the inequality.

By the same arguments as above there are c-good sides and c-good roots, and any pair of them gives rise, by the same process, to one solution of the equation. The solutions obtained now can also be divergent, but if we choose at each step the maximum among the numbers $\mu - s$ which give c-good sides (whose corresponding sides are called principal ones) the solution is convergent.

To prove the convergence we consider the solution constructed above and let $Q = (a, b)$ be its pivot point, reached at the step r_0. Since obviously $b > 0$, there are two possibilities:

(1) **b** ≥ 2. In this case we consider the curve

$$g(x, y) = \frac{\partial^{b-1} f}{\partial y^{b-2} \partial y'} = 0$$

and, as we do with f, let us define

$$g_0 = g, \; g_i = g_{i-1}[c_i x^{\mu_i} + y].$$

Then

$$g_i(x, y) = \frac{\partial^{b-1} f_i}{\partial y^{b-2} \partial y'}, \quad \text{for all } i \geq 0$$

and consequently

$$(i, j) \in \Delta(g_r) \Rightarrow (i - 1, j + b - 1) \in \Delta(f_r), \text{ and } j \geq 0.$$

Since $L(f, \mu_r)$ is principal, the coefficient of $x^{a+1} y^{b-1} y'$ in f_r is different from zero, if $r \geq r_0$, then, $(a + 1, 1) \in \Delta(g_r)$ for $r \geq r_0$, and in consequence is over $L(g_r, \mu_r)$. Then the solution of f is also a branch of g and is convergent.

(2) **b = 1**. In this case, after a ramification we may suppose that the solution z is a formal series in x, and the pivot point is $(a, 1)$. Now the coefficient of $x^{a+1}y'$ in f_r, $r \geq r_0$ is fixed and different from zero, then

$$\nu_x\left(\frac{\partial f}{\partial y'}[z]\right) = a + 1, \ \nu_x\left(\frac{\partial f}{\partial y}[z]\right) \geq a,$$

and the differential operator

$$\left(\frac{\partial f}{\partial y'}[z]\right)\partial + \frac{\partial f}{\partial y}[z]$$

has a regular singularity at the origin, and then z converges (Malgrange [91]).

Remark 6.3.5 It is an interesting problem to study the character of (divergent) formal power series solutions, $\sum a_i x^i$, of convergent equations. All of them have the so-called *Gevrey character*, which means that there is a real positive number s, such that

$$\sum \frac{a_i}{(i!)^s} x^i$$

converges. We will say that a series with this property has Gevrey index s. If we take a differential equation of first order and first degree with Gevrey character, all their solutions have Gevrey character (in general with greater Gevrey index), as can be proved by using a modification of the last proof.

□

6.4 Codimension-One Foliations in a Dimension-Three Space: Thom's Conjecture

In this section we will expose the general ideas of the reduction of singularities of differential forms in an ambient space of dimension three; we will follow the papers of F. Cano [21–23, 25]. As in the case of dimension two, which we have seen in some detail, there are two stages, reduction to presimple singularities, stable under blowing-up, and destruction of resonances to obtain simple singularities.

Reduction of singularities of differential forms (codimension-one foliations) is still a conjecture for dimension $n \geq 4$; the main difficulties are in the reduction to presimple singularities, specially in the local control. As F. Cano pointed out, it should be possible to imagine reasonable global strategies, inspired by Giraud's ideas [52, 53], similar the ones used in the positive characteristic results for varieties, but is not clear how to generalize the differential idealistic exponents, and mainly the use of the characteristic polyhedra [64], which becomes very complicated in higher dimensions.

There is a theorem of Malgrange (Frobenius theorem with singularities [92]) proving that if ω is a germ of integrable holomorphic form and the codimension of $\mathrm{Sing}(\omega)$ is greater or equal than 3, there is a germ of holomorphic function (first integral) such that $\omega = df$.

The reduction to presimple singularities of foliations given by forms $\omega = df$ is similar to the reduction of singularities of the hypersurface $f = 0$. We are able to obtain, after finitely many blow-ups, a total transform of f, locally defined by a monomial, with at most one irreducible component outside the exceptional divisor; this is a presimple singularity.

Then to study reduction of singularities of codimension-one foliations, we need to consider only the case $codim(\mathrm{Sing}(\omega)) = 2$. But even with this limitation, for ω in general the problem is much more complicated than the one for hypersurfaces, mainly in the dicritical case, because there exists nothing similar to Hilbert–Samuel functions, and so appears the problem of resonances close to the typical characteristic p problems.

As usual we have a set of invariants that can be described in any dimension. There are two types of them: invariants of transversality (vertical invariants) and invariants of resonance. Before we define the invariants we need to prepare the equation of the generating differential form, by adding to the problem a divisor with normal crossing and adapting the equation to the divisor. The reason, as we have seen in the dimension-two case, is that it is important to know the position of ω with respect to the exceptional divisor in a sequence of blow-ups.

Let M be a complex analytic variety of pure dimension n, D a normal crossing divisor, and $\mathscr{F}$ a codimension-one singular holomorphic foliation on M. Locally at a point $\mathbf{p} \in M$ there exists a system of coordinates, *adapted to D*, $(x_1, \dots, x_n)$, i.e. such that there is a subset $A \subset \{1, 2, \dots, n\}$ with D locally defined at $\mathbf{p}$ by the equation:

$$\prod_{i \in A} x_i = 0.$$

Then the sheaf of logarithmic 1-forms along D, $\Omega_M[D]$ is a locally free $\mathscr{O}_M$ -module whose basis at $\mathbf{p}$ is:

$$\left\{ \frac{dx_i}{x_i} \right\}_{i \in A} \cup \{dx_j\}_{j \in \{1,\dots,n\} \setminus A}.$$

- A singular integrable hyperplane field adapted to D is an invertible submodule $\mathscr{H}$ of $\Omega_M[D]$ locally generated by a form:

$$\omega = \sum_{i \in A} a_i \frac{dx_i}{x_i} + \sum_{i \in \{1,\dots,n\} \setminus A} a_i dx_i, \ a_i \in \mathscr{O}_{M\mathbf{p}}$$

  such that $\omega \wedge d\omega = 0$.

- A foliation over M adapted to D is a singular integrable hyperplane field adapted to D, $\mathscr{H}$, such that $\Omega_M[D]/\mathscr{H}$ has no torsion.
- Locally, that means $gcd(a_1, \ldots, a_n) = 1$ there is a transformation, *the saturation*, consisting locally in the division by $gcd(a_1, \ldots, a_n)$, that acting over a integrable hyperplane field $\mathscr{H}$ gives rise to the minimum foliation adapted to D, $Sat(\mathscr{H}, D)$, containing $\mathscr{H}$.
- Globally $\mathscr{H} = Sat(\mathscr{H}, D)\mathscr{I}_{(\mathscr{H},D)}$, $\mathscr{I}_{(\mathscr{H},D)}$ being the ideal locally generated by $gcd(a_1, \ldots, a_n)$. If $D = \varnothing$, for any singular integrable hyperplane field $\mathscr{H}$, $Sat(\mathscr{H}, \varnothing)$ will be called the *foliation associated* with $\mathscr{H}$.
- A *non-dicritical component* of D is an irreducible component of D which is an integral hypersurface of $\mathscr{F}$. Then $Nd(D, \mathscr{F})$ is the union of the non-dicritical components of D.
- We will denote by $\Theta_M[D]$ the sheaf of vector fields tangent to D; it is locally generated by:

$$\left\{x_i \frac{\partial}{\partial x_i}\right\}_{i \in A} \cup \left\{\frac{\partial}{\partial x_i}\right\}_{i \in \{1,\ldots,n\} \setminus A}$$

  and is the dual of $\Omega_M[D]$.
- For any singular integrable hyperplane field adapted to D, $\mathscr{H}$, there is a coherent ideal sheaf $\mathscr{U}(\mathscr{H}, D) \subset \mathscr{O}_M$ given by:

$$\mathscr{U}(\mathscr{H}, D)_{\mathbf{p}} = \{\theta(X) \mid \theta \in \mathscr{H}_{\mathbf{p}},\ X \in \Theta_M[D]_{\mathbf{p}}\},$$

  i.e. locally generated by the coefficients $\{a_1, \ldots, a_n\}$ of a generator ω of $\mathscr{H}$.

If we consider a divisor $E = Nd(E, \mathscr{F})$, i.e. with all its components non-dicritical, and E is defined by $\prod_{i \in A} x_i = 0$, $\mathscr{F}$ can be generated by a form:

$$\omega = \prod_{i \in A} x_i (\sum_{i \in A} b_i \frac{dx_i}{x_i} + \sum_{i \in \{1,\ldots,n\} \setminus A} b_i dx_i),\ b_i \in \mathscr{O}_{M\mathbf{p}},\ gcd(b_1, \ldots, b_n) = 1.$$

With this notation the vertical invariants are:

(1) The *dimensional type* $\tau(\mathscr{F}, \mathbf{p})$ codimension in $T_{\mathbf{p}}(M)$ of the vector subspace:

$$\mathscr{D}(\omega)(\mathbf{p}) = \{X(\mathbf{p}) \mid X \in \Theta_{M\mathbf{p}},\ \omega(X) = 0\}.$$

(2) The *adapted order* $\nu(\mathscr{F}, E; \mathbf{p}) = min(\{\nu_{\mathbf{p}}(b_i)\}_{1 \le i \le n})$.
(3) The *adapted multiplicity* $\mu(\mathscr{F}, E; \mathbf{p})$ order at $\mathbf{p}$ of the ideal generated by:

$$\{b_i\}_{i \in A} \cup \{x_j b_i\}_{i \in \{1,\ldots,n\} \setminus A,\ 1 \le j \le n}.$$

If the dimensional type of $\mathscr{F}$ is t, there is a local system of coordinates at $\mathbf{p}$, $(x_1, \ldots, x_n)$, such that $\mathscr{F}$ can be generated by a form:

$$\omega = \sum_{i=1}^{t} a_i(x_1, \dots, x_t)dx_i.$$

Then the foliation is locally trivial along a germ of foliation of $(\mathbb{C}^t, \mathbf{0})$. The adapted order $\nu(\mathscr{F}, E; \mathbf{p})$ is the order at $\mathbf{p}$ of the coherent ideal $\mathscr{U}(\mathscr{H}, E)$ and as a consequence is analytically upper semicontinuous; then the sets:

$$S_r((\mathscr{F}, E) = \{\mathbf{p} \in M \mid \nu(\mathscr{F}, E; \mathbf{p}) \geq r\}$$

are closed analytic subsets of M.

If Y is a closed irreducible analytic subspace of M, we may generalize the adapted order to the *adapted order along* Y, $\nu(\mathscr{F}, E; Y)$, defined as the order with respect to Y of $\mathscr{U}(\mathscr{H}, E)$ along Y.

Let us consider a closed connected non-singular subspace of M with normal crossings with the divisor D, then there exists at each point of M a coordinate system (*adapted to* D, Y), $(x_1, \dots, x_n)$ and two subsets A, B of $\{1, 2, \dots, n\}$ such that D and Y are respectively locally defined by:

$$D : \prod_{i \in A} x_i = 0, \;\; Y : \{x_i = 0\}_{i \in B}.$$

Now let $\pi : M' \longrightarrow M$ be the blow-up of M with center Y. $D' = \pi^{-1}(D \cup Y)$ is a normal crossing divisor, and for a foliation $\mathscr{F}$ locally defined by ω, the total transform of $\mathscr{F}$ is the singular integrable hyperplane field $\pi^{-1}(\mathscr{F})$ generated by the pull-back $\pi^*(\omega)$. Then we define the *multiplicity* of $\mathscr{F}$ along Y by:

$$\mu(\mathscr{F}, Y) = \nu_{\pi^{-1}(Y)}(\mathscr{I}_{\pi^{-1}(\mathscr{F}, D')}) = \nu_{\pi^{-1}(Y)}(\mathscr{U}(\pi^{-1}(\mathscr{F}), D')).$$

and the *adapted multiplicity* by:

$$\mu(\mathscr{F}, D; Y) = \mu(Sat(\mathscr{F}, D)Y).$$

Then:

- If $Y = \{\mathbf{p}\}$, $\mu(\mathscr{F}, D; Y)$ is the adapted multiplicity defined before.
- The foliation $\mathscr{F}'$ associated with $\pi^{-1}(\mathscr{F})$ will be called the strict transform of $\mathscr{F}$ by π.
- $\mathscr{F}'$ is the unique singular foliation over M' such that $\mathscr{F}'|_{M' - \pi^{-1}(Y)}$ is transformed in $\mathscr{F}|_{M-Y}$ by the isomorphism $\pi|_{M' - \pi^{-1}(Y)}$.
- $Sat(\mathscr{F}', D') = \mathscr{I}_{\pi^{-1}(Y)}^{-\mu(\mathscr{F}, D; Y)} Sat(\mathscr{F}, D)$.
- $\nu(\mathscr{F}, D; \mathbf{p}) \leq \mu(\mathscr{F}, D; \mathbf{p}) \leq \nu(\mathscr{F}, D; \mathbf{p}) + 1$.

Let us observe that the multiplicity is not semicontinuous.

With these invariants we may define the final forms of the reduction process and the permissible centers of blowing-up.

Definition 6.4.1 A point $\mathbf{p} \in \mathrm{Sing}(\mathscr{F})$ is a presimple singularity for $\mathscr{F}$ adapted to E if and only if it verifies one of the following properties:

(1) $\nu(\mathscr{F}, E; \mathbf{p}) = 0$.
(2) $\nu(\mathscr{F}, E; \mathbf{p}) = \mu(\mathscr{F}, E; \mathbf{p}) = 1$ and there is a coefficient $b_i,\ i \in A$ such that the linear part of b_i does not depend only on the coordinates $\{x_j\}_{j\in A}$.

Remark 6.4.1 There are some interesting properties that assures the vertical stability of presimple singularities:

- If $\mathbf{p}$ is a presimple singularity of $\mathscr{F}$, and the number of components of E at $\mathbf{p}$ is $e(\mathscr{F}, \mathbf{p}) = e$, the dimensional type $t = \tau(\mathscr{F}, \mathbf{p})$ verifies:

$$t = \nu(\mathscr{F}, \varnothing; \mathbf{p}) + 1,\ e - 1 \leq t \leq e.$$

- Let us assume that $\nu(\mathscr{F}, E; \mathbf{p}) = \mu(\mathscr{F}, E; \mathbf{p}) = 1$ (then $t = e - 1$) and that $(x_1, \ldots, x_n)$ is a coordinate system adapted to $\mathscr{F}$ and E with E defined by $\prod_{i=1}^{t-1} x_i = 0$. Then a local generator of $\mathscr{F}$ can be written as:

$$\omega = \prod_{i=1}^{t-1} x_i (\sum_{i=1}^{t-1} b_i \frac{dx_i}{x_i} + \sum_{i=t}^{n} b_i dx_i,\ b_i \in \mathscr{O}_{M\mathbf{p}},$$

$$gcd(b_1, \ldots, b_n) = 1.$$

  Then there exists a formal hypersurface $\widehat{H}$ not a component of E that is a formal integral hypersurface for $\mathscr{F}$ with normal crossings with E.
- We may modify biholomorphically $(x_t, \ldots, x_n)$ in such a way that the equation of $\widehat{H}$ becomes, for any index $k,\ 1 \leq k \leq t-1$:

$$x_t = \phi(x_1, \ldots, x_{t-1}) = \sum_{j=1}^{\infty} \phi_j^{(k)}(x_1, \ldots, x_{k-1}, x_{k+1}, \ldots, x_{t-1} x_k^j)$$

  with the $\phi_j^{(k)}$ convergent series in $t-2$ variables with a common domain of convergence.
- The linear part of the b_i's, $\{B_i\}_{1\leq i\leq t-1}$ is non-resonant, i.e. there do not exist natural numbers $r_1, \ldots, r_{t-1}$ such that $\sum_{i=1}^{t-1} r_i B_i = 0$.

Definition 6.4.2 Given a foliation $\mathscr{F}$ and a normal crossings divisor D over a complex-analytic variety M, we will say that a subspace $Y \subset M$ is *a permissible center for $\mathscr{F}$ adapted to D at a point* $\mathbf{p} \in Y$ if:

(1) Y is a closed non-singular analytic subspace tangent to $\mathscr{F}$.
(2) Y has normal crossings with D.

(3) For $\mathcal{F}_D = Sat(\mathcal{F}, D)$:

$$\nu_{\mathbf{p}}(\mathcal{U}(\mathcal{F}_D, D; Y)) = \nu_Y(\mathcal{U}(\mathcal{F}_D, D; Y))(= \mu(\mathcal{F}, D; Y)).$$

If in addition we have:

$$\mu(\mathcal{F}, D; Y) = \mu(\mathcal{F}, D; \mathbf{p})$$

we will say that Y is *an appropriate center for* $\mathcal{F}$ at $\mathbf{p}$. A *permissible center* is a center Y permissible at any point $\mathbf{p} \in Y$.

The condition of permissibility is horizontally stable and is useful for the control of generic points. The condition of appropriate center is not horizontally stable and we will need it to control the bad points that appear mainly in the dicritical situations.

Then the invariants order and multiplicity and the presimple condition have good behavior under permissible blow-ups. F. Cano proves that:

Proposition 6.4.1 *Let $\pi M' \longrightarrow M$ be a permissible blow-up with center Y for a foliation $\mathcal{F}$ adapted to D, let $\mathcal{F}'$ be the strict transform of $\mathcal{F}$, $D' = \pi^{-1}(D \cup Y)$ and $\mathbf{p} \in Y$. Then:*

(1) $\exists \mathbf{q}' \in \pi^{-1}(\mathbf{p})$ *such taht* $\nu(\mathcal{F}', D'; \mathbf{q}') = 0$.
(2) $\forall \mathbf{p}' \in \pi^{-1}(\mathbf{p}),\ \nu(\mathcal{F}', D'; \mathbf{p}') \le \nu(\mathcal{F}, D; \mathbf{p})$.
(3) *If* $e(D', \mathbf{p}') \ge e(D, \mathbf{p})$ *then* $\mu(\mathcal{F}', D'; \mathbf{p}') \le \mu(\mathcal{F}, D; \mathbf{p})$.

Proposition 6.4.2 *With the same notation, if all the singularities of $\mathcal{F}$ adapted to D are presimples, the singularities of $\mathcal{F}'$ adapted to D' are all presimples.*

There are also invariants related to the directrix (differential version of Hironaka's strict tangent space). Let D be a normal crossings divisor, $\mathbf{p} \in D$, let F be the intersection of the components of D through $\mathbf{p}$ and let Y be a non-singular closed analytic subspace with normal crossing with D at $\mathbf{p}$. We may choose a coordinate system centered at $\mathbf{p}$, $(x_1, \ldots, x_n)$ in such a way that there are subsets $A,\ B$ of $\{1, \ldots, n\}$ such that:

$$\mathcal{I}_{D,\mathbf{p}} = \left(\prod_{i \in A} x_i\right) \mathcal{O}_{M\mathbf{p}},\ \mathcal{I}_{F,\mathbf{p}} = \sum_{i \in A} x_i \mathcal{O}_{M\mathbf{p}},\ \mathcal{I}_{Y,\mathbf{p}} = \sum_{i \in B} x_i \mathcal{O}_{M\mathbf{p}}.$$

If we write

$$\mathcal{O} = \mathcal{O}_{M\mathbf{p}},\ \mathcal{M} = m_{M\mathbf{p}},\ \widetilde{\mathcal{O}} = \mathcal{O}/\mathcal{I}_{F,\mathbf{p}},\ \widetilde{\mathcal{M}} = \mathcal{M}/\mathcal{I}_{F,\mathbf{p}},$$

we have graded rings:

$$Gr_{\mathbf{p}}(M) = Gr_{\mathcal{M}}(\mathcal{O}) \simeq \mathbb{C}[T_1, \ldots, T_n],\ Gr_{\widetilde{\mathcal{M}}}(\widetilde{\mathcal{O}}) \simeq \mathbb{C}[\{T_i\}_{i \notin A}]$$

with $T_i = x_i + \mathcal{M}^2$, $\widetilde{T}_i = x_i + \widetilde{\mathcal{M}}^2$ and a canonical morphism:

$$\xi : Gr_{\mathcal{M}}(\mathcal{O} \to Gr_{\widetilde{\mathcal{M}}}(\widetilde{\mathcal{O}}).$$

Given a homogeneous $h \in Gr_{\mathcal{M}}(\mathcal{O}$ the directrix of h, $Dir(h)$ is the largest linear subspace of $T_{\mathbf{p}}(M)$ that leaves $h = 0$ acting by translation, which means that the ideal of the directrix $\mathcal{I}Dir(h)$ in $Gr_{\mathcal{M}}(\mathcal{O}$ is generated by independent linear forms $\theta_1, \ldots, \theta_r$, such that $h = P(\theta_1, \ldots, \theta_r)$ for a certain polynomial P. If $\mathcal{J}$ is a homogeneous ideal of $Gr_{\mathcal{M}}(\mathcal{O}$, we will denote by $\mathcal{I}Dir(\mathcal{J})$ the ideal generated by $\{\mathcal{I}Dir(h)\}_{h\in\mathcal{J}}$ and by $Dir(\mathcal{J})$ the corresponding linear subspace of the tangent space to M at $\mathbf{p}$.

Now for any s, $\nu_{\mathbf{p}}(\mathcal{J}) \geq s$ we will write:

- $In^s(\mathcal{J})$, ideal of $Gr_{\mathcal{M}}(\mathcal{O}$ generated by $\{f + \mathcal{M}^{s+1}\}_{f\in\mathcal{J}}$.
- $\mathcal{I}Dir^s(\mathcal{J}) = \mathcal{I}Dir(In^s(\mathcal{J}))$, $Dir^s(\mathcal{J}) = Dir(In^s(\mathcal{J}))$.
- The *adapted codimensions*:

$$d^s(\mathcal{J}, D) = dim_{\mathbb{C}}\left(\xi(\mathcal{I}Dir^s(\mathcal{J})) \cap Gr^1_{\widetilde{\mathcal{M}}}(\widetilde{\mathcal{O}})\right),$$

$$t^s(\mathcal{J}, D) = dim_{\mathbb{C}}\left(\xi(In^s(\mathcal{J})) \cap Gr^1_{\widetilde{\mathcal{M}}}(\widetilde{\mathcal{O}})\right).$$

- If $\mathcal{F}$ is a foliation and $\widetilde{\mathcal{F}}_D = Sat(\mathcal{F}, D)$, and we let $r = \nu(\mathcal{F}, D; \mathbf{p})$, $m = \mu(\mathcal{F}, D; \mathbf{p})$ and:

  – $\mathcal{U} = \mathcal{U}(\mathcal{F}_D, D)$ if $m = r + 1$,
  – $\mathcal{U} = \mathcal{U}(\mathcal{F}_D, D, \mathbf{p})$ if $m = r$,

  we define the directrix of $\mathcal{F}$ adapted to D at $\mathbf{p}$ as:

$$Dir(\mathcal{F}, D; \mathbf{p}) = Dir^r(\mathcal{U}),\ d(\mathcal{F}, D; \mathbf{p}) = d^r(\mathcal{U}, D)$$

  and its ideal:

$$\mathcal{I}Dir(\mathcal{F}, D; \mathbf{p}) = \mathcal{I}Dir^r(\mathcal{U}),\ t(\mathcal{F}, D; \mathbf{p}) = t^r(\mathcal{U}, D).$$

  Also $l(\mathcal{F}, D; \mathbf{p}) = dim_{\mathbb{C}}(Dir(\mathcal{F}, D; \mathbf{p}))$.

The more significant of the invariants attached to the directrix is the transversal codimension of the directrix (invariant d); this invariant gives the number of transversally independent (modulo D) linear forms appearing in the ideal of the directrix.

In our case the codimension of the directrix is not stable by blowing-up, but the transversal codimension d has a well-controlled vertical behavior, except if after blowing up $\mathbf{p}$ the exceptional divisor is a dicritical component for the strict transform of the foliation (radially dicritic point). With this exception, after blowing up, $d' \geq d$. The most difficult case is $d = 0$ but in this case one can show that we are in a *maximal contact* situation and we may drop the dimension.

The invariants attached to the directrix are complementary invariants of the essential one, the pair (r, m^*), where $m^* = m$, except if $r = m$ and there is a resonance between the B_i's defined in Remark 6.4.1; in this case $m^* = m + 1$. In the non-dicritical case is the only invariant we need.

The origin of pathologies is the resonances, as in the reduction of singularities of algebraic varieties in a positive characteristic. There is another difference between the resolution of singularities of complex-analytic spaces and the reduction of singularities of foliations. The vertical behavior of the invariants we have described, lexicographically ordered as usual, is good, but they do not have horizontal stability. They are not always useful either for the construction of permissible centers or for the control of propagation of singularities.

This is the point of difficulty for work in dimensions greater than three. In dimension three the good points, i.e. points in the generic part of the analytic set of non-presimple singularities, can be controlled by our invariants simultaneously in horizontal and vertical ways, and there are only a finite number of isolated bad points; that can be destroyed, but that does not happen in dimensions greater than three; in this case there may exist too many bad points.

The destruction of bad points can be controlled by the maximum lexicographic of the values of (r, m^*) over these points. We need blow-ups with centers appropriate with respect to the bad points and a *global preparation* to prevent cycles.

More precisely, if we call $\mathrm{Sing}^*(\mathscr{F}, D)$ the set of singular not presimple singularities of $\mathscr{F}$ adapted to D, we will say that $\mathbf{p}$ is a *good point* if there is a *generic equireduction* around $\mathbf{p}$, i.e. if there is a neighborhood U of $\mathbf{p}$ and a finite sequence of blow-ups:

$$U = U_0 \xleftarrow{\pi_1} U_1 \xleftarrow{\pi_2} U_2 \cdots \xleftarrow{\pi_k} U_k$$

such that, working in U:

- If $(\mathscr{F}, D) = (\mathscr{F}_0, D_0)$ and $(\mathscr{F}_j, D_j)$, $j > 0$, are the strict transforms by π_j of $(\mathscr{F}_{j-1}, D_{j-1})$, $\mathrm{Sing}^*(\mathscr{F}, D)$ is non-singular of codimension two, and, for $j \geq 1$ $\mathrm{Sing}^*(\mathscr{F}_j, D_j)$ is non-singular of codimension two and etalé over $\mathrm{Sing}^*(\mathscr{F}, D)$.
- π_j is the blow-up of U_{j-1} with center $\mathrm{Sing}^*((\mathscr{F}_{j-1}, D_{j-1}))$.
- $\mathrm{Sing}^*(\mathscr{F}_k, D_k) = \varnothing$.

Then the bad (not good) points form a closed analytic set of codimension at least three, that is they are isolated points if the ambient space is tridimensional. Then:

(1) If there are bad points with the maximum value of (r, m^*), it can be proved that there are permissible centers, appropriate at the bad points and well prepared to avoid cycles.
(2) If a bad point $\mathbf{p}$ is radially dicritic we make first a blow-up with center $\mathbf{p}$; if not we choose the largest center, provided it accomplishes the conditions of (1).

(3) After a finite number of blow-ups with centers fulfilling (1) and (2) necessarily a bad point is modified.

By following these criteria we have a sequence of global blow-ups; if it stops, i.e. the bad points of the initial (r, m^*)-stratum disappear, we are on the way to achieving a situation of generic equireduction. Then we must prove that it is impossible to have an infinite sequence. If this infinite sequence exists, we have an infinite sequence of local blow-ups fulfilling the conditions of (1) and (2):

$$\begin{array}{ccccccc} \cdots U_k & \xrightarrow{\pi_k} & \cdots U_2 & \xrightarrow{\pi_2} & U_1 & \xrightarrow{\pi_1} & U_0 = U, \\ \cdots P_k & \longmapsto & \cdots P_2 & \longmapsto & P_1 & \longmapsto & P_0, \end{array}$$

where all the P_i's are bad points with the same (r, m^*) and this sequence cannot exist. There are two possibilities:

(1) *The stable case*: The adapted multiplicity m remains constant at all the points P_i, then as we saw in the dimension-two case, if we are in a *maximal contact situation*, i.e. if we have a non-singular hypersurface H such that its strict transforms along the sequence of blow-ups contains the centers of all the π_i's, we have a concept of an idealistic exponent that allows us to reduce the ambient space to H. If we do not have maximal contact, we need to get it, and this is a difficult task. It is necessary to construct new invariants attached to a characteristic polygon in the spirit of Hironaka's characteristic polyhedron [64].
(2) *The unstable case*: It corresponds to infinite repetitions of $m = r,\ m = r + 1$, i.e. infinite succession of resonances, the directrix jumps to infinity many times. In dimension three it is possible to solve the problems by an easy but very long process, but in dimensions greater than three the situation apparently becomes out of control.

Then by following the described strategy we may reduce all the singularities to pre-simple ones, and we have the following theorem:

Theorem 6.4.1 (F. Cano [21]) *Let $\mathscr{F}$ be a codimension-one singular foliation over an ambient space M and let D be a normal crossings divisor on M. Assume that $dim(M) = 3$ and M is a germ along a compact analytic subset Z. Then there is a morphism $\pi : M' \longrightarrow M$ such that:*

(1) *The morphism π is the composition of a finite sequence of blow-ups with non-singular closed permissible analytic centers.*
(2) *If $\mathscr{F}'$ is the strict transform of $\mathscr{F}$ by π and D' is the total transform of D, then $\mathscr{F}'$ has at most presimple singularities adapted to D'.*

We need a better end for the reduction process in order to find analytic integral hypersurfaces of the foliation. In the presimple singularities remain resonances that may be killed by using more convenient blow-ups and at the end we may have *simple singularities*.

The difference between presimple and simple singularities is visible after Jordanization; it consists in the non-existence of resonances between proper values of the initial form, as we did in the dimension-two case. The formal normal forms for presimple singularities are described in [21, 27], by following ideas of Martinet [93]. We will point out some interesting properties of simple singularities and the theorem of reduction.

Proposition 6.4.3 (F. Cano [25]) *Let* $\mathbf{p} \in \mathrm{Sing}(\mathscr{F})$ *be a simple singularity adapted to* D, *let* $t = \tau(\mathscr{F}, D)$ *be the dimensional type of the adapted foliation at* $\mathbf{p}$, *and* $e = e(D, \mathbf{p})$ *the number of components of* D *at* $\mathbf{P}$, *then there are two possibilities:*

(1) Simple corner*: Corresponds to* $e = t$, *in this case the irreducible components of* D *are the only integral hypersurfaces of* $\mathscr{F}$ *at the point* $\mathbf{p}$.
(2) Trace simple singularity*: Corresponds to* $e = t - 1$ *then there is exactly one transversally formal hypersurface* $\widehat{H}$ *at* $\mathbf{p}$ *having normal crossings with* D *and such that the only integral hypersurfaces of* $\mathscr{F}$ *at the point* $\mathbf{p}$ *are* $\widehat{H}$ *and the irreducible components of* D *at* $\mathbf{p}$.

Also if Y *is a permissible center for* $\mathscr{F}$ *adapted to* D, $\mathbf{p} \in Y$, *and* $\pi : M' \to M$ *is the blow-up with center* Y, *then:*

(1) π *is non-dicritical.*
(2) *Any singular point* $\mathbf{p}' \in \pi^{-1}(\mathbf{p}) \cap \mathrm{Sing}(\mathscr{F}')$ *is a simple singularity of* $\mathscr{F}'$ *adapted to* D'.
(3) *If* $\mathbf{p}$ *is a simple corner any point in* $\pi^{-1}(\mathbf{p}) \cap \mathrm{Sing}(\mathscr{F}')$ *is a simple corner.*
(4) *If* $\mathbf{p}$ *is a trace simple singularity,* $\pi^{-1}(\mathbf{p}) \cap \mathrm{Sing}(\mathscr{F}') \subset \widehat{H}'$.

Theorem 6.4.2 (F. Cano [21]) *Let* $\mathscr{F}$ *be a codimension-one singular foliation over an ambient space* M *and let* D *be a normal crossings divisor on* M *whose irreducible components are non-dicritical for* $\mathscr{F}$. *Assume that* M *is a germ along a compact analytic subset* $C \subset M$, *and that all the singularities of* $\mathscr{F}$ *are presimples. Then there is a morphism* $\pi : M' \longrightarrow M$ *such that:*

(1) *The morphism* π *is the composition of a finite sequence of blow-ups with non-singular closed permissible analytic centers.*
(2) *If* $\mathscr{F}'$ *is the strict transform of* $\mathscr{F}$ *by* π *and* D' *is the divisor composed by the non-dicritical components of the total transform of* D. *Then* $\mathscr{F}'$ *has at most simple singularities adapted to* D'.

The proof of the theorem consists, as in the two-dimensional case, in the destruction of resonances, in a global coherent way, by using a characteristic polyhedron and a global strategy derived from that of Spivakovsky to solve the *game of Hironaka* [109].

Thom's conjecture about integral hypersurfaces is as follows:

Conjecture (R. Thom) Let $\mathscr{F}$ be a germ of foliation in $(\mathbb{C}^n, \mathbf{0})$; there is at least one germ of convergent integral hypersurface of $\mathscr{F}$.

Let us remember that, with the usual notation, an irreducible component of D is non-dicritical for $\mathcal{F}$ if it is an integral hypersurface, and then to be dicritiucal means be transversal to the leaves of $\mathcal{F}$. Then saying that the germ of foliation $\mathcal{F}$ in $(\mathbb{C}^n, \mathbf{0})$ is non-dicritical means that for any finite sequence of local permissible blow-ups:

$$(\mathbb{C}^n, \mathbf{0}) = M_0 \xleftarrow{\pi_1} M_1 \xleftarrow{\pi_2} M_2 \cdots \xleftarrow{\pi_k} M_k = M'$$

all the irreducible components of the total exceptional divisor D' on M' are non-dicritical for the transform $\mathcal{F}'$ of $\mathcal{F}$. Then in dimension two:

- If the foliation $\mathcal{F}$ is non-dicritical there is a bijection between the set of formal integral curves of $\mathcal{F}$ and the set of trace simple singularities after reduction of singularities, and to prove the conjecture we need to prove that at least one of the formal integral curves is convergent.
- The foliation is dicritical if and only if it has infinitely many integral curves, obtained by projecting the integral curves transversal to one dicritical component of the exceptional divisor after reduction of singularities, and in fact in this case there are infinite convergent integral curves for the foliation.

And in dimension two as we have already seen, Camacho and Sad [19], and later J. Cano by other methods [36, 38], prove Thom's conjecture.

The situation in dimension $n \geq 3$ is more difficult, because a germ of hypersurface at a point obtained after some blow-ups does not necessarily project in a hypersuface in the original space. For this reason there are counterexamples to the conjecture. For instance, we may consider the dicritical integrable form in $\mathbb{C}^3$:

$$\omega = (x^m y - z^{m+1})dx + (y^m z - x^{m+1})dy + (z^m x - y^{m+1})dz, \ m \geq 2.$$

the blow-up with center the origin of ω is in the first chart:

$$\omega' = (y^m z - 1)dy + (z^m - y^{m+1})dz.$$

Then the exceptional divisor is a projective plane transverse to the strict transform of the foliation, Jouanolou [76] proves that the possible leaves of the foliation generated by ω would be cones whose projectivization would be the algebraic integral curves of ω'. He proves that ω' has no algebraic integral curves in the projective plane, then ω gives a counterexample to the conjecture of Thom.

The counterexample being a dicritical foliation, one may modify the conjecture and Cano, Cerveau and Mattei [28, 29], prove for dimension $n \geq 3$ the following theorem:

Theorem 6.4.3 *Let $\mathcal{F}$ be a germ of non-dicritical foliation in $(\mathbb{C}^n, \mathbf{0})$; there is at least one germ of convergent integral hypersurface of $\mathcal{F}$.*

In dimension three the proof of Cano and Cerveau uses the reduction to simple singularities, which exists in this dimension. In the corner singularities the only integral hypersurfaces are the components of the divisor; then we need to trace simple singularities. After reduction of singularities to simple ones: Beginning with a foliation $\mathscr{F}$, in the final ambient space M' the trace simple singularities of its strict transform $\mathscr{F}'$ forms a closed analytic space, whose connected components are in biunivocal correspondence with the formal integral hypersurfaces of $\mathscr{F}$; any of them is a projection of a formal integral hypersurface of $\mathscr{F}'$ in M'. Moreover, convergent integral hypersurfaces of $\mathscr{F}$ are projections of convergent integral hypersurfaces of $\mathscr{F}'$. Then it is necessary only to prove the existence of a convergent integral hypersurface over a component of the space of trace simple singularities, and the existence may be tested at a well-chosen point (generic point of dimensional type 2). The proof follows an argument of transversality by Mattei and Moussu [94].

For the case $n > 3$, Cano and Mattei use the result of Cano and Cerveau in dimension three, the generic equireduction, and an argument of cohomological triviality [108] that works by going from three to n, but not from two to n.

6.5 Vector Fields and Valuations

In this chapter, we have so far studied only codimension-one foliations; to finish this epilogue we will say some things about dimension-one foliations, i.e. foliations by curves, and about vector fields as generators of dimension-one foliations.

In ambient dimension-two, codimension-one and dimension-one foliations are the same objects, and both are covered by Seidenberg's result [105]. The first results about vector fields in dimension three are in F. Cano's thesis [20], which gives a winner strategy for Hironaka's game attached to the problem of reduction of singularities of vector fields.

To explain Cano's result, we will define firstly the invariants: In codimension-one foliations we used as the first invariant the adapted multiplicity; for foliations by lines the "dual concept" of adapted multiplicity is the logarithmic multiplicity. We may consider the foliation by curves as a locally free dimension-one submodule of the tangent sheaf; then, fixing a normal crossing divisor D, a germ of dimension-one foliation $\mathscr{L}$ at a point $\mathbf{p}$ can be generated by a vector field, described in a local system of coordinates $(\mathbf{z}) = (z_1, \ldots, z_n)$, centered at $\mathbf{p}$, by:

$$\xi = \sum_{i=1}^{e} a_i(\mathbf{z}) z_i \frac{\partial}{\partial z_i} + \sum_{i=e+1}^{n} a_i(\mathbf{z}) \frac{\partial}{\partial z_i}, \quad gcd(\{a_i(\mathbf{z})\}_{1 \le i \le n}) = 1,$$

provided D is locally defined at $\mathbf{p}$ by the equation $\prod_{i=1}^{e} z_i = 0$.

Then the logarithmic multiplicity at $\mathbf{p}$ of $\mathscr{L}$ with respect to D is the minimum of the multiplicities of the coefficients a_i, and we will say that $\mathbf{p}$ is a *log-elementary* singularity of $\mathscr{L}$ with respect to D if the logarithmic multiplicity at $\mathbf{p}$ is less than or equal to one.

The game is as follows:

There are two players A and B and a board (a germ of foliation by lines $\mathscr{L}$ at a point $\mathbf{p}$):

- If $\mathbf{p}$ is log-elementary A wins. If not A choose a formal center of blowing-up.
- B chooses a point $\mathbf{p}'$ in the pre-image of $\mathbf{p}$ by blowing-up.
- The game recommences for A with $\mathbf{p}'$ and the strict transform $\mathscr{L}'$ of $\mathscr{L}$ at $\mathbf{p}'$.

Find a winning strategy for A by fixing decision criteria to finish the game in a finite number of steps.

In connection with the reduction of singularities, the game has two problems: The centers of the blow-ups are not necessarily convergent, and the process is strictly local. Also it is possible to strengthen the conditions for the victory of the player A demanding that at the end of the game the condition for $\mathbf{p}$ be *elementary singularity*.

A vector field ξ over the variety M gives at any point $\mathbf{p} \in M$ a derivation of the local ring $\mathscr{O}_{M\mathbf{p}}$. The point $\mathbf{p}$ is a singular point, if an only if $\xi(m_{M\mathbf{p}}) \subset m_{M\mathbf{p}}$, then ξ induces a linear endomorphism of the tangent space of M at $\mathbf{p}$ that is called the *jacobian map* or the *linear part* of the vector field. The point $\mathbf{p}$ is elementary if either it is non-singular or the linear part at $\mathbf{p}$ of the vector field is not nilpotent. In [26], F. Cano gives a winning strategy for the stronger game consisting in finalizing with elementary singularities.

The first step toward a reduction of singularities of vector fields would be the following result:

Proposition 6.5.1 *Let ξ be a germ of a vector field, and let D be a germ of a divisor with normal crossings both on $(\mathbb{C}^3, \mathbf{0})$. Let $\mathscr{L}$ be the foliation generated by ξ. Then there is a finite sequence of local blow-ups:*

$$(\mathbb{C}^3, \mathbf{0}) = (M_0, \mathbf{p}_0) \xleftarrow{\pi_1} (M_1, \mathbf{p}_1) \xleftarrow{\pi_2} \cdots \xleftarrow{\pi_k} (M_k, \mathbf{p}_k)$$

such that, being $\mathscr{L}_i$, D_i the transformed of $\mathscr{L}_{i-1}$, D_{i-1} by π_i:

(1) *The center Y_{i-1} of π_i is either a point or a germ of a non-singular analytic curve invariant by $\mathscr{L}_{i-1}$ and with normal crossings with D_{i-1}.*
(2) *$\mathscr{L}_k$ is generated by a vector field with an elementary singularity at $\mathbf{p}_k$.*

But the proposition is false. There are counterexamples of a specially bad case:

- There is a totally transcendental (i.e. not contained in any germ of an analytic surface) germ of formal curve invariant by ξ, $\widehat{\Gamma}$ suth that $\widehat{\Gamma}_i$ being the strict transform by π_i of $\widehat{\Gamma}_{i-1}$ $\forall i$, $\mathbf{p}_i \in \widehat{\Gamma}_i$, $\forall i$.

In this case there are examples by F. Sanz and F. Sancho (see [34] for details) of vector fields ξ such that it is impossible to find a sequence of blow-ups fulfilling the proposition. These vector fields are:

$$\xi_{\alpha,\beta,\lambda} = x_1 \left(x_1 \frac{\partial}{\partial x_1} - \alpha x_2 \frac{\partial}{\partial x_2} - \beta x_3 \frac{\partial}{\partial x_3} \right) + x_1 x_3 \frac{\partial}{\partial x_2} + (x_2 - \lambda x_1) \frac{\partial}{\partial x_3},$$

with $\alpha, \beta \in \mathbb{R}_{\geq 0}$, $\lambda \in \mathbb{R}_{>0}$.

Then to transform Proposition 6.5.1 into a true proposition there are three possibilities:

1. Add a restrictive hypothesis to the vector field: Cano and Roche [34] prove that in dimension three with the additional hypothesis:

   (F) There exists a germ of a codimension-one holomorphic foliation $\mathscr{F}$ tangent to $\mathscr{L}$.

   Proposition 6.5.1 becomes true.
2. Use more general blow-ups: Panazzolo [99] uses weighted blow-ups, i.e. combinations of blow-ups and ramifications to prove Proposition 6.5.1, in the real setting and in dimension three.
3. Fewer conditions for the final singularities: If one demands log-elementary singularities as unique final forms, Cano, Roche and Spivakovsky prove in [35] the existence of a birational reduction of vector field singularities to log-elementary ones.

We will look briefly at the three possibilities. To add a supplementary hypothesis in order to give an affirmative answer to the problem, Cano and Roche analyze the Sanz–Sancho example and prove that $\xi_{\alpha,\beta,\lambda}$ is not tangent to any codimension-one foliation. After that they observe that for a vector field ξ, the property of not admitting a resolution is equivalent to some specific properties of the Sanz–Sancho example that motivate the nonexistence of a tangent codimension-one analytic foliation.

In the second possibility, Panazzolo uses weighted blow-ups of $\mathbb{R}^3$, i.e. combinations of blow-ups and ramifications, defined as follows:

For a vector $\mathbf{n} = (n_1, n_2, n_3) \in \mathbb{Z}^3_{>0}$ the $\mathbf{n}$-weighted blow-up centered at the origin is the proper analytic surjective map:

$$\begin{array}{rcl} \Phi_{\mathbf{n}} : \mathbb{S}^2 \times \mathbb{R}^+ & \to & \mathbb{R}^3 \\ ((\overline{x}, \overline{y}, \overline{z}), t) & \mapsto & (t^{n_1}x, t^{n_2}y, t^{n_3}z) \end{array}$$

and for $\mathbf{n} = (n_1, n_2) \in \mathbb{Z}^2_{>0}$ the $\mathbf{n}$-weighted blow-up centered on the line $x = y = 0$ is:

$$\begin{array}{rcl} \Phi^z_{\mathbf{n}} : \mathbb{S}^1 \times \mathbb{R}^+ \times \mathbb{R} & \to & \mathbb{R}^3 \\ ((\overline{x}, \overline{y}), t, z) & \mapsto & (t^{n_1}x, t^{n_2}y, z) \end{array}$$

By using weighted blow-ups and a strategy similar to that of Cano's paper [21], Panazzolo is able to prove the reduction theorem for one-dimensional analytic foliations in $\mathbb{R}$. But as McQuillan and Panazzolo observe in [88]: "*the only caveat to use this techniques in the complex space is that there is no suitable 1-category where weighted blowing-up will preserve smoothness of the ambient space*". Then to extend Panazzolo's result to the complex case they needed the hard machinery of the theory of algebraic stacks.

In the third possibility there is a novelty in the approach to the problem, Cano, Roche and Spivakowsky change the ambient space from analytic to algebraic, trying to obtain a birational reduction of singularities by following Zariski's method [127], i.e. by proving a result of local uniformization and using a patching method developed by O. Piltant [100]. The theorem of local uniformization has also been proved by Fernández and Duque [43] for codimension-one foliations in any ambient dimension.

From the algebro-geometric point of view, the ambient space is a projective algebraic variety M_0 over $\mathbb{C}$ of dimension n, $K(M_0) = K$ is the field of rational functions over M_0, $\omega_{K/\mathbb{C}}$ is the K-vector space of Kähler differentials of K with respect to $\mathbb{C}$, and $Der_{\mathbb{C}}K$ the K-vector space of $\mathbb{C}$-derivations of K. Then a codimension-one foliation over M_0 is a one-dimensional K-vector subspace $\mathscr{F} \subset \Omega_{K/\mathbb{C}}$ generated by ω such that $\omega \wedge d\omega = 0$, and one foliation by lines over M_0 is a one-dimensional K-vector subspace $\mathscr{L}$ of $Der_{\mathbb{C}}K$.

Let $\mathbf{p}$ be a non-singular point in a projective model M of K, and let $\mathscr{O}$ denote $\mathscr{O}_{M\mathbf{p}}$ and $\mathfrak{m}$ its maximal ideal. Then, $\Omega_{\mathscr{O}/\mathbb{C}}$ and $Der_{\mathbb{C}}\mathscr{O}$ are free rank-n $\mathscr{O}$-modules. Now:

- For a codimension-one foliation $\mathscr{F}$, $\mathscr{F}_{M\mathbf{p}} = \mathscr{F} \cap \Omega_{\mathscr{O}/\mathbb{C}}$ is a free submodule of rank 1 of $\Omega_{\mathscr{O}/\mathbb{C}}$ and we will call it the local foliation induced by $\mathscr{F}$.
- For a foliation by lines $\mathscr{L}$, $\mathscr{L}_{M\mathbf{p}} = \mathscr{L} \cap Der_{\mathbb{C}}\mathscr{O}$ is a free submodule of rank 1 of $Der_{\mathbb{C}}\mathscr{O}$ and we will call it the local foliation induced by $\mathscr{L}$.

Then:

- We will say that $\mathscr{F}$ (resp. $\mathscr{L}$) is singular at $\mathbf{p}$ if

$$\mathscr{F}_{M\mathbf{p}} \subset \mathfrak{m}\Omega_{\mathscr{O}/\mathbb{C}}, \text{ resp. } \mathscr{L}_{M\mathbf{p}} \subset \mathfrak{m}Der_{\mathbb{C}}\mathscr{O}.$$

- We will say that $\mathscr{F}$ (resp. $\mathscr{L}$) is log-elementary at $\mathbf{p}$ if there exists a regular system of parameters of $\mathscr{O}$ $(\mathbf{z}) = (z_1, \ldots, z_n)$, an integer e, $0 \le e \le n$, and $\omega \in \mathscr{F}_{M\mathbf{p}}$ (resp $\xi \in \mathscr{L}_{M\mathbf{p}}$), in such a way that:

$$\omega = \sum_{i=1}^{e} a_i \frac{dz_i}{z_i} + \sum_{i=e+1}^{n} a_i dz_i,$$

respectively:

$$\xi = \sum_{i=1}^{e} a_i z_i \frac{\partial}{\partial z_i} + \sum_{i=e+1}^{n} a_i \frac{\partial}{\partial z_i},$$

with $a_i \in \mathscr{O}$, $1 \le i \le n$ and $\exists j, a_j \notin \mathfrak{m}^2$.

Then it is possible to enunciate two theorems:

Theorem 6.5.1 (Local uniformization) *Let $\mathscr{F} \subset \Omega_{K/\mathbb{C}}$ be a codimension-one foliation (resp. $\mathscr{L} \subset Der_{\mathbb{C}}K$ a foliation by lines), given a $\mathbb{C}$-valuation ν of K and a projective model of K M_0, then there is a composition of a finite sequence of blow-ups $M \to M_0$ such that $\mathscr{F}$, (resp. $\mathscr{L}$) is log-elementary at the center of ν in M.*

Theorem 6.5.2 (Reduction of singularities) *Let $\mathscr{F} \subset \Omega_{K/\mathbb{C}}$ be a codimension-one foliation (resp. $\mathscr{L} \subset Der_{\mathbb{C}}K$ a foliation by lines), given a projective model of K M_0, then there is a composition of a finite sequence of blow-ups $M \to M_0$ such that $\mathscr{F}$, (resp. $\mathscr{L}$) is log-elementary at all the points of M.*

Both theorems are conjectures in all their generality, but:

(1) In ambient dimension three for one-dimensional foliations, Cano, Roche and Spivakovski [35] prove local uniformization, and by using Piltant's axiomatic theory of patching [100] they prove also reduction theorem.
(2) In ambient dimension three for codimension-one foliations it is possible to adapt Cano's proof for the analytic case [21].
(3) In ambient dimension n for codimension-one foliations, Fernàndez Duque [43] proves the local uniformization for rational archimedean $\mathbb{C}$-valuations of K.

Let us remark that the hypothesis of rational archimedean for the valuation in the theorem by Fernández Duque is not too restrictive; local uniformization for these valuations is the most difficult case in the classical theory of Zariski [123], and it seems that it is possible to derive general uniformization from this case. The proof of the local uniformization for these valuations is by induction in the rational rank, and avoids the difficulties of extending valuations to the formal completion by using a promising technique of truncation.

6.6 Valuations and Solutions of Differential Equations

The problem of finding solutions of differential equations is a "nonsense" problem. There is an *algebraic theory of differential equations* by Ritt [101] similar to the classical theory of algebraic equations. Ritt proves that for any differential equation there is a differential algebra where the equation has a solution. The true problem is to find solutions in a given algebra. In the classical theory of the algebra of formal series, these solutions can be convergent; another method, asymptotic analysis, allows the construction of some types of solution of functions. We will try to explain how valuation theory and techniques in this book may help to prove the existence of a generalized series solution of ordinary differential equations.

As we said in Sects. 6.2 and 6.3, there exist analytic solutions for any first order and first degree analytic equation, or equivalently convergent integral curves for any analytic foliation in the complex plane. We may propose also the problem of existence of integral curves for foliations over possibly singular complex surfaces. This is not the same problem, as we pointed out in Sect. 6.1, because we can solve the singularities of the surface and we have the problem already solved, a foliation in the germ $(\mathbb{C}^2, \mathbf{0})$, but we have a non-empty divisor, the exceptional divisor of the resolution, from the origin of the process. We also need an integral curve out of this divisor. Camacho [18] gives an example of analytic foliation over a normal surface without convergent integral curves.

Also by the Newton polygon method, we know that the equation:

$$f(x, y, y_1) = y_1^3 + x^2 y_1^2 - 7xyy_1 + 12y^2 + x^8, \; y_1 = \frac{dy}{dx}$$

has no Puiseux series solution, but this equation can be interpreted as the differential form $\omega = zdx - dy$ over the singular surface:

$$z^3 + x^2 z^2 - 7xyz + 12y^2 + x^8 = 0.$$

Then we have an analytic form over a singular surface that has not even formal integral curves.

For codimension-one non-dicritical foliations we saw in Sect. 6.4 the existence of convergent integral hypersurfaces, but in the case of foliations by lines there are families of examples by Gómez-Mont and Luengo [55] for dimension three and by Luengo and Olivares [86] for any dimension $n > 2$ of vector fields without convergent integral curves. But there are more general series than the Puiseux ones, and we will try to find more general series solutions.

As in the case of foliations, the initial idea can be found also in Seidenberg. In his paper [106], Seidenberg wrote: "Roughly, derivations are related to contact, and so are valuations, so one may ask for a study connecting derivations and valuations", and to show his ideas about the nature of the connection, he proved an algebraic version of Picard's theorem of existence and uniqueness:

Theorem 6.6.1 (Seidenberg) *Let K be a field (zero characteristic) and let $\mathcal{O}$ be a finite integral K-algebra. If $\mathfrak{m} \in Spec\mathcal{O}$ and $D \in Der_K(\mathcal{O})$ is such that $D(\mathcal{O}_\mathfrak{m}) \subset \mathcal{O}_\mathfrak{m}$ then there is a valuation of $Frac(\mathcal{O})$ centered in $\mathfrak{m}$ with valuation ring D-invariant.*

From a geometric point of view we may consider the derivation as a rational vector field over an algebraic variety M, defined in a neighborhood of a closed point P corresponding to a maximal ideal $\mathfrak{m}$. Then there are two possibilities:

- $D(\mathfrak{m}.\mathcal{O}_{:\mathfrak{m}}) \not\subseteq \mathfrak{m}.\mathcal{O}_\mathfrak{m}$: That is, P is a regular point of the vector field, and the valuation is the valuation defined by the integral formal curve of the vector field passing through P.
- $D(\mathfrak{m}.\mathcal{O}_\mathfrak{m}) \subseteq \mathfrak{m}.\mathcal{O}_\mathfrak{m}$: Now P is singular and there may exist more than one valuation with local ring stable by the Lie derivative attached to the vector field; those valuations do not come from integral curves.

Seidenberg chose a property that gives the connection between valuations and derivations: "*valuation ring invariant for the derivation (Lie derivative attached to the vector field)*".

This is not a good criterion to characterize the valuations corresponding to integral curves of a vector field X, because:

- It is a condition not stable by product of rational functions, i.e. it is a condition for the derivation and not for the attached foliation by lines.
- It does not work at the singular points.
- It is not stable by blowing-up.

After Seidenberg, there have been several proposals of compatibility criteria between valuations and derivations by Ritt [101], Kolchin [80], Morrison [96], etc. But we will say some things only about Rosenlicht's approach, the nearest to reduction of singularities. Rosenlicht's L'Hopital property [102], is the classical L'Hopital property for the place attached to the valuation.

With the notation:

- k : characteristic zero field,
- $\mathcal{O}$: local integral k-álgebra of maximal ideal $\mathfrak{m}$ and field of fractions K,
- ∂ : a derivation of $\mathcal{O}$, with $\partial|_k = 0$, ν a valuation of K/k,

Rosenlicht defines:

Definition 6.6.1 ν is a l'Hopital valuation with respect to ∂ if it verifies one of the following equivalent conditions:

(1) For all $a, b \in K^*$ such that $\nu(a) \geq \nu(b) > 0,\ \partial(b) \neq 0$,

$$\nu\left(\frac{a}{b} - \frac{\partial a}{\partial b}\right) > 0.$$

(2) For all $a, b \in K^*$ such that $\nu(a) \geq 0,\ \nu(b) > 0,\ \partial(b) \neq 0$,

$$\nu\left(\frac{\partial ab}{\partial b}\right) > 0.$$

Then:

- For rational valuations of K/k, i.e. with k as residual field, the L'Hopital condition is equivalent to:

$$\nu(a) \geq \nu(b) \Leftrightarrow \nu(\partial a) \geq \nu(\partial b).$$

- L'Hopital $\Rightarrow$ Morrison's continuity [96].
- The monomial valuations are in general continuous but not necessarily L'Hopital, for instance the order in $\mathbb{C}(x, y)$ is a monomial rank-1 discrete valuation,

continuous with respect to the derivation $\partial(x) = 1/x, \ \partial y = y^2$ and is not l'Hopital.

- If $K = \mathbb{R}(\{x\})$, the l'Hopital property is exactly the classical l'Hopital property with respect to the usual derivative.

Let $\mathscr{O}$ be the ring of germs of analytic functions at the origin $\mathbf{0}$ in $\mathbb{C}^2$ with maximal ideal $\mathfrak{m}$ and field of fractions $\mathscr{M}$. A valuation ν, different from the order valuation, of $\mathscr{M}$ centered in $\mathscr{O}$, gives a point $\mathbf{p}_\nu$ in the exceptional divisor of the blow-up of $\mathbb{C}^2$, then if ν is L'Hopital for a germ of foliation ω with a singularity at $\mathbf{0}$, the point $\mathbf{p}_\nu$ still being singular of the strict transform of the foliation. Then for the different types of valuations of $\mathscr{M}$, Fortuny [46, 47] proves that:

- If ν is the contact valuation with a formal, non-analytic curve $\widehat{\Gamma}$, ν is L'Hopital for ω if and only if $\widehat{\Gamma}$ is a formal integral curve of ω.
- If ν is divisorial, i.e. after a finite sequence of blow-ups gives rise to the order valuation, ν is L'Hopital for ω if and only if $\mathbf{0}$ is a dicritical singularity for ω.
- The valuations with infinite Puiseux pairs, or with an irrational Puiseux exponent cannot be L'Hopital.
- If ν is the contact valuation with an analytic curve Γ, this is a rank-two valuation, L'Hopital condition is not equivalent to the condition "Γ is a integral curve of ω".

But for the valuations of contact with curves, both formal an convergent, Fortuny proves the following:

Theorem 6.6.2 (Fortuny [46]) *Let ω be a germ of foliation at $(\mathbb{C}^2, \mathbf{0})$, let ν be the contact with a curve Γ valuation of $\mathscr{M}$, i.e.:*

- *either $\exists x(t), y(t) \in t.\mathbb{C}[[t]], f(x(t), y(t)) \neq 0, \ \forall f \in \mathscr{O}, \ \nu(h) = ord_0(h(x(t), y(t))$,*
- *or, $\exists x(t), y(t) \in t.\mathbb{C}[[t]], \ \exists f \in \mathscr{O}$ irreducible such that $f(x(t), y(t)) = 0$ and $\forall h \in \mathscr{O}, \ h = f^r.\overline{h}, \ f \nmid \overline{h}$ then $\nu(h) = (r, ord(\overline{h}(x(t), y(t)) \in \mathbb{Z}^2$.*

Then Γ is an integral curve for ω if and only if ν is weakly L'Hopital, i.e.:

- *Δ_1 being the first non-trivial isolated subgroup of the group of values of ν, the L'Hopital condition is verified only for $\nu(a), \nu(b) \in \Gamma_1$.*

Fortuny's results open the possibility to consider the problem in a new, but too classic, frame: the Hardy fields theory.

Definition 6.6.2 A Hardy field, $\mathscr{H}$, is a subfield of ring of germs of derivable real functions at infinity in $\mathbb{R}$, $\mathfrak{H}_\infty$, closed under derivation.

This means:

- $\forall [f] \in \mathscr{H}. [f]$ is a germ of C^∞ function.
- $\forall [f] \in \mathscr{H}^*, \ \exists a \in \mathbb{R}$ with either $f(x) > 0, \ \forall x \geq a$ or $f(x) < 0, \ \forall x \geq a$, then $\mathscr{H}$ is an ordered field.

- $\forall [f] \in \mathscr{H}$, $\exists \lim_{x\to\infty} f(x) \in \mathbb{R} \cup \{\infty\}$.
- The map $\lim_{x\to\infty}(-) : \mathscr{H} \to \mathbb{R} \cup \{\infty\}$ is a place in the field $\mathscr{H}$.

Then, X being an analytic vector field with an isolated singularity at the origin O in $\mathbb{R}^n$, outside of the singularity, X has integral analytic curves. Let Γ be a integral curve of X with ω-limit set $\{O\}$, i.e. defined outside O by an analytic function:

$$\gamma : [0, +\infty) \to \mathbb{R}^n, \quad \lim_{t\to+\infty} \gamma(t) = O.$$

Now, attached to Γ we have a correspondence of substitution: between the ring of germs of real analytic functions $\mathbb{R}\{x_1, \ldots, x_n\}$ and the ring of germs of real C^∞ functions at infinity $\mathscr{H}_{infty}$:

$$S_\gamma : \mathbb{R}\{x_1, \ldots, x_n\} \to \mathscr{H}_\infty, \; S_\gamma(f) = f \circ \gamma.$$

$S_\gamma(f)$ is not defined if $\Gamma \cap \{f = 0\}$ is an infinite set that accumulates at O. Cano, Moussu, and Sanz [31, 32] define a non-oscillating curve Γ as before, as a curve such that for any $f \in \mathbb{R}\{x_1, \ldots, x_n\}$, $\Gamma \cap \{f = 0\}$ is not an infinite set that accumulates at O.

Then if Γ is a non-oscillating curve, S_γ is a ring's homomorphism, whose kernel is composed by the equations of all the germs of analytic hypersurfaces containing Γ. Also Γ being a integral curve of X, if we call L_X the Lie derivative attached to X:

$$\frac{d(f \circ \gamma(t))}{dt} = \langle grad_{\gamma(t)}, \frac{d(\gamma(t))}{dt} \rangle = \langle grad_{\gamma(t)}, X(\gamma(t)) \rangle = L_X(f) \circ \gamma(t),$$

as a consequence, $Ker\, S_\gamma$ is invariant by L_X and the substitution gives a differential isomorphism of differential domains:

$$S_\gamma : (\mathbb{R}\{x_1, \ldots, x_n\}/Ker(S_\gamma), L_X) \to \left(Im(S_\gamma), \frac{d}{dt}\right)$$

that may be extended to their fields of fractions.

The field of fractions of $(Im(S_\gamma), \frac{d}{dt})$ has a natural structure of a Hardy field; its valuation gives rise to a valuation of the field of fractions of $(\mathbb{R}\{x_1, \ldots, x_n\}/Ker(S_\gamma)$. This valuation verifies compatibility conditions with the derivation L_X (L'Hopital for instance). For the totally transcendent curves $(Ker\, S_\gamma = \{0\})$ this field is the field of germs of real meromorphic functions at $O \in \mathbb{R}^n$.

In the complex case, given an analytic vector field X in $\mathbb{C}^n, \mathbf{0}$ and a formal, totally transcendental, integral curve Γ of X defined by

$$\gamma : \mathbb{C}\{x_1, \ldots, x_n\} \to \mathbb{C}[[t]], \; \gamma(x_i) = x_i(t)$$

X induces a derivation of $\mathbb{C}(\{x_1, \ldots, x_n\})$ and γ an immersion:

$$\psi_\gamma \mathbb{C}(\{x_1, \ldots, x_n\}) \hookrightarrow \mathbb{C}((t))$$

such that the diagram:

$$\begin{array}{ccc} \mathbb{C}\{x_1, \ldots, x_n\} & \xrightarrow{\psi_\gamma} & \mathbb{C}((t)) \\ X \downarrow & & \downarrow \partial_t \\ \mathbb{C}\{x_1, \ldots, x_n\} & \xrightarrow{\psi_\gamma} & \mathbb{C}((t)) \end{array}$$

is commutative.

Also γ defines a discrete rank-1 valuation of $\mathbb{C}(\{x_1, \ldots, x_n\}),\ \nu_\gamma$, and ψ_γ gives an isomorphism of differential valued fields:

$$\psi_\gamma : (\mathbb{C}\{x_1, \ldots, x_n\},\ \nu_\gamma,\ X) \to (L,\ \nu_0,\ \partial_t)$$

where L is a differential subfield of$(\mathbb{C}((t)),\ \nu_0,\ \partial_t)$.

The program is do the same thing, with a valuation in general, instead of the discrete valuation attached to a formal curve.

For every field k and every ordered group Δ one may construct the field of generalized series with coefficients in k and exponents in Δ (see Zariski and Samuel [132]). Then if we are working in zero characteristic and with valuations over a field K whose residual field is a subfield of K, Kaplansky's theorem [78] proves the existence of an immersion of K in the field of generalized series with coefficients in the residual field and exponents in the group of values, $k_\nu((t^\Delta))$, endowed with the order valuation. The problem is that the proof of Kaplansky's theorem is not constructive, and that there is neither a unique nor a canonical immersion of (K, ν) in a generalized series field. Also there are two problems:

- how to construct derivations of generalized series fields;
- given a derivation in $(\mathcal{M}, \nu)$, determining compatibility conditions with the valuation (L'Hopital and something more) to have an extension to $k_\nu((t^\Delta))$.

Solving both problems, we will have a more general notion of formal solution, as a valuation fulfilling compatibility conditions, and this opens the door to a theory of arc spaces of solutions as part of a Riemann–Zariski surface.

Bibliography

1. Abhyankar, S.S.: Current status of the resolution problem. In: Lecture notes prepared in connection with the Summer Institute on Algebraic Geometry held at the Whitney State (Woods Hole, Mass 1964). American Mathematical Society (1964)
2. Abhyankar, S.S.: On the problem of resolution of singularities. Proc. Int. Congress Math. Moscow **1**, 469–481 (1966)
3. Abhyankar, S.S.: Analytic desingularization in characteristic zero. Appendix to the second edition of the book: Resolution of singularities of embedded algebraic surfaces. Springer (1998)
4. Abramovich, D., de Jong, J.: Smoothness, semi-stability and toroidal geometry. J. Algebraic Geom. **6**, 789–801 (1997)
5. Aroca, F.: Metodos algebraicos en ecuaciones diferenciales. Tesis Doctoral, University of Valladolid (2000)
6. Aroca, J.M., Hironaka, H., Vicente, J.L.: The theory of the maximal contact. Memorias de Matemática del Instituto Jorge Juan, No. 29. Consejo Superior de Investigaciones Científicas, Madrid (1975)
7. Aroca, J.M., Hironaka, H., Vicente, J.L.: Desingularization theorems. Memorias de Matemática del Instituto Jorge Juan, No. 30. Consejo Superior de Investigaciones Científicas, Madrid (1977)
8. Bennett, B.-M.: On the characteristic function of a local ring. Ann. Math. **91**, 25–87 (1970)
9. Berthelot, P.: Altérations de variétés algébriques (d'après A.J. de Jong). Sém. Bourbaki, exp. 815 (1995/1996)
10. Bierstone, E., Milman, P.: Canonical desingularization in characteristic zero by blowing up the maximum strata of a local invariant. Invent. Math. **128**, 207–302 (1997)
11. Bierstone, E., Milman, P.: Uniformization of analytic spaces. J. Am. Math. Soc. **2**, 801–836 (1989)
12. Bierstone, E., Milman, P.: A simple constructive proof of canonical resolution of singularities. In: Mora, T., Traverso, C. (eds.) Effective Methods in Algebraic Geometry. Progress in Mathematics, vol. 94, pp. 11–30. Birkhäuser, Boston (1991)
13. Bierstone, E., Milman, P.: Relations among analytic functions I. Ann. Inst. Fourier **37**, 187–239 (1987)
14. Bogomolov, F., Pantev, T.: Weak Hironaka theorem. Math. Res. Lett. **3**, 299–307 (1996)
15. Bourbaki, N.: Algèbre commutative. Chapitre 5: Entiers. Chapitre 6: Valuations. Hermann, Paris (1964)
16. Briançon, J.: Weierstrass prepare à la Hironaka. Astèrisque **7 et 8**, 67 (1973)

J. M. Aroca et al., *Complex Analytic Desingularization*,
https://doi.org/10.1007/978-4-431-49822-3

17. Briot, C.A., Bouquet, J.C.: Propriétés des fonctions définies par les équations différentielles. Journal de l'Ecole Polytechnique **36**, 133–198 (1856)
18. Camacho, C.: Quadratic forms and holomorphic foliations on singular surfaces. Math. Annalen **282**, 177–184 (1988)
19. Camacho, C., Sad, P.: Invariant varieties through singularities of holomorphic vector fields. Ann. Math. **115**, 579–595 (1982)
20. Cano, F.: Desingularization Strategies for Three Dimensional Vector Fields. Lecture Notes in Mathematics, vol. 1259. Springer, Berlin/New York (1987)
21. Cano, F.: Reduction of singularities of codimension one foliations in dimension three. Ann. Math. **160**, 907–1011 (2004)
22. Cano, F.: Reduction of the singularities of non-dicritical singular foliations. Dimension three. Am. J. Math. **115**, 509–588 (1993)
23. Cano, F.: Dicriticalness of a singular foliation. In: Gómez-Mont, X., Seade, J., Verjovski, A. (eds.) Holomorphic Dynamics. Proceedings, 1986, México. Lecture Notes in Mathematics, vol. 1345, pp. 73–95 (1988)
24. Cano, F.: Foliaciones singulares dicríticas. Mem. Real Acad. Ciencias Exactas, Físicas y Naturales, serie Ciencias Exactas, t. XXIV (1989)
25. Cano, F.: Reduction of the singularities of foliations and applications. Singularities symposium–Lojasievicz 70, vol. 44. Banach Center Publications, Warsaw (1998)
26. Cano, F.: Final forms for a three-dimensional vector field under blowing-up. Annales de linstitut Fourier **37**, 151–193 (1987)
27. Cano, F., Cerveau, D.: Desingularization of non-dicritical holomorphic foliations and existence of separatrices. Acta Math. **169**, 1–103 (1992)
28. Cano, F., Cerveau, D., Deserti, J.: Thorie lmentaire des feuilletages holomorphes singuliers. Belin, Paris (2013)
29. Cano, F., Mattei, J.F.: Hypersurfaces intégrales des feuilletages holomorphes. Ann. Inst. Fourier **42**, 49–72 (1992)
30. Cano, F., Moussu, R., Rolin, J-P.: Non-oscillating integral curves and valuations. J. Reine Angew. Math. **582**, 107–141 (2005)
31. Cano, F., Moussu, R., Sanz, F.: Oscillation, spiralement, tourbillonnement. Commentarii Mathematici Helvetici **75**, 284–318 (2000)
32. Cano, F., Moussu, R., Sanz, F.: Nonoscillating projections for trajectories of vector fields. J. Dyn. Control Syst. **13**, 173–176 (2007)
33. Cano, F., Piedra, R.: Characteristic polygon of surface singularities. In: Aroca, J.-M., Scannchez-Giralda, T., Vincente, J.-L. (eds.) Géométrie algébrique et applications II. Proceedings of Conference on Singularities, Hermann. La Rábida 1984 (1987)
34. Cano, F., Roche C.: Vector fields tangent to foliations and blow-ups. J. Sing. **9, 7**, 43–49 (2014)
35. Cano, F., Roche, C., Spivakovsky, M.: Reduction of singularities of three dimensional line foliations. arXiv: 1009.1348
36. Cano, J.: An extension of the Newton-Puiseux polygon construction to give solutions of Pfaffian forms. Ann. Inst. Fourier **43**, 125–142 (1993)
37. Cano, J.: On the series defined by differential equations with an extension of the Puiseux polygon construction to these equations. Analysis **3**, 103–117 (1993)
38. Cano, J.: Construction of invariant curves for holomorphic vector fields. Proc. Am. Math. Soc. **125**, 2649–2650 (1997)
39. Cossart, V.: Polyèdre carastéristique d'une singularité. Thèse d'Etat, Orsay (1987)
40. Cossart, V.: Uniformisation et désingularisation des surfaces d'après Zariski. Progr. Math. **181**, 239–258 (2000)
41. Cossart, V., Giraud, J., Orbanz, U.: Resolution of Surface Singularities. Lecture Notes in Mathematics, vol. 1101. Springer, Berlin (1984)
42. Farto, M.: Resolucion efectiva de ecuaciones diferendiales. Tesis Doctoral, University of Valladolid (1995)

43. Fernandez Duque, M.: Local uniformization of codimension one foliations. Rational archimedean valuations. Tesis Doctoral, University of Valladolid (2016)
44. Fine, J.: On the functions defined by differential equations with an extension of the Puiseux polygon construction to these equations. Am. J. Math. **XI**, 317–328 (1889)
45. Fine, J.: Singular solutions of ordinary differential equations. Am. J. Math. **XII**, 293–322 (1890)
46. Fortuny, P.: L'Hopital. Tesis doctoral, Valladolid (1999)
47. Fortuny, P.: The valuative theory of foliations. Canad. J. Math. **54**, 897–915 (2002)
48. Galligo, A.: A propos du théorème de préparation de Weierstrass. In: Lecture Notes in Mathematics, vol. 409, pp. 543–579. Springer, Berlin/Heidelberg (1974)
49. Giraud, J.: Etude locale des singularitiés. Cours de 3e Cycle, Orsay (1971/1972)
50. Giraud, J.: Sur la théorie du contact maximal. Math. Z. **137**, 285–310 (1974)
51. Giraud, J.: Contact maximal en caractéristique positive. Ann. Scient. E. N. S. **8**, 201–234 (1975)
52. Giraud, J: Forme normale d'une fonction sur une surface de caractéristique positive. Bull. Soc. Math. France **111**, 109–124 (1983)
53. Giraud, J.: Condition de Jung pour les revêtements radiciels de hauteur un. Proc. Algr. Geom. Tokyo/Kyoto 1982. Lecture Notes in Mathematics, vol. 1016. Springer, Heidelberg/Berlin/New York (1983)
54. Goldin, R, Tessier, B.: Resolving singularities of plane analytic branches with one toric morphism. Progr. Math. **181**, 315–340 (2000)
55. Gómez-Mont, X., Luengo, I.: Germs of holomorphic vector fields in $\mathbb{C}^3$ without a separatrix. Invent. Math. **109**, 211–219 (1992)
56. Grauert, H.: Ein Theorem der analytischen Garbentheorie und die Modulrume komplexer Strukturen. Publications Mathmatiques de LInstitut des Hautes Scientifiques **5**(1), 5 (1960)
57. Grigoriev, D.Y., Singer, M.: Solving ordinary differential equations in terms of series with real exponents. Trans. A.M.S. **327**, 339–351 (1991)
58. Grothendieck, A.: Travaux de Heisuké Hironaka sur la résolution des singularités. Actes, Congrès Intern. Math., Nice, pp. 7–9 (1970)
59. Hardy, G.H.: Orders of Infinity. Cambridge Tracts in Mathematics, vol. 12. Cambridge University Press, Cambridge (1910)
60. Hensel, K.: Über eine neue Theorie der algebrischen Funktionen zweier Variablen. Acta Math. **23**, 339–416 (1900)
61. Hauser, H.: Seventeen obstacles for resolution of singularities. In: Festband in Honour of Brieskorn's 60th Birthday. Birkhäuser, Basel/Boston/Berlin (1997)
62. Herrmann, M., Ikeda, S., Orbanz, U.: Equimultiplicity and Blowing Up. Springer, Berlin (1988)
63. Hironaka, H.: Resolution of singularities of an algebraic variety over a field characteristic zero: I and II. Ann. Math. **79**, 109–326 (1964)
64. Hironaka, H.: Characteristic polyhedra of singularities. J. Math. Kyoto Univ. **7**, 251–293 (1967)
65. Hironaka, H.: La voûte étoilée. Singularités à Cargèse. Astérisque 7–8 (1973)
66. Hironaka, H.: Desingularization of excellent surfaces. Notes by B. Bennett at the Conference on Algebraic Geometry, Bowdoin, 1967. Reprinted in: Cossart, V., Giraud, J., Orbanz, U.: Resolution of Surface Singularities. Lecture Notes in Mathematics, vol. 1101. Springer, Berlin (1984)
67. Hironaka, H.: Additive groups associated with points of a projective space. Ann. Math. **92**, 327–334 (1970)
68. Hironaka, H.: Idealistic exponents of singularity. In: Algebraic Geometry. The Johns Hopkins Centennial Lectures. Johns Hopkins University Press, Baltimore (1977)
69. Hironaka, H.: Schemes etc. In: Oort, F. (ed.) 5th Nordic Summer School in Mathematics, Oslo, pp. 291–313 (1970)
70. Hironaka, H.: Desingularization of complex analytic varieties. Actes du Congrès Intern. des Math., Nice, pp. 627–631 (1970)

71. Hironaka, H.: Gardening of infinitely near singularities. In: Proceedings of Nordic Summer School in Mathematics, Oslo, pp. 315–332 (1970)
72. Hironaka, H.: Bimerromorphic smoothing of complex-analytic spaces. University of Warwick (1971). reproduit par le Centre de Mathematique de l'Ecole polytecnique, Paris
73. Hironaka, H.: Introduction to the Theory of Infinitely Near Singular Points. Mem. de Mat. del Inst. Jorge Juan., Madrid, vol. 28 (1974)
74. Ince, P.: Ordinary Differential Equations. Dover, New York (1956)
75. de Jong, A.: Smoothness, semi-stability and alterations. Inst. Hautes Études Sci. Publ. Math. **83**, 51–93 (1996)
76. Jouanolou, J.P.: Equations de Pfaff algébriques. Lecture Notes in Mathematics, vol. 708. Springer, Berlin/New York (1979)
77. Jung, H.: Darstellung der Funktionen eines algebraischen Körpers zweier unabhängigen Veränderlichen x, y in der Umgeburg einer Stelle $x = a, y = b$. J. Reine Angew. Math. **133**, 289–314 (1908)
78. Kaplansky, I.: Maximal fields with valuations. Duke Math. J. **9**, 303–321 (1942)
79. Kiyek, K.H., Vicente, J.L.: On the Jung–Abhyankar theorem. Arch. Math. **83**, 123–134 (2004)
80. Kolchin, E.R.: Differential Algebra and Algebraic Groups. Academic Press, New York/London (1973)
81. Laufer, H.: Normal Two-Dimensional Singularities. Annals of Mathematics Studies, vol. 71. Princeton University Press, Princeton (1971)
82. Lê, D.T., Oka, M.: On resolution complexity of plane curves. Kodai Math. J. **18**, 1–36 (1995)
83. Lejeune, M., Teissier, B.: Contribution à l'étude des singularitiés du point de vue du polygone de Newton. Thèse d'Etat (1973)
84. Lipman, J.: Introduction to resolution of singularities. In: Proceedings Symposium on Pure Applied Mathematics, vol. 29, pp. 187–230. American Mathematical Society, Providence, Rhode Island (1975)
85. Luengo, I.: A new proof of the Jung–Abhyankar theorem. J. Algebra. **85**, 399–409 (1983)
86. Luengo, I., Olivares, J.: Germs of holomorphic vector fields in $\mathbb{C}^n$ without a separatrix. Trans. Am. Math. Soc. **352**, 5511–5524 (2000)
87. McDonald, J.: Fiber polytopes and fractional power series. J. Pure Appl. Algebra **104**, 213–233 (1995)
88. McQuillan, M., Panazzolo, D.: Almost étale resolution of foliations. IHES preprint 2009, elect.av. at preprints.ihes.fr/2009/M/M-09-51.pdf
89. Mahler, K.: On formal power series defined by algebraic differential equations. Rend. Linc. **L**, 76–89 (1971)
90. Maillet, E.: Sur las séries divergentes et les équations différentielles. L'ens. Mat. **20**, 487–518 (1903)
91. Malgrange, B.: Sur le théorème de Maillet. Assymp. Anal. **2**, 1–4 (1989)
92. Malgrange, B.: Frobenius avec singularités 1. Codimension un. Publ. Math. I.H.E.S. **46**, 162–173 (1976)
93. Martinet, J.: Normalisation des champs de vecteurs holomorphes (d'apres A.D. Brjuno). Sem. Bourbaki, 1980/1981. Lecture Notes on Mathematics, vol. 901. Springer, Berlin (1982)
94. Mattei, J.F., Moussu, R.: Holonomie et intgrales premires. Ann. Sci. cole Norm. Sup. **13**, 469–523 (1980)
95. Moh, T.-T.: Quasi-canonical uniformization of hypersurface singularities of characteristic zero. Commun. Alg. **20**, 3207–3249 (1992)
96. Morrison, S.: Continuous derivations. J. Algebra **110**, 468–479 (1987)
97. Oka, K.: Sur les fonctions analytiques de plusieurs variables. VII. Sur quelques notions arithmétiques. Bulletin de la Société Math. de France **78**, 1–27 (1950)
98. Orlik, P., Wagreich, P.: Equivariant resolution of singularities with C^* action. In: Proceedings of the 2nd Conference on Compact Transformation Groups. Lecture Notes in Mathematics, vol. 298, pp. 270–290. Springer, Berlin/Heidelberg (1972)
99. Panazzolo, D.: Resolution of singularities of real-analytic vector fields in dimension three. Acta Math. **197**, 167–289 (2009)

100. Piltant, O.: An axiomatic version of Zariski's patching theorem. University of Valladolid (2008, preprint)
101. Ritt J.F.: Differential Equations from the Algebraic Standpoint, vol. XIV. A. M. S. Colloquium publications, New York (1932)
102. Rosenlicht, M.: Differential valuations. Pac. J. Math. **86**, 301–319 (1980)
103. Saeki, A.: Foliations on complex spaces. Funkcialaj Ekvacioj **38**, 121–157 (1995)
104. Saito, K.: On a generalization of de Rham Lemma. Ann. Inst. Fourier **26**, 165–170 (1976)
105. Seidenberg, A.: Reduction of the singularities of the differential equation $Ady = Bdx$. Am. J. Math. **90**, 248–269 (1968)
106. Seidenberg, A.: Derivations and valuations rings. Contributions to Algebra. Papers in Honor of E. Kolchin. Academic Press, New York (1977)
107. Serre, J.P.: Algèbre locale. Lecture Notes in Mathematics, vol. 11. Springer, Berlin (1965)
108. Siu, Y.T.: Techniques of Extension of Analytic Objects. Lecture Notes in Pure and Applied Mathematics, vol. 8. Marcel Decker, New York (1974)
109. Spivakovsky, M.: A solution to Hironaka's polyhedra game. In: Artin, M., Tate, J. (eds.) Arithmetic and Geometry. Papers Dedicated to I.R. Shafarevich on the Occasion of His Sixtieth Birthday: Geometry, vol. II. Progress in Mathematics, vol. 36, pp. 419–432. Birkhäuser, Boston (1983)
110. Spivakovsky, M.: A counterexample to Hironaka's "hard" polyhedra game. Publ. RIMS, Kyoto Univ. **18**, 1009–1012 (1982)
111. Spivakovsky, M.: A counterexample to the theorem of Beppo Levi in three dimensions. Invent. Math. **96**, 181–183 (1989)
112. Sussmann, H.J.: Real analytic desingularization and subanalytic sets: an elementary approach. Trans. Am. Math. Soc. **317**, 417–461 (1990)
113. Suwa, T.: Indices of Vector Fields and Residues of Singular Holomorphic Foliations. Hermann, Paris (1988)
114. Van der Hoeven, J.: Transseries an Real Differential Algebra. Lecture Notes in Mathematics, vol. 1888. Springer, Berlin (2006)
115. Villamayor, U.O.: Constructiveness of Hironaka's resolution. Ann Sci. Éc. Norm. Sup. Paris **22**, 1–32 (1989)
116. Villamayor, U.O.: Patching local uniformizations. Ann. Sci. Éc. Norm. Sup. Paris **25**, 629–677 (1992)
117. Villamayor, U.O.: Introduction to the algorithm of resolution. In: Campillo Lopez, A., Narváez Macarro, L. (eds.) Algebraic Geometry and Singularities. Proceedings of Conference on Singularities La Rábida, pp. 123–154. Birkhäuser, Basel/Berlin (1996)
118. Walker, R.J.: Reduction of the singularities of an algebraic surface. Ann. Math. **36**, 336–365 (1935)
119. Walker, R.J.: Algebraic Curves. Dover Publications Inc., New York (1962)
120. Youssin, B.: Newton polyhedra without coordinates. Mem. AMS **433**, 1–74, 75–99 (1990)
121. Zariski, O.: Algebraic Surfaces. Ergebnisse der Mathematik und ihrer Grenzgebiete, vol. 3. Springer, New York (1971)
122. Zariski, O.: The reduction of singularities of an algebraic surface. Ann. Math. **40**, 639–689 (1939)
123. Zariski, O.: Local uniformization on algebraic varieties. Ann. Math. **41**, 852–896 (1940)
124. Zariski, O.: A simplified proof for the resolution of singularities of an algebraic surface. Ann. Math. **43**, 583–593 (1942)
125. Zariski, O.: Foundations of a general theory of birational correspondences. Trans. Am. Math. Soc. **53**, 490–542 (1943)
126. Zariski, O.: The compactness of the Riemann manifold of an abstract field of algebraic functions. Bull. Am. Math. Soc. **50**, 683–691 (1944)
127. Zariski, O.: Reduction of the singularities of algebraic three dimensional varieties. Ann. Math. **45**, 472–542 (1944)
128. Zariski, O.: Exceptional singularities of an algebroid surface and their reduction. Atti Accad. Naz. Lincei Rend. Cl. Sci. Fis. Mat. Natur. **43**, 135–146 (1967)

129. Zariski, O.: A new proof of the total embedded resolution theorem for algebraic surfaces. Am. J. Math. **100**, 411–442 (1978)
130. Zariski, O., Muhly, M.T.: The resolution of singularities of an algebraic curve. Am. J. Math. **61**, 107–114 (1939)
131. Zariski, O., Samuel, P.: Commutative Algebra, Volume I. Graduate Texts in Mathematics, vol. 28. Springer, New York (1960)
132. Zariski, O., Samuel, P.: Commutative Algebra, Volume II. Graduate Texts in Mathematics, vol. 29, Springer, New York (1960)

Index

J. M. Aroca et al., *Complex Analytic Desingularization*,
https://doi.org/10.1007/978-4-431-49822-3

P

Q

R

S

T

Printed by Printforce, the Netherlands